Deep Reinforcement Learning with Python

RLHF for Chatbots and Large Language Models

Second Edition

Nimish Sanghi

Apress®

Deep Reinforcement Learning with Python: RLHF for Chatbots and Large Language Models, Second Edition

Nimish Sanghi
Bangalore, India

ISBN-13 (pbk): 979-8-8688-0272-0   ISBN-13 (electronic): 979-8-8688-0273-7
https://doi.org/10.1007/979-8-8688-0273-7

Managing Director, Apress Media LLC: Welmoed Spahr
Acquisitions Editor: Celestin Suresh John
Development Editor: James Markham
Editorial Assistant: Gryffin Winkler
Copy Editor: Kezia Endsley

Cover designed by eStudioCalamar
Cover image designed by Freepik (www.freepik.com)

Distributed to the book trade worldwide by Springer Science+Business Media New York, 1 New York Plaza, Suite 4600, New York, NY 10004-1562, USA. Phone 1-800-SPRINGER, fax (201) 348-4505, e-mail orders-ny@springer-sbm.com, or visit www.springeronline.com. Apress Media, LLC is a California LLC and the sole member (owner) is Springer Science + Business Media Finance Inc (SSBM Finance Inc). SSBM Finance Inc is a **Delaware** corporation.

For information on translations, please e-mail booktranslations@springernature.com; for reprint, paperback, or audio rights, please e-mail bookpermissions@springernature.com.

Apress titles may be purchased in bulk for academic, corporate, or promotional use. eBook versions and licenses are also available for most titles. For more information, reference our Print and eBook Bulk Sales web page at http://www.apress.com/bulk-sales.

Any source code or other supplementary material referenced by the author in this book is available to readers on GitHub. For more detailed information, please visit https://www.apress.com/gp/services/source-code.

If disposing of this product, please recycle the paper

This book is dedicated to my wife Suman, and our kids, Harsh and Yash, who selflessly sacrificed our shared moments to allow these pages to come to life.

Table of Contents

About the Author

Nimish Sanghi is a seasoned entrepreneur and an angel investor, with a rich portfolio of AI, automation, and SaaS tech ventures across India, the United States, and Singapore. He has over 30 years of work experience. Nimish ventured into entrepreneurship in 2006 after holding leadership roles at global corporations like PwC, IBM, and Oracle.

Nimish holds an MBA from Indian Institute of Management, Ahmedabad, India (IIMA), and a Bachelor of Technology in electrical engineering from Indian Institute of Technology, Kanpur, India (IITK).

About the Technical Reviewer

Akshay Kulkarni is an AI and machine learning evangelist and thought leader. He has consulted several Fortune 500 and global enterprises to drive AI and data science-led strategic transformations. He is a Google developer expert, author, and regular speaker at major AI and data science conferences (including Strata, O'Reilly AI Conf, and GIDS). He has also been a visiting faculty member at some of the top graduate institutes in India. In 2019, he was featured as one of the top 40 under 40 data scientists in India. In his spare time, he enjoys reading, writing, coding, and building next-gen AI products

Acknowledgments

A huge thank you to the Apress team—particularly Sowmya Thodur, Celestin John, Mark Powers, and Joseph Quatela—for their guidance and help in making this publication possible. A huge shoutout to Akshay Kulkarni for meticulously reviewing the drafts to ensure accuracy and completeness and for verifying the code. My gratitude also extends to all the Apress staff who contributed in various ways.

A special mention to my wife, Suman, who once again became the pillar of our family, shielding me from distractions and providing me the peace and space needed to write this second edition. Your unwavering support was crucial.

I must also express my gratitude to the readers of the first edition, whose constructive feedback was instrumental in shaping this edition. I eagerly await your honest opinions of this book—did it meet your expectations, and was it worth your time? Your feedback is invaluable, and I'm available at nimish.sanghi@gmail.com or https://github.com/nsanghi for any comments or suggestions.

Introduction

This book is about reinforcement learning, taking the readers through the basics to advanced topics. Although this book assumes no prior knowledge of the field of reinforcement learning, it expects the readers to be familiar with the basics of machine learning. Have you coded in Python? Are you comfortable working with libraries like NumPy and scikit-learn? Have you heard of deep learning and have you explored the basic build blocks of training simple models in PyTorch? You should answer yes to these questions to get the most out of this book. If not, I suggest you learn a bit about these concepts first. Nothing too deep—any introductory online tutorial or book from Apress on these topics will be sufficient.

In this second edition, I have made some major changes while keeping most of the content from the first edition. The main additions are related to the new developments in the field of Large Language Models (LLM) and Multimodal Generative AI, which have revolutionized the world since late 2022. Reinforcement learning (RL) has played a crucial role in enabling this through Reinforcement Learning from Human Feedback (RLHF). This edition has a new chapter dedicated to this topic. It gives the reader a high-level overview of transformers, LLMs, and related topics like prompt engineering, Retrieval Augmented Generation (RAG), parameter efficient fine-tuning (PEFT), and chaining of LLMs and LLM-based auto agents, followed by a detailed explanation of the concept of RLHF. In the same chapter, you'll also explore Proximal Policy Optimization (PPO), which is a popular state-of-the-art RL based algorithm that was used by OpenAI for the RLHF fine-tuning of ChatGPT.

Another addition is a chapter on multi-agent RL (MARL) and deep MARL (DMARL), which deals with more than one agent cooperating or competing in the same environment. In this chapter, I start with the introduction and go all the way to a working example. I limit the discussion to introducing the key concepts, enabling interested readers to follow specialized texts on MARL for further exploration.

This edition also covers additional topics, like hyperparameter tuning. It includes an overview of other topics like curiosity learning, use of transformers in RL in various ways, emerging areas such as sample efficient offline RL, decision transformers, automated curriculum learning, zero-shot RL, and various other advances in the field since the first edition. The chapter on Deep Q networks has been split into two to provide better organization to the topic.

On the code front, there have been significant changes too. I cover a lot more about RL environments. I introduce Gymnasium, the successor library of OpenAI Gym. I also introduce other environments, like FinRL, which is dedicated to applying RL in market trading, other robotic environments, and environments for MARL. I cover various other RL libraries that have gained popularity since the first edition. The way code runs in 2024 is very different. Therefore, immediately in Chapter 1 the book gives readers step-by-step installation instructions to be able to run the accompanying code on local machines and/or cloud providers, including cloud-based monitoring and tracking of training.

This book walks readers through the basics of reinforcement learning, spending a lot of time explaining the concepts in the initial chapters. A reader with prior knowledge of reinforcement learning can go through the first four chapters quickly. Chapter 5 picks up pace and starts exploring advanced topics, combining deep learning with reinforcement learning. The accompanying code hosted on GitHub is an integral part of this book. While the book has listings of the relevant code, Jupyter notebooks in the code repository provide additional insights and practical tips on coding these algorithms. The reader is best served by first reading the chapter and going through the explanations followed by working through the code in the Jupyter notebooks. The reader is also encouraged to try and rewrite the code on their own training agents for different additional environments, as discussed in the book.

For a subject like this, math is unavoidable. However, I have tried to keep it minimal. The book quotes a lot of research papers giving short explanations of the approach taken. Readers wanting a deeper understanding of the theory should go through these research papers. This book's purpose is to introduce practitioners to the motivation and high-level approach behind many of the latest techniques in this field. However, by no means it is meant to provide a complete theoretical understanding of these techniques, which is best gained by reading the original papers.

The book is organized into 13 chapters.

- Chapter 1, Introduction to Reinforcement Learning: Readers will learn about the basics and the applications of reinforcement learning in the real world. This chapter also provides the steps for setting up the code that comes with this book on different platforms, such as a local computer or the cloud.

- Chapter 2, The Foundation: Markov Decision Processes: This chapter describes the problem that reinforcement learning aims to solve. I introduce the components of an RL system—the agent, environment, rewards, value functions, model, and policy. I explain the Markov process and its different variants, along with backup equations by Richard Bellman.

- Chapter 3, Model-Based Approaches: This chapter looks at the scenario where the agent has a model and plans its action for the best result. I also introduce Gymnasium, a popular RL environment library. Lastly, I examine value and policy iteration methods for planning, including generalized policy iteration and asynchronous backups.

- Chapter 4, Model-Free Approaches: This chapter explores the model-free learning methods that the agent can use when it cannot access the environment/model's dynamics. It focuses on the Monte Carlo (MC) and Temporal Difference (TD) approaches to learning. It first examines them separately and then merges them under the concept of n-step returns and eligibility traces.

- Chapter 5, Function Approximation and Deep Learning: Explores continuous valued states instead of discrete states as in the previous chapters. It starts with the conventional method of designing function approximation by hand, especially the linear ones. At the end, the chapter presents the idea of using deep learning-based models as non-linear function approximators.

- Chapter 6, Deep Q-Learning (DQN): Dives into DQN, an approach that successfully demonstrated the use of deep learning together with reinforcement learning. It introduces the popular libraries for hyperparameter optimization and experiment tracking. Finally, it does a broad survey of RL environments from robotics, finance and trading, and so on.

- Chapter 7, Improvements to DQN: This is an optional chapter that discusses enhancements to the DQN. For each of these, theoretical discussions are followed by complete implementation in Python and PyTorch. It covers topics like Prioritized Replay, Double DQN, Dueling DQN, NoisyNets DQN, C51 DQN, Qunatile DQN, and Hindsight Experience Replay.

- Chapter 8, Policy Gradient Algorithms: This chapter shifts gears from value-based methods to direct policy learning. After establishing the foundations, it discusses various approaches, including recent and highly successful ones like Trust Region Policy Optimization and Proximal Policy Optimization, complete with implementations in PyTorch. It also looks at curiosity-driven learning.

- Chapter 9, Combining Policy Gradients and Q-Learning: This chapter looks at DQN with policy gradients methods to leverage the advantages of both approaches, which also allows for continuous action spaces, which are not easily feasible otherwise. It looks at three very popular ones—Deep Deterministic Policy Gradients (DDPG), Twin Delayed DDPG (TD3), and Soft Actor-Critic (SAC).

- Chapter 10, Integrated Planning and Learning: This chapter is all about combing the model-based approach from Chapter 3 and the model-free approach from Chapters 4 through 9. It discusses a general framework called Dyna along with some variants. Finally, it looks at Monte Carlo Tree Search (MCTS) along with its application for training AlphaGo that can beat champion human Go players.

- Chapter 11, Proximal Policy Optimization (PPO) and RLHF: This chapter looks at natural policy gradients that motivated PPO along with a line-by-line walkthrough of PPO implementation. The chapter introduces Large Language Models (LLM) and related concepts. This is followed by an in-depth discussion on RLHF along with a working example.

- Chapter 12, Multiagent RL (MARL): This chapter introduces the concept of multiple agents in the same environment, the theoretical foundations, the new taxonomy for MARL, its linkage to game theory, the new challenges, and the RL solution approaches for MARL.

- Chapter 13, Additional Topics and Recent Advances: This chapter surveys various other extensions of reinforcement learning. It touches on concepts like world models, decision transformers, automatic curriculum learning imitation and inverse learning, derivative-free methods, transfer and multi-task learning, meta learning, unsupervised zero-shot RL, and a lot more.

CHAPTER 1

Introduction to Reinforcement Learning

Reinforcement learning is a fast-growing discipline and is helping to make AI real, especially when it comes to robots and autonomous vehicles. Combining deep learning with reinforcement learning has led to many significant advances that are increasingly getting machines closer to acting the way humans do. Recently, deep reinforcement learning has been applied to Large Language Models like ChatGPT and others to make them follow human instructions and produce output that's favored by humans. This is known as *Reinforcement Learning from Human Feedback* (RLHF).

This book starts with the basics and finishes up discussing some of the most recent developments in the field. There is a good mix of theory (with minimal mathematics) and code implementations using PyTorch and other libraries.

This chapter sets the context and gets everything prepared for you to follow along in the rest of the book.

Reinforcement Learning

All intelligent beings start with some knowledge. However, as they interact with the world and gain experience, they learn to adapt to the environment and become better at doing things. To quote a 1994 op-ed statement in the *Wall Street Journal*,[1] intelligence can be defined as follows:

---

[1] https://en.wikipedia.org/wiki/Mainstream_Science_on_Intelligence

© Nimish Sanghi 2024
N. Sanghi, *Deep Reinforcement Learning with Python*, https://doi.org/10.1007/979-8-8688-0273-7_1

A very general mental capability that, among other things, involves the ability to reason, plan, solve problems, think abstractly, comprehend complex ideas, learn quickly, and learn from experience. It is not merely book learning, a narrow academic skill, or test-taking smarts. Rather, it reflects a broader and deeper capability for comprehending our surroundings—"catching on," "making sense" of things, or "figuring out" what to do.

Intelligence in the context of machines is called *artificial intelligence*. Oxford Languages defines artificial intelligence (AI) as follows:

The theory and development of computer systems able to perform tasks normally requiring human intelligence, such as visual perception, speech recognition, decision-making, and translation between languages.

This is what you study in this book: the theory and design of algorithms that help machines (agents) gain the capability to perform tasks by interacting with the environment and continually learning from their successes, failures, and rewards. Originally AI revolved around designing solutions as a list of formal rules that could be expressed using logic and mathematical notations. These rules consisted of a collection of information codified into a knowledge base. The design of these AI systems also consisted of an inference engine that enabled users to query the knowledge base and combine the individual strands of rules/knowledge to make inferences. These systems were also known as *expert systems, decision support systems,* and so on. However, soon people realized that these systems were too brittle. As the complexity of problems grew, it became exponentially harder to codify the knowledge or to build an effective inference system.

The modern concept of reinforcement learning is a combination of two different threads through their individual development. First is the concept of optimal control. Among many approaches to the problem of optimal control, in 1950 Richard Bellman came up with the discipline of *dynamic programming*, which is used extensively in this book. Dynamic programming, however, does not involve learning. It is all about planning through the space of various options using Bellman recursive equations. Chapters 2 and 3 have a lot to say about these equations.

The second thread is that of learning by trial and error, which finds its origin in the psychology of animal training. Edward Thorndike was the first one to express the concept of *trial and error* in clear terms. In his words:

Of several responses made to the same situation, those which are accompanied or closely followed by satisfaction to the animal will, other things being equal, be more firmly connected with the situation, so that, when it recurs, they will be more likely to recur; those which are accompanied or closely followed by discomfort to the animal will, other things being equal, have their connections with that situation weakened, so that, when it recurs, they will be less likely to occur. The greater the satisfaction or discomfort, the greater the strengthening or weakening of the bond.

The concept of increasing the occurrence of good outcomes and decreasing the occurrence of bad outcomes is something you will see in use in Chapter 8, on policy gradients.

In the 1980s, these two fields merged to give rise to the field of modern reinforcement learning. In the last decade, with the emergence of powerful deep learning methods, reinforcement learning (when combined with deep learning) is giving rise to very powerful algorithms that could make artificial intelligence real in coming times. Today's reinforcement systems interact with the world to acquire experience and learn to optimize their actions based on the outcomes of their interactions with the world, by generalizing their experience. There is no explicit coding of expert knowledge.

Machine Learning Branches

Machine learning (ML) involves learning from the data presented to the system so that the system can perform a specified task. The system is not explicitly told how to do the task. Rather, it is presented with the data, and the system learns to carry out a task based on a defined objective. I do not say more about it, as I assume that you are familiar with the concept of machine learning. Machine learning approaches are traditionally divided into three broad categories, as shown in Figure 1-1.

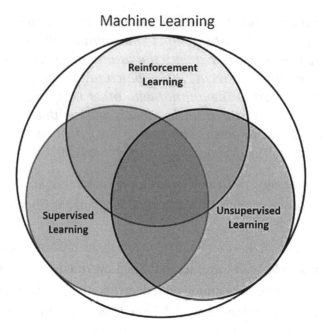

Figure 1-1. *Branches of machine learning*

The three branches of machine learning differ in the sense of the "feedback" available to the learning system. They are discussed in the following sections. The distinction between supervised and unsupervised learning is blurring, with many nuanced approaches that mix and match the two paradigms in different ways. It has given rise to additional sub-categories, such as semi-supervised learning, self-supervised learning, and Generative AI.

Supervised Learning

In supervised learning, the system is presented with the labeled data, and the objective is to generalize the knowledge so that new, unlabeled data can be labeled. Consider images of cats and dogs presented to a system, along with labels indicating whether the image shows a cat or a dog. The input data is represented as a set of data, $D = (x_1, y_1)$, $(x_2, y_2), ..., (x_n, y_n)$, where $x_1, x_2, ..., x_n$ are the vectors of pixel values of individual images and $y_1, y_2, ..., y_n$ are the labels of the respective images—say, a value of 0 for a cat and a value of 1 for a dog. The system/model takes this input and learns a mapping from image x_i to label y_i. Once trained, the system is presented with a new image x' to get a prediction of the label $y' = 0$ or 1, depending on whether the image shows a cat or a dog.

This is a classification problem where the system learns to classify an input into the correct class. The other problem type is that of regression, such as when you'd want to predict the price of a house based on its features. The training data is again represented as $D = (x_1, y_1), (x_2, y_2), ..., (x_n, y_n)$. The inputs are $x_1, x_2, ..., x_n$ where each input x_i is a vector of certain attributes—for example, the number of rooms in a house, its area, the size of the front lawn, and so on. The system is given a label y_i, which represents the market value of the house. The system uses the input data from many houses to learn to map the input features, x_i to the value of a house, y_i. The trained model is then presented with a vector x', which consists of the features of a new house, and the model predicts the market value y' of this new house.

Unsupervised Learning

Unsupervised learning has no labels. It only has the inputs $D = x_1, x_2, ..., x_n$ and no labels. The system uses this data to learn the hidden structure of the data so that it can cluster/categorize the data into broad categories. Post learning, when the system is presented with a new data point x', it can match the new data point to one of the learned clusters. Unlike supervised learning, there is no well-defined meaning to each category. Once the data is clustered into a category, based on the most common attributes within a cluster, you can assign some meaning to it. The other use of unsupervised learning is to leverage underlying input data to learn the data distribution so that the system can be subsequently queried to produce a new synthetic data point. This latter approach has been fueling the growth of image- and text generation. The new name for this approach is Generative AI. I elaborate on this in a section to follow.

Many times, unsupervised learning is used for feature extraction, which is then fed to a supervised learning system. By clustering data, you can first identify the hidden structure and remap the data to a lower dimensional form. With this lower dimensional data, supervised learning can learn faster. In this case, unsupervised learning is used as a feature extractor.

There is yet another way to leverage the unsupervised learning approach. Consider a case in which you have a small amount of labeled data and a large amount of unlabeled data. The labeled data and the unlabeled data are first clustered together. Next, in each such cluster, the unlabeled data is assigned labels based on the strength of labeled data in that cluster. You are basically leveraging labeled data to assign labels to the unlabeled data. The completely labeled data is next fed into the supervised learning algorithm to train a classification system. This approach of combining supervised and unsupervised learning together is known as *semi-supervised learning*. Semi-supervised learning is a good option when a large amount of unlabeled data has been collected and you want to avoid the additional workload of having to label all of it.

Reinforcement Learning

Now that you've learned about the supervised and unsupervised approaches, this section turns to the third broad category of machine learning—reinforcement learning. It is distinct from all the approaches discussed so far.

Let's first look at an example. Say you are trying to design an autonomous vehicle that can drive on its own. You have a car called the *agent;* that is, a system or an algorithm that is learning to drive on its own. It is learning the behavior of driving. Its current coordinates, speed, and direction of motion when combined as a vector of numbers is known as its *current state*. The agent uses its current state to make a decision to either apply the brake or press on the gas pedal. It also uses this information to turn the steering to change the direction of the car's motion. The combined decision of "breaking/accelerating" and "steering the car" is known as an *action*. The mapping of a specific current state to a specific action is known as a *policy*. When the agent's action is good, it will yield a happy outcome, and when an action is bad, it will result in an unhappy outcome. The agent uses this feedback of the outcome to assess the effectiveness of its action. The outcome as feedback is known as a *reward* that the agent gets for acting in a particular way in a specific state. Based on the current state and its action, the car reaches a new set of coordinates, speed, and direction. This is the *new state* that the agent finds itself in based on how it acted in the previous step. Who provides this outcome and decides the new state? It is the surroundings of the car, and it is something that the car/agent has no control over. Everything else that the agent does not control is known as the *environment*. I have a lot more to say about each of these terms throughout the book.

In this setup, the data provided by the system in the form of state vectors, actions taken, and rewards obtained is sequential and correlated. Based on the action the agent takes, the next state and reward obtained from the environment could change drastically. In the previous example of the autonomous car, imagine a pedestrian who is crossing the road in front of the car. In this situation, the action of accelerating versus breaking would have very different outcomes. Accelerating the car could lead to injuring the pedestrian as well as damaging to the car and harming its occupants. Breaking could lead to avoiding any damage and the car continuing on its journey after the road is clear.

In reinforcement learning, the agent does not have prior knowledge of the system. It gathers feedback and uses that feedback to plan/learn actions to maximize a specific objective. As it does not have enough information about the environment initially, it must explore to gather insights. Once it gathers "enough" knowledge, it needs to exploit that knowledge and adjust its behavior to maximize the objective it is chasing. The difficult part is that there is no way to know when the exploration is "enough." If the agent continues to explore even after it has perfect knowledge, it is wasting resources trying to gather new information, of which there is none. On the other hand, if the agent prematurely assumes that it has gathered enough knowledge, it may end up optimizing based on incomplete information and may perform poorly. This dilemma of when to explore and when to exploit is a core recurring theme of reinforcement learning algorithms. As you learn about different behavior optimization algorithms in this book, you will see this issue play out again and again.

Emerging Sub-branches

As the discipline of ML is seeing rapid advances and as ML-based solutions are becoming more and more sophisticated, the boundaries between the three traditional branches of machine learning are blurring. The last four to five years has seen innovative combinations of these three approaches, which has given rise to some common patterns, especially ones involving large volumes of unlabeled data that are used to develop models that act as supervised models. I discuss now the most prominent sub-branches.

Self-Supervised Learning

Self-supervised learning is a sub-branch of machine learning that aims to learn useful representations from unlabeled data without human supervision. Unlike supervised learning, where the data is labeled with the correct output beforehand, self-supervised learning creates its own labels from the data. For example, a self-supervised learning system can generate labels by masking some parts of the data and trying to predict them from the rest of the data. This way, the system can learn general features and patterns that can be useful for various downstream tasks. Some examples of self-supervised learning are as follows:

Masked language modeling: This method randomly masks some words in a sentence and tries to predict them from the remaining words. For example, given the sentence "She went to the [MASK] yesterday", the system tries to fill in the blank with the most likely word. This way, the system can learn the syntax and semantics of natural language and the relationships between words. A popular model that uses this method is called BERT (Bidirectional Encoder Representations from Transformers).

Next sentence prediction: This method takes two sentences as input and tries to predict whether they are consecutive or not. For example, given the sentences "He bought a new car" and "It was very expensive", the system tries to determine if they are from the same paragraph or not. This way, the system can learn the coherence and logic of natural language and the structure of paragraphs. This method is also used by BERT, along with masked language modeling.

Denoising autoencoding: This method randomly corrupts some parts of the input data and tries to reconstruct the original data from the corrupted version. For example, given the sentence "She wnt to the park yestrday" (words misspelled), the system tries to correct the spelling errors and output the sentence "She went to the park yesterday". This way, the system can learn to recover information and reduce the noise in the data. A popular model that uses this method is called GPT-3 (Generative Pre-trained Transformer 3).

Another domain where self-supervised learning is widely used is computer vision, where large amounts of image data are available but not labeled. Self-supervised learning methods can leverage the visual features and patterns in images to learn robust image representations that can be useful for various downstream tasks. Some examples of self-supervised learning methods in computer vision are:

Contrastive learning: This method takes pairs of images as input and tries to determine whether they are similar or different. For example, given two images of the same object from different angles, the system tries to recognize that they are related.

Conversely, given two images of different objects, the system tries to distinguish that they are not related. This way, the system can learn the invariant and discriminative features of images and the similarity and differences between images. A popular model that uses this method is called SimCLR (Simple Contrastive Learning of Visual Representations).

Generative modeling: This method takes an image as input and tries to generate a new image that is similar to the input but not identical. For example, given an image of a face, the system tries to generate a new face that has different facial features but still looks realistic. This way, the system can learn the generative and creative aspects and the diversity and variability of images. A popular model that uses this method is called StyleGAN (Style-Based Generator Architecture for Generative Adversarial Networks).

Self-distillation: This method takes a large and complex model as input and tries to compress it into a smaller and simpler model that can perform the same task. For example, given a large image classification model that can recognize thousands of classes of images, the system tries to create a smaller model that can achieve the same accuracy with less computational resources. This way, the system can learn to optimize and simplify the model and retain the essential information and functionality of the model. A popular model that uses this method is called DistilBERT (Distilled BERT).

The field of self-supervised learning is fueling the explosive growth of AI post COVID-19. In a matter of a few years, it has opened up unimaginable human quality models that were unthinkable a few years back. The approaches of generating novel images and text based on instructions and reference context have been combined under a new category called Generative AI.

Generative AI

Generative AI is a sub-branch of artificial intelligence that focuses on creating novel content or data from scratch, such as images, text, music, or speech. Generative AI models learn from a large amount of data and use statistical techniques to generate new data. This new data resembles the original data distribution but is not an exact copy of any existing sample. Generative AI can be used for various purposes, such as data augmentation, artistic expression, content creation, data synthesis, and anomaly detection.

One of the most popular Generative AI techniques is generative adversarial networks (GANs), which consists of two neural networks: a generator and a discriminator. The generator tries to create fake data that can fool the discriminator, while the discriminator tries to distinguish between real and fake data. The two networks compete with each other and improve over time, until the generator can produce realistic data that can deceive the discriminator. GANs have been used to generate high-quality images of faces, animals, landscapes, and objects, as well as to manipulate and enhance existing images, such as face aging, style transfer, super-resolution, and inpainting.

Another Generative AI technique is variational autoencoders (VAEs), which are a type of autoencoder that learns to encode the data into a latent space and then decodes it back to the original space. The latent space is a lower-dimensional representation of the data that captures its essential features. VAEs also impose a probabilistic distribution on the latent space, such as a Gaussian distribution, and then sample from it to generate new data. VAEs can generate diverse and smooth data, such as images of handwritten digits, faces, and flowers, as well as interpolate between different data points in the latent space.

A third Generative AI technique is autoregressive models, which are a type of neural network that generates data sequentially, by predicting the next element in the sequence based on the previous elements. Autoregressive models can capture long-term dependencies and complex patterns in the data, such as natural language, music, and speech. Autoregressive models have been used to generate coherent and fluent text, such as stories, articles, summaries, and translations, as well as to generate realistic and expressive music and speech, such as songs, melodies, and voices. The GPT family of text-generation Large Language Models follow this paradigm.

A fourth Generative AI technique is deep learning diffusion models, which are a type of probabilistic model that learns to reverse the process of adding random noise to the data. The idea is to start with a noisy version of the data and gradually remove the noise until the original data is recovered. By doing so, the model learns to capture the distribution of the data and can generate new data by sampling from it. Deep learning diffusion models can generate high-fidelity and diverse data, such as images of faces, animals, and objects, as well as conditional generation based on text or class labels.

Generative AI vs Other Learning Paradigms

One way to categorize Generative AI models is based on how they learn from the data. As you have learned, there are three main types of learning paradigms: supervised, unsupervised, and self-supervised.

Supervised learning is when the model learns from labeled data, that is, data that has some ground truth or desired output associated with it.

Unsupervised learning is when the model learns from unlabeled data, that is, data is grouped based on some similarity or difference criteria.

Self-supervised learning is when the model learns from unlabeled data by creating its own labels or pseudo-labels based on some pretext task or objective. Self-supervised learning is useful for tasks that require learning rich and generalizable representations of the data that can be used for downstream tasks.

Another way to categorize Generative AI models is based on how they generate new data or content. There are two main types of generative models: explicit and implicit.

Explicit generative models are those that explicitly model the probability distribution of the data or the likelihood of generating a given data point. For example, a variational autoencoder (VAE) is an explicit generative model that learns to encode the data into a latent space and decode it back into the data space while maximizing the likelihood of the data given the latent variables. Explicit generative models can be used for tasks that require estimating the probability of the data or sampling from the data distribution.

Implicit generative models are those that implicitly learn the data distribution by optimizing a different objective function than the likelihood function. For example, a generative adversarial network (GAN) is an implicit generative model that learns to generate realistic data by playing a minimax game between a generator and a discriminator.

Semi-supervised learning is a hybrid approach that combines supervised and unsupervised learning methods.

To summarize, *Generative AI* can be seen as a subset of unsupervised learning or self-supervised learning, depending on how the model learns from the data. Generative AI can also be seen as a distinct type of learning paradigm, depending on how the model generates new data or content. Generative AI can be compared with other learning paradigms based on their goals, methods, and applications.

The next section returns to the topic of reinforcement learning (RL).

Core Elements of RL

A reinforcement learning system can be broken into four key components: policy, rewards, value functions, and model of the environment.

Policy is what forms the intelligence of the agent. An agent gets to interact with the environment to sense the current state of the environment, for example, a robot getting visual and other sensory inputs from the system, also known as the *current state* of the environment or the current observation data being perceived by the robot. The robot, like an intelligent entity, uses this current information and possibly the past history to decide what to do next, that is, what *action* to perform. The *policy* maps the state to the action that the agent takes. Policies can be *deterministic*. In other words, for a given state of environment, there is a single fixed action that the agent takes. Sometimes the policies can be *stochastic*; in other words, for a given state, there are multiple possible actions that the agent can take.

Reward refers to the goal/objective that the agent is trying to achieve. Consider a robot trying to go from Point A to Point B. It senses the current position and takes an action. If that action brings it near to its goal B, the reward should be positive. If it takes the robot away from Point B, it is an unfavorable outcome, the reward should be negative. In other words, the reward is a numerical value indicating the "goodness" of the action taken by the agent based on the objective/goal it is trying to achieve. The reward is the primary way for the agent to assess if the action was good or bad and use this information to adjust its behavior, thus optimizing the policy that it is learning.

Rewards are an intrinsic property of the environment. The reward obtained is a function of the current state the agent is in and the action it takes while in that state. Rewards and the policy being followed by the agent define the *value functions*.

- *Value* in a state is the total cumulative reward an agent is expected to get, based on the state it is in and the current policy it is following.

- *Rewards* are immediate feedback from the environment based on the state and action taken in that state. Unlike value, rewards do not change based on an agent's actions. Taking a specific action in a specific state will always produce the same reward.

Value functions are like long-term rewards that are influenced not only by the environment but also by the policy the agent is following. Value exists because of the rewards. The agent accumulates rewards as it follows a policy and uses these cumulative rewards to assess the value in a state. It then makes changes to its policy to increase the value of the state.

You can connect this idea to the *exploration-exploitation dilemma* discussed earlier. There may be certain states in which an optimal action may bring an immediate negative reward. However, such an action may still be optimal, as it may put the agent in a new state from which it can get to its goal a lot faster. An example is reaching a goal by crossing over a hump that includes a shorter length path versus taking a detour with a longer path and avoiding the hump. In many cases, the detour, although longer, may be an easier path overall.

Unless the agent explores enough, it may not uncover these optimal paths and may end up settling for a less than optimal path. However, having discovered the better path, it has no way of knowing whether it still needs more exploration to find yet another quicker path to a goal or whether it's better to just exploit its prior knowledge to race toward the goal.

Chapters 2 to 7 focus on algorithms that use the previously described value functions to find optimal behavior/policy.

The last component is the model of the *environment*. In some approaches of finding optimal behavior, agents use interactions with the environment to form an internal model of the environment. Such an internal model helps the agent plan, that is, consider one or more chains of actions to assess the best sequence of actions. This method is called *model-based* learning. At the same time, there are other methods that are completely based on trial and error. Such methods do not form any model of the environment. Hence, these are called *model-free* methods. The majority of the agents use a combination of model-based and model-free methods to find the optimal policy.

Deep Learning with Reinforcement Learning

In recent years, a sub-branch of machine learning involving models based on neural networks has exploded. With the advent of powerful computers, the abundance of data, and new algorithms, it is now possible to train models to generalize based on raw inputs like images, text, and voice, similar to the way humans operate. The need for domain-specific handcrafted features to train the models is being replaced with powerful neural-network-based models under the sub-branch of deep learning.

In 2014, DeepMind successfully combined deep learning techniques with reinforcement learning to learn from the raw data collected from an environment without any domain-specific processing of the raw input. Its first success was converting the conventional Q-learning algorithm under reinforcement learning into a deep

Q-learning approach that was named Deep Q Networks (DQN). Q-learning involves an agent following some policy to gather experiences of its actions in the form of a tuple of the current state, the action it took, the reward it got, and the next state it found itself in. The agent then uses these experiences with Bellman equations in an iterative loop to find an optimal policy so that the value function (as explained earlier) of each state increases.

Earlier attempts to combine deep learning with reinforcement learning had not been successful due to the unstable performance of the combined approach. DeepMind made some interesting and smart changes to overcome instability issues. It first applied the combined approach of traditional reinforcement learning and deep learning for developing game-playing agents for Atari games. The agent would get a snapshot of the game and have no prior knowledge of the rules of the game. The agent would use this raw visual data to learn to play Atari video games. In many cases, it achieved human-level performance. The company subsequently extended the approach to develop agents that could defeat champion human players in a game like Go. The use of deep learning with reinforcement learning has given rise to robots that act a lot more intelligently, without the need to handcraft domain-specific knowledge.

And most recently it has been applied to Large Language Models (LLMs) in the form of RLHF (Reinforcement Learning from Human Feedback) to make LLMs follow instructions and generate human-quality text based on the prompts. This is what made the world take notice of chatGPT3.5 and the emergence of new opportunities that arose.

This is an exciting and fast-growing field. Chapter 6 covers deep learning with RL. Most of the algorithms from Chapter 6 onward involve a combination of deep learning and reinforcement learning.

Examples and Case Studies

To motivate you, this section looks at various uses of reinforcement learning and explains how it is helping solve some real-world problems today.

Autonomous Vehicles

This section looks at the field of autonomous vehicles (AVs). AVs have sensors like LiDAR, radar, cameras, and so on, that sense their nearby environment. These sensors are then used to carry out object detection, lane detection, and so on. The raw sensory data and object detection are combined to get a unified scene representation that is used for planning a path to a destination. The planned path is next used to feed inputs to the controls to make the system/agent follow that path. The motion planning is the part in which trajectories are planned.

A concept like inverse reinforcement learning in which one observes an expert and learns the implied objective/reward based on the expert's interaction can be used to optimize cost functions to come up with smooth trajectories. Actions such as overtaking, lane changing, and automated parking also leverage various parts of reinforcement learning to build intelligence into the behavior. The alternative would be to handcraft various rules, and that can never be exhaustive or flexible.

Robots

Using computer vision and natural language processing or speech recognition with deep learning techniques has added human-like perceptions to autonomous robots. Further, combined deep learning and reinforcement learning methods have resulted in teaching robots to learn human-like gaits to walk, pick up and manipulate objects, or observe human behavior through cameras and learn to perform like humans.

Recommendation Systems

Today recommender systems are everywhere. Video sharing/hosting applications, YouTube, TikTok, and Facebook suggest videos that you might like to watch based on your viewing history. When you visit any e-commerce site, based on the current product you are viewing and your past purchase patterns or based on the way other users have acted, you are presented with other similar product recommendations.

All such recommender engines are increasingly driven by reinforcement learning-based systems. These systems continually learn from the way users respond to the suggestions presented by the engine. A user acting on the recommendation reinforces these actions as good actions given the context.

Finance and Trading

Because of its sequential action optimization focus, wherein past states and actions influence the future outcomes, reinforcement learning finds significant use in time-series analysis, especially in the field of finance and stock trading. Many automated trading strategies use a reinforcement learning approach to continually improve and fine-tune the trading algorithms based on the feedback from past actions. Banks and financial institutions use chatbots that interact with users to provide effective, low-cost user support and engagement. These bots again use reinforcement learning to fine-tune their behavior. Portfolio risk optimization and credit scoring systems have also benefitted from RL-based approaches.

Healthcare

Reinforcement learning finds significant use in healthcare, be it generating predictive signals and enabling medical intervention at early stages or be it robot-assisted surgeries or managing the medical and patient data. It is also used to refine the interpretation of imaging data, which is dynamic in nature. RL-based systems provide recommendations learned from its experiences and continually evolve.

Large Language Models and Generative AI

One of the applications of reinforcement learning in Generative AI is to create Large Language Models that can generate coherent and diverse text. Reinforcement learning helps optimize the language models for specific objectives, such as relevance, informativeness, or creativity. For example, RLHF (Reinforcement Learning from Human Feedback) is a technique that uses human ratings as rewards to fine-tune the language models and improve their quality and fluency. RLHF has been used to generate more engaging and personalized dialogues, summaries, and stories.

Game Playing

Finally, I cannot stress enough the way RL-based agents are able to beat human players in many board games. While it may seem wasteful to design agents that can play games, there is a reason for this. Games offer a simpler idealized world, making it easier to design, train, and compare approaches. Approaches learned under such idealized environments/setups can be subsequently enhanced to make agents perform well in real-world situations. Games provide a well-controlled environment to research deeper into the field.

As stated, deep reinforcement learning is a fascinating and rapidly evolving field, and I hope to provide you with a solid foundation to get started in your journey to master this field.

Libraries and Environment Setup

All the code examples in this book are in Python and they use various Python packages/libraries, including PyTorch, TensorFlow, Gymnasium RL environments, Stable Baselines 3, and a few other ones. All the accompanying code has been set up as Jupyter notebooks to make the code execution interactive, with detailed explanations accompanying the code. Jupyter Notebook is a great way to mix the code and explanations with rich format. There are many ways to set up the environment and I will talk about a few common ways with step-by-step instructions. Note that the code is self-contained and can be run in any type of Python environment, local or cloud.

As the book covers quite a few variations, readers should be able to use these instructions as guidance to set up code execution environments on other platforms of their choice. However, unless there is a reason, I suggest you choose one of the two recommended ways to go about setting it up—either a local environment or cloud-based Google Colab. Most of the notebooks have been designed to run on a local computer with just CPU. To leverage GPUs, you may need to do some platform-specific installs and make some minor code changes in relevant notebooks requiring GPU processing.

Local Install (Recommended for a Local Option)

I first walk through the steps required for a local installation. This is the preferred approach for running the code locally. Follow these steps:

1. I recommend Python version 3.9, as the code has been tested on this specific version of Python. Having said that, the code should work on most of the recent Python versions, including 3.10 and 3.11.

    On a Windows machine, I recommend using the WSL2-based Ubuntu distribution. On a Mac or Linux (especially the latest Ubuntu distributions), you should already have everything ready. If you are new to WSL2, refer to the https://learn.microsoft.com/en-us/windows/wsl/install link to get it ready. All the code in the book has been tested using WSL2 with Ubuntu 22.04. It should also work flawlessly on most other versions of Linux.

2. Having identified the platform on which you will set up the local environment, the next step is to install a specific version of Python. This section explains two approaches. One is creating a virtual environment based on venv, the approach I take in the book. The other one is to set up a Miniconda-based virtual environment. Step 3 walks you through the venv setup and Step 4 details the miniconda based setup.

3. VENV-based virtual environment: On Windows WSL and Ubuntu, you can use the apt-get command on your WSL2/Ubuntu shell to install the specific version of Python, 3.9.x in this case. Run the following set of commands to do so:

```
sudo add-apt-repository ppa:deadsnakes/ppa
sudo apt-get update
sudo apt-get install python3.9
```

Next, create and activate a virtual environment based on Python 3.9 using the following commands:

```
# Install venv package for python 3.9
sudo apt install python3.9-venv

# Make a folder for venv virtual environments
mkdir ~/.venvs

# Create a new virtual environment
python3.9 -m venv ~/.venvs/drl

# Activate the new venv
source ~/.venvs/my-venv-name/bin/activate
```

At this point, the prompt in the shell window should show (drl) as a prefix on the command line. You can check the version of Python installed by using the following command on the shell prompt: python -V. It should print something like: Python 3.9.18 with "18" in the end replaced with the latest sub-version 3.9 has at the point in time you do the installation. You can also check the pip version with the command: `pip3 --version`. This should print the version number of whatever pip is linked in your virtual environment that you just installed.

To deactivate the virtual environment, you can use the `deactivate` command, which will deactivate the virtual environment and return you back to the original system default version of Python. Do not do this right now, as you need to be in the activate Python3.9.x version to follow the remaining steps of the installation.

At this point, you can move to Step 5. Step 4 uses Conda environment management to install a specific Python version and creates an isolated environment for the execution of code in this book.

Make sure that all the commands that I ask you to run in the upcoming steps are carried out in the same terminal where you activated the new venv environment.

4. Conda-based environment: Conda provides package dependency and environment management for any language. You can visit the `https://conda.io/projects/conda/en/latest/index.html` link for more details on what problem Conda solves, as well as its advantages and disadvantages. This section works through the steps for installing Conda on WSL2/UNIX. The links also have specific instructions for installing the same on macOS, both in the native way or with the help of Homebrew, a popular package manager for macOS.

   a. Visit `https://docs.conda.io/en/latest/miniconda.html` and download the Miniconda install for your platform. Choose the latest Python3.x version. If you already have Anaconda or Miniconda installed, you can skip this step. Run the downloaded program to install Miniconda on your local machine.

   b. Next, create a specific environment with Python3.9 version. Open a shell window using the following command:

```
conda create -n drl python=3.9
```

where `drl` is the name of the environment. Answer Yes to all the prompts.

c. Switch to the new environment you created using the following command:

```
conda activate drl
```

Make sure that all the commands that I ask you to run in the upcoming steps are carried out in the same terminal where you activated the new Conda environment.

Having installed a specific Python version, you can continue to Step 5.

5. With venv (Step 3) or Conda (Step 4) activated, follow along to complete the rest of the steps. Create a local folder and download or clone the source code provided in the links to the book. A mirror repository maintained by the author with the latest updates can be found at `https://github.com/nsanghi/drl-2ed`.

Change the directory (`cd`) to the newly created local folder and run the following command to make a local copy of the source code accompanying the book:

```
git clone https://github.com/nsanghi/drl-2ed.git
```

This command will create another subfolder named `drl-2ed` in the current folder where you ran the command. You can also visit the publisher's source code link to clone the repository or download and unzip the code.

6. Setting up the dependencies: You are now ready to install all the required libraries with the help of two commands, as follows:

```
# First install some required packages - For Linux and
Windows-wsl users
apt-get install swig cmake ffmpeg freeglut3-dev xvfb \
git-lfs

git lfs install
```

These commands install the required system packages, which are required by some of the Python packages you will be installing next. On macOS, I recommend using the brew command from Homebrew to install these packages.

```
# For mac-OS users
brew install swig cmake ffmpeg freeglut3 git-lfs
git lfs install
```

Next, install the required Python packages. These have been provided as a requirements.txt file, which is a preferred and reproducible way to share environments in the Python ecosystem. Make sure that you are currently in the drl-2ed sub-folder, which was created with the git clone command or the copy and unzip commands in Step 5. Once you have confirmed the same, you can run the following command to install all the required dependencies. Note that it may take five to ten minutes for the installation to finish. Time for you to grab a cup of coffee while the installation finishes.

```
# install python packages from requirements.txt file
pip install -r requirements.txt
```

On macOS if you face installation or build errors while running the above command, please run "pip install --use-pep517 pymunk" and then rerun the above command. After this command is executed and completed, you have everything required to explore and execute the code.

Steps 1 to 6 need to be run only once. After the initial installation, you do not need to run these again unless you want to do a reinstall.

7. As part of the Python installation, you have also installed the JupyterLab package. JupyterLab is the latest web-based interactive development environment for notebooks, code, and data. Its flexible interface allows users to configure and arrange workflows in data science, scientific computing, computational journalism, and machine learning. You can use JupyterLab interface to open a notebook or even a command shell.

Make sure that your venv or Conda environment is active and you are currently inside the drl-2ed directory, where you cloned/copied the code. You can now start the JupyterLab session with this command:

```
jupyter lab
```

It will start the Jupyter session and will print the following instructions somewhere in the middle of the output created by this command:

```
To access the server, open this file in a browser:
    file:///home/nsanghi/.local/share/jupyter/runtime/jpserver-1608-
    open.html
Or copy and paste one of these URLs:
    http://localhost:8888/lab?token=<random string>
    http://127.0.0.1:8888/lab?token=<random string>
```

Click any of the links to open your default browser. You should see a page similar to the one in Figure 1-2. You may navigate to any of the chapter specific folders and double-click the specific notebooks (files with the .ipynb extension) to open them. You are ready to follow the notebook instructions and code execution.

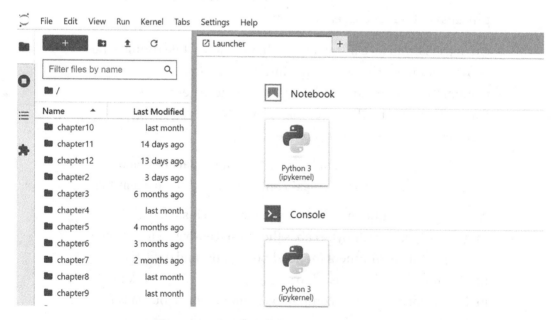

Figure 1-2. *JupyterLab landing page*

Local Install with VS Code

If you want to use VS Code, you can still follow all the steps for "local Install" and then use VS Code to open, run, and execute the notebooks. It provides a similar experience to a local install, but with some additional capabilities. You may also use this if you prefer to use the richer GUI experience VS Code offers and if VS Code is already a part of your developer experience. The high-level steps to follow are as follows:

1. Install VS Code on your local machine. Refer to the `https://code.visualstudio.com/` link for detailed setup instructions.

2. Next, you need to enable a few plugins for Python: Python, Python Debugger, PyLance, Jupyter, Jupyter Notebook Renderers and Jupyter Keymap, Jupyter Cell Tags, and Jupyter Slide Show. Some of these are optional but installing all of them will give you a richer experience. You can read through each individual plugin and decide which ones you want to skip.

3. Navigate to the `drl-2ed` directory where you cloned/downloaded the source code. Run the following command:

   ```
   code .
   ```

   Do not forget the period (`.`) at the end of the command. This command instructs the system to open VS Code in the current folder.

4. You may see a popup in the bottom-right corner with the message "Folder contains Dev Container configuration file. Reopen folder to develop in a container". **Do not click the** "Reopen in Container" button. Instead click the cross at the top right of the popup to close it. See Figure 1-3.

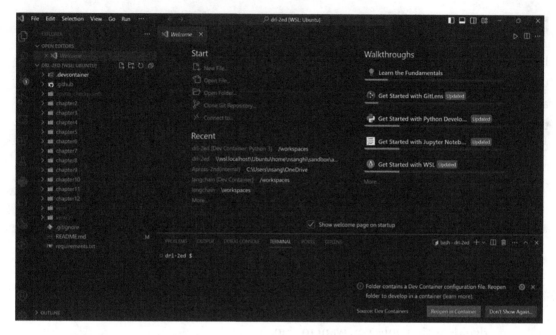

Figure 1-3. *Opening in VS Code*

5.   Next, press Control+Shift+P or Command+Shift+P to open the
     command palette in VS Code and type **Python** to see a list of
     commands applicable to Python in VS Code. Choose the "Python:
     Select Interpreter" option, as shown in Figure 1-4.

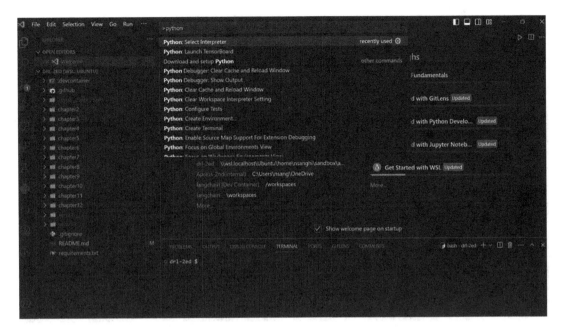

Figure 1-4. *Selecting the Python environment in VS Code*

It will show all the Python environments found in your system.
Choose the venv or Conda environment you created for the local
install. Either you will find this option listed in the command
palette or you can add it using the "Enter Interpreter
Path…" option.

6. You may now open any notebook. At times, even after you have
   chosen the Python environment in the previous step, you again
   may need to select the same environment once more as the
   notebook kernel, before you can execute the code in the notebook.
   You will find that option on the top-right side of the notebook;
   it's called "Select Kernel". Click this and select the same Python
   environment that you created for the local install and the one you
   selected for the Python Interpreter in Step 5. See Figure 1-5.

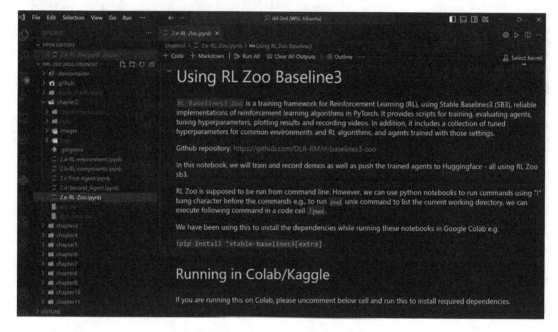

Figure 1-5. *Selecting the Notebook Kernel in VS Code*

You are ready to explore the notebooks and execute the code cells contained in Jupyter Notebook.

Running on Google Colab (Recommended for a Cloud Option)

Google Colab is a free online platform that allows you to write and run Python code in your browser. It is based on Jupyter Notebook. Google Colab provides some advantages over Jupyter Notebook, such as:

- You can access Google Colab from any device with an Internet connection, without installing anything on your local machine.

- You can use the computing power of Google's servers, including GPUs and TPUs, to speed up your code execution and save resources on your own device.

- You can easily share your notebooks with others and collaborate on them in real time.

- You can integrate your notebooks with Google Drive and other Google services, such as Google Sheets, Google Forms, and Google Maps.

To use Google Colab, you need a Google account and a web browser. You can create a new notebook by visiting `https://colab.research.google.com/` or opening an existing notebook from Google Drive. You can also upload a notebook file from your local machine or import one from GitHub or other sources. Once you have a notebook open, you can write and execute code cells, add text and images, and use various tools and features provided by Google Colab. You can also customize your notebook settings, such as the runtime type, the theme, and the keyboard shortcuts.

This book uses the GitHub option to directly open notebooks from the GitHub repository. It is just a matter of personal preference. If you have cloned the source code repository in GitHub to your GitHub account, you will also be able to make changes and issue commits to your repo, thus giving you a seamless developer experience without any local install on your computer. Here are the steps:

1. Navigate to `https://colab.research.google.com/`, which will open the web page with a dialog box giving you various options to open a notebook. For the current workflow, choose the GitHub option.

2. If this is the first time you are using GitHub from inside the Colaboratory (Google Colab), clicking the GitHub option will take you through an OAuth workflow to grant Google Colab access to your GitHub account.

3. After the authentication is done and Google Colab is connected to your GitHub account, you can use the dialog box to open a specific notebook of interest or you can directly type the URL of the GitHub account housing the repository, such as `https://github.com/nsanghi/`.

4. Click the repository dropdown to navigate to the repository. Keep the branch selection to "main" or whatever else is applicable for the repository you are trying to open. A sample of this dialog is shown in Figure 1-6.

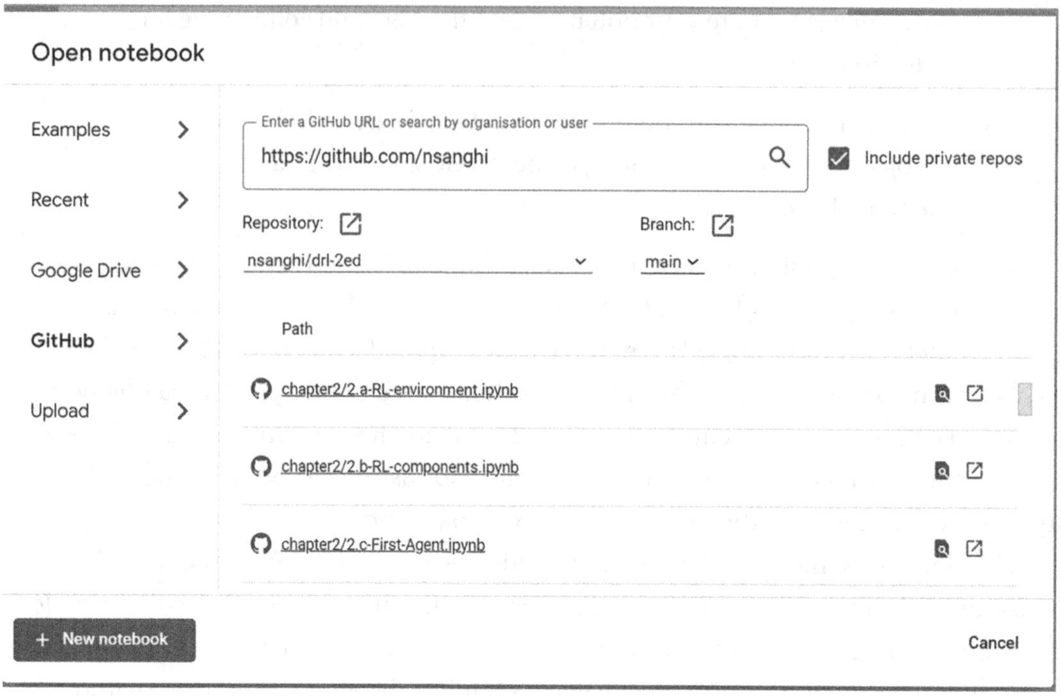

Figure 1-6. *Selecting the repository and notebook in Google Colab*

5. Select the notebook, for example `chapter2/2.a-RL-envrionment.ipynb`. It will close the dialog and open the notebook in the browser with an interface very similar to JupyterLab's. There are some Google Colab-specific enhancements, but most are not relevant to this case. The only thing to note is the runtime type, which you can choose from the "Connect" dropdown on the top-right side of the Colab web page.

This example mostly uses the CPU environment, as the free tier of Google Colab provides limited access to the GPU/TPU environment. Most of the notebooks can be run on a CPU environment. However, from Chapter 6 onward, once you enter the realm of deep reinforcement learning and start training agents on either large text data (such as RLHF) or images (such as Atari games), you'll run those notebooks with GPU enabled.

6. Google Colab comes with most of the libraries preinstalled. However, I have designed the notebooks so that you can run some additional commands from inside the notebook to install system-level packages and the required Python libraries. One example is the 2.e-RL-Zoo.ipynb notebook. Open this notebook in Google Colab following the instructions in Step 4. Once it is opened, you will see two code execution cells that have code commented out, as shown in Figure 1-7.

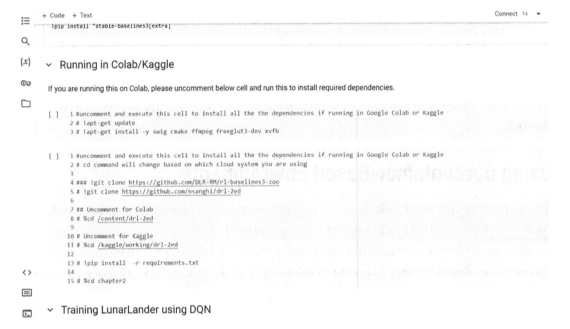

Figure 1-7. *Running code cells for package installation in Google Colab*

The first cell installs the Ubuntu system-level packages. The second cell clones the GitHub repository and runs the requirements.txt file from the cloned repository to install all the dependencies.

Uncomment and run these cells before proceeding with the notebook.

Note that you do not need to run the commented code cells for any kind of local installation-based execution.

Running on Kaggle

You can use a functionality offered by Kaggle that is similar to Google Colab.

Kaggle hosts data science and machine learning competitions, where you can find and explore various datasets, learn new skills, and share your solutions with other users. Kaggle also provides a cloud-based service called Kaggle Kernel, which allows you to run notebooks online without installing anything on your local machine. You can use Kaggle Kernel to access the datasets from the competitions, run code in Python or R, and publish your results as interactive web pages. Kaggle Kernel is a convenient way to experiment with different models and techniques, collaborate with other users, and showcase your work. You can also fork and edit existing notebooks from other users and use them as a starting point for your own projects.

Running the notebook from this book on Kaggle is pretty similar to the way you run it on Google Colab. You can read more about it in the Kaggle documentation and try this out.

Using devcontainer-Based Environments

Another way to run notebooks is to use *devcontainer*, which stands for development container. A `.devcontainer` is a folder in a repository that contains a Docker file and a configuration file that specifies the tools and settings for a development environment. You can use `.devcontainer` to create a consistent and reproducible workspace that can run on any machine that supports Docker and Visual Studio Code (VS Code).

To use devcontainer, you need to have Docker and VS Code installed on your local machine or cloud service. You also need to install the Remote Development extension pack for VS Code, which enables you to work with remote environments. Then, you can find a repository that has a `.devcontainer` folder, which indicates that it supports devcontainer.

Once you have a devcontainer repository, you can clone it to your local machine or cloud service and open it with VS Code. VS Code will detect the `.devcontainer` folder and prompt you to reopen the folder in a container. This will build the Docker image and launch the container with the specified tools and settings. You can then open the notebook file and run it inside the container. You can also modify the code and push the changes back to the repository.

Using devcontainer is a convenient way to work with notebooks that require specific dependencies or configurations, without affecting your local machine or cloud service. You can also share your devcontainer setup with other users and ensure that they have the same development environment as you. Devcontainer is a useful feature for collaborating and deploying code. You can read more about it in the VS Code documentation and try it out.

The source code repository for this book has this capability enabled. This section looks at two ways to use this. One is using a local system, and another is using GitHub's Codespaces, which provides a cloud environment to run code.

Running devcontainer Locally

As discussed previously, you need to have Docker and VS Code installed locally on your machine. To install Docker, head over to the `https://www.docker.com/products/docker-desktop/` link and install the Docker Desktop version for your system. Download and run the binary for your platform and follow all the prompts, choosing the recommended/default options. VS Code can be installed from `https://code.visualstudio.com/`. Once VS Code is installed, also install the Remote Development extension pack for VS Code. After these steps, your local machine will be ready for local deployment of a Docker container. The advantage of this approach is that your development environment is housed in a Docker container, completely isolating your development environment from your local machine. Regardless of your local machine's operating system, the code will run the same Docker configuration—in this case, it's a Ubuntu image with all the required dependencies. The following steps walk through downloading the source code and running it locally:

1. Navigate to a folder of your choice on the local machine. If you are on a Windows machine, you can choose the WSL2 or Windows subsystem.

2. Clone the source code repository locally with this command:

    `git clone https://github.com/nsanghi/drl-2ed.git`

    This command will create another subfolder named `drl-2ed` in the current folder where you ran this command. You can also visit the publisher's source code link to clone the repository or download and unzip the code.

3. Navigate to the `drl-2ed` folder created with the `git` command and run VS Code from there with this command:

```
code .
```

Do not forget the period (`.`) at the end of the command. This command instructs the system to open VS Code in the current folder.

4. You may see a popup in the right corner with the message "Folder contains Dev Container configuration file. Reopen folder to develop in a container". Click the "Reopen in Container" button to kick in the Docker build. It will use the development environment image specified inside the `.devcontainer` folder of the repository to create and start the Docker container. This is opposite of the instructions you had when running on VS Code locally, where you did not click the "Reopen in Container" button.

The first time you do this, it has to download the Docker image as well as install the required Ubuntu packages and run the `pip install requirements.txt` command to set up the Python environment. It may take a while—say about ten minutes or so to run the whole setup process. It will save the final Docker container locally and make it visible in your Docker desktop.

The subsequent runs will be very fast, as they will use the saved Docker container. If you need to do a rebuild of the Docker environment, the VS Code command palette has options to do so.

When the code is opened in a .devcontainer specified Docker image, you will see a status like "Dev Container: Python 3" in the bottom-left side of the VS Code, as shown in Figure 1-8. This confirms that you are now working inside a Docker-based dev container.

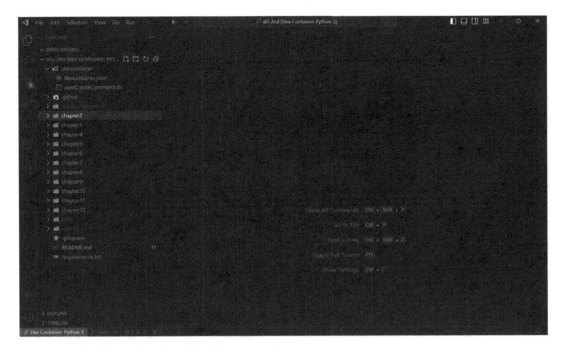

Figure 1-8. *Running in a Docker container*

5. Next, press Control+Shift+P or Command+Shift+P to open the
    command palette in VS Code and type **Python** to see a list of
    commands for Python in VS Code. Choose the "Python: Select
    Interpreter" option, as shown in Figure 1-9. This time the Python
    Interpreter options will be different. Choose the Python 3.9.18
    version, as highlighted in Figure 1-9.

Figure 1-9. *Selecting the Python environment in VS Code*

6. You can open any notebook. You again may need to select the
    same environment as the notebook kernel before you can execute
    the code in the notebook. You will find that option on the top-
    right side of the notebook; it's called "Select Kernel". Click this and
    select the same Python environment with Python 3.9.18.

You can now execute the code just as you would in a JupyterLab environment.

Running on GitHub Codespaces

GitHub Codespaces is a cloud-based development environment that allows you to
create, edit, and run code from any device. You can access your codespaces from Visual
Studio Code, a browser-based editor, or github.com. GitHub Codespaces integrates
seamlessly with GitHub features, such as pull requests, issues, actions, and packages.
You can also customize your codespaces with your own dotfiles, extensions, and settings.
The key advantages of using something like Codespaces are:

- You can start coding quickly without setting up a local environment or installing dependencies.

- You can work on multiple projects and switch between them easily without cluttering your machine.

- You can collaborate with others in real time and share your codespaces with teammates or contributors.

- You can use the same tools and workflows that you are familiar with, such as VS Code, Git, and terminal.

- You can leverage the power and scalability of the cloud and run your code on fast and secure servers.

You can start with Codespaces on the free tier, which will get you 60 hours per month of two-core machines. Here are the steps involved in using Codespaces:

1. Navigate to the repository, for example, `https://github.com/nsanghi/drl-2ed/`.

2. Make sure you are logged into your GitHub account and click the dropdown arrow next to the blue "Code" button. Next, click the "Create Codespace on Main" button, as shown in Figure 1-10. This will open a new tab and start setting up Codespaces using the Docker container specifications provided in the `.devcontainer` folder of the repository.

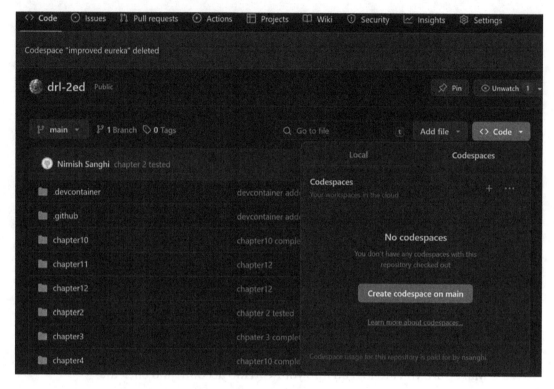

Figure 1-10. *Setting up GitHub Codespaces*

3. Once the container is set up, it will open the repository in a cloud
    version of VS Code and run the pip install to install all the Python
    packages. Like the local Docker container setup, the first time
    this is set up on Codespaces, it may take about five to ten mins.
    However, subsequent runs will be faster, as it will reuse the cloud-
    saved Docker container it created during the first run.

4. Similar to the option of running on a local Docker, you need to
    select the right Python Interpreter with Python 3.9.18, as shown in
    Figure 1-9. You also need to select the same Python version as the
    kernel while running the notebooks.

5. You are now ready to run the code on GitHub Codespaces. This
    option gives you a full developer experience without any local
    install.

6. Because the Docker image is stored in Codespaces and is tied
    to your GitHub account, the next time you visit the GitHub
    repository and click the dropdown next to the "Code" button, you
    will see the previously created codespace, as shown in Figure 1-11.
    You can use that codespace to run the code in subsequent visits.
    It will use the saved Docker container and, in a matter of seconds,
    get your environment running.

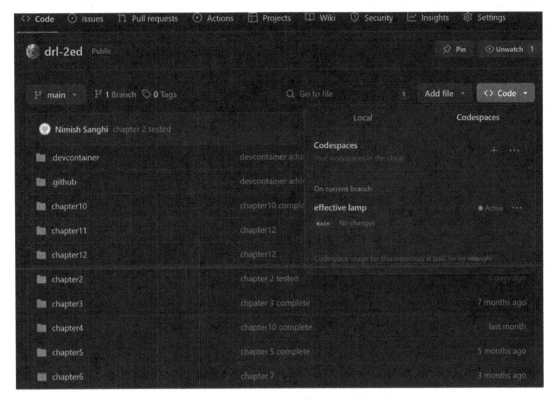

Figure 1-11. *Setting up GitHub Codespaces*

7. You can click the three dots on the right side of the container
    name and open the codespace container. You can also use the
    menu to open the cloud-hosted codespace container inside your
    local VS Code.

Running on AWS Studio Lab

AWS Studio Lab is a free tier version from AWS and can be used as an alternative to Google Colab. It provides you with certain hours of free CPU and GPU access. You need to create an account by navigating to the `https://studiolab.sagemaker.aws/` link.

Once you have created the account and logged in, you can clone the GitHub repository, create a Conda environment, and run the setup to install the Ubuntu packages as well as the Python libraries specified in the `requirements.txt` file. The steps involved are as follows:

1. Once you are logged in, click the "Start Runtime" button to start a Jupyter instance.

2. After the runtime is running, click the "Open Project" circle button. This will open a new tab with the JupyterLab interface.

3. Next you need to clone the repository. You can open a terminal from the JupyterLab interface or choose the Git ➤ Clone Git Repository menu option. Follow the instructions given in the dialog box to complete this step.

4. The next step is to create the Conda environment with Python 3.9, install the Ubuntu packages, and run the pip install. First open a terminal from the JupyterLab interface and run these commands:

```
# create conda environment and activate it
conda create -n drl python=3.9.18
conda activate drl

# install required Linux packages
conda install -c conda-forge swig cmake ffmpeg
conda install conda-forge::xvfbwrapper
conda install conda-forge::freeglut

# Navigate to the drl-2ed folder and
# install the required Python libraries
pip install requirements.txt
```

5. You can now open any of the notebooks from the explorer on the left side. When presented with the option to choose the kernel, choose drl, which is the Conda environment created in a previous step. If you do not see the option, you may need to choose the Amazon SageMaker Studio Lab ➤ Restart JupyterLab menu option.

6. After selecting the right kernel, you may start executing the notebook code cells. The Conda environment you created persists between sessions, making subsequent runs very fast.

Using SageMaker Studio Lab allows you to run the code on-cloud, just like a few other options I have talked about in earlier sections. It requires no local install at all. Figure 1-12 shows a screenshot of AWS Studio Lab in action.

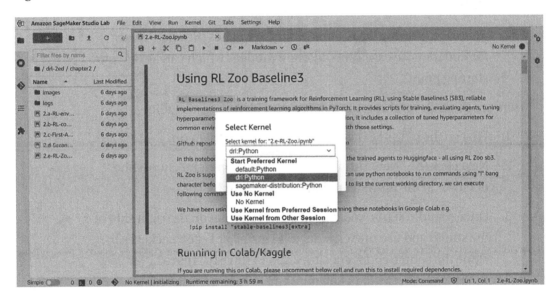

Figure 1-12. *AWS Studio Lab*

Running Using Lightning.ai

Lightning AI Studio (https://lightning.ai/) is a cloud-based AI development platform (similar to Google Colab and Amazon Sage maker Studio Lab) that aims to eliminate the hassle of setting up local environments for machine learning projects. Some key features of Lightning AI Studio include:

- It integrates popular machine learning tools into a single interface. This allows you to build scalable AI apps and endpoints more easily.

- There is no environment setup required. You can code in the browser or connect your local IDE (VS Code or PyCharm). You can also easily switch between CPU and GPU with no environment changes.

- It allows you to host and share AI apps built with Streamlit, Gradio, React JS, and so on. It also enables multi-user collaboration by coding together.

- It provides unlimited storage and the ability to upload and share files as well as connect S3 buckets.

- It enables training models at a massive scale using thousands of GPUs under paid plans.

- There is a growing list of community templates (Studios) ready for different use cases and different models, which can act as your starting point.

Here are the high-level steps to run code on Lightning.ai studio:

1. First go the `https://lightning.ai/` link and sign up.

2. Next click the "New Studio" button on the top right, choose the defaults on the prompt, and then click "Start".

3. Once a studio starts, you can click the terminal on the right side of the studio and use the terminal to a) install `apt-get` packages b) clone the source code repository and c) run `pip install -r requirements.txt` to install required Python packages and libraries.

4. At this point you are ready to start executing Jupyter Notebook. At the time of writing, the free tier version offers a four CPU configuration, making it a powerful experimentation platform in free tier offering. You also get access to GPU resources.

Other Options to Run Code

There are many more choices. If you want to run the code on GPU machines for graphic games or 3D environments, you can use many paid choices. A few of them are listed here. Depending on your need, you should refer to the documentation of the relevant platform to set them up for code execution:

1. Google Vertex AI Notebooks

2. AWS SageMaker

3. MS Azure ML Studio

4. Paperspace

5. Lambda Labs

6. Jarvis Labs

There is a growing list of cloud GPU and CPU providers for deep learning workloads. For the purpose of running the source code of this book, any entry-level configuration is good enough.

Summary

This chapter started by introducing the field of reinforcement learning and the history of how it has evolved from a rigid rule-based decision-making system to a flexible optimal behavior learning system. It did this by learning on its own from prior experiences.

The chapter explained the three sub-branches of machine learning—supervised learning, unsupervised learning, and reinforcement learning. In the context of the supervised and unsupervised methods, the chapter also discussed emerging hybrid approaches such as semi-supervised and self-supervised learning as Generative AI.

I compared the three core approaches of supervised, unsupervised, and reinforcement learning to elaborate on the context in which each of these makes sense. I also talked about the subcomponents and terms that comprise a reinforcement learning setup. These are *agent, behavior, state, action, policy, reward,* and *environment.* I used the example of a car and a robot to show how these subcomponents interact and what each of these terms means.

I also talked about the concept of reward and value functions. The chapter discussed that rewards are short-term feedback and that value functions are long-term feedback of the agent's behavior. Finally, I introduced model-based and model-free learning approaches. Next, I talked about the influence of deep learning in the field of reinforcement learning and explained how DQN started the trend of combining deep learning with reinforcement learning. I also discussed how the combined approach has resulted in scalable learning, including from unstructured inputs such as images, text, and voice.

The chapter moved on to talk about the various use cases of reinforcement learning, citing examples from the fields of autonomous vehicles, intelligent robots, recommender systems, Large Language Models and stock trading, healthcare, and video/board games.

Finally, I walked through various ways to set up a Python environment and be able to run the Jupiter notebooks accompanying this book.

## CHAPTER 2

The Foundation: Markov Decision Processes

As discussed in Chapter 1, reinforcement learning involves sequential decision-making. This chapter formalizes the notion of using stochastic processes under the branch of probability that models sequential decision-making behavior. Although most of the problems you'll study in reinforcement learning are modeled as *Markov decision processes* (MDP), this chapter starts by introducing Markov chains (MC) followed by Markov reward processes (MRP). Next, the chapter discusses MDP in-depth while covering model setup and the assumptions behind MDP.

The chapter then discusses related concepts, such as value functions of state and action value functions of state-action pairs. It follows with an in-depth discussion of the various forms of Bellman equations, like Bellman expectation equations and Bellman optimality equations. It also includes a quick introduction to various types of learning algorithms. With the framework of MDP in place, the chapter switches gears to train a couple of agents, along the way covering a high-level overview of the software libraries you will using in this book. While the chapter's focus is the theoretical foundations of reinforcement learning, it includes examples and exercises to help cement these concepts. There is no better way to learn than to code yourself, which is a recurring theme of this book.

This chapter finishes with a brief discussion of the different approaches to RL training.

Definition of Reinforcement Learning

The previous chapter explained the cycle of an agent interacting with the environment by taking an action based on its current state, getting a numerical reward, and finding itself in a new state. Figure 2-1 illustrates this concept.

© Nimish Sanghi 2024
N. Sanghi, *Deep Reinforcement Learning with Python*, https://doi.org/10.1007/979-8-8688-0273-7_2

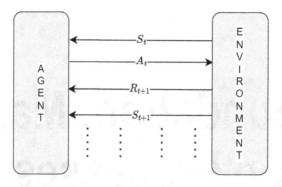

Figure 2-1. *The cycle of agent-environment interaction*

The agent at time t is in a state S_t. From the set of actions, the agent can take in this state it takes a specific action, A_t. At this point, the system transitions to the next time tick $(t + 1)$. The environment responds to the agent with a numerical reward, R_{t+1} as well as puts the agent in a new state, S_{t+1}. The cycle of "state to action to reward and next state" goes on until the time agent reaches some end-goal state, like the end of a game, the completion of a task, or the end of a specified number of time steps.

$$S_t \xrightarrow{\text{Agent Acts}} A_t \xrightarrow{\text{Environment Reacts}} R_{t+1}, S_{t+1}$$

current state   action taken   reward & new state

$$S_0, A_0, R_1, S_1, A_1, R_2, S_2, A_2, R_3, S_3, A_3, R_4, S_4, \ldots.$$

There is a cycle of state, action, reward, and next state $(S_t, A_t, R_{t+1}, S_{t+1})$.

The agent takes an action based on the state it finds itself in, that is, the agent "acts" by taking an action. The environment "reacts" to the agent's action by rewarding the agent with some numeric reward and transitioning the agent to a new state. The agent's objective is to take a sequence of actions that maximizes the sum total of rewards it gets from the environment.

The purpose of reinforcement learning is to have the agent learn the best possible action for each of the states it could find itself in, keeping in mind the cumulative reward maximization objective.

As an example, consider the game of chess. The position of a piece on the board could form the current state (S_t). The agent (player) takes action (A_t) by moving a chess piece. The agent gets a reward (R_{t+1})—let's say 0 for a safe move and -1 for a move leading to checkmate. The game also moves to a new state, (S_{t+1}).

In the literature, states are sometimes referred to as *observations* to distinguish between the fact that in some situations the agent may get to see only partial details of the actual state. This partially observable state is known as *observation*. The agent uses the full or partial state information, as may be available, to make decisions with respect to the action it should take. It is an important aspect in real-life implementations to understand what will be observed by the agent and how detailed it will be. The choice and theoretical guarantee of the learning algorithm can be significantly influenced by the level of partial observability. This chapter initially focuses on situations where state and observations are the same; in other words, the agent knows every possible detail of the current state. But starting in Chapter 5, you will start looking at situations where the state is either not fully known or, even if fully known, needs to be summarized using some kind of approximation.

Let's now go through a couple of examples to understand in depth the cycle of state/observation to action to reward to the next state. The book uses a Python library by *Gymnasium,* which is a maintained fork of the *Gym* library by OpenAI. Gym by OpenAI has stopped new development and all the additions now happen in the Gymnasium library. Gymnasium implements some common simple environments. Let's look at the first environment, called `MountainCar-v0,` from the Gymnasium library.

The first step is to start your Jupyter Notebook and navigate to `2.a-RL-environment.ipynb` or you can open the notebook in Google Colab. Refer to Chapter 1 for how to set up a local environment or the steps to follow to run the notebook directly on Google Colab.

Once you have opened the notebook, read further to explore the environment and related code. In the `MountainCar` environment, there is a hill that the car is trying to climb with the end goal of reaching the flag on the top-right side of the hill. The car is not powerful enough, so the car needs to swing to the left and then accelerate to the right to reach the goal. This back-and-forth swing needs to happen multiple times so that the car can gain enough momentum and reach the flag at the right-top side of the valley. See Figure 2-2.

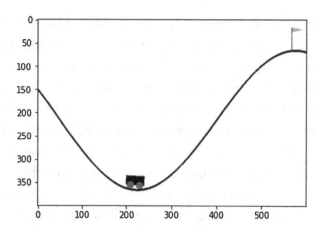

Figure 2-2. *MountainCar-v0 environment. This environment has a two-dimensional state and a set of three discrete actions*

Listing 2-1 shows the code needed to test the MountainCar environment.

Listing 2-1. MountainCar Environment – "2.a-RL-environment.ipynb"

```
import gymnasium as gym

env = gym.make('MountainCar-v0', render_mode="rgb_array")
obs = env.reset()

# It is a random policy. As we progress
# we will use different RL algorithms
# to teach agent to take long-term reward
# maximizing actions
def policy(observation):
    return env.action_space.sample()

# take steps
for _ in range(5):

    # to render environment for visual inspection
    # when you train, you can skip rendering to speed up
    plt.imshow(env.render())

    # policy - i.e. agent uses state to take an action
    action = policy(obs)
```

```
    # pass the action to env which will return back
    # a set of values - new state, reward
    # and a few other things
    obs, reward, terminated, truncated, info = env.step(action)

# close the environment
env.close()
```

Let's walk through the code line by line. You first import the Gymnasium library by using import gymnasium as gym. Gymnasium has multiple environments implemented for reinforcement learning. You will be using some of these environments as you go through the chapters of this book.

Moving on, you instantiate the MountainCar environment with env = gym. make('MountainCar-v0',...). This is followed by initializing the environment through obs = env.reset(), which returns the initial observation. In the case of MountainCar, the observation is a tuple of two values: (x-position, x-velocity). The agent uses observation to find the best action: action = policy(obs).

In Listing 2-1, the agent is taking random actions. However, as you progress through the book, you will learn different algorithms that can be used to find reward-maximizing policies. For the MountainCar environment, there are three possible actions: accelerate left, do nothing, and accelerate right. The agent passes the action to the environment. At this point, the system conceptually takes a time step, moving from time t to time $t + 1$.

The environment executes the action and returns a tuple of five values: *new observation* at time $(t + 1)$, *reward* (r_{t+1}), *terminated* flag, *truncated* flag, and some debug info. Note that the Gym environment from OpenAI had a *done* flag, which has now been split into two flags in the Gymnasium library—*truncated* and *terminated*. The reason for this split is to provide better control. For now, just think of the old done flag as a combination of truncated and terminated flags with the expression (done = terminated or truncated).

Next, the agent uses the new observation to get back a new tuple of five values—*next state, reward, terminated, truncated, and debug* info. This cycle of "state to action to reward to new state" goes on until the game ends (i.e., terminated) or until it times out (i.e., truncated). In this setup, the agent gets a reward of -1 at each time step, with a reward of 0 when the game ends. Accordingly, the agent's objective is to reach the flag in the shortest possible number of steps.

Let's look at yet another environment. Replace the line env = gym. make('MountainCar-v0', ...) with env = gym.make('CartPole-v1', render_ mode="rgb_array") and again run the code in Listing 2-1. See Figure 2-3.

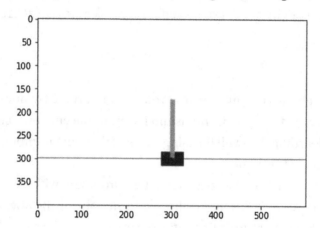

Figure 2-3. *The CartPole-v1 environment. The objective is to keep the pole upright as long as possible*

The CartPole environment has an observation space consisting of four values. The cart's position on the x-axis and velocity along the x-axis are the first two values. The tilt angle of the pole is the third value in the observation tuple and it has to be between −24° to 24°. The fourth value in the observation tuple is the angular velocity of the pole.

Possible actions are 0 and 1, which push the cart to the left or right, respectively. The agent keeps getting the reward of 1 at each time step, thereby incentivizing the agent to balance the pole for the longest possible duration and earn as many points as possible. The game ends if the pole tilts more than 12° on either side or if the cart moves more than 2.4 on either side, that is, the cart reaches either ends or if 200 steps have been taken.

You may have noticed that the code setup is the same in both cases. I just changed one code line to instantiate different environments. The rest of the code to step through the environment and receive feedback remains the same. The abstraction provided by the Gymnasium library makes it easier to test a particular algorithm across many environments. Further, you can also create your own custom environment based on the problem at hand.

This section covered a brief definition of reinforcement learning and the basics of the Gymnasium environment that you will be using. In the upcoming sections, you will learn about the different components of a problem set up as reinforcement learning (RL) in detail.

Agent and Environment

The setup of the agent and environment is a flexible one. The agent is a closed system and it gets the state/observation details from outside the system. It uses policy that could be handcrafted or learned, to take an action with a goal to maximize the rewards the agent is seeking. The agent gets a reward from the environment based on the current observation/state and the action the agent takes. The agent has no direct control over what those rewards will be—rather, the agent can influence it by taking good actions given the state it finds itself in. The agent also does not control the transition from state to state. The transition depends on the environment, and the agent can only influence the outcome indirectly by deciding which action to take in each state it finds itself in.

On the other hand, the environment is everything that is outside the agent. In other words, it could be the rest of whole world. However, the environment is usually defined in a more narrow way: to comprise information that could influence the reward. The environment receives the action the agent wants to take. It provides feedback to the agent in the form of reward and transitions to a new state based on the action the agent took. Next, the environment provides part/all of the revised state information to the agent, which becomes the agent's new observation/state.

The boundary between the agent and the environment is an abstract one. It is defined based on the task at hand and the goal you are trying to achieve in a particular case. Let's look at some examples.

Consider the case of an autonomous car. The agent state could be the visuals from multiple cameras, light detection and ranging (LiDAR), and the other sensor readings as well as geocoordinates. While the "environment" is everything outside the agent, that is, the whole world, the agent's relevant state/observation is only those parts of the world that are relevant for the agent to take an action. The position and actions of a pedestrian two blocks away may not be relevant for an autonomous car to make decisions and hence do not need to be part of the observation/state of the agent. The action space of an autonomous car could be defined in terms of gas-pedal values, breaks, and steering controls. The actions taken by the vehicle result in transitioning the car into a new observation state. This cycle goes on. The agent (i.e., the autonomous car) is expected to take actions that maximize the reward based on a specific goal, for example, going from Point A to Point B.

Let's consider another example of a robot trying to solve a Rubik's Cube. The observation/state in this case would the configuration of the Rubik Cube's six faces, and the actions would be the manipulations that can be performed on the Rubik's Cube.

The reward could be -1 for each time step and 0 at the end of a successful solution, that is, on termination. Such a reward setup will incentivize the agent to find the least number of manipulations to solve the puzzle.

In this setup where you will be using the Gymnasium environment, observations will always be a tuple of various values with the exact composition dependent on the specific Gymnasium environment. Actions will be dependent on the specific environment. Rewards will be provided by the environment with the exact real number dependent on the specific Gymnasium environment. The agent in this case will be the software program that you write. The agent's (software program) main job will be to receive observation from the Gymnasium environment, take an action, and wait for the environment to provide feedback in terms of the reward and the next state.

Although this describes discrete steps—that is, the agent in state (S_t) taking an action (A_t) and, in the next time step, receiving a reward (R_{t+1}) and the next state (S_{t+1})—many times the nature of the real-life problem would be a continuous one. In such cases, you could conceptually divide the time into small discrete time steps, thereby modeling the problem back into the discrete time-step environment and solving it using the familiar setup shown earlier.

At the most general level, when the agent finds itself in state S, it can take one of the many possible actions with some probability distribution over the space of actions. Such policies are known as *stochastic policies*. Further, in some cases the agent may take only one specific action for a given state every time the agent finds itself in that state. Such policies are known as *deterministic policies*. A policy is defined as follows:

$$\pi = p(a|s) \tag{2-1}$$

That is, the probability of taking an action a when the agent is in state s.

Similarly, at a most general level, the reward received and the next state of the agent would be a probability distribution over the possible values of rewards and next states. This is known as *transition dynamics*.

$$p(s',r) = \Pr\{S_t = s', R_t = r \mid S_{t-1} = s, A_{t-1} = a\} \tag{2-2}$$

where S_t and S_{t-1} belong to all possible states, also called state/observation space. A_t belongs to all possible actions, also called action space, and reward r is a numerical value. The previous equation defines the probability of the next state being s' and the reward being *r* when the last state was *s* and the agent took an action, *a*. Figure 2-4 shows the relationship between the state, the action, and feedback from the environment in the form of the next state and reward.

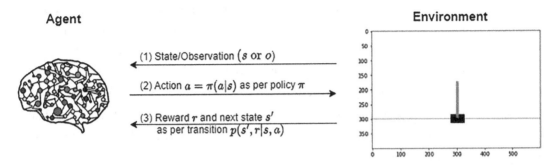

Figure 2-4. *Agent and environment interaction*

Rewards

In reinforcement learning, a reward is a signal from the environment to the agent to let the agent know how good or bad the action was. The agent uses this feedback to fine-tune its knowledge and learn to take good actions that maximize the rewards. It gives rise to some important questions, such as what do you maximize, the immediate reward for the last action or the reward over the complete life history? What happens when the agent does not know enough about the environment? How much should it explore the environment by taking some random steps before it starts? This dilemma is known as the *exploration versus exploitation dilemma*. I will keep coming back to this point as I go through various algorithms. The objective of the agent to maximize the cumulative total rewards is known as the *reward hypothesis*.

To reiterate, a reward is a signal, or a scalar numerical value, that the environment sends back to the agent to inform the agent of the quality of the agent's action and the quality of the resulting state/observation due to that action. Note that the observations could be multidimensional—such as two-dimensional for `MountainCar` and four-dimensional for `CartPole`. Similarly, actions can be multidimensional, for example, acceleration value and steering angle for the autonomous car scenario. However, in each such case, the reward is always a scalar real number. It may seem limiting to have only a single value, but it is not. At the end, the agent is trained for the purpose of reaching a goal, and rewards codify that progress.

Now you'll look at a maze example where the agent is trying to find its way out. You could formulate the reward as the agent getting a reward of -1 at each time step and a reward of 0 at the end of episode. Such a reward configuration incentivizes the agent to come out of the maze with the least possible number of steps and minimize the sum of the negative ones (-1). An alternate reward configuration could be an agent getting a reward of 0 in all the time steps and a reward of 1 at the end of episode when the agent is out of the maze. What do you think will happen to the agent's behavior in the latter setup? The agent has a reason to get out of the maze to collect a reward of +1, but it is in no hurry. It will get the same +1 whether it comes out after 5 steps or 500 steps. How do you modify the situation to push the agent to not focus just on reward collection but to do so in the fastest possible time?

This question leads naturally to the concept of *discounting*. What carries more utility—a reward x after 5 time steps or the same reward x after 500 time steps? Of course, the earlier the better, and hence a reward of +1 after 5 steps is more valuable as compared to +1 reward after 500 steps. You can induce this behavior by discounting the rewards from the future to present. A reward of R from the next time step is discounted to the current time by a discount factor, γ (gamma). The discount factor is a value between 0 and 1. In the maze example, the reward with maze completion in five steps would mean a reward of $\gamma^5 \cdot (+1)$ versus a reward of $\gamma^{500} \cdot (+1)$ for 500 steps to complete. The "return" at time t is defined as follows:

$$G_t = R_{t+1} + \gamma \cdot R_{t+2} + \gamma^2 \cdot R_{t+3} + \cdots + \gamma^{T-t+1} \cdot R_T \tag{2-3}$$

The discount factor is similar to what you see in the financial world. It is the same concept as money now being more valuable than money later.

The discount factor also serves an important mathematical purpose of making sure that total return, G_t, is bounded for continuing tasks. Consider a scenario of a continuing task where each state gives some positive reward. As it is a continuing task that has no logical end, the total reward will keep adding and exploding to infinity. However, with a discounting factor, the total cumulative reward will be capped. Therefore, discounting is always introduced in continuing tasks and is optional in episodic tasks.

It is interesting to note the effect of the discounting factor on the horizon over which the agent tries to maximize the cumulative reward. Consider a discount factor of 0. If you use this discount value in Equation 2-3, you will see that cumulative reward is equal to just the reward at the next instant reward. It would lead the agent to become short-sighted and greedy with respect to the next time step reward, without any focus on future rewards beyond the next time step. Next, consider the other extreme with a discount factor approaching a value of 1. In this case, the agent will become more and more far-sighted because, using a discount factor of 1, you can see that the cumulative reward as defined in Equation 2-3 will give equal importance to all the future rewards. The whole sequence of actions, from the current time step until the end, becomes important and not just the immediate next time step reward.

The previous discussion should emphasize the importance of designing an appropriate reward based on the behavior the agent needs to optimize. The reward is the signal that the agent uses to decide between a good or bad state and/or action. For example, in a game of chess, if you design the reward as the number of opponent pieces captured, the agent may learn to play dangerous moves just to maximize rewards from immediate actions instead of sacrificing one of its own pieces to get into a position of strength and a possible win in a future move. The field of reward design is an open area of research. Having said that, the examples in this book use fairly simple and intuitive reward definitions.

The reward design is not an easy task, especially in the field of continuous control and robotics. Consider the task of a humanoid where the objective of training is, say, to make the agent learn to run for as long as possible. How will the agent know the way the arms and legs need to be moved to learn the task of running? Specific measures, such as the distance of the center of gravity of the agent from the ground, the energy spent in making action, and the torso angle to the ground are combined with trial and error to make the agent learn a good policy. In the absence of a good reward signal shaping the behavior you want the agent to learn, it will take a long time for agent to learn. Even worse, the agent sometimes may learn counterintuitive behavior. A good example is OpenAI's agent to play the video game Coast-Runners that had the objective of finishing the boat race quickly and ahead of other players. The game provided a score for hitting the targets in the way, and there was no explicit score for finishing the game. Under such reward configuration, the agent learned a hilarious and destructive behavior of repeatedly hitting a set of targets and not progressing through the race but scoring 20 percent higher than human players. You can read more about it in the OpenAI blog[1] and see a video of the behavior in action. Rewards need to be designed with care to ensure that autonomous RL agents do not learn potentially dangerous behaviors.

---

[1]https://openai.com/blog/faulty-reward-functions/

In other cases, it is not at all clear how to model the reward function. Consider a robot trying to learn human-like behavior, such as pouring water in a glass from a jug without spilling water or breaking the glass tumbler due to excessive grip force. In such cases, an extension of reinforcement learning, known as *inverse reinforcement learning*, is used to learn the implicit reward function based on observations from watching a human expert perform the task. Chapter 13 briefly talks about this. However, a detailed study of reward shaping and discovery could take a book of its own.

So, in a nutshell, just as the quality of data is important for supervised learning, an appropriate reward function is important to make algorithms train the agent for a desired behavior.

RL has recently found interesting use in the field of Large Language Models. This approach is known as RLHF—Reinforcement Learning from Human Feedback. It was introduced by OpenAI in 2017 and was first used by OpenAI in 2022 to make Large Language Models safer in their InstructGPT model. To learn more, you can read the paper "Training Language Models to Follow Instructions with Human Feedback," by OpenAI Jan 2022.[2] The reward is the human preference of one sentence over another for safety. Recently, Meta has used two reward functions to steer the Llama2 model introduced in July 2023. One reward function models the human preference for safety and another reward function models the human preference for helpfulness. You can read more about it in Meta's bog titled "Llama 2: Open Foundation and Fine-Tuned Chat Models".[3] Specific details of RLHF are mentioned in Section 3.2 of that paper.

Markov Processes

The field of reinforcement learning is based on the formalism of Markov processes. Before diving deep into learning (behavior optimization) algorithms, you need to have a good grasp of the Markov Processes construct. This section goes over Markov Chains followed by Markov Reward Processes and lastly Markov Decision Processes.

---

[2] https://arxiv.org/abs/2203.02155

[3] https://ai.meta.com/research/publications/llama-2-open-foundation-and-fine-tuned-chat-models/

Markov Chains

Let's first talk about what a Markov Property is. Consider the diagram in Figure 2-5. It is trying to model the daily rain status of a city. It has two states—on any given day, whether it rains or not. The arrows from one state to another indicate the probability of being in one of the two states the next day based on the current day's state. For example, if it rains today, the chance of rain the next day is 0.3, and the chance it will not rain the next day is 0.7. Similarly, if there is no rain today, the chance it will continue to be dry tomorrow is 0.8, while there is a 0.2 chance that it will rain tomorrow.

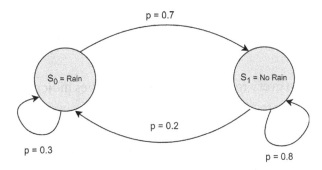

Figure 2-5. *A simple two-state Markov Chain. One state is "rain," and second state is "no rain"*

This model assumes that rain on a given day depends on the state of the previous day, that is, the chance of rain tomorrow is 0.3 if it also rained today, and the chance of rain tomorrow is 0.2 if it did not rain today. Whether it rained yesterday or before that has no impact on the probability of rain tomorrow. This is an important concept. It is known as *independence*, that is, knowing the present (present state at time t) makes the future (future state at time $t + 1$) independent of the past (all past states at time 0, 1, $\cdots$, $t - 1$). Mathematically, you can express this as follows:

$$P(S_{t+1} = s' \mid S_0, S_1, \ldots \ldots S_{t-1}, S_t = \mathrm{s}) = P(S_{t+1} = s' \mid S_t = \mathrm{s}) \tag{2-4}$$

It simply means that the probability of being in a state at time $t + 1$ depends only on the state at time t. The state at time $(t + 1)$ has no dependence on the states before time (t), that is, no dependence on the states S_0 to S_{t-1}.

If the environment provides observations to the agent that is detailed enough, the agent will know what is to be known from its current state and will not need to remember the chain of states/events it had to go through in the past to reach the present. This *Markov independence* is an important assumption that is required to prove the theoretical soundness of reinforcing learning algorithms. In practice, many times you'll still get fairly good results for non-Markov systems. However, it is important to remember that in the absence of the Markov Property, the results cannot be evaluated for theoretical worst-case bounds.

Going back to the Markov Chain diagram in Figure 2-5, you can define transition probability as the probability of moving to state S_{t+1} from state S_t in the prior time step. If a system has m states, the transition probabilities will be a square matrix with m rows and m columns. The transition probability matrix for Figure 2-5 will look like this:

$$P = \begin{bmatrix} 0.3 & 0.7 \\ 0.2 & 0.8 \end{bmatrix}$$

The sum of values in every row will be 1. The row values indicate the probability of going from one given state to all the states in the system. For example, row 1 indicates that the probability from S_1 to S_1 is 0.3 and from S_1 to S_2 is 0.7.

The previous Markov Chain will have a steady state in which there is a defined probability of being in one of the two states on a given day. Let's say that the probability of being in state S_1 and S_2 is given by a vector $S = [S_1\ S_2]^T$. From Figure 2-5, you can see that

$$S_1 = 0.3 \cdot S_1 + 0.2 \cdot S_2 \cdots (a)$$

$$S_2 = 0.7 \cdot S_1 + 0.8 \cdot S_2 \cdots (b)$$

As the system must be in one of the two states at any point in time, you also have:

$$S_1 + S_2 = 1 \cdots (c)$$

The equations in (a), (b), and (c) form a consistent and solvable system of equations. From (a),

$$0.7 \cdot S_1 = 0.2 \cdot S_2 \text{ or, } S_1 = \frac{0.2}{0.7} \cdot S_2$$

Substituting the value of s_1 in (c), you get:

$$\frac{0.2}{0.7} \cdot S_2 + S_2 = 1$$

Which gives:

$$S_2 = \frac{0.7}{0.9} = 0.78$$

Substituting this in (c), gives

$$S_1 = \frac{0.2}{0.9} = 0.22$$

In vector algebra notations, you can specify the relationship at steady state as follows:

$$S^T = S^T \cdot P$$

This relationship can be used to solve steady-state probability iteratively. Listing 2-2 shows a code snippet of this.

Listing 2-2. Markov Chain Example and Its Solution by Iterative Method-"2.b-RL-components.ipynb"

```
# MC with no end

# import numpy library to do vector algebra
import numpy as np

# define transition matrix
P = np.array([[0.3, 0.7], [0.2, 0.8]])
print("Transition Matrix:\n", P)

# define a random starting solution for state probabilities
# Here we assume equal probabilities for all the states
S = np.array([0.5, 0.5])

# run through 10 iterations to calculate steady state
# transition probabilities
for i in range(10):
    S = np.dot(S, P)
    if i % 5 == 0:
        print("\nIter {0}. State Probability vector S = {1}".format(i, S))

print("\nFinal Vector S={0}".format(S))
```

When you run the Python notebook `2.b-RL-components.ipynb`, the output produced is given here, which matches with the values you got from solving equations (*a*), (*b*), and (*c*) together:

`Final Vector S=[0.22222222 0.77777778]`

The formulation in Figure 2-5 has no start or end state. This is an example of a continuing task. There is another class of formulation that has one or more end states. Look at Figure 2-6. This is known as an *episodic task,* in which the agent starts in some state and goes through many transitions to eventually reach an end state. There could be one or more end states with different outcomes. In Figure 2-6, the end state is the successful completion of an exam resulting in a certificate. In a game of chess, there could be three end states: win, loss, or draw.

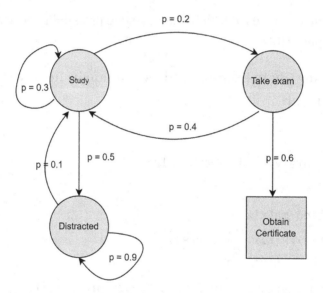

Figure 2-6. *An example of an episodic Markov Chain with one end state depicted by way of a square box*

Like the continuing formulation, there is a concept of a transition probability matrix. First you define the states with a variable name:

S_1 = "Study"; S_2 = "Distracted", S_3 = "Take Exam" ; S_4 = "Obtain Certificate"

The Transition matrix will look like this:

$$P = \begin{bmatrix} 0.3 & 0.5 & 0.2 & 0 \\ 0.1 & 0.9 & 0 & 0 \\ 0.4 & 0 & 0 & 0.6 \\ 0 & 0 & 0 & 1 \end{bmatrix}$$

In the case of the episodic task as shown earlier, you can look at multiple runs, with each run called an *episode*. Let the start state always be S_1. As you have only one end state—"Obtain Certificate"—the episodes will always end in S_4, as shown here:

$$S_1, S_2, S_2, S_1, S_3, S_4$$

$$S_1, S_2, S_2, S_1, S_3, S_1, S_2, S_3, S_4$$

$$S_1, S_2, S_2, S_2, S_2, S_2, S_2, S_2, S_2, S_2, S_2, S_2, S_2, S_1, S_3, S_4$$

Episodic tasks do not have a steady state. Eventually the system will transition to one of the end states. The path/trajectory from state to end state could go through multiple hoops and other states, but eventually the agent will find an end state and the episode will end.

Markov Reward Processes

Moving on to the Markov Reward Process, I now introduce the concept of rewards. Look at the modified state diagrams in Figures 2-7 and 2-8. They are the same problems from the previous section (Figures 2-5 and 2-6, respectively) with the addition of rewards at each transition.

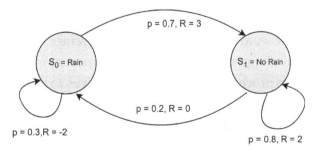

Figure 2-7. *Continuing Markov Reward Process. This is similar to the Markov Chain in Figure 2-5, with an addition of reward R for each transition arrow*

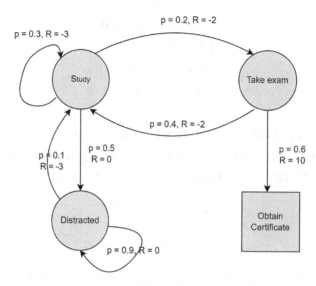

Figure 2-8. *The episodic Markov reward process similar to Figure 2-6, with additional rewards R for each transition*

In these two MRP setups, you can compute transition probabilities similar to that in MC, as in the preceding section. In addition, you can calculate the value of a state $V(S)$, which is the cumulative reward that the agent gets when it is in state $S = s$ at time t and then it follows the dynamics of the system.

$$V(s) = E[G_t \mid S_t = s] \tag{2-5}$$

Where G_t is defined in Equation 2-3:

$$G_t = R_{t+1} + \gamma R_{t+2} + \gamma^2 R_{t+3} + \dots$$

The notation $E[G_t \mid S_t = s]$ is read as an expectation of returning G_t when the starting state at time t is $S_t = s$. The word *expectation* implies the average value of G_t when the simulation is performed a large number of times. The expectation operator ($E[\bullet]$) is used to derive formulae and to prove theoretical results. However, in practice, it is replaced by averages over many sample simulations and is also known as the *Monte Carlo simulation*. In the Monte Carlo simulation, you have the agent start at state $S_t = s$ and follow the trajectory collecting the rewards, calculating G_t each time, and finally taking the average of G_t across such multiple repeats. The average $\bar{G_t}$ so calculated works as an approximation of the theoretical expectation $V(s) = E[G_t \mid S_t = s]$.

Note the use of γ as the discount factor. As explained earlier, γ captures the notion that the reward today is better than the reward tomorrow. It is also important mathematically to avoid unbounded returns for continuing tasks. I do not go into the mathematical details here beyond mentioning this fact.

A value of $\gamma = 1$ implies there is no discount, and that the agent is far-sighted. It cares as much about future rewards as it does for immediate rewards. A value of $\gamma = 0$ implies that an agent is short-sighted. It now cares only about immediate reward in the next time step. You can check this concept by putting different values of γ in the $G_t = R_{t+1} + \gamma R_{t+2} + \gamma^2 R_{t+3} + \dots$ equation.

To summarize, up to now I have introduced the concept of transition probability P, return G_t, and value of a state $V(s) = E[G_t \mid S_t = s]$.

Markov Decision Processes

Markov Decision Processes (MDP) extend reward processes by bringing the additional concept of "action". In MRP, the agent had no control over the outcome. Everything was governed by the environment. However, under MDP regime, agents can choose actions based on the current state/observation. The agent can learn to take actions that maximize the cumulative reward, that is, total return G_t.

For example, look at an extension of episodic MRP from Figure 2-8. Looking at Figure 2-9, you can see that in the "Learning" state, the agent can take one of two actions—study more or take the exam. The agent chooses the two actions with equal probability of 0.5. Choosing actions with certain probabilities based on the current state is called a *policy* that the agent learns in order to optimize the long-term rewards. The action influences the probabilities of the next state and the reward values that the agent will see. If an agent decides to "study" more, there is a 0.7 chance that agent will get "distracted" and turn to social media, while there is 0.3 chance that they will continue to stay focused on "learning". If the agent, while in state of "learning," decides to "take exam" instead of "study", the outcome of the "take exam" action would lead to two outcomes—the agent fails with probability of 0.4, going back to "Learning" with a reward of -2 or (with a probability of 0.6) the agent successfully completes the exam, thus obtaining a certificate and getting a reward of 10.

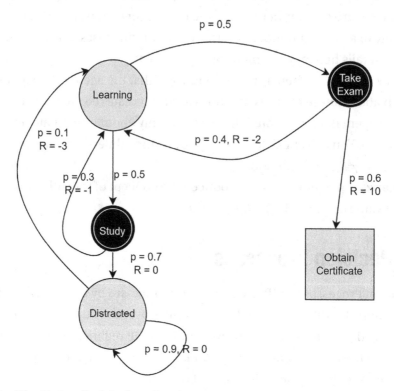

Figure 2-9. *The Episodic Markov Decision Process as an extension of the Markov Reward Process shown in Figure 2-8. The black filled circles are the decisions that the agent can take*

The transition function is now a mapping from the current state and action to the next state and reward. It defines the probability of getting a reward r with a transition to the next state s' whenever the agent takes action a in state s.

$$p(s',r \mid s,a) = \Pr\{S_t = s', R_t = r \mid S_{t-1} = s, A_{t-1} = a\} \qquad (2\text{-}6)$$

You can use Equation 2-6 to derive many useful relations. Transition probability can be derived from the previous transition function. Transition probability defines the probability of the agent finding itself in state s' when it takes action a while in state s. It is defined as follows:

$$p(s' \mid s,a) = \Pr\{S_t = s' \mid S_{t-1} = s, A_{t-1} = a\} = \sum_{r \in R} p(s',r \mid s,a) \qquad (2\text{-}7)$$

Equation 2-7 is derived by averaging all the possible rewards that the agent could get while transitioning from (s, a) to s'.

Let's look at a different example of MDP, one that is a continuing task. Think of an electric cart at the airport that ferries passengers from one terminal to another. The cart has two states, "High Charge" and "Low Charge." In each state, the cart can stay idle, recharge by connecting to the charging station, or ferry passengers. If the cart ferries passengers, it gets a reward of b, but there is a chance that in the "Low Charge" state, ferrying passengers may drain the battery completely and the cart would need to be rescued, resulting in a reward of -10. Figure 2-10 shows the MDP.

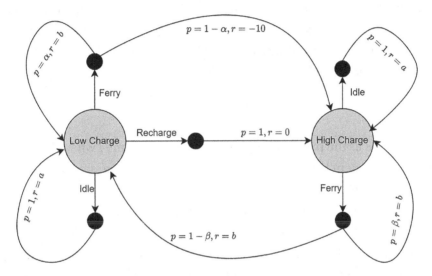

Figure 2-10. *Electric cart: A continuing Markov Decision Process with two states. Each state has three actions: idle, ferry, or recharge*

In the "Low Charge" state, the cart can take an action from the set of possible actions, {recharge, idle, ferry}. In the "High Charge" state, the cart can take an action from {idle, ferry}. All other transitions are 0. With $p(s', r \mid s, a)$ given as shown earlier, $p(s' \mid s, a)$ can be calculated. In this case, it is the same as $p(s', r \mid s, a)$. This is because for each transition from (s, a) to (s') there is only one fixed reward. In other words, there is no uncertainty or a probability distribution for rewards after an action is taken in a given state. The reward is the same fixed value every time the agent in state s takes an action a and finds itself in the next state s'. This is one of the most common setups encountered in practical problems. However, theoretically in the most general case, the reward could be a probability distribution like a "ferry" action reward being linked to the number of passengers being ferried by the cart.

The following sections of this chapter continue to use this example.

Policies and Value Functions

As shown earlier, MDP has states, and the agent can take actions that transition the agent from the current state to the next state. In addition, the agent gets feedback from the environment in the form of a reward. The dynamics of MDP are defined as $p(S_t = s'$, $R_t = r | S_{t-1} = s, A_{t-1} = a)$. You have also seen "cumulative return," G_t, which is defined as the sum of all rewards received from time t. The agent has no control over transition dynamics. It is outside the agent's control. However, the agent can control the decision, that is, what action to take in which state.

This is exactly what the agent learns based on the transition dynamics of the system. The agent does so with an objective to maximize the G_t for each state S_t that can be expected on average across multiple runs (the expected value). The mapping of states to actions is known as *policy*. It is formally defined as follows:

$$\pi\left(A_t = a | S_t = s \right) \tag{2-8}$$

Policy is defined as the probability of taking action a at time t when the agent is in state s at time t. The agent learns the mapping function from the state to actions with a goal to maximize the total return.

Policies can be of two types. See Figure 2-11. The first type is a *stochastic policy* in which $\pi(a | s)$ is a probability function. For a given state, there are multiple actions that the agent can take, and the probability of taking each such action is defined by $\pi(a | s)$. The second type is a *deterministic policy* where there is only one unique action for a given state. In other words, the probability function $\pi(A_t = a | S_t = s)$ becomes a simple mapping function with the value of a function being 1 for some action $A_t = a$ and 0 for all other actions A_t.

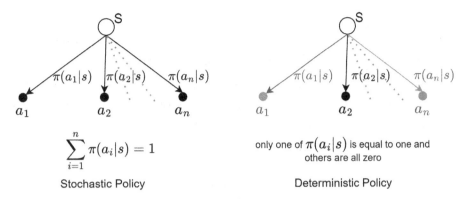

$$\sum_{i=1}^{n} \pi(a_i|s) = 1$$

Stochastic Policy

only one of $\pi(a_i|s)$ is equal to one and
others are all zero

Deterministic Policy

Figure 2-11. *Types of policies. (a) Stochastic policy in which the agent can take one of the many actions based on a probability distribution. (b) Deterministic policy in which the agent learns to take only one specific action that is optimal*

The cumulative reward G_t (i.e., return) that the agent can get at time t while in state S_t is dependent on the state S_t, and the policy agent is the following. It is known as a *state value function*. The value of G_t is dependent on the trajectory of states that the agent will see after time t, and trajectories in turn depend on the policy the agent is going to follow. Hence, the value function is always defined in the context of the policy the agent is following. It is also referred to as the agent's *behavior*. It is formally defined as follows:

$$v_\pi(s) = E_\pi[G_t|S_t = s] \tag{2-9}$$

Let's try to break this down a little bit. $v_\pi(s)$ specifies the "state value" of state s when the agent is following a policy π. $E_\pi[\cdot]$ signifies that whatever comes inside the square bracket is an average taken over many samples. While it is mathematically known as the *expectation* of the expression inside the square brackets under policy π, in reality you'll usually calculate these values using simulation. You carry out the calculation over multiple iterations and then average the value. As per one of the fundamental laws of statistics, averages converge to expectations under general conditions. This concept is regularly used in computers while estimating an expectation and is called *Monte Carlo simulations*. Coming to the final part, the expression inside the square brackets is $(G_t | S_t = s)$. That's the average return G_t over many runs that the agent can see in repeated trials at time t when the agent is in state s at time t and the agent's behavior is that of following policy π.

Backup diagrams show the path from time t when the agent is in state S_t to its successor states that the agent can find itself in at time $t + 1$. It depends on the action the agent takes at time t, that is, $\pi(A_t = a | S_t = s)$. Further, it depends on the environment/model transition function $Pr\{S_{t+1} = s', R_{t+1} = r | S_t = s, A_t = a\}$, which transitions the agent to state S_{t+1} with a reward of R_{t+1} based on the action A_t it took while in state S_t. Pictorially, a one-step transition from the current state to the possible successor states is called a *backup diagram* and looks like Figure 2-12.

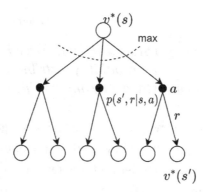

Figure 2-12. *Backup diagram starting from state, taking an action. Empty circles signify states, and dark circles signify actions*

You will make extensive use of backup diagrams, especially in the next section about Bellman equations. Backup diagrams are helpful in conceptualizing and reasoning about the equations as well as working out the proofs for various RL algorithms. They are also helpful in reasoning about data collection required to train agents.

Another related concept to *state value functions* (also called *value functions* in short) is the concept of *action value functions*. A value function is the expected cumulative reward the agent gets when it is in a specific state and takes actions based on the policy π. However, suppose that the agent is free to take any action at this first-time step t with the condition that it must follow policy π on all the following time steps from $t + 1$ onward. The expected return that the agent gets at time t is now known as *action-value-function* $q_\pi(s, a)$. Formally, it is defined as follows:

$$q_\pi(s,a) = E_\pi[G_t | S_t = s, A_t = a] \tag{2-10}$$

There is a simple and subtle relationship between v-value (state value) and q-values (state action value), which is explored in detail in the next section.

This pretty much completes the definition of various components of an MDP. The next section explores the recursive relationship between v-value of the states and q-values of state action at time t to that of the successor values at time $t+1$. Almost all the RL algorithms exploit this recursive relationship.

Bellman Equations

Bellman equations are a fundamental concept in the field of reinforcement learning (RL). They formalize the recursive relationship between the current and successor values in a sequential decision-making setup like RL. Almost all the algorithms in RL are based on variations of these equations with the purpose of maximizing the value functions under different assumptions and approximations. Bellman equations provide a powerful tool for understanding and solving Markov Decision Processes (MDP) in RL.

Let's look again at Equation 2-9, which defines the value function. Also let's look at the definition of G_t defined in Equation 2-3; both are reproduced here:

$$G_t = R_{t+1} + \gamma \cdot R_{t+2} + \gamma^2 \cdot R_{t+3} + \cdots + \gamma^{T-t+1} \cdot R_T \qquad (2\text{-}11)$$

$$v_\pi(s) = E_\pi[G_t | S_t = s] \qquad (2\text{-}12)$$

Comparing the two equations, you can see that the value of a state is the expectation/average of the cumulative reward if an agent in that state S follows a policy π. The state that the agent finds itself in and the reward it gets from the environment are dependent on the policy it is following, that is, the action it takes in a given state. There is a recursive relationship in which the expression for G_t can be written in terms of G_{t+1}.

$$G_t = R_{t+1} + \gamma \left[R_{t+2} + \gamma R_{t+3} + \gamma^2 R_{t+4} + \cdots + \gamma^{T-t+1-1} R_T \right] \qquad (2\text{-}13)$$

Now extract the expression inside the square brackets:

$$R_{t+2} + \gamma R_{t+3} + \gamma^2 R_{t+4} + \cdots + \gamma^{T-t+1-1} R_T$$

In this expression, you do a change of variables $t' = t+1$ changing the above expression to:

$$R_{t'+1} + \gamma R_{t'+2} + \gamma^2 R_{t'+3} + \cdots + \gamma^{T-t'+1} R_T \qquad (2\text{-}14)$$

Comparing Equation 2-14 to the expression of G_t as given in Equation 2-11, you can see that:

$$R_{t'+1} + \gamma R_{t'+2} + \gamma^2 R_{t'+3} + \cdots + \gamma^{T-t'+1} R_T = G_{t'} = G_{t+1} \tag{2-15}$$

Next, substitute Equation 2-15 in 2-13 to get:

$$G_t = R_{t+1} + \gamma G_{t+1} \tag{2-16}$$

You can now substitute this recursive definition of G_t in the expression for $v_\pi(s)$ in Equation 2-12 to get:

$$v_\pi(s) = E_\pi\left[R_{t+1} + \gamma G_{t+1} | S_t = s\right] \tag{2-17}$$

The expectation E_π is over all possible actions a that the agent can take in state $S_t = s$ as well as all over the new states the environment transitions the agent to defined by the transition function $p(s', r | s, a)$. The expanded form of expectation leads to the revised expression of $v_\pi(s)$ as follows:

$$v_\pi(s) = \sum_a \pi(a|s) \sum_{s',r} p(s',r | s,a)\left[r + \gamma v_\pi(s')\right] \tag{2-18}$$

The way to interpret this equation is that the state value for s is the average over all the rewards and state values for successor states s'. The averaging is done based on the policy $\pi(a|s)$ of taking an action a in state s followed by the environment transition probability $p(s', r|s, a)$ of the agent which moves the agents to s' with reward r based on the initial state-action pair (s, a). Equation 2-18 shows the recursive nature of linking the state value of the current state (s) with the state values of the successor states (s').

A similar relationship exists for action value functions. Let's start from Equation 2-10 and walk through the derivation of the recursive relationship between q-values.

$$q_\pi(s, a) = E_\pi[G_t | S_t = s, A_t = a]$$

$$= E_\pi[R_t + \gamma G_{t+1} | S_t = s, A_t = a]$$

Expanding the expectation with summing over all possibilities, you get this:

$$q_\pi(s, a) = \sum_{s',r} p(s', r|s, a)\left[r + \gamma v_\pi(s')\right] \tag{2-19}$$

Comparing Equations 2-18 and 2-19, you will notice that the expression for $q_\pi(s, a)$ has the outer summation over action a missing with the inner summation of (Equation 2-18) matching the summation in Equation 2-19. This shows the relationship of $v_\pi(s)$ and $q_\pi(s, a)$. They are related through the policy $\pi(s|a)$ the agent is following. The q-value is the value of the tuple (s, a), and the state value is the value for a state (s). The policy links the state to the possible set of actions through a probability distribution. The agent in state s takes an action a using the policy $\pi(s|a)$ and the value of that state action pair is defined by the action-value function $q_\pi(s, a)$:

$$v_\pi\left(s\right) = \sum_a \pi(a|s).q_\pi\left(s,a\right) \tag{2-20}$$

You can also infer this relationship from visual inspection of Equations 2-18 and 2-19. You can write the relationship between $v_\pi(s)$ in terms of $q_\pi(s, a)$, wherein you replace the inner summation on the right side in Equation 2-18 with $q_\pi(s, a)$ as per Equation 2-19.

Just like Equation 2-18 gives a recursive relationship of $v_\pi(s)$ in terms of the state values of the next state $v_\pi(s')$, you could also express $q_\pi(s, a)$ in Equation 2-19 in terms of the pairs $q_\pi(s', a')$. This is achieved by replacing $v_\pi(s')$ in Equation 2-19 with the expression in Equation 2-20. This manipulation gives a recursive relationship between q-values.

$$q_\pi\left(s, a\right) = \sum_{s', r} p(s', r|s, a)\left[r + \gamma \sum_{a'} \pi\left(a'|s'\right) q\left(s', a'\right)\right] \tag{2-21}$$

Equations 2-18 and 2-21 link the current state-values/q-values with successive state-value/q-values and can be represented by way of backup diagrams. To show this relationship clearly, let's expand the backup diagram to cover the q-value transitions also, as shown in Figure 2-13. The diagram follows the standard convention: states s are shown as outlined circles, and action nodes representing a are shown as filled black circles.

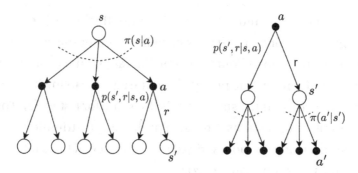

Figure 2-13. *Backup diagram for state values and action values. Outline circles represent states, and filled circles represent actions*

Equation 2-18 is the Bellman equation for $v_\pi(s)$, which is reflected in the left side of Figure 2-13. Equation 2-21 is the Bellman equation for $q_\pi(s, a)$, which is reflected in right side of Figure 2-13. These equations formalize the recursive relationship in a sequential decision-making setup like reinforcement learning. All the algorithms in reinforcement learning are based on variations of these two equations with the purpose of maximizing the value functions under different assumptions and approximations. As I go through these algorithms, I will keep highlighting the parts of the equation being approximated under certain assumptions along with the pros and cons of individual approaches.

A large part of expertise as a practitioner of reinforcement learning will revolve first on formulating a real-life problem into an RL setup, choosing what the state is, how actions are defined, and what will constitute reward. Secondly, expertise will involve choosing the right set of algorithms based on constraints and assumptions. Accordingly, the emphasis of this book is on the assumptions required to make a technique work best. When to choose which technique based on the problem statement, data availability, and other factors along with pros and cons of different approaches is also discussed as I introduce different algorithms under RL learning. I will present mathematical equations to formalize the relationship, but the core focus is to help you gain an intuitive sense of what is going on and when a given approach/algorithm makes sense.

Optimality Bellman Equations

Solving a reinforcement learning problem means finding (learning) a policy that maximizes the state value functions. Suppose you have a set of policies. The objective then would be to choose the policy that maximizes the state value $v_\pi(s)$. The optimal value function for a state is defined as $v^*(s)$. The relation for optimal state value can be stated as follows:

$$v^*(s) = \max_{\pi} v_{\pi}(s) \tag{2-22}$$

The previous equation shows that the optimal state value is the maximum state value that can be obtained across all possible policies π. Suppose this optimal policy is denoted by a superscript (*) to the policy, that is, π^*. If an agent is following the optimal policy, then the agent in state (s) will take that action (a), which maximizes the $q(s, a)$ obtained under the optimal policy. In other words, Equation 2-20 is modified from being an expectation to a maximization as follows:

$$v^*(s) = \max_{a} q_{\pi^*}(s, a) \tag{2-23}$$

Replacing the expression of $q_{\pi^*}(s, a)$ from Equation 2-19, the revised expression of $v^*(s)$ as given here, in which $v^*(s)$ is recursively linked to the optimal value of the next states:

$$v^*(s) = \max_{a} \sum_{s', r} [p(s', r|s, a)[r + \gamma \cdot v^*(s')]\} \tag{2-24}$$

Equation 2-24 links the optimal state value in state s with the optimal state values s' in the next time step. Next, let's derive a recursive relationship between the q-values. Start with Equation 2-19 and substitute the $v(s')$ with the expression in 2-23, which links v^* with q^*. This substitution leads to the revised recursive relation between q-values:

$$q^*(s, a) = \sum_{s', r} p(s', r|s, a) \cdot \left[r + \gamma \cdot \max_{a'} q^*(s', a') \right] \tag{2-25}$$

You will notice the optimal equation for state value in Equation 2-24 is a tough one to solve, as the max operator comes outside the summation side. At the same time, the optimal value of the q state action values in Equation 2-25 is far easier to solve due to the max operation being restricted to the individual following states s' instead of the max over summation. These optimal equations can be represented using backup diagrams, as shown in Figure 2-14, which highlights the recursive relationship between the current and successor values.

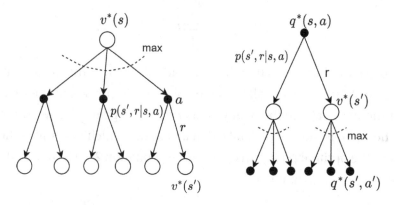

Figure 2-14. *Backup diagram for optimal state values and action values. Policy π is replaced by the "max" operation*

Once you have $v^*(s)$ using Equation 2-24, it is straightforward to derive the optimal policy. You use a one-step backup diagram, as shown in Figure 2-14, to find the a^* that resulted in the optimal value v^*. It can be seen as a one-step search. Once you have optimal $q^*(s, a)$ using Equation 2-25, finding the optimal policy is even easier. In a state s, you just choose the action a, which has the highest value of q. This is evident from right-side backup diagram in Figure 2-14. Let's apply these concepts to the problem of the electric van introduced in Figure 2-10, which is reproduced in Figure 2-15.

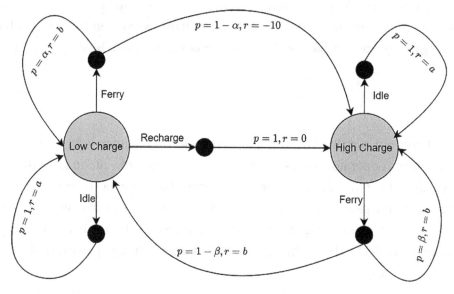

Figure 2-15. *Electric van: continuing Markov Decision Process with two states. Each state has three actions: idle, ferry, or recharge. Reproduction of Figure 2-10*

Let's draw the table depicting various values, as shown in Table 2-1. This table lists all the *possible combinations of* (s, a, s', r) as well as probability $p(s', r \mid s, a)$.

Table 2-1. *System Dynamics for MDP in Figure 2-5*

State (s)	Action (a)	New State (s')	Reward (r)	$p(s', r \mid s, a)$
Low battery	Idle	Low battery	a	1.0
Low battery	Charge	High battery	0	1.0
Low battery	Ferry	Low battery	b	α
Low battery	Ferry	High battery	-10	1- α
High battery	Idle	High battery	a	1.0
High battery	Ferry	High battery	b	β
High battery	Ferry	Low battery	b	1-β

You can use Table 2-1 to calculate the optimal state values using Equation 2-24.

$$v^*\left(low\right) = max \begin{cases} a + \gamma \, v^*\left(low\right) \\ \gamma \, v^*\left(high\right) \\ \alpha\left[b + \gamma v^*\left(low\right)\right] + (1-\alpha).\left[-10 + \gamma \, v^*\left(high\right)\right] \end{cases}$$

$$v^*\left(high\right) = max \begin{cases} a + \gamma \, v^*\left(high\right) \\ \beta\left[b + \gamma \, v^*\left(high\right)\right] + (1-\beta).\left[b + \gamma \, v^*\left(low\right)\right] \end{cases}$$

For given set of values of a, b, α, β, γ, there will be a unique solution for $v^*(high)$ and $v^*(low)$ that will satisfy the previous equations. However, explicitly solving equations is feasible only for simple toy problems. Real-life bigger problems require the use of other methods that are scalable. This is exactly what you are going to study in the rest of the book: various approaches and related algorithms to solve Bellman equations for an optimal policy.

Train Your First Agent

This section looks at two Gymnasium environments and two algorithms to train agents for these two environments. Along the way, you will learn about different RL libraries in Python that you will be using throughout the book. As this is the first introduction to RL training, you will be looking at the training algorithms as black boxes with a focus to get familiar with the libraries and different components required for carrying out RL training. However, as you progress with the book, you will be opening up the black boxes and writing your own versions of these algorithms to gain deeper understanding.

First Agent

This section uses an environment called LunarLander, which is a classic rocket trajectory optimization problem (see Figure 2-16). The aim is to land the craft on the landing pad as marked by two flags. There are four possible actions (a) to control the craft:

- 0: Do nothing

- 1: Fire left orientation engine

- 2: Fire main engine

- 3: Fire right orientation engine

The observation (o), that is, the part of the state environment exposed to the agent consists of eight-dimensional vector:

- x, y coordinates of the lander

- Linear velocities in the x and y directions

- The craft's angle and its angular velocity

- Two flags indicating respectively if the left and right leg are in contact with the ground

The rewards are awarded at each time step based on how far the craft is to the landing pad, whether it is tilted or straight, whether the engines are firing, and whether either leg is in contact with the ground. Landing safely or crashing earns extra +ve / -ve rewards. You can read more about it in the Gymnasium documentation.[4]

---

[4] https://gymnasium.farama.org/environments/box2d/lunar_lander/

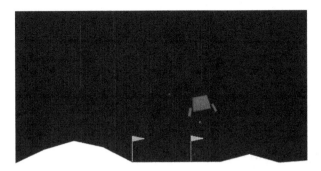

Figure 2-16. *LunarLander environment from the Gymnasium library -* `https://` `gymnasium.farama.org/environments/box2d/lunar_lander/`

This example uses DQN, or Deep Q-Network, which is a reinforcement learning algorithm that combines the traditional Q-learning algorithm with the power of deep neural networks. It is used to approximate the optimal action-value function, which represents the maximum expected future rewards for taking a particular action in a given state. DQN has been successfully applied to a wide range of problems, including game playing and control tasks, and has been shown to achieve human-level performance in many cases.

Chapter 6 and 7 are devoted to a deep dive of DQN and its variants. Therefore, this chapter simply shows an example without getting into the specifics of how DQN works.

This example uses another library called `stable_baseline3` to create a small neural network that learns the agent policy $\pi(a|s)$, that is, the mapping of state to action. Input to the network is the eight-dimensional observation vector and action is a value from the discrete set of four actions {0, 1, 2, 3}. DQN agents need to explore the environment in a random way to have good coverage of possible *(state, action)* ➤ *next_action* transitions. This is specified by setting `exploration_final_eps=0.1` The verbosity is defined as 1 in order to get appropriate level of logs. Refer to the notebook `2.c-First-Agent.ipynb` for details. Note that you can either run the notebook locally in an environment you created following the instructions in Chapter 1, or you can run the notebooks directly on Google Colab, which is a web-hosted Python environment with a free tier and paid tiers. For the book, the free tier of Colab will mostly suffice. The excerpt where DQN is instantiated is shown in Listing 2-3.

Listing 2-3. Initializing DQN agent-"2.c-First-Agent.ipynb"

```
import gymnasium as gym
from stable_baselines3 import DQN
model = DQN(
    "MlpPolicy", "LunarLander-v2",
    verbose=1, exploration_final_eps=0.1,
    target_update_interval=250)
```

Next, train the agent and save it using the built-in DQN algorithm from the stable_baselines3 library. Training is done by calling the model.learn function with total_timesteps=int(1e5) to run training for 100,000 steps, log_interval=400 to log the progress of training every 400 episodes (and not 400 steps), and finally setting progress_bar=True to observe the progress of training with the help of a progress bar.

Listing 2-4. Train the agent-"2.c-First-Agent.ipynb"

```
# Train the agent
model.learn(total_timesteps=int(1e5), log_interval=400, progress_bar=True)
# Save the agent
model.save("dqn_lunar")
```

You then evaluate the trained agent to check the performance using the following code. The agent, before training, produced a mean reward of -600, which improves to about +100 after 100,000 steps of training. See Listing 2-5.

Listing 2-5. Evaluate Trained Agent--"2.c-First-Agent.ipynb"

```
# Evaluate the trained agent
mean_reward, std_reward = evaluate_policy(model, eval_env, n_eval_
episodes=10, deterministic=True)

print(f"mean_reward={mean_reward:.2f} +/- {std_reward}")
-----------
Output:
mean_reward=94.54 +/- 87.28950452232776
```

The notebook `2.c-First-Agent.ipynb` also shows the way you can record the video of the trained agent. After you are satisfied with the result of training, you can use `huggingface` to upload the model, which then can be shared with others. In order to upload the model to `huggingface`, you will need to generate a token to enable notebook access to your `huggingface` account. First create an account at `huggingface.co` if you do not have one already. You can create/find a `huggingface` token at `https://huggingface.co/settings/tokens`. Once your model has been pushed to `huggingface`, you can navigate to the link where it is hosted. The URL is printed at the end by the code used for pushing the model to the `huggingface` hub. In my case, it looks like this: `https://huggingface.co/nsanghi/dqn-LunarLander-v2`. The `nsanghi` corresponds to my `huggingface` login and `dnq-LunarLander-v2` refers to the space/repository on `huggingface` under my account where the model was pushed, based on the name I defined in the `package_to_hub()` function call in the notebook. The call to this function is shown in Listing 2-6. Change the `repo_id` to refer to your `huggingface` account login before you run this code.

Listing 2-6. Package Trained Agent and Video to Hugging Face hub--"2.c-First-Agent.ipynb"

```
package_to_hub(model=model, # Our trained model
            model_name="dqn-LunarLander-v2", # The name of our
            trained model
            model_architecture="DQN", # The model architecture we used:
            in our case PPO
            env_id="LunarLander-v2", # Name of the environment
            eval_env=eval_env, # Evaluation Environment

    # id of the model repository from the Hugging Face Hub
    # (repo_id = {organization}/{repo_name} for instance
    #nsanghi/dqn-LunarLander-v2
            repo_id="nsanghi/dqn-LunarLander-v2",

            commit_message="Push to Hub")
```

You can share this link with your friends and family to show the performance of the agent you have trained using just a few lines of code.

Walkthrough of Common Libraries Used

This section takes a quick tour of the libraries used in training an agent on the LunarLander environment, as well as a few other common libraries you will be using in other chapters.

Environments: Gymnasium and OpenAI Gym

As described in the first section in this chapter on RL definitions, every RL problem must have a concept of environment, which is about states, actions, and rewards, that is, what a state is, what actions are possible by the agent, and what kinds of rewards the agent can get. The agent receives a state, uses a policy that it learns outside of environment to take an action, and passes this action to the environment. The environment takes the action as input and steps the agent to the next time instance, returning a bunch of things: the next state, the reward, the terminated flag, the truncated flag, and some debug info. The Gymnasium[5] library, which is a port of the OpenAI Gym library, provides a standard interface to all the RL environments. This makes it modular, whereby the learning algorithm code is abstracted away from the code of the environment dynamics/model. Gymnasium, in addition to providing a standard interface, also provides a wide range of environments:

- **Classic Control:** This is classic reinforcement learning based on real-world problems and physics.

- **Box2D:** These environments all involve toy games based around physics control, using Box2D-based physics and PyGame-based rendering.

- **Toy Text:** These environments are designed to be extremely simple, with small discrete state and action spaces, and hence are easy to learn. They are suitable for debugging implementations of reinforcement learning algorithms.

- **MuJoCo:** Physics engine based environments with multi-joint controls, which are more complex than the Box2D environments.

---

[5] https://gymnasium.farama.org/

- **Atari:** A set of 57 Atari 2600 environments simulated through Stella and the Arcade Learning Environment that have a high range of complexity for agents to learn.

- **Third-party options:** A number of environments have been created that are compatible with the Gymnasium API.

Gymnasium is a new port of the original OpenAI Gym. While it has mostly retained the function signatures of the OpenAI Gym, there are some subtle differences. The popular RL libraries are still in the process of adding support to Gymnasium. While installing these other libraries, be sure to install the version appropriate for the Gymnasium port.

Stable Baselines3 (SB3)

Stable Baselines3[6] (SB3) is a set of reliable implementations of reinforcement learning algorithms in PyTorch. It is the next major version of Stable Baselines version 2. You can read a detailed presentation of Stable Baselines3 in the JMLR paper[7]. It implements seven common algorithms:

- Advantage Actor Critic(A2C)

- Deep Deterministic Policy Gradient (DDPG)

- Deep Q Network (DQN)

- Hindsight Experience Replay (HER), which works with DQN, SAC, TD3, and DDPG

- Proximal Policy Optimization (PPO)

- Soft Actor Critic (SAC)

- Twin Delayed DDPG (TD3)

The implementations have been benchmarked against reference codebases, and automated unit tests cover 95% of the code. The algorithms follow a consistent interface.

---

[6] https://stable-baselines3.readthedocs.io/
[7] https://jmlr.org/papers/volume22/20-1364/20-1364.pdf

RL Baselines3 Zoo

RL Baselines3 Zoo[8] is a training framework for reinforcement learning, using Stable Baselines3. It provides scripts for training, evaluating agents, tuning hyperparameters, plotting results, and recording videos.

In addition, it includes a collection of tuned hyperparameters for common environments and RL algorithms, and agents trained with those settings.

Hugging Face

Hugging Face[9] has an extensive set of Python libraries around Transformers, which was introduced in a seminar paper titled "Attention is all you need"[10] in 2017. This resulted in an explosion of various derivatives of this idea for solving natural language processing (NLP) related learning methods. All the recent language models are increasing based on the variations of this architecture—including the OpenAI GPT family, Google's BERT and BARD, Meta's Llama and Llama2, and many more. Hugging Face came into the scene to provide standard, easy-to-use implementations of these language models—the model code, the infrastructure including data management, training runs, and space to host trained models. It enabled collaboration in the open-source community and now covers all other kinds of deep learning models, including ones related to vision, voice, and reinforcement learning. I use this library extensively when I touch upon the topic of RLHF (Reinforcement Learning from Human Feedback), which is an approach introduced by OpenAI in developing InstructGPT and ChatGPT. Since then, it has been adopted by almost all LLM-based chat and instruction following models. A section in one of the later chapters is devoted to RLHF.

However, even before I go to RLHF, I will be using the Hugging Face hub to host the trained models so that the same can be shared with others, including a video of the agent in action post training. This is what I did in the section titled "Train First Agent".

---

[8] https://stable-baselines3.readthedocs.io/en/master/guide/rl_zoo.html
[9] https://huggingface.co/docs
[10] https://arxiv.org/abs/1706.03762

Second Agent

Now you'll see how to train an agent to play the Atari Game of Pong. This example uses an environment from the Atari Simulator. Atari environments are simulated via the Arcade Learning Environment ALE[11] through Stella.[12]

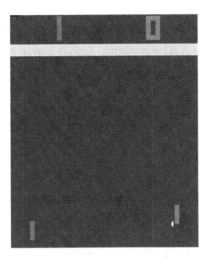

Figure 2-17.   *Atari Pong from the Gymnasium library -* `https://gymnasium.farama.org/environments/atari/pong/`

This example uses a specific variant of Pong from the Atari game simulator, `PongNoFrameskip-v4`. Atari simulator 1600 has 18 actions, of which Pong uses 6. The observation space is an RGB image of the game state in the form of 210x160x3 three-color channel pixel values, each ranging from 0-255, that is, uint8 values.

As a single image of the paddles and ball does not tell you anything about the direction in which the ball is travelling, you have to take four consecutive images of the game and stack them together to form a single observation and capture the movements in the game. The observation is hence a stacked image consisting of four images. The policy that agent learns is to take these stacked four-image observation vectors

---

[11] MG Bellemare, Y Naddaf, J Veness, and M Bowling. "The arcade learning environment: An evaluation platform for general agents." Journal of Artificial Intelligence Research (2012). `https://github.com/mgbellemare/Arcade-Learning-Environment`

[12] `https://github.com/stella-emu/stella`

as input and produce the probability of making one of the six actions possible. This is mathematically depicted as learning a policy $p_{\pi(\theta)}(a|o)$ where the policy is a neural network $\pi(\theta)$ with stacked image vector (o) as input and six actions (a) as output.

This example uses a `CnnPolicy` class of a neural network for the policy. The inputs are stacked images and a CNN class of networks are best suited for image processing in neural networks.

In the previous case of LunarLander, the agent's policy was based on DQN. However, for the game of Pong, I use the A2C algorithm. A2C is a synchronous, deterministic variant of Asynchronous Advantage Actor Critic (A3C).[13]

Listing 2-7. Train Atari Pong Using the A2C Algorithm-"2.d-Second-Agent.ipynb"

```
env_id = "PongNoFrameskip-v4"

# `n_envs=4` enables multiworker training using 4 environments
vec_env = make_atari_env(env_id, n_envs=4, seed=0)

# `n_stack=4` - Frame-stacking with 4 frames
vec_env = VecFrameStack(vec_env, n_stack=4)

# instantiate the A2C based agent using stablebaselines3
model = A2C("CnnPolicy", vec_env, verbose=1)

# If you have a GPU, you can increase the `total_timesteps` to something
like 1_000_000 i.e. one million and it would take about 45 mins to train
# Trian the agent i.e. the Neural network CNN based model which learns
the policy
model.learn(total_timesteps=20_000, log_interval=500, progress_bar=True)
```

The rest of the steps of creating a video to check the performance and pushing a trained model to Hugging Face are all very similar to what you did in the previous example of LunarLander. You can refer to the accompanying Python notebook `2.d-Second-Agent.ipynb` for details.

---

[13] https://stable-baselines3.readthedocs.io/en/master/modules/a2c.html

RL Zoo Baselines3

RL Baselines3 Zoo[14] is a training framework for reinforcement learning. Using Stable Baselines3 (SB3), reliable implementations of reinforcement learning algorithms in PyTorch are provided in this library. It provides scripts for training, evaluating agents, tuning hyperparameters, plotting results, and recording videos. In addition, it includes a collection of tuned hyperparameters for common environments and RL algorithms, and agents trained with those settings.

RL Zoo is supposed to be run from the command line. However, you can use Python notebooks to run commands using the ! bang character before the commands. For example, to run the pwd UNIX command to list the current working directory, you can execute the !pwd command in a code cell.

In this section, you learn how to train agents, record videos of the trained agents, as well as push the trained agents to Hugging Face—all using RL Zoo SB3. You will first train the same LunarLander agent using the DQN algorithm. Notebook 2.e-RL-Zoo.ipynb contains the code for running the RL Zoo commands.

Listing 2-8. Using RL Zoo to Launch Training-"2.e-RL-Zoo.ipynb"

```
!python -m rl_zoo3.train --algo dqn --env LunarLander-v2 --n-timesteps
100000 --log-interval 400 –progress
```

At the start of script run, it prints the various hyperparameters that control the training. You can refer to the RL Zoo documentation for more details. As an example, the default values for the DQN agent for LunarLander are given in the location; see the footnote.[15] You can modify these values at the time of calling the training script using an extended list of command-line arguments given in the RL Zoo SB3 documentation, under the section "Train an Agent".[16]

Once the agent is trained, you can run additional command-line scripts to a) evaluate the performance of the trained agent; b) record a video on the local machine; and c) push the trained agent model, performance, and video to the Hugging Face hub. These additional commands are shown in Listing 2-9.

---

[14] https://rl-baselines3-zoo.readthedocs.io/en/master/

[15] https://github.com/DLR-RM/rl-baselines3-zoo/blob/master/hyperparams/dqn.yml

[16] https://rl-baselines3-zoo.readthedocs.io/en/master/guide/train.html

Listing 2-9. Additional Commands in RL Zoo-"2.e-RL-Zoo.ipynb"

```
#evaluate trained agent
!python -m rl_zoo3.enjoy --algo dqn --env LunarLander-v2 --no-render --
n-timesteps 5000 --folder logs/

#record a video
!python -m rl_zoo3.record_video --algo dqn --env LunarLander-v2 --exp-id
0 -f logs/ -n 1000

#push all artefacts to Hugging Face Hub
!python -m rl_zoo3.push_to_hub --algo dqn --env LunarLander-v2 -f logs/ -orga
nsanghi -m "Initial commit"
```

Now it is your turn to see if you can replicate these steps for the second agent—the Atari Pong game playing an agent trained using A2C.

This ends the exploration of various libraries and sample code to load environments, train agents, record performance, and push things to Hugging Face for sharing. While the approach taken here is a high-level overview of libraries involved and many introductory books on Practical RL stop at this, the endeavor in this book is to master both—a theoretical code-level understanding of various training algorithms used in RL as well as common production-level libraries involved in RL. Accordingly, the rest of the chapters have an equally strong emphasis on coding your own working algorithms.

Solution Approaches with a Mind Map

Now that you have a background in reinforcement learning configuration, related blocks, Bellman equations, and a sample of algorithms, it is time to look at the complete landscape of algorithms in the reinforcement learning world. Figure 2-18 from OpenAI shows a high-level landscape of the various types of learning algorithms in the RL space.

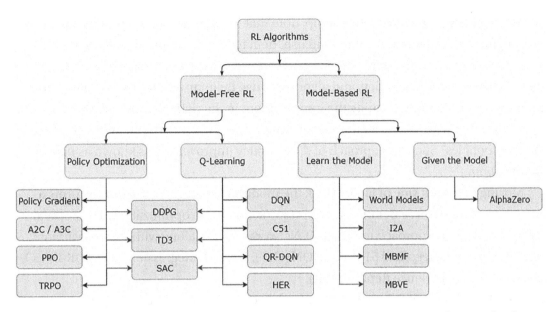

Figure 2-18. *Mind map of algorithms in reinforcement learning. This is a high-level map showing only broad categorizations (refer to* https://spinningup. openai.com/en/latest/spinningup/rl_intro2.html)

As seen in Bellman equations, the system transition dynamics $p(s', r \mid s, a)$ forms a central part. Transition dynamics describe the model of the environment, that is, based on the current state and action taken by the agent, what is the response of the environment in terms of the reward given back to the agent and the new state the agent is transitioned into? However, transition dynamics are not always known. Accordingly, the first broad categorization of learning algorithms can be done based on the knowledge (or lack thereof) of this transition model. The algorithms where the transition dynamics are known or learned explicitly are categorized as *model-based algorithms* and the ones where transition dynamics are not known or not used directly are categorized as *model-free algorithms.*

Model-based algorithms can be further classified into two categories: one where you are given the model, for example, the game of Go or chess, and the second category being the one where the agent needs to explore and learn the model. Some popular approaches under "learn the model" are World Models, Imagination Augmented Agents (I2A), Model-Based RL with Model-Free fine tuning (MBMF), and Model-Based Value Exploration (MBVE).

Moving back to the model-free setup, note that Bellman equations provide you with a way to find the state/action values and use them to find the optimal policies. Most of these algorithms take an approach of iterative improvement. Eventually you want the agent to have an optimal policy, and algorithms use Bellman equations to evaluate the goodness of a policy and to guide the improvement in the right direction. However, there is another way. Why not improve the policy directly instead of the indirect way of values to policies? Such an approach of direct policy improvement is known as *policy optimization*.

Moving to the left side of Figure 2-18, the model-free category, *Q-learning* forms a major part of model-free Bellman-driven state/action value optimization. The popular variants under this approach are Deep Q-Networks (DQN) along with various variants of DQN, such as Categorical 51-Atom DQN (C51), Quantile Regression DQN (QR-DQN), and Hindsight Experience Replay (HER).

The algorithms that follow the path of policy optimization directly are Policy Gradient, Actor Critic and its variations (A2C/A3C), Proximal Policy Optimization (PPO), and Trust Region Policy Optimization (TRPO). PPO is the state of the art when it comes to policy-based algorithms. Recently, PPO has been used with Large Language Models such as the GPT family, Llama, and other variants to make the model generate safe and helpful human outputs, under an approach called Reinforcement Learning from Human Feedback (RLHF).

Finally, there is a set of algorithms that lie midway between Q-learning and Policy Optimization. The popular ones in this category are deep deterministic policy gradient (DDPG), twin delayed DDPG (TD3), and soft actor-critic (SAC).

These categorizations are just to help you appreciate the different approaches and popular algorithms. However, the list and categorizations are not exhaustive. The field of reinforcement learning is evolving rapidly, with new approaches being added on a regular basis. Use the previous mind map only as high-level guidance.

Summary

This chapter sets the context for what reinforcement learning is, the setup, and the various definitions. The chapter first talked about the high-level setup and the mathematical foundations of various Markov formulations. This was followed by a discussion of Bellman equations, the derivation of their recursive nature, and the optimality equations. State/action value functions and backup diagrams formalized this notion.

The chapter then looked at a couple of agent training workflows using Stable Baselines3. It dove deep into the related libraries and software frameworks in RL and the ones you will be using significantly during the course of this book. The chapter then concluded with a view of the algorithm landscape.

In the upcoming portions of the book, I delve deeper into many of these algorithms, starting with dynamic programming in the next chapter.

CHAPTER 3

Model-Based Approaches

Chapter 2 talked about the parts of the setup that form the agent and the part that forms the environment. To recap, the agent gets the state $S_t = s$ and follows a policy $\pi(s|a)$ that maps states to actions. The agent uses this policy to take an action $A_t = a$ when in state $S_t = s$. The system transitions to the next time instant of $t + 1$. The environment responds to the action $(A_t = a)$ by putting the agent in a new state of $S_{t+1} = s'$ and providing feedback to the agent in terms of a reward, R_{t+1}. The agent has no control over what the new state S_{t+1} and reward R_{t+1} will be. The transition from $(S_t = s, A_t = a) \rightarrow (R_{t+1} = r, S_{t+1} = s')$ is governed by the environment. This is known as *transition dynamics*. For a given pair of (s, a), there could be one or more pairs of (r, s'). In a deterministic world, you would have a single pair of (r, s') for a fixed combination of (s, a). However, in stochastic environments, that is, environments with uncertain outcomes, you could have many pairs of (r, s') for a given (s, a). Chapter 2 also looked at the computation of state values $V(s)$ and state-action values $Q(s, a)$ and the recursive relationship between current state/state-action values and next state/state-action values (in Equations 2-19 and 2-21).

RL agents learn optimal actions for any given state, which maximizes the total reward the agent gets by following an optimal set of actions as the agent transitions from one state to another. Equations 2-19 and 2-21 clearly indicate that $V(s)$ and $Q(s, a)$ depend on two components, the transition dynamics and the next state/state-action values. To lay the foundations of RL learning, this chapter starts with the simplest setup—one in which the transition dynamics $\Pr\{ S_{t+1} = s', R_{t+1} = r \mid S_t = s, A_t = a\}$ are known. It also assumes that the number of states and actions possible in a given state form a closed, small set of discrete values. Such a simplifying assumption will help you develop the foundations of RL learning algorithms.

In real life, these assumptions are not valid. Consider an example of a robotic arm (or any physical system). Each time the arm is moved, due to the mechanical parts involved, the movement may be imprecise and therefore the next state is only approximately defined in a strict sense. Further, consider the case of a self-driving car, where the state

89

© Nimish Sanghi 2024
N. Sanghi, *Deep Reinforcement Learning with Python*, https://doi.org/10.1007/979-8-8688-0273-7_3

at the next instant depends on many other stochastic elements like the sudden move by another car or a pedestrian, which may vary from time to time. This shows that prior knowledge of the transition function is an assumption that can be made only in very simple scenarios. The second assumption of a finite set of discrete states and actions is not valid for most of the physical systems that live in the world of continuous states and actions.

Hence, the algorithms you will study in later chapters use various techniques to either estimate these transition functions using sampling and other related approaches under *model-based RL* or side-step the need to explicitly estimate the transition functions under *model-Free RL*. I discussed these in the section on "Solution Approaches with a Mind Map" toward the end of Chapter 2. These approaches help remove both these constraints and I will discuss these in detail as you progress through the book.

Coming back to the focus of this chapter, the agents in this chapter use the transition knowledge to "plan" a policy that maximizes the cumulative return of the state value $v_\pi(s)$ or $q_\pi(s, a)$. All these algorithms are based on dynamic programming, which allows you to break the problem into smaller subproblems and use the recursive relationship of Bellman equations explained in Chapter 2. Next, I will extend the approach to come up with a more general framework of policy improvement.

Grid World Environment

The coding exercises are based on a simple grid world environment—an idealized maze world of 16 squares arranged in a 4x4 format, as shown in Figure 3-1.

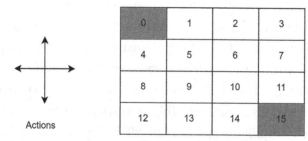

Figure 3-1. *Grid world environment. It is a 4×4 grid with terminal states at the top left and bottom right. Numbers in the grid represent the state S*

The top-left and bottom-right positions are terminal states shown as shaded cells in the figure. In a given cell, the agent can move in any of the four directions: UP, RIGHT, DOWN, and LEFT. The actions deterministically move the agent in the direction of the action unless there is a wall. In the case of hitting a wall, the agent stays in the current position. The agent gets a reward of -1 every time step until it reaches the terminal state.

I will keep the implementation of the grid-world environment to bare bones, just implementing the core capabilities. For a detailed way to create custom environments in the Gymnasium library, refer to the documentation of Gymnasium.

You'll create a new class called GridworldEnv that implements the transition dynamics $\Pr\{ S_{t+1} = s', R_{t+1} = r \mid S_t = s, A_t = a \}$ as a dictionary P in which P[s][a] gives a list of tuples with the values (probability, next_state, reward, done). In other words, for a given state s and action a, it gives a list of tuples consisting of the possible next states s', reward r, and probability $p(s', r \mid s, a)$. The fourth value in the tuple is a Boolean flag called done that indicates if the next state s' is a terminal state or not.

The transition dynamics P are known under the current setup of model-based learnings, which is the focus of this chapter. However, P is unknown in real-world problems, whether model-based or model-free. Algorithms either learn without the knowledge of the model (transition dynamics) or build an approximation of the model dynamics to solve the RL optimization problem. You will study these algorithms in subsequent chapters. Listing 3-1 shows the transition function implementation of grid world, which is contained in the script file gridworld.py.

Listing 3-1. The Grid World Environment

```
def limit_coordinates(self, coord):
    """

    Prevent the agent from falling out of the grid world
    :param coord: a tuple(x,y) position on the grid
    :return: new coordinates ensuring that they are within the grid world
    """

    coord[0] = min(coord[0], self.shape[0] - 1)
    coord[0] = max(coord[0], 0)
    coord[1] = min(coord[1], self.shape[1] - 1)
    coord[1] = max(coord[1], 0)
    return coord
```

```
def transition_prob(self, current, delta):
    """

    Model Transitions. Prob is always 1.0.
    :param current: Current position on the grid as (row, col)
    :param delta: Change in position for transition
    :return: [(1.0, new_state, reward, done)]
    """

    # if stuck in terminal state
    current_state = np.ravel_multi_index(tuple(current), self.shape)
    if current_state == 0 or current_state == self.nS - 1:
        return [(1.0, current_state, 0, True)]

    new_position = np.array(current) + np.array(delta)
    new_position = self.limit_coordinates(new_position).astype(int)
    new_state = np.ravel_multi_index(tuple(new_position), self.shape)

    is_done = new_state == 0 or new_state == self.nS - 1
    return [(1.0, new_state, -1, is_done)]
```

The limit_coordinates function makes sure that the agent does not use all of the grid. If you are in one of the left-most cells, an action to move "left" does not let the agent get out of the grid. The agent in such a case stays unmoved, as there is no way to go left while in a left-most cell.

The transition_prob function takes the current position of the agent and an action such as UP or DOWN. It moves the agent as per the action, taking care to not let the agent fall off the grid. It returns a tuple of prob_of_new_state, next state value, reward and a done flag. In the current environment, you are modeling a deterministic world, which means that the move based on the action is certain and the outcome is only a single new state with a probability of 1.0. Therefore, prob_of_new_state in the returned tuple is always 1 and only one tuple is returned for each state-action pair. However, in a stochastic world, for each state, action pair would get a list of tuples, whereby each tuple refers to one of the possible states the agent could find itself in after the action.

Let's now focus on the model-based algorithms, which is the focus of this chapter. I will first explain the concept of dynamic programming.

Dynamic Programming

Dynamic programming is an optimization technique that was developed by Richard Bellman in the 1950s. In the context of algorithmic thinking, dynamic programming is equated with the concept of avoiding repeated work by remembering partial results. Stated another way, it refers to simplifying a complex problem into simpler sub-parts in a recursive way. This recursive breaking down into simpler problems stops once you reach a level of simplification where the problem can be solved trivially. An example is the calculation of a Fibonacci series.

A Fibonacci series is defined as $F_i = F_{i-1} + F_{i-2}$ with the base case of $F_1 = F_2 = 1$. Let's look at how to calculate F_{51}. $F_{51} = F_{50} + F_{49}$. But F_{50} itself has to be calculated first, using $F_{49} + F_{48}$. As you can see, to calculate F_{51}, you need to calculate F_{49} twice, first as part of F_{50} and then directly as the second term in the equation of F_{51}. Using dynamic programming, you can calculate F_{49} only once and store it the first time it is calculated. Any subsequent call to calculate F_{49} would return the previously stored value instead of calculating it again. There are two ways to implement this. A top-down approach of starting with the big problem and recursively breaking it down into calculations of smaller sub-problems (when implemented in a naïve way) would lead to exponential growth of runtime due to repeated calculations of a sub-problem multiple times, similar to what I talked about in the context of F_{49} and illustrated in Figure 3-2.

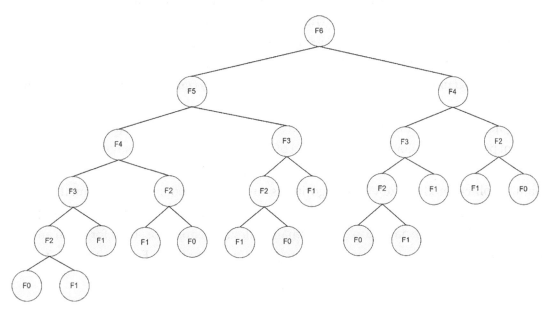

Figure 3-2. *Recursion diagram of calculating Fibonacci F_6*

You can see from Figure 3-2 that to calculate F_6, you need to calculate F_5 once, F_4 twice, F_3 three times, F_2 five times, and F_1 seven times. As you calculate the Fibonacci value of higher numbers, the lower number's Fibonacci values need to be calculated/accessed multiple times. The way to avoid this exponential growth is to a) store the values as you go from the top node in Figure 3-2 to the leaf nodes, in a *top-down approach*, or b) start the calculation from leaves and climb up the tree to reach the root node at the top, in a *bottom-up approach*. The technique of saving values as you calculate and using them the next time when they are needed is called *memoization* and should not be confused with memorization.

Dynamic programing in the world of mathematics optimization refers to the simplification of a calculation into a sequence of steps over time—that is, calculating current time step values from the values of the previous time step. This is the approach you will be using to build a time-indexed recursion for calculating the state values $v_\pi(s)$. Take a look at the Bellman equation, which expresses the value of a state $v_\pi(s)$ in terms of the policy $\pi(a|s)$, system dynamics $p(s', r|s, a)$, and state value of successor states $v_\pi(s')$.

$$v_\pi(s) = \sum_a \pi(a|s) \sum_{s',r} p(s', r|s, a) \big[r + \gamma\, v_\pi(s') \big]$$

This expresses the value $v_\pi(s)$ in terms of other state values $v_\pi(s')$, all of which are unknown. If you could somehow get all the successor state values for the current state, you would be able to calculate $v_\pi(s)$. This shows the recursive nature of the equation.

Note that a particular value $v_\pi(s')$ would be needed multiple times wherever s' is a successor state of some state s. Because of this, you can cache (i.e., store) the value $v_\pi(s')$ and use it multiple times to avoid recalculating $v_\pi(s')$ again and again, every time it is needed.

Dynamic programming is an extensively used optimization technique for a varied class of problems; it allows decomposition of complex problems into smaller problems. Some common applications are scheduling algorithms, graph algorithms like shortest path, graphical models like Viterbi algorithms, and lattice models in bioinformatics. As this book is about reinforcement learning, I restrict the use of dynamic programming for solving the Bellman expectation and Bellman optimality equations, both for value and action-value functions. These equations are given in Chapter 2, in Equations 2-18, 2-21, 2-24, and 2-25.

Equation 2-18 shows the recursive nature of state values $v_\pi(s)$ in terms of successor state values $v_\pi(s')$ as reproduced here:

$$v_\pi\left(s\right)=\sum_a \pi\left(a|s\right)\sum_{s',r} p(s',r|s,a)\left[r+\gamma\,v_\pi\left(s'\right)\right] \tag{3-1}$$

Equation 2-21 showed the recursive nature state-action values $q_\pi(s,a)$ in terms of successor state-action values $q_\pi(s'|a')$ as reproduced here:

$$q_\pi\left(s,a\right)=\sum_{s',r} p(s',r|s,a)\left[r+\gamma\sum_{a'}\pi\left(a'|s'\right)q\left(s',a'\right)\right] \tag{3-2}$$

Equation 2-24 showed the recursive relationship of state-value $v_*(s)$ in terms of successor state-values $v_*(s')$ as shown in Equation 3-3. I remove the outer summation in Equation 3-1 and replace it with a *max* function. This is the Bellman Optimality Equation:

$$v_*\left(s\right)=\max_a\sum_{s',r} p(s',r|s,a)\left[r+\gamma\,v_*\left(s'\right)\right] \tag{3-3}$$

Equation 2-25 showed the recursive relationship between the state-action values $q_*(s,a)$ in terms of successor state-action values $q_*(s'|a')$ under the Bellman optimality construction, where the inner sum in Equation 3-2 is replaced with the *max* function.

$$q_*\left(s,a\right)=\sum_{s',r} p(s',r|s,a)\left[r+\gamma\max_{a'}q_*\left(s',a'\right)\right] \tag{3-4}$$

Each of these four equations represents the v or q value of a state or state-action pair in terms of the value of successor states or state-actions meeting the recursive nature of dynamic programming.

Remember that each of the summations in these four equations is nothing but the expectation operation $E[\cdot]$ over all possible values reward and the next state in the next time period, which is a function of the current state and action taken in that state.

In the following sections, you will first use the expectation in Equation 3-1 and/or 3-2 to evaluate a policy, which is known as an *evaluation* or *prediction*. You will then use optimality Equation 3-3 and/or 3-4 to find an optimal policy that maximizes the state values and state-action values, which is known as *policy improvement*.

Once you understand this concept, you will look at a generalized setup, an extensively used generalized framework for policy improvement. I conclude the chapter by talking about the practical challenges in large-scale problem setups and various approaches to optimizing dynamic programming in that context.

As discussed at the beginning of this chapter, I mostly focus on the class of problems whereby you have a finite set of states the agent can find itself in as well as a finite set of actions in each state. Problems with continuous states and continuous actions can technically be solved using dynamic programming by first discretizing the states and actions. You will see an example of such an approach toward the end of Chapter 4. It also forms the bulk of Chapter 5.

Policy Evaluation/Prediction

This section uses Equation 3-1 to derive the state values using its iterative nature and the concepts of dynamic programming. Equation 3-1 represents the state value for a state s in terms of its successor states. The values of states also depend on the policy the agent is following, which is defined as policy $\pi(a|s)$ in Equation 3-1. As you may have noticed, because of this dependence of value on policy, all the state values are subscripted with π to signify that state values in Equation 3-1 are the ones obtained by following a specific policy $\pi(a|s)$. To stress once again the importance of a policy π, note that changing the policy π will produce a different set of values, $v_\pi(s)$ and $q_\pi(s, a)$.

The relationship in Equation 3-1 can be graphically represented by a backup diagram, as shown in Figure 3-3.

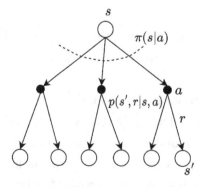

Figure 3-3. *Backup diagram of Bellman expectation equation for state-value functions. Empty circles signify states, and dark circles signify actions*

The agent starts in state s. It takes an action a based on its current policy $\pi(a|s)$. As a result of this action, the environment transitions the agent to a new state s' along with the reward r based on the system dynamics $p(s',r|s,a)$. It should be evident that Equation 3-1 forms a system of equations, one equation for each state. If there are N states, you will have N such equations. The number of equations equals the total number of states, mathematically represented as $N = |S|$, and is the same as the number of unknown state values $v(s)$, one for each state $S = s$. Thus, Equation 3-1 represents a system of $|S|$ equations with $|S|$ unknowns.

You could solve the system of equations using any linear programming technique. The recursive set of N equations can be represented in Vector Algebra notation. While you will not be using this approach in real-life problems for reasons soon to be discussed, I will go over the approach to vectorization of Equation 3-1. This is optional and you may skip it if you are not very mathematically inclined. You first rewrite Equation 3-1 by merging the outer and inner summations in two separate summations:

$$v_\pi(s) = \sum_{a,s',r} \pi(a|s) . p(s',r|s,a) . r + \gamma . \sum_{a,\,s',r} \pi(a|s) . p(s',r|s,a) . v_\pi(s') \qquad (3\text{-}5)$$

Now take the summation over successor state s' out of the summation:

$$v_\pi(s) = \sum_{a,s',r} \pi(a|s) . p(s',r|s,a) . r + \gamma . \sum_{s'} \sum_{a,r} \pi(a|s) . p(s',r|s,a) . v_\pi(s')$$

The first term on the right—$\pi(a|s) . p(s',r|s,a)$ with the summation over all possible s', a, r is nothing but the expectation, $E[.]$ due to a specific policy, π. Also note that you will have N such equations, one for each state $S = s$. If you write all the N equations, one below the other, you can specify all $v_\pi(s)$, all N of them as a single vector of size $1xN$ denoted as V_π. In a similar fashion, the first term on the right side can be expressed as a vector of rewards of size $1xN$ depicted as R_π, thereby denoting the expected reward for each state based on the policy that the agent follows in each of the state. The summation $\sum_{a,r} \pi(a|s) . p(s',r|s,a)$ in the second expression on the right can be expressed as P_π and is the transition dynamics of the system under a specific policy. This will be a matrix of size NxN with a number of rows and number of columns equal to the total number of states N. Making these changes, you can represent the previous equation as follows:

$$V_\pi = R_\pi + \gamma \cdot P_\pi \cdot V_\pi$$

Collecting all the terms with V_π on left, you get:

$$V_\pi \cdot (I + \gamma \cdot P_\pi) = R_\pi$$

You can now rewrite the vector equation as follows:

$$V_\pi = (I + \gamma \cdot P_\pi)^{-1} \cdot R_\pi$$

As you can see, calculating V_π involves inverting an $N \times N$ matrix $(I + \gamma \cdot P_\pi)$, which is not feasible for a large value of N. And this is the primary reason that you would not use the linear algebra approach to solve this system of N linear equations.

Instead, I will resort to the use of iterative solutions. This is achieved by starting with some random state values $v_0(s)$ at first iteration $k = 0$ and using them on the right side of Equation 3-1 to obtain the state values at the next iteration step.

$$v_{k+1}(s) \leftarrow \sum_a \pi(a|s) \sum_{s',r} p(s', r|s, a) \left[r + \gamma\, v_k(s') \right] \tag{3-6}$$

Notice the change of subscript from π to (k) and $(k + 1)$. Also notice the change of equality (=) to assignment ($\leftarrow$). You are now representing the state values at iteration $(k + 1)$ in terms of the state values in the previous iteration k, and there will be N (total number of states) such updates in each iteration. It can be shown that, as the iteration index k increases and goes to infinity (∞), v_k will converge to v_π. This approach of finding the values of all the states for a given policy is known as *policy evaluation*. You start from an arbitrarily chosen value of v_o at $k = 0$ and iterate over the state values using Equation 3-6, until the state values v_k stop changing. The other name for policy evaluation is *prediction*—that is, predicting the state values for a given policy. The convergence of v_k to v_π is guaranteed because $v_k = v_\pi$ is a fixed point and it flows from a contraction mapping theorem which you can look up in advanced texts.

While updating all the N state values v_{k+1}, one for each of the N states, from the previous iteration values of states (N of the v_k values), there are two approaches. The first such approach is known as the *synchronous version*. In this, you maintain two arrays for the values of the states—one from the previous iteration k which is used to update the second array to hold updated state values in the current iteration $k + 1$. At the end of the current iteration, the values from the second array to the first array and the next cycle of iteration from $k + 1$ to $k + 2$ is initiated.

The second approach is known as *asynchronous version*, because there is only one array and updates are carried out in place. This is shown to display faster convergence while requiring memory for only one array. In the approach, there are many ways to choose which state value to update in what sequence. I will discuss this in detail in a subsequent section in this chapter.

Figure 3-4 shows the pseudocode for iterative policy evaluation.

ITERATIVE POLICY EVALUATION

Input π, the policy to be evaluated and convergence threshold θ

Initialize state values $v\,(s) = 0$ or to any arbitrary values for all state s $\in$ S. However terminal state values should always be initialized to 0

Make a copy: $v'\,(s) \leftarrow v(s)$ for all s
Loop:
$\quad \Delta = 0$
$\quad$ Loop for each $s \in S$
$$v'(s) \leftarrow \sum_a \pi(a|s) \sum_{s',r} p(s',r|s,a)[r + \gamma v(s')]$$
$$\Delta = max(\Delta, |v(s) - v'(s)|)$$
$\quad\quad v(s) \rightarrow v(s')$ for all $s \in S$, i.e., make a copy of $v(s)$
Until $\Delta < \theta$

Figure 3-4. *Iterative policy evaluation algorithm*

Let's now apply the algorithm to the grid world given in Figure 3-1. This assumes a random policy $\pi(a| s)$ where each of the four actions (UP, RIGHT, DOWN, LEFT) have an equal probability of 0.25. Listing 3-2 shows relevant portions of the code for iterative policy evaluation applied to the grid world. This is from the file 3.a-policy-evaluation.ipynb.

Listing 3-2. Policy Evaluation/Policy Planning: "3.a-policy-evaluation.ipynb"

```
# Start with a (all 0) value function
k = 0
V = np.zeros(env.nS)
V_new = np.copy(V)
```

```
while True:
    k += 1
    delta = 0
    # For each state, perform a "backup"
    for s in range(env.nS):
        v = 0
        # Look at the possible next actions
        for a, pi_a in enumerate(policy[s]):
            # For each action, look at the possible next states...
            for prob, next_state, reward, done in env.P[s][a]:
                # Calculate the expected value as per backup diagram
                v += pi_a * prob * \
                    (reward + discount_factor * V[next_state])
        # How much our value function changed (across any states)
        V_new[s] = v
        delta = max(delta, np.abs(V_new[s] - V[s]))

    V = np.copy(V_new)
    # print grid for specified iteration values
    if k in print_at:
        grid_print(V.reshape(env.shape), k=k)
    # Stop if change is below a threshold
    if delta < theta:
        break
grid_print(V.reshape(env.shape), k=k)
return np.array(V)
```

You start with k=0, the index of current iteration, and two arrays V and V_new to hold the values of the states in the grid. The inner for loop for s in range(env.nS) loops over all the states to update the array V_new with $v_{k+1}(s)$ from the previous state values $v_k(s)$ contained in the array V using Equation 3-1. Along the way, it records the maximum change in value across all the states and stores this maximum change in the delta variable. Next, it copies the values from V_new to V and checks if the maximum change in the current iteration, as stored in delta, is less than the initial threshold, called theta. If the check is true, the program terminates printing the grid and its state values.

If delta is more than theta, the program loops back to the next iteration, incrementing the iteration counter k by 1. The final values of the states in the grid for following this random policy produced by the program are shown in Figure 3-5.

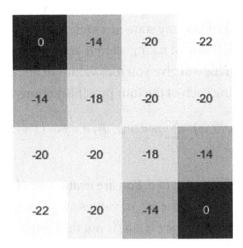

Figure 3-5. *Policy evaluation: $v_\pi(s)$ for the grid world from Figure 3-1 with the agent following a random policy. Each of the four actions (UP, DOWN, LEFT, and RIGHT) have an equal probability of 0.25*

Note that the values have converged. Let's look at the third row in the last column with a state value of $v_\pi(s) = -14$. In this state, the UP action will take the agent to a cell with a state value of -20, the LEFT action takes the agent to a cell with a state value of -18, the DOWN action takes the agent to a terminal state with a value of 0, and the RIGHT action hits the wall, leaving the agent in the same state. Let's apply Equation 3-1. I will expand the right side of Equation 3-1 with the actions applied in order—TOP, RIGHT, DOWN, and LEFT:

-14 = 0.25*(-1+(-20)) + 0.25*(-1+(-14)) + 0.25*(-1+0) + 0.25*(-1+(-18))

-14 = -14

The values on both sides are matching, which confirms convergence. Accordingly, the values shown in Figure 3-5 are the state values when the agent follows a random policy. Note that this considers a discount factor of $\gamma = 1.0$.

This section considered a random policy with equal probability of action in all four directions. However, this probably may not be an optimal policy to follow. There could be one or more optimal policies producing higher state values. The next section looks at a way to devise the algorithm to find a better policy.

Policy Improvement and Iterations

The previous section discussed an algorithm to get state values $v_\pi(s)$ for a given policy. You could use this information to improve upon the policy. In the grid world, you have four actions that you can take from any state. Instead of following a random policy $\pi(a|s)$, you'll now look at the value of taking all four actions individually and then follow the policy π after that step. This will give you four values of $q(s, a)$ action values, which are the action values of taking each of the four possible actions in the grid world.

$$q(s, a) = \sum_{s',r} p(s', r|s, a)\left[r + \gamma\, v_\pi(s')\right]$$

Note that $q(s, a)$ has no subscript of π. You are evaluating $q(s, a)$ for all possible actions while in state $S = s$. If any of the $q(s, a)$ values are greater than the current state value $v_\pi(s)$, it means that current policy $\pi(a|s)$ is not taking the optimal action. You can improve the current policy in current state $S = s$ by taking the q-value, maximizing action $A = a$ and defining this to be the policy in state $S = s$. It will give a higher state value compared to the current policy $\pi(a|s)$. This new policy is defined as $\pi'(a|s)$. In other words, you define the following:

$$\pi'(a|s) = \operatorname*{argmax}_a q(s, a) \tag{3-7}$$

There is a general result called *policy improvement theorem,* which states:
Let there be two deterministic policies π and π' such that, for all $s \in S$,

$$q_\pi\left(s, \pi'(s)\right) \geq v_\pi(s)$$

Then the policy π' must be as good as or better than π. In other words:

$$v'_\pi(s) \geq v_\pi(s)\ s \in S$$

The values for all states under the new policy π' are going to be equal to or greater than the state values under previous policy π.

Stated differently, while choosing a maximizing action in a specific state $S = s$ improves the state value of that state, it cannot reduce the values of other states. It can either leave them unchanged or improve those other states that depend on $S = s$.

The previous maximizing step could be applied to all the states based on their current q-values. It can be seen as a greedy step since you are maximizing the state value of a particular state without considering anything else. Extending this greedy step action across all states in the MDP is known as a *greedy policy*. The recursive state value relationship is given by the Bellman's optimality Equations 3-3 and 3-4.

You now have a framework to improve the policy. For a given MDP, you first do policy evaluation iteratively to obtain state values $v(s)$, and then you apply the greedy selection of action by maximizing the q values as per Equation 3-7. This results in state values going out of sync with Bellman equations, as the maximization step is applied to each individual state without flowing it through all the successor states. Therefore, you again carry out the policy iteration under the new policy π' to find the state/action values under the improved policy. Once the values are obtained, the maximizing action from Equation 3-7 is applied again to further improve the policy to π''. The cycle goes on until no further improvement is observed. This sequence of action can be depicted as follows:

$$\pi_0 \xrightarrow{\text{evaluate}} v_{\pi_0} \xrightarrow{\text{improve}} \pi_1 \xrightarrow{\text{evaluate}} v_{\pi_1} \xrightarrow{\text{improve}} \pi_2 \ldots \ldots \xrightarrow{\text{improve}} \pi_* \xrightarrow{\text{evaluate}} v_*$$

From the policy improvement theorem, you know that each swipe of greedy improvement and policy evaluation ($v_\pi \xrightarrow{\text{improve}} \pi' \xrightarrow{\text{evaluate}} v_\pi$) gives you a policy that is better than the previous one with $v'_\pi(s) \geq v_\pi(s)$ $s \in S$. For an MDP that has a finite number of discrete states along with a finite number of actions in each state and with each swipe leading to an improvement, an optimal policy would be found once you stop observing any further improvement in state values. This is bound to happen within a finite number of cycles of improvement.

This approach to finding an optimal policy is called *policy iteration*. Figure 3-6 shows the pseudocode for policy iteration.

POLICY ITERATION

Initialize state values $v(s)$ and policy π arbitrarily
e.g. $v(s) = 0, \forall s \in S$, and $\pi(a|s)$ as random

define convergence threshold θ

Policy-Evaluation-Step
Loop:
$\quad \Delta = 0$
$\quad$ Loop for each $s \in S$
$$v'(s) \leftarrow \sum_a \pi(a|s) \sum_{s',r} p(s',r|s,a)[r + \gamma v(s')]$$
$$\Delta = \max(\Delta, |v(s) - v'(s)|)$$
$\quad v(s) \leftarrow v'(s)$ for all $s \in S$, i.e. make a copy of $v(s)$
Until $\Delta < \theta$

Policy-Improvement-Step
policy-changed $\leftarrow$ false
Loop for each $s \in S$
$\quad$ old-action $\leftarrow \pi(s)$
$$\pi(s) \leftarrow \operatorname*{argmax}_a \sum_{s',r} p(s',r|s,a)[r + \gamma v(s')]$$
$\quad$ If $\pi(s) \neq$ old-action then policy-changed = true
If policy-changed = true: then go back to Policy-Evaluation step followe by Policy Improvement
otherwise return $v(s)$ as v_* and $\pi(s)$ as π_*

Figure 3-6. *Policy iteration algorithm for finite MDP*

Now apply the policy-iteration algorithm to the grid world from Figure 3-1. Listing 3-3 shows the code for the policy iteration applied to the grid world. The complete code is given in `3-b-policy-iteration.ipynb`. The `policy_evaluation` function remains the same as Listing 3-2. There is a new function, called `policy_improvement`, which applies greedy maximization to return a policy that improves upon the existing policy. `policy_iteration` is a function that runs `policy_evaluation` followed by `policy_improvement` in a loop until the state values stop increasing and converge to a fixed point.

Listing 3-3. Policy Improvement: 3-b-policy-iteration.ipynb

```python
def policy_improvement(policy, V, env, discount_factor=1.0):

    def argmax_a(arr):
        """
        Return idxs of all max values in an array.
        """
        max_idx = []
        max_val = float('-inf')
        for idx, elem in enumerate(arr):
            if elem == max_val:
                max_idx.append(idx)
            elif elem > max_val:
                max_idx = [idx]
                max_val = elem
        return max_idx

    policy_changed = False
    Q = np.zeros([env.nS, env.nA])
    new_policy = np.zeros([env.nS, env.nA])

    # For each state, perform a "greedy improvement"
    for s in range(env.nS):
        old_action = np.array(policy[s])
        for a in range(env.nA):
            for prob, next_state, reward, done in env.P[s][a]:
                # Calculate the expected value as per backup diagram
                Q[s, a] += prob * (reward + discount_factor *
                V[next_state])

        # get maximizing actions and set new policy for state s
        best_actions = argmax_a(Q[s])
        new_policy[s, best_actions] = 1.0 / len(best_actions)

    if not np.allclose(new_policy[s], policy[s]):
        policy_changed = True

    return new_policy, policy_changed
```

In the policy improvement, you loop over all states and do greedy maximization for each state, as shown in the for loop for s in range(env.nS). In a given state *s*, you first calculate the current state-action values Q[s,a] and pick the action *a* with maximum Q value. This action becomes the new policy in that given state. If there is more than one action with same maximum state value, all such actions are taken as a set of best actions in that state with equal probabilities.

The policy iteration step, as shown in Listing 3-4, iterates over policy evaluation and policy improvement in a loop until the values converge and there is no further improvement observed.

Listing 3-4. Policy Iteration: 3-b-policy-iteration.ipynb

```
def policy_iteration(env, discount_factor=1.0, theta=0.00001):

    # initialize a random policy
    policy = np.ones([env.nS, env.nA]) / env.nA
    while True:
        V = policy_evaluation(policy, env, discount_factor, theta)
        policy, changed = policy_improvement(policy, V, env,
        discount_factor)
        if not changed:  # terminate iteration once no improvement is
        observed
            V_optimal = policy_evaluation(policy, env, discount_
            factor, theta)
            print("Optimal Policy\n", policy)
            return np.array(V_optimal)
```

Figure 3-7 shows the state values for each grid cell as a result of running policy_iteration on the grid world.

State Values

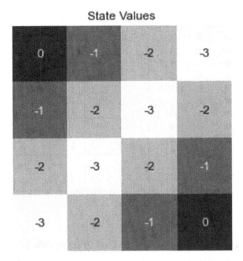

Figure 3-7. *Policy iteration $v_*(s)$ for the grid world from Figure 3-1. The agent is following an optimal policy as found by applying policy_iteration from Listing 3-4*

Note that the optimal state values are the number of steps required to reach the closest terminal state with a -ve sign. As the reward is -1 for each time step until the agent reaches the terminal state, the optimal policy would take the agent to the terminal state in the minimal number of possible steps multiplied with -1. For some states, more than one action could lead to the same number of steps to reach the terminal state. For example, looking at the top-right state with the state value = -3, it takes three steps to reach the terminal state at the top left and also three steps to reach the terminal state at the bottom right. In other words, the state value is the negative of the Manhattan distance between the state and nearest terminal state.

You can also extract the optimal policy, as shown in Figure 3-8. The left side of the figure shows the policy as extracted from the code in Listing 3-4, and the right side of the diagram shows the same policy graphically superimposed on the grid.

```
Optimal Policy
[[0.25 0.25 0.25 0.25]
 [0.   0.   0.   1.  ]
 [0.   0.   0.   1.  ]
 [0.   0.   0.5  0.5 ]
 [1.   0.   0.   0.  ]
 [0.5  0.   0.   0.5 ]
 [0.   0.   0.5  0.5 ]
 [0.   0.   1.   0.  ]
 [1.   0.   0.   0.  ]
 [0.5  0.5  0.   0.  ]
 [0.   1.   0.   0.  ]
 [0.   0.   1.   0.  ]
 [1.   0.   0.   0.  ]
 [0.   1.   0.   0.  ]
 [0.   1.   0.   0.  ]
 [0.25 0.25 0.25 0.25]]
```

Figure 3-8. *Optimal state values v*(s) for the grid world as shown in Figure 3-7. Left: optimal policy with probability of action for each cell in the grid. Right: grid superimposed with the optimal policy*

As discussed earlier, policy evaluation is also known as *prediction,* as you are trying to find the state values consistent with the current policy of the agent. Similarly, using policy iteration to find an optimal policy is also known as *control*—to control the agent and find an optimal policy.

Value Iteration

This section looks at policy iteration and computes how many passes you need to find an optimal policy. Policy iteration has two steps. The first step is policy evaluation, which is run for the current policy. It requires multiple passes over the state space so that state values converge and become consistent with the current policy. The second part of the loop is policy improvement, which requires one pass over the state space to find the best action for each state—that is, greedy improvement with respect to the current state action values. From this, it is evident that a large part of the time is spent in policy evaluation and letting values converge.

An alternative approach is to truncate the loop inside policy evaluation well before the state values have converged. When you truncate the loop inside policy evaluation to only one loop, you are using an approach known as *value iteration*. Similar to the approach of Equation 3-6, you take the Bellman optimality from Equation 3-3 for the state value and convert it into an assignment with iteration. The revised equation is as follows:

$$v_{k+1}(s) \leftarrow \max_a \sum_{s',r} p(s',r|s,a)\left[r + \gamma\, v_k(s')\right] \qquad (3\text{-}8)$$

As you iterate, the state values will keep improving and will converge to v*, which are the optimal values.

$$v_0 \rightarrow v_1 \rightarrow \ldots v_k \rightarrow v_{k+1} \rightarrow \ldots \rightarrow v_*$$

Once the values converge to optimal state values, you can use a one-step backup diagram to find the optimal policy.

$$\pi_*(a|s) = argmax_a \sum_{s',r} p(s',r|s,a)\left[r + \gamma\, v_*(s')\right] \qquad (3\text{-}9)$$

The previous procedure to iterate over values by taking a max at each step is known as *value iteration*. Figure 3-9 shows the pseudocode for value iteration.

```
VALUE ITERATION

Initialize state values v(s) (Terminal states are always initialized to 0)
e.g. v(s) = 0, ∀s ∈ S
define convergence threshold θ
Make a copy: v'(s) ← v(s) for all s

Loop:
    Δ = 0
    Loop for each s ∈ S
        v'(s) ← maxₐ Σ p(s',r|s,a)[r + γv(s')]
                     s',r
        Δ = max(Δ, |v(s) − v'(s)|)
        v(s) ← v'(s) for all s ∈ S, i.e. make a copy of v(s)
Until Δ < θ


Output a deterministic policy, breaking ties deterministically.
Initialize π(s), an array of length |S|
Loop for each s ∈ S
    π(s) ← arg maxₐ Σ p(s',r|s,a)[r + γv(s')]
                      s',r
```

Figure 3-9. *Value iteration algorithm for a finite MDP*

You can now apply the previous value iteration algorithm to the grid world given in Figure 3-1. Listing 3-5 contains the code for value iteration applied to the grid world. Check out the `3.c-value-iteration.ipynb` file for a detailed implementation. The `value_iteration` function is the straight implementation of the pseudocode in Figure 3-9.

Listing 3-5. Value Iteration: 3.c-value-iteration.ipynb

```
def value_iteration(env, discount_factor=1.0, theta=0.00001):
    """

    Carry out Value iteration given an environment and a full description
    of the environment's dynamics.

    def argmax_a(arr):
        # Return idx of max element in an array.
        ........<see ipynb notebook for omitted code> ........
        return max_idx

    optimal_policy = np.zeros([env.nS, env.nA])
    V = np.zeros(env.nS)
    V_new = np.copy(V)

    while True:
        delta = 0
        for s in range(env.nS):
            q = np.zeros(env.nA)
            for a in range(env.nA):
                for prob, next_state, reward, done in env.P[s][a]:
                    if not done:
                        q[a] += prob*(reward + discount_factor *
                        V[next_state])
                    else:
                        q[a] += prob * reward
            V_new[s] = q.max()
            delta = max(delta, np.abs(V_new[s] - V[s]))
        V = np.copy(V_new)
        if delta < theta:
            break
```

```
# V(s) has optimal values. Use these values and one step backup
# to calculate optimal policy
for s in range(env.nS):
    q = np.zeros(env.nA)
    for a in range(env.nA):
        for prob, next_state, reward, done in env.P[s][a]:
            if not done:
                q[a] += prob * (reward + discount_factor *
                V[next_state])
            else:
                q[a] += prob * reward
    # find the optimal actions
    best_actions = argmax_a(q)
    optimal_policy[s, best_actions] = 1.0 / len(best_actions)
return optimal_policy, V
```

The output of running the value algorithm against the grid world will produce the optimal state values $v_*(s)$ and optimal policy $v_{\pi^*}(a|s)$ with the same values and policy, as shown in Figures 3-7 and 3-8.

Before moving forward, let's summarize. What you have learned until now are classified as *synchronous dynamic programming algorithms*, as summarized in Table 3-1.

Table 3-1. *Synchronous Dynamic Programming Algorithms*

Algorithm	Bellman Equation	Type of Problem
Iterative policy evaluation	Expectation equations	Prediction
Policy iteration	Expectation equations and greedy improvement	Control
Value iteration	Optimality equations	Control

Generalized Policy Iteration

The policy iteration described earlier has two steps. The first step is *policy evaluation,* which gets the state values in sync with the current policy the agent is following. It requires multiple passes over all the states for state values to converge to v_π. The second step is greedy action selection, which improves the policy. As explained, the second step of improvement results in the current state of values going out of sync with the new policy. Therefore, you have to carry out yet another round of policy evaluation to bring the state values back into sync with the new policy. The cycle of *evaluation* followed by *improvement* stops when there is no further change to the state values. This happens when the agent has reached an optimal policy, when the state values are optimal and in sync with the optimal policy. The convergence to the optimal policy (fixed point) v_π can be depicted visually, as shown in Figure 3-10.

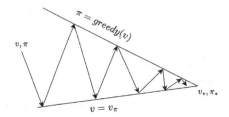

Figure 3-10. *Iteration between the two steps. The first step is evaluating to get the state values in sync with the policy being followed. The second step is improving the policy to do a greedy maximization for actions*

You have seen two extremes for the number of loops to iterate inside the policy evaluation step. Also, every iteration, be it policy evaluation or policy improvement, covered all the states in the model. But there can be many variations.

First, even within a single iteration of *evaluation + improvement*, you can visit only a partial set of states to evaluate as well as a partial set of the state-actions in a given state to improve with greedy maximization. The *policy improvement theorem* guarantees that even partial coverage of the states or partial coverage of state-action maximization will lead to an improvement unless the agent is already following an optimal policy.

Secondly, you do not have to carry out the evaluation step over multiple iterations until the state values converge. You could stop early. The state value sync does not need to be complete. It may terminate midway.

All of this results in the arrows in Figure 3-10 stopping short of touching the bottom line of $v = v_\pi$. The arrow diagram in this scenario is shown in Figure 3-11.

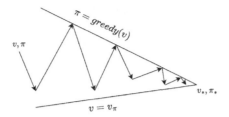

Figure 3-11. *Iterating just enough to cover all the states in a single iteration leads to not touching the bottom line $v = v_\pi$*

Similarly, the step of policy improvement may not carry out improvement for all the states, which this time leads to arrows in Figure 3-10 stopping short of the upper line of $\pi = greedy(v)$. Figure 3-12 shows the arrow diagram for such a case.

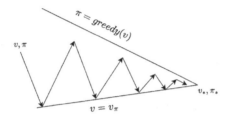

Figure 3-12. *Iterating when the policy improvement step is not carried out over all the states, leading to not touching the top line $\pi = greedy(v)$*

In summary, these two steps of policy evaluation and policy improvement—with all their variations—lead to convergence as long as each state is visited enough times in both evaluation and improvement. This is known as *generalized policy iteration* (GPI). Most of the algorithms that you will study in this book can be classified as some form of GPI. When you go through the various algorithms, keep in mind the images shown in Figures 3-10 through 3-12.

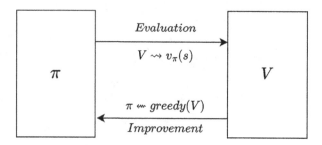

Figure 3-13. *Iteration between the two steps. The first step is evaluation to get state values in sync with the policy being followed. The second step is that of policy improvement to do a greedy maximization for actions*

Asynchronous Backups

Dynamic programming–based algorithms have scalability issues. The dynamic programming approach is much better and more scalable than direct solution methods such as linear programing, which involve solving a matrix equation. However, dynamic programming still does not scale well with real-life problems. Consider a single sweep under policy iteration. It requires a visit to each state, and under each state, you need to consider all the possible actions. Further, each action involves a calculation that may theoretically involve all the states again, depending on the state transition function $p(s', r|s, a)$. In other words, every iteration has a complexity of $O(|A| * |S|^2)$, one $|S|$ for covering each state and $|A|$ for visiting each action in a given state, taking the computation to $O(|A| * |S|)$. Then one more $|S|$ is added due to the backup equation, where for a given state and one specific action in that state, you need to visit all possible states the system may transition to as a result of the action, taking the overall complexity to $O(|A| * |S|^2)$.

You started with policy iteration under which you carried out multiple iterations as part of evaluation step for the state values to converge. The second control method was value iteration. You then reduced the evaluation iterations to only one step by leveraging Bellman optimality equations. All these were synchronous dynamic programming algorithms under which all states are updated using Bellmen backup Equations 3-1 to 3-4.

However, there is no need to update each state in every iteration. You could update and/or optimize in any order covering only partial parts of the total states in the system. All such methods to sweep over states yield optimal state values and optimal policies as long as each state is visited often enough. There are various approaches to making a sweep, as shown in Table 3-2.

Table 3-2. *Flavors of Dynamic Programming Sweeps*

Approach	Pros	Cons
Dynamic Programming	Scales better than linear programming	Requires multiple passes for convergence
In-Place Dynamic Programming	Lower memory footprint	Same scaling issues as DP
Dynamic Programming with Prioritized Sweeping	Faster convergence	Some states may take longer to converge due to infrequent updates
Real-time Dynamic Programming	Most scalable	Offers only asymptotic convergence if each state is updated enough times (infinite in asymptotic limit)

The first one is *in-place dynamic programming*. Up until now, you have maintained two copies of the states. The first copy holds the existing state values, and the second copy holds the new state values after the update. An *in-place* strategy uses only a copy of the state value array. The same array is used for reading old and updating new state values. As an example, let's look at the value iteration in Equation 3-8 and rewrite the version for an in-place update. Note the subtle difference in the subindices on state values on the right and left sides of original value iteration equation, as compared to those in-place versions. The original version uses the array V_k to update a new array V_{k+1}, while the in-place edit updates the same array.

Here is the original update Equation 3-8:

$$v_{k+1}(s) \leftarrow \max_a \sum_{s',r} p(s', r \mid s, a) \left[r + \gamma v_k(s') \right]$$

The in-place version of the same is shown here. It's same array of $v(s)$ on both sides of the arrow.

$$v(s) \leftarrow \max_a \sum_{s',r} p(s', r \mid s, a) \left[r + \gamma v(s') \right]$$

Experiments have shown that in-place edits offer faster convergence as the values move up even midway of an iteration.

The second idea revolves around the order in which states should be updated. In synchronous programing, you update all the states in a single iteration. However, values may converge faster if you use *prioritized sweeping*. Prioritized sweeping requires knowledge of the predecessors of a state. Suppose you have just updated a state $S = s$ and the value changed by an amount Δ. All the predecessor states of state $S = s$ are added to a priority queue with a priority of Δ. If a specific predecessor state is already there in the priority queue with a priority more than Δ, it is left untouched. In the next iteration, a new state with the highest priority is taken out of the queue and updated, taking you back to the beginning of the loop. The strategy of priority sweeping *requires knowledge of reverse dynamics*—that is, the predecessors of a given state.

The third idea is *real-time dynamic programming*. With this approach, you only update the value of the states that the agent has currently seen—that is, the states that are relevant to the agent and use its current path of exploration to prioritize the updates. The approach avoids wasteful updates to the states that are not on the horizon of the agent's current path and hence mostly irrelevant.

Dynamic programming, both synchronous and asynchronous, uses full-width backups like the one shown in Figure 3-3. For a given state, you need to know every action possible and every successor state $S = s'$. You also need to know the environment dynamics $p(s', r \mid s, a)$. However, using an asynchronous approach does not completely solve the scalability issue. It only extends the scalability a bit. In other words, dynamic programming is feasible only for midsize problems even with asynchronous updates.

The big assumption that breaks down for most of the real-life problems is that of having prior knowledge of the transition function or environment dynamics $p(s', r \mid s, a)$. This chapter assumed knowledge of dynamics to introduce Bellman equations and the core foundation of how dynamic programming is used for solving problems in RL. Starting with the next chapter, I explain a more scalable approach to solving reinforcement learning problems using sample-based methods. Under sample-based methods, you drop the assumption of prior and complete knowledge of the environment dynamics. Also, to build scalability, you'll find efficient ways to avoid carrying out full-width sweeps.

Summary

This chapter introduced the concept of dynamics programming and explained how it is applied to the field of reinforcement learning. In order to facilitate this discussion, I made an assumption of knowing the dynamics of the system $p(s', r| s, a)$. The chapter first looked at policy evaluation for prediction—that is, to calculate the state values for a given policy. Next, the chapter looked at policy iteration for control—that is, to find a better policy than the current one if the current policy is not already optimal. This was followed by exploring the concept of value iteration, which was obtained by combining evaluation and improvement into a single equation.

These approaches then led to the topic of generalized policy iteration. You looked at the convergence path that generalized policy iteration follows. Throughout the chapter, I used a running example of grid world, which had a 4x4 grid with simple dynamics. Finally, the chapter covered various shortcuts you can take to speed up the algorithms under the topic of asynchronous dynamic programming.

CHAPTER 4

Model-Free Approaches

The previous chapter looked at dynamic programming, used when you know the model dynamics $p(s', r| s, a)$, and the knowledge is used to "plan" the optimal actions. This is also known as the *planning problem*. This chapter shifts focus and looks at *learning problems*—that is, setups where the model dynamics (aka transition dynamics) are not known. You will learn to calculate/learn state values and state-action values by *sampling*—that is, collecting experience by following some policy in the real world or running the agent through a policy in simulation. There is another class of problems where the model-free approach is more applicable. In some problems, it is easier to sample than to calculate the transition dynamics, such as a problem of finding the best policy to play a game like blackjack. There are many combinations to reach a score that depend on the cards seen so far and the cards still in the deck. It is almost impossible to calculate the exact transition probability from one state to another, but it is easy to sample states from an environment. To summarize, you use model-free methods when either you do not know the model dynamics or you know the model, but it is much more practical to sample than to calculate the transition dynamics.

This chapter looks at two broad classes of model-free learning: Monte Carlo (MC) methods and temporal difference (TD) methods. You will first learn how policy evaluation works in a model-free setup and then extend this understanding to look at control—that is, finding the optimal policy. I also touch upon the important concepts of bootstrapping and the exploration-versus-exploitation dilemma as well as off-policy versus on-policy learning. Initially, the focus is to look at the MC and TD methods individually. Subsequently, I cover additional concepts like n-step returns, importance sampling, and eligibility traces in a bid to combine the MC and TD methods into a common, more generic approach, called TD(λ).

© Nimish Sanghi 2024
N. Sanghi, *Deep Reinforcement Learning with Python*, https://doi.org/10.1007/979-8-8688-0273-7_4

Estimation/Prediction with Monte Carlo

When you do not know the model dynamics, what do you do? Think back to a situation when you did not know something about a problem. What did you do in that situation? You experiment. In the context of reinforcement learning, it could mean having an agent take actions randomly and seeing how the system responds. For example, say you want to find out if a playing die or a coin is biased or not. You toss the coin or throw the die multiple times, observe the outcomes, and use that to form your opinion. In other words, you sample. In the case of coin, say you toss the coin 10,000 times and calculate the fraction of times it shows heads. If that fraction is close to 0.5, you can surmise that the coin is unbiased with a high degree of confidence. If the experiment is conducted 100,000 times with fraction of heads still hovering around 0.5, your confidence in the unbiasedness of the coin will be even higher. Why does taking the average of outcomes from samples and using it as an estimate work?

The law of large numbers (LLN) from statistics tells us that the average of samples is a good estimate for the actual unknown quantity and that this estimate becomes better and better as the number of trials of the experiment (samples) increases. Further, these estimates formed by averaging the outcomes from samples become better as the number of samples increase—stated in mathematical language the error as measured by the variance between the average of samples and the actual unobserved value decreases as the number of trial n increases and with n going to infinity, the estimate from trials becomes equal to the actual unobserved value. This law is one of the very important laws in the field of statistics. Having said that, do remember that there are some cases where such convergence does not happen. However, thankfully you will not be encountering such pathological cases in most real-world RL examples.

If you examine the Bellman equations in the previous chapter, you will notice that there is an expectation operator $E[\cdot]$ in those equations; for example, the value of a state being $v(s) = E[G_t | S_t = s]$. Further, to calculate $v(s)$, you used dynamic programming requiring the transition dynamics $p(s', r | s, a)$. In the absence of the model dynamics knowledge, what can you do?

Following the law of large numbers, you sample from the model. You fix a starting state $S_t = s$ and fix a policy π for which you want to calculate $v_\pi(s)$. You let the agent start from this state $S_t = s$, follow the policy π to take actions, and keep doing so until termination. You record the total sum of rewards—that is, the return the agent gets in the current single sequence/episode. This sequence of state, action, reward until termination is also known as an *episode* or a single *rollout* of the policy. You repeat this experiment to collect individual returns from each such rollout and average these as an estimate of $v_\pi(s)$ for the policy π. This, in a nutshell, is the approach of *Monte Carlo (MC) methods*: replace expected returns with the average of sample returns.

There are a few points to note. The MC approach does not require knowledge of the model. The only thing required is that you should be able to sample from it. The ability to sample or play out the agent depends a lot on the problem. If it is a board game whose laws can be easily modeled in a program or if there is a simulator available for a real-world system, you can use such a program or simulator to sample very quickly and cheaply. However, if a rollout of an episode will involve interaction with the agent in the real world and if the real-world agent is something like an expensive robot or a self-driving car, such interaction may not be feasible, especially when the agent is still not trained enough and is bound to make mistakes. In some cases, a mistake could be very costly, leading to the destruction of the machine or endangering the public safety. Therefore, the MC approach to sample the return from a complete rollout may not always work, especially if it involves real-world interactions. There is another scenario where MC may not work—in the case of non-terminating environments where there is no terminal state. You need to know the return of starting from a state until termination, and hence you can use MC methods only on episodic MDPs in which every run finally terminates. It will not work on non-terminating environments.

The second point is that for a large MDP, you can keep the focus on sampling only that part of the MDP that is relevant and avoid exploring irrelevant parts of the MDP. Such an approach makes MC methods highly scalable for very large problems. As stated earlier, the MC approach can be useful for the class of problems in which listing all states and/or all actions in a state makes the problem intractable and where there could be large parts of the state and state-action combinations not of interest in a specific context. A very good example was demonstrated by OpenAI, when they used a variant of the MC method called Monte Carlo tree search (MCTS) to train a Go game-playing agent.

The third point is about Markov assumption—that is, the past is fully encoded in the present state. In other words, knowing the present makes the future independent of the past. As more formally stated the transition to state S_{t+1} in time $t+1$ is only dependent on the current state S_t and action A_t taken in time t. This transition is completely independent of the states and actions that came before S_t, A_t. I talked about this property in Chapter 2. Markov independence forms the basis of the recursive nature of Bellman equations, where a state just depends on the successor state values. However, under the

MC approach, you observe the full return starting from a state S going until termination. You are not depending in any way on the value of the successor states to calculate the current state value. There is no Markov property assumption being made here. A lack of Markov assumption in MC methods make them much more feasible for the class of MDPs known as POMDPs (for "partially observable MDPs"). In a POMDP environment, you get only partial-state information, which is known as an observation.

We have been talking about using MC methods to estimate the state and state-action values without making any Markov property assumption or without worrying about the transition dynamics. However, you can also use the MC method to roll out multiple episodes, and then use these rollouts in turn to calculate the dynamics $p(s', r|s, a)$. Such an approach will take you back to the world of model-based RLs by first using MC rollouts to estimate system dynamics and then using those system dynamics in a DP setup to find the optimal policies. I will talk about those in detail in a future chapter. For now, the chapter focuses on the model-free approach.

Moving on, let's look at a formal way to estimate the state values for a given policy. You let the agent start a new episode and observe the stepwise rewards from the time agent enters state $S = s$ for the first time in that specific episode going until the termination of the episode. The cumulative reward from first time $S = s$ until termination is called the reward G. Many episode runs are carried out, every time calculating the return from $S = s$ to termination. The average of return across episodes is taken as an estimate of $v_\pi(S = s)$. This is known as the *first-visit MC method*. Note that, depending on the dynamics, an agent may visit the same state $S = s$ in some later step within the same episode before termination. In the first-visit MC method, you take the total return only from the first visit in an episode until the end of the episode. There is another variant in which you take the average of returns from every visit to that state until the end of the episode. This is known as the *every-visit MC* method.

Let's see how the value of a state under a specific policy can be estimated under first-visit MC method. Figure 4-1 shows the backup diagram for the MC method. In DP, in a given state you take a full swipe to cover all the actions possible from a state and all the possible transitions to a new state from the state-action pair $(S = s, A = a)$. You go only one level deep from state $S = s$ to $A = a$ and then to the next state $S = s'$. Compared to this, in the MC method, the backup covers a full sample trajectory from the current state $S = s$ to the terminal state. However, It does not cover all the branching possibilities; rather, it covers only one single path that was sampled from the start state $S = s$ to the terminal state.

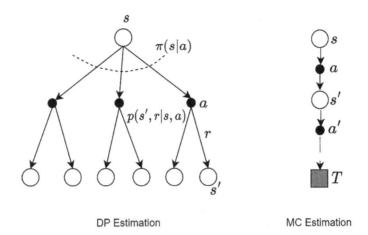

Figure 4-1. *Backup diagram of MC methods as compared to Bellman equation–based DP backup*

Now look at the pseudocode for the first-visit version, which is shown in Figure 4-2. You input the agent's current policy π. You initialize two arrays: one to hold the current estimate of $V(s)$ and the second to hold the number of visits $N(s)$ to a state $S = s$. Multiple episodes are executed, and you update $V(s)$ as well as $N(s)$ for each of the states the agent visits in every episode, taking care to do the update for the first visit in the "first-visit" variant and updating every visit in the "every-visit" variant. The pseudocode covers only the "first visit" variant. The "every visit" variant is easy to implement by just dropping the condition toward the end of the pseudocode, which reads as "Unless S_t appears in S_0, S_1, ..." As per the law of large numbers, the statistical law on which Monte Carlo simulations are based, $V(s)$, will converge to true $v_\pi(s)$ for all states under the condition that the number of visits to each state approaches infinity.

FIRST VISIT MC PREDICTION

Input:
    π, the policy to be evaluated

Initialize:
    Cumulative state values $S(s) = 0$ for all $s \in S$
    Estimated state values $V(s) = 0$ for all $s \in S$
    Visit count $N(s) = 0$ for all $s \in S$

Loop:
    sample episode following policy $\pi : S_0, A_0, R_1, S_1, A_1, R_2, \ldots\ldots A_{t-1}, R_T, S_T$
    $G \leftarrow 0$
    Loop backwards for each step of episode, $t = T - 1, T - 2, \ldots, 1, 0$
        $G \leftarrow \gamma.G + R_{t+1}$
        Unless S_t appears in $S_0, S_1, \ldots, S_{t-1}$
            $N(s) \leftarrow N(s) + 1$
            $S(s) \leftarrow S(s) + G$
            $V(s) \leftarrow S(s)/N(s)$

Figure 4-2. *First-visit MC prediction for estimating $v\pi(s)$. The pseudocode uses the online version to update the values as samples are received*

You store the cumulative total and the count of first visits to a state. The average is calculated by dividing the total by the count. Every time a state is visited, the following update is carried out:

$$N(s) = N(s) + 1$$

$$S(s) = S(s) + G$$

$$V(s) = \frac{S(s)}{N(s)}$$

With a minor change, you could look at the update in another way, without needing the cumulative totals $S(s)$. In this alternate formulation, you have only one array $V(s)$ that is updated for every visit directly without needing to divide the total by the count. This is an important computational optimization that will occur many times, whenever you need a running average of the numbers that are coming one at a time in some kind of online fashion. If you have two numbers, instead of remembering them all, you just keep the average. When a third number arrives, you update the average of the two numbers with the third number so that the new average is now an average of the three numbers. The derivation of this updated rule is as follows.

In the following derivation, $N(s)$ is the visit count, which is also the notation used in Figure 4-2. $S(s)$ is the cumulative sum of returns that the agent sees in state s and $V(s)$ is the value function of state s. The cumulative sum $S(s)$ and state value $V(s)$ are specific to the policy the agent is following, which is fixed and provided as input in the beginning, as shown in Figure 4-2. You also use subscript n or $n + 1$ to indicate the nth or $n + 1^{th}$ visit to the specific state s.

$$N(s)_{n+1} = N(S)_n + 1$$

$$V(s)_{n+1} = \frac{S(s)_n + G}{N(s)_{n+1}}$$

$$\Rightarrow V(s)_{n+1} = \frac{V(s)_n . N(s)_n + G}{N(s)_{n+1}}$$

$$\Rightarrow V(s)_{n+1} = \frac{V(s)_n \cdot \left[N(s)_{n+1} - 1 \right] + G}{N(s)_{n+1}}$$

$$\Rightarrow V(s)_{n+1} = \frac{V(s)_n . N(s)_{n+1} + \left[G - V(s)_n \right]}{N(s)_{n+1}}$$

$$\Rightarrow V(s)_{n+1} = V(s)_n + \frac{1}{N(s)_{n+1}} \cdot \left[G - V(s)_n \right]$$

$$\Rightarrow V(s)_{n+1} = V(s)_n + \frac{1}{N(s)_{n+1}} \cdot \left[G - V(s)_n \right] \qquad (4\text{-}1)$$

The difference $[G - V(s)_n]$ can be viewed as an error between the latest sampled value G and the current estimate $V(s)_n$. The difference/error is then used to update the current estimate by adding $\frac{1}{N}.error$ to the current estimate. As the number of visits increases, the new sample has increasingly low influence in revising the estimate of $V(s)$. This is because the multiplication factor $1/N$ reduces to zero as N becomes very large.

Sometimes, instead of using a diminishing factor $1/N$, you can use a constant α as the multiplication factor in front of $[G - V(s)_n]$. You use this approach for non-stationary environments specifically. As the environment could be changing, you want to always give enough importance to the latest error with the help of a contact, α.

$$V_{n+1} = V_n + \alpha \left(G - V_n \right) \text{ for all states } s \in S \qquad (4\text{-}2)$$

Note that in Equation 4-2, I have removed the function arguments s from V_{n+1} and V_n.

The constant multiplication factor approach is more appropriate for problems that are non-stationary or when you want to give a constant weight to all the errors. A situation like this can happen when the old estimate V_n may not be very accurate.

Having looked at the pseudocode, now it is time to look at the Python implementation of the first visit MC and run the logic on the same grid world you saw in the previous chapter. In Chapter 3, you learned how to derive the state values for a random walk policy using DP. Now you will arrive at the same state value estimates using the MC method. Remember that with MC methods, you do not know or use the knowledge of transition dynamics. Rather, you just sample the rollouts for multiple episodes and use the rewards collected in those rollouts to estimate the state values for all the states in the grid for the random walk policy agent is following.

For this, you need to update the implementation of grid world from Chapter 3. The Python code in gridworld.py inside the Chapter 4 code includes revised code for the grid world environment. There is a section toward the end of this chapter devoted to the things you need to do to implement your own Gymnasium environment. For now, it suffices to know that you need to implement two new functions in your environment.

The first function is env.reset(), which randomly initializes the environment and returns the starting state of the agent plus some debug info about the environment. The second function is env.step(action) and it takes an action as input, transitions the agent to the new state as per the internal dynamics of the system, and returns a tuple of five values (new_state, reward, terminated, truncated, info).

Let's now look at the implementation of the MC value prediction. Listing 4-1 shows the code, and the full code is available in the 4.a-mc_estimation.ipynb file.

Listing 4-1. MC Value Prediction Algorithm for Estimation – "4.a-mc_estimation.ipynb"

```
# MC Policy Evaluation
def mc_policy_eval(policy, env, discount_factor=1.0, episode_count=100):
    """

    Evaluate a policy given an environment.

    Args:
        policy: [S, A] shaped matrix representing the policy. Random in
        our case
```

```
    env: GridWorld env. use step(a) to take an action and receive
        a tuple of (s', r, terminated, terminated, info)
        env.nS is number of states in the environment.
        env.nA is number of actions in the environment.
    episode_count: Number of episodes:
    discount_factor: Gamma discount factor.

Returns:
    Vector of length env.nS representing the value function.
"""

# Start with (all 0) state value array and a visit count of zero
V = np.zeros(env.nS)
N = np.zeros(env.nS)
i = 0

# run multiple episodes
while i < episode_count:

    # collect samples for one episode
    episode_states = []
    episode_returns = []
    state, _ = env.reset()
    episode_states.append(state)
    while True:
        action = np.random.choice(env.nA, p=policy[state])
        state, reward, done, _, _ = env.step(action)
        episode_returns.append(reward)
        if not done:
            episode_states.append(state)
        else:
            break
    # update state values
    G = 0
    count = len(episode_states)
    for t in range(count-1, -1, -1):
        s, r = episode_states[t], episode_returns[t]
        G = discount_factor * G + r
```

```
            if s not in episode_states[:t]:
                N[s] += 1
                V[s] = V[s] + 1/N[s] * (G-V[s])

        i = i+1

    return np.array(V)
```

The code in Listing 4-1 is a straight implementation of the pseudocode in Figure 4-2. The code implements the online version of the update, as explained in Equation 4-1— that is, `N[s]+=1; V[s]=V[s]+1/N[s]*(G-V[s])`. The code also implements the "first-visit" version but can be converted to "every visit" with a very small tweak. You just need to drop the `if` check—that is, "`if s not in episode_states[:t]`" and move the code under this condition as an unconditional code to carry out the updates at every step.

To obtain convergence to the true state values, you need to ensure each state is visited enough times, ideally infinite times in limit. As shown in Figure 4-3, state values do not converge very well for 100 episodes. However, with 10,000 episodes, the values have converged well and match those produced with the DP method.

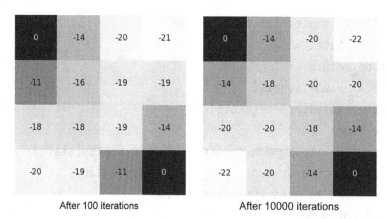

After 100 iterations   After 10000 iterations

Figure 4-3. *Policy evaluation: $v_\pi(s)$ computed using the MC method for the grid world when the agent is following a random policy. Each of the four actions (UP, DOWN, LEFT, and RIGHT) have an equal probability of 0.25. The left figure shows the values after 100 iterations and the right figure shows the state values after 10,000 iterations*

A key aspect of Monte Carlo techniques is that the predictions for each state stand alone. One state's estimate doesn't rely on the estimate of another, unlike in dynamic programming (DP). Put simply, Monte Carlo methods don't use bootstrapping as described in the earlier chapter.

Related to the same point is that of computational expense in estimating a specific state value. The computational expense incurred for calculating a single state's value is independent of the total number of states and independent of the branching factor of possible actions in that specific state. In the case of DP, you need to consider all actions in a specific state as well as all following possible successor states for each action. This is expressed by the two summation signs in Equation 3-6. However, in the MC method, you just consider the rollouts that happen without worrying about an exhaustive list of possible actions and successor states. Therefore, if the interest is in calculating a specific state's value, you just need the agent to start from that state and follow the policy being evaluated until termination. And you do so for many episodes to take the average value of each episode's return as an estimate of specific starting state's value. In the MC approach, you are not worried about all other states and actions that you do not see in the sample rollouts.

The next subsection dives deeper into the two versions of MC methods: *first-visit* vs *every-visit*.

Bias and Variance of MC Predication Methods

Let's now look at the pros and cons of "first visit" versus "every visit." Do both converge to the true underlying $V(s)$? Do they fluctuate a lot while converging? Does one converge faster to the true value? Before I answer these questions, let's first review the basic concept of bias-variance tradeoff that you see in all statistical model estimations, for example , in supervised learning.

Bias refers to the property of the model to converge to the true underlying value that you are trying to estimate—in this case, $v_\pi(s)$. Some estimators are biased, meaning they are not able to converge to the true value due to their inherent lack of flexibility—that is, estimation approach being too simple or restricted for a given true value. The bias of a model goes down as you increase the number of samples in an estimate or as you make the estimation process more complex and hence more flexible. A biased estimation approach or a model will consistently over- or under-predict the true values. For example, a model that is biased toward predicting higher values will consistently predict that new data points will have higher values than they do.

Variance refers to the model estimate being sensitive to the specific sample data being used. This means the estimate value may fluctuate a lot. Mostly all estimation processes show lower variance as you increase the number of samples. However, for a given number of samples, some approaches may have lower variance, and some may have more. Usually, the models that are more complex show higher variance as they can fit better for a given data. Therefore, as you collect different sets of samples with multiple runs of an experiment, the estimated value from each run will fit well to the data seen in that run. This in turn leads to higher variance of these estimated values across multiple runs. The model with higher complexity or flexibility will show lower bias but will show more variance.

A model that is trained on a small dataset may have a high variance, because the predictions will be sensitive to the specific data points that are included in the training set.

If you have unlimited data, you could always choose a very complex model and still drive down the variance by using an increasingly large volume of data to form the estimate. However, in all practical problems, you do not have unlimited data, or compute, or time to process unlimited data even if it was available.

This fundamental tradeoff between the bias and variance is a tradeoff between the accuracy and variability of an estimation approach, especially for a given and fixed volume of data.

The bias-variance tradeoff can be illustrated by the following example. Suppose you are trying to estimate the average height of all adults in the world. You can collect samples of adults and measure their heights. Then, you can use the sample mean to estimate the population mean height. "Population mean" refers to the true average height of all the adults in the world, which is unknown, and which is what you are trying to estimate with the help of samples.

The sample mean is an unbiased estimator of the population mean height. This means that the average of all possible sample means will be equal to the population mean height. However, the sample mean is also a variable estimator. This means that the value of the sample mean will vary from sample to sample.

The variance of the sample mean is inversely proportional to the sample size. This means that the larger the sample size, the smaller the variance of the sample mean. In other words, the more data you have, the more accurate your estimate of the population mean height will be.

However, there is a limit to how much you can reduce the variance of the sample mean by increasing the sample size. This is because the sample mean is still a biased estimator of the population mean height. The bias of the sample mean is due to the fact that you are only sampling a small fraction of the population.

If you want to reduce the bias of the sample mean, you need to use a more complex model. For example, you could use a regression model to predict the height of an individual based on their age, gender, and other factors. A regression model will typically have a lower bias than the sample mean, but it will also have a higher variance.

The bias-variance tradeoff is important because it is impossible to have a model that has both zero bias and zero variance. If a model has zero bias, then it will perfectly fit the training data. However, this means that the model is likely to overfit the training data and will not generalize well to new data. On the other hand, if a model has zero variance, then it will produce the same prediction for all data points. This means that the model is not able to capture any of the complexity in the data. The models, which are very flexible, have low bias as they can fit the model to any configuration of a dataset. At the same time, due to flexibility, they can be overfit to the data, making the estimates vary a lot as the training data changes. On the other hand, models that are simpler have high bias. Such models, due to the inherent simplicity and restrictions, may not be able to represent the true underlying model. But they will also have low variance as they do not overfit.

Consider that you have some data points (say 100). You are trying to fit a regression model with a polynomial of degree n and assume that underlying true curve on which the data points lie follow some unknown equation. Further, assume that each data point may not lie exactly on this unknown ground truth curve due to noise in data. If you take a polynomial of lower degree (less complex model), say a degree of 1, it will do a poor fit and hence will have a higher bias. However, the fitted curve will not change much even if you were to carry out the fitting exercise with another set of 100 data points generated from the game ground truth—that is, it will show lower variance. Now consider the other extreme of using a polynomial of degree=100. For any given 100 data points, the curve will exactly fit all the data points—that is, lower bias. However, in this case another set of 100 data points may produce a fitted curve very different from the previous curve. In other words, the fitted curve will fluctuate wildly in shape based on which 100 sample points are used for fitting the curve. Such a situation is known as high variance. As the model complexity increases the bias reduces but variance increases. Based on the problem, there is a range in the middle that balances bias and variance. Readers may refer to any standard text on machine learning for an in-depth discussion.

This tradeoff, known as *bias-variance tradeoff*, is presented in the graph in Figure 4-4.

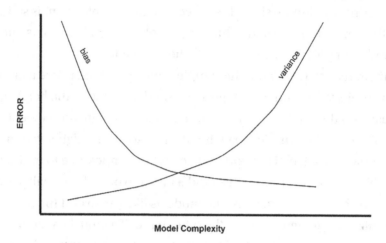

Figure 4-4. *Bias variance tradeoff. Model complexity is increasing toward the right on the x-axis. Bias starts off high when the model is restricted and drops off as the model becomes flexible. Variance shows a reverse trend with model complexity*

In the field of reinforcement learning, until a certain point, you can reduce variance without increasing the bias by making model simpler on time domain. This is because of causality. The action agent takes in current state—that is, the policy it follows in the current state does not impact the rewards seen in the past transition. Current action can only impact the future rewards. You will see variations of these when you study policy gradient methods in later chapters.

With this background, now turn your attention back to comparing "first visit" to "every visit". The *first visit* is unbiased but has high variance. *Every visit* has bias that goes down to zero as the number of trials increases. In addition, *every visit* has low variance and usually converges to the true value estimates faster than the *first visit*.

Control with Monte Carlo

Let's now talk about control in a model-free setup. You need to find the optimal policy in this setup without knowing the model dynamics. As a refresher, let's look at the generalized policy iteration (GPI) that was introduced in Chapter 3. In GPI, you iterate between two steps. The first step is to find the state values for a given policy, and the second step is to improve the policy using greedy optimization. You will follow the same GPI approach for control under MC. You will have some tweaks, though, to account for the fact that you are in a model-free world with no access/knowledge of transition dynamics.

Chapter 3 looked at state values, $v(s)$. However, in the absence of transition dynamics, state values alone will not be sufficient. For the greedy improvement step, you need access to the action values, $q(s, a)$. You need to know the q-values for all possible actions—that is, all $q(S = s, a)$ for all possible actions a in state $S = s$. Only with that information will you be able to apply a greedy maximization to pick the best action—that is, $argmax_a q(s, a)$.

There is another complication when compared to DP. The agent follows a policy at the time of generating the samples. However, such a policy may result in many state-action pairs never being visited, and even more so if the policy is a deterministic one. If the agent does not visit a state-action pairs for a given state, it does not know all $q(s, a)$ for that state. Hence it cannot find the action yielding maximum q-value. One way to solve the issue is to ensure enough exploration by *exploring starts*—that is, ensuring that the agent starts an episode from a random state-action pair and over the course of many episodes covers each state-action pair enough times as a starting point. Ideally, an infinite number of times in the limit.

Figure 4-5 shows the GPI diagram with one change, v-values being replaced by q-values. The evaluation step now is the MC prediction step that I talked about in the previous section. Once the q-values stabilize, greedy maximization can be applied to obtain a new policy. The policy improvement theorem ensures that the new policy will be better or at least as good as the old policy.

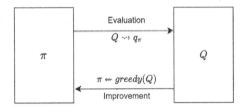

Figure 4-5. *Iteration between the two steps. The first step evaluates to get the state-action values in sync with the policy being followed. The second step performs policy improvement to do a greedy maximization for actions*

This approach of GPI will be a recurring theme. Based on the setup, the evaluation steps will change, and the improvement step invariably will continue to be greedy maximization.

The assumption of "exploring starts" is not very practical and efficient. It's not practical, as in many scenarios, because the agent does not get to choose the starting condition, for example, while training a self-driving car. It may not be feasible, and it may also be wasteful for the agent to visit each state-action pair infinite times in limit. However, you still need continued exploration for the agent to visit all the actions in the states that are visited by the current policy. This is achieved by using an ε-greedy policy.

In an ε-greedy policy, the agent takes the action with the maximum q-value with the probability $1 - \epsilon$, and it takes any action randomly with probability $\frac{\epsilon}{|A|}$. In other words, the remaining probability of ϵ is divided equally across all actions to ensure that the agent continues to explore non-maximizing actions. In other words, the agent *exploits* the knowledge with probability 1-ε, and it *explores* with probability ε. You will see this concept of exploration and exploitation being repeated in RL.

$$\pi(a|s) = \begin{cases} 1 - \varepsilon + \dfrac{\varepsilon}{|A|} & \text{for } a = \text{argmax}_a Q(s,a) \\ \dfrac{\varepsilon}{|A|} & \text{otherwise} \end{cases} \tag{4-3}$$

The action with a maximum q-value gets picked with probability $1 - \epsilon$ from greedy max. All actions are picked up with a probability ε/|A|.

While ε-greedy solves the problem of exploring all possible pairs of s, a, it is a compromise. The final policy learned by the agent is also an ε-greedy one. Imagine such a final policy for an autonomous robot. The probability ε/ | A| of some random action could put the robot in a destructive state or could cause some harm to objects/people around it. So how do you fix it? An easy fix is to reduce the value of ε as the episode count increases and in limit you let ε → 0. This lets the agent learn the optimal greedy policy in the limit.

Let's make one more refinement to the estimation/prediction step of the iteration. In the previous section, you saw that even for a simple 4×4 grid MDP, you needed an order of 10,000 episodes for the values to converge. Refer to Figure 3-10, which shows the convergence of GPI: the MC prediction/evaluation step makes the v-values or q-values in sync with the current policy. However, in the previous chapter, you also saw that there was no need for the agent to go all the way to the convergence of values. This was shown in Figure 3-11. Similarly, you could run MC prediction followed by policy improvement on an episode-by-episode basis. This approach will remove the need for a large number of iterations in the estimation/prediction step, thereby making the approach scalable for large MDPs. Similar to Figure 3-11, such an approach will produce convergence, as shown in Figure 4-6.

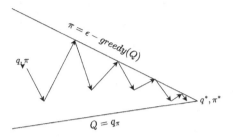

Figure 4-6. *Iteration between the two steps. The first step is the MC prediction/ evaluation for a single step to move the q-values in the direction of the current policy. The second step is that of policy improvement to do a ε-greedy maximization for actions*

Combining the two tweaks, you now have a practical algorithm for Monte Carlo-based control for optimal policy learning. This is called *greedy in the limit with infinite exploration* (GLIE). You will use the *every-visit* version in the *q*-value prediction step. To make it first-visit, you just need a minor tweak, like the one you saw in MC prediction in Figure 4-2. The GLIE pseudocode is given in Figure 4-7.

GLIE FOR POLICY OPTIMIZATION

Initialize:
    State-action values $Q(s, a) = 0$ for all $s \in S$ and $a \in A$.
    Visit count $N(s, a) = 0$ for all $s \in S$ and $a \in A$.
    Policy π with enough exploration e.g., random policy

Loop:
    sample episode(k) following policy π : $S_0, A_0, R_1, S_1, A_1, R_2, \ldots \ldots A_{t-1}, R_T, S_T$
    $G \leftarrow 0$
    Loop backwards for each step of episode, $t = T - 1, T - 2, \ldots, 1, 0$
        $G \leftarrow \gamma.G + R_{t+1}$
        $N(s, a) \leftarrow N(s, a) + 1$
        $Q(s, a) \leftarrow Q(s, a) + 1/N(s, a) * [G - Q(s, a)]$

    Reduce ε using $\varepsilon = 1/k$
    Update policy with ε-greedy using revised $Q(s, a)$

Figure 4-7. *Every-visit (GLIE) MC control for policy optimization*

Start with q-values $Q(s, a) = 0$ and visit counts $N(s, a) = 0$ for all (s, a) pairs. Then you initialize a random policy like a policy with equal probability for all actions in a state. The main thing to note is that you can use any random policy as long as every action in a specific state has a non-zero probability. The main loop is to carry rollouts of the policy multiple times and in each iteration of the loop use the rolled out state, action, reward tuples to first do one step of prediction followed by greedy maximization, as shown in the pseudocode.

Let's now look at the actual implementation given in 4.b-mc-control.ipynb. Listing 4-2 shows the Python code implementation of the main loop in GLIE. The implementation follows the pseudocode. The argmax_a() function finds the action with max Q(s,a) for a given state. Another function, get_action(state), returns an ε-greedy action for the current policy. As a coding exercise, you may want to make tweaks and convert it into the "first-visit" variant. You can also compare the result of the "first-visit" with the "every-visit" MC control.

Listing 4-2. GLIE MC Control Algorithm

```
# collect samples for one episode
episode_states = []
episode_actions = []
episode_returns = []

state,_ = env.reset()
episode_states.append(state)

while True:
    action = get_action(state)
    episode_actions.append(action)
    state, reward, done, _, _ = env.step(action)
    episode_returns.append(reward)
    if not done:
        episode_states.append(state)
    else:
        break

# update state-action values
G = 0
count = len(episode_states)
```

```
for t in range(count-1, -1, -1):
    s, a, r = episode_states[t], episode_actions[t], episode_returns[t]
    G = discount_factor * G + r
    N[s, a] += 1
    Q[s, a] = Q[s, a] + 1/N[s, a] * (G-Q[s, a])

# Update policy and optimal value
k = k+1
eps = 1/k

for s in range(env.nS):
    best_action = argmax_a(Q[s])
    policy[s] = best_action
```

So far in this chapter, you have studied on-policy algorithms for prediction and control. In other words, the same policy is used to generate samples as well as do policy improvement. Policy improvement is done with respect to q-values of the same ϵ-greedy policies. You need exploration to find $Q(s, a)$ for all state-action pairs, which is achieved by using ϵ-greedy policies. However, as you progress, the ϵ-greedy policy in a limit is suboptimal. You need to exploit more as your understanding of the environment grows with more and more episodes. You have also seen that while theoretically the policy will converge to an optimal policy, at times it requires careful control of the ϵ value. The other big disadvantage of MC methods is that you need each episode to finish before you can update q-values and carry out policy optimization. Therefore, MC methods can only be applied to environments that are episodic and where the episodes terminate. Starting in the next section, you will start looking at a different class of algorithms called *temporal difference* methods that combine the advantages of DP (single time-step updates) and MC (i.e., no need to know system dynamics). However, before I dive deep into TD methods, let's go through a short section to introduce and then compare on-policy versus off-policy learning.

Off-Policy MC Control

In GLIE, you saw that to explore enough, you needed to use ε-greedy policies so that all state actions are visited often enough in limit. The policy learned at the end of the loop is used to generate the episodes for the next iteration of the loop. You are using the same policy to explore as the one that is being maximized. Such an approach is called *on-policy,* where samples are generated from the same policy that is being optimized.

In many cases this is not very efficient. The episodes you have collected in the past have to be thrown away and a new set of episodes must be generated. In the real world or where generating episodes is expensive, the throwing away of old episodes leads to inefficiency or slow process of training. The other issue is that the ability to explore unexplored parts is dependent on the policy being optimized. You have no separate ability to control exploration. Even the ε in the ε-greedy policy needs to be reduced to zero as training progresses, making exploration harder.

Stated differently, the optimal policy control method is seeking to learn action-values conditional on optimal behavior from the starting (state, action) pair but the agent also needs to behave sub-optimally in order to explore all actions in all states so that it can find the maximizing actions in all the states seen in the rollouts.

There is another approach in which the samples are generated using a policy that is more exploratory with a higher ε, while the policy being optimized is the one that may have a lower ε or could even be a fully deterministic one. Such an approach of using a different policy to learn than the one being optimized is called *off-policy* learning. The policy being used to generate the samples is called the *behavior policy* usually denoted with subscript "*b*" and the one being learned (maximized) is called the *target policy,* which is denoted with subscript "π". This is contrast to the on-policy approach, where the policy being used to generate samples is the same as the one which is optimized. In the on-policy approach, the samples are discarded after each policy update as the samples become stale and are no more on-policy after the policy update.

This approach makes the process more complex and needs careful mathematical handing in the form of *importance sampling,* discussed next. You will first derive a concise expression for importance sampling and then see its use for off-policy approach.

Importance Sampling

Mathematically, if you want to estimate the expected value of a function f under a distribution p, but you have samples from distribution q, the importance sampling estimator is given by:

$$E_p[f] \approx \frac{1}{N} \sum_{i=1}^{N} \frac{p(x_i)}{q(x_i)} f(x_i)$$

Where $E_p[f]$ is the expected value of f under distribution p. x_i are the samples

drawn from distribution q and N is the number of samples. The term $\dfrac{p(x_i)}{q(x_i)}$ is called the

importance weight or importance ratio.

Imagine you're a teacher and you want to know the average grade of all the students in the school. However, you only have access to the grades of students in the math club. The students in the math club might be better at math than the average student, so their grades might be higher. To get an estimate of the average grade of all students using only the math club's grades, you can use importance sampling.

Here, p is the distribution of grades for all students. q is the distribution of grades for the math club students. f is the grade itself. You would weigh each math club student's grade by the ratio of the probability that a random student gets that grade to the probability that a math club student gets that grade.

Having seen this introduction, let's now look at importance sampling in the context of Monte Carlo sampling in reinforcement learning. Consider a starting state S_t and the trajectory of action and state sequences until the end of the episode, $A_t,\ S_{t+1},\ A_{t+1},\ \cdots,\ S_T$. The probability of observing the sequence under a policy π is given by the following expression:

$$\Pr_{\pi}\{\text{trajectory}\} = \pi(A_t|S_t) \cdot p(S_{t+1}|S_t, A_t) \cdot \pi(A_{t+1}|S_{t+1})\cdots p(S_T|S_{T-1}, A_{T-1})$$

Similarly, the probability of observing the sequence under a behavior policy b is given by the following expression:

$$\Pr_{b}\{\text{trajectory}\} = b(A_t|S_t) \cdot p(S_{t+1}|S_t, A_t) \cdot b(A_{t+1}|S_{t+1})\cdots p(S_T|S_{T-1}, A_{T-1})$$

The importance sampling ratio is the ratio of the probability of trajectory under the target policy π and the probability of trajectory under the behavior policy b.

$$\rho_{t:T-1} = \frac{\Pr_{\pi}\{\text{trajectory}\}}{\Pr_{b}\{\text{trajectory}\}}$$

Substituting the expressions for trajectory probabilities for target policy π and behavior policy b:

$$\rho_{t:T-1} = \frac{\pi(A_t|S_t) \cdot p(S_{t+1}|S_t, A_t) \cdot \pi(A_{t+1}|S_{t+1})\cdots p(S_T|S_{T-1}, A_{T-1})}{b(A_t|S_t) \cdot p(S_{t+1}|S_t, A_t) \cdot b(A_{t+1}|S_{t+1})\cdots p(S_T|S_{T-1}, A_{T-1})}$$

You see that the transition probabilities $p(S_{t+1}|S_t, A_t)$ occur in both the numerator and denominator. Also, these are dependent on the environment and have nothing to do with the policy agent follows. You can cancel then from both numerator and denominator. This gives you the final form of the importance sampling ratio:

$$\rho_{t:T-1} = \frac{\Pr_{\pi}\{\text{trajectory}\}}{\Pr_{b}\{\text{trajectory}\}} = \prod_{k=t}^{T-1} \frac{\pi(A_k|S_k)}{b(A_k|S_k)} \tag{4-4}$$

Having derived the expression for importance sampling in the context of a trajectory rollout, let's look at why you need it. What is it that you are trying to estimate in MC methods? You are using average of sample rollouts from a starting state/action and these rollouts are observed under a target policy π:

$$Q_{\pi}(s,a) = E_{\pi}[G_t|S_t = s, A_T = a]$$

Remember that G_t is the return of the episode starting from S_t, A_t and continuing until termination where the actions are taken based on the policy π. But when you use a behavior policy b to do the same rollout, you do not see this specific trajectory with same frequency as you would under the target policy π. Taking a simple average of returns across all rollouts under behavior policy b will give you the following:

$$Q_b(s,a) = E_b[G_t|S_t = s, A_T = a]$$

The q-value you get while following policy b is a q-value applicable for policy b and not π. The importance sampling ratio ensures that the return of a trajectory observed under the behavior policy is adjusted up or down based on the relative chance of observing a trajectory under the target policy versus the chance of observing the same trajectory under the behavior policy. One single rollout estimate of $Q(S_t, A_t)$ under π is given as:

$$Q_{\pi}(S_t, A_t) = \rho_{t:T-1} \cdot G_t$$

Where G_t is the return observed in a single episode when agent in state S_t takes action A_t and then follows behavior policy b until termination.

You observe multiple rollouts under policy b from stating point (S_t, A_t), multiply each rollout return by the relevant importance sampling ratio $\rho_{t:T-1}$ to map the returns back to the target policy, and then take the average of all the adjusted returns to form an estimate of state-action values under that policy.

When taking the average, you have two options. The first option is to add all the returns and divide that by the number of rollouts. This approach is called *ordinary importance sampling*.

$$Q_\pi(S_t, A_t) = \frac{\sum \rho_{t:T-1} G_t}{N}$$

The second option is to divide the sum of the importance sampling ratios across all the rollouts. This version is called *weighted importance sampling*:

$$Q_\pi(S_t, A_t) = \frac{\sum \rho_{t:T-1} G_t}{\sum \rho_{t:T-1}}$$

Both the versions have pros and cons. Nothing comes for free, which is the case with importance sampling. The importance sampling ratio can cause wide variance. Plus, these are not computationally efficient. There are many advanced techniques like *discount-aware importance sampling* and *per-decision importance sampling*, which look at the importance sampling and rewards in various different ways to reduce the variance and to make these algorithms efficient. For more details, you can refer to RL book by Sutton and Barto.[1]

The key point to keep in mind is that the weighted-importance sampling method has a much lower error when the number of episodes is smaller and the weighted sampling approach has a bias that is higher initially but goes to zero as numbers of episodes increase.

Similar to the *first-visit* MC prediction, you can use a formula to calculate running average. Suppose you have a sequence of returns G_1, G_2, $\cdots$, G_{n-1} all in the same starting state and action and each has a corresponding weight W_i ($W_i = \rho_{t:T-1}$). The weighted average is calculated as:

$$Q_n = \frac{\sum_{k=1}^{n-1} W_k G_k}{\sum_{k=1}^{n-1} W_k}$$

---

[1] http://incompleteideas.net/book/the-book.html

Now define $C_{n-1} = \sum_{k=1}^{n-1} W_k$. The revised formula for Q_n will be:

$$Q_n = \frac{\sum_{k=1}^{n-1} W_k G_k}{C_{n-1}}$$

A new return G_n arrives. The updated estimate and running total of weights will be:

$$Q_{n+1} = \frac{\sum_{k=1}^{n} W_k G_k}{C_n} = q_n + \frac{W_n}{C_n}[G_n - Q_n]$$

$$C_n = C_{n-1} + W_n$$

Figure 4-8 shows the pseudocode of the off-policy MC control employing the previous idea.

OFF POLICY MC CONTROL OPTIMIZATION

Initialize, for all $s \in S$ and $a \in A(s)$:
    State-action values $Q(s, a) \in \mathbb{R}$ (arbitrarily)
    $C(s, a) \leftarrow 0$
    Policy $\pi = \arg\max_a Q(s, a)$

Loop for each episode:
    $b \leftarrow$ a "behavior policy" with enough exploration
    sample episode(k) following policy $\pi : S_0, A_0, R_1, S_1, A_1, R_2, \ldots\ldots A_{t-1}, R_T, S_T$
    $G \leftarrow 0$
    $W \leftarrow 1$
    Loop backwards for each step of episode, $t = T - 1, T - 2, \ldots, 1, 0$
        $G \leftarrow \gamma.G + R_{t+1}$
        $C(S_t, A_t) \leftarrow C(S_t, A_t) + W$
        $Q(S_t, A_t) \leftarrow Q(S_t, A_t) + \frac{W}{C(S_t, A_t)}[G - Q(S_t, A_t)]$
        $\pi(S_t) \leftarrow \arg\max_a Q(S_t, a)$
        If $A_t \neq \pi(S_t)$ exit inner loop
        $W \leftarrow W \frac{1}{b(A_t|S_t)}$

Figure 4-8. *Off-policy MC control for policy optimization*

This completes the topic of the MC approach. The biggest issue with the MC approach is that you need to wait for an episode to terminate before you can update the q-values. The next sections looks at another approach that takes care of this issue.

Temporal Difference Learning Methods

Refer to Figure 4-1 to study the backup diagrams of the DP and MC methods. In DP, you back up the values over only one step using values from the successor states to estimate the current state value. You also take an expectation over action probabilities based on the policy being followed and then from the (s, a) pair to all possible rewards and successor states. This is as expressed in the equation for state value calculation:

$$v_{\pi}(s) = \sum_{a} \pi(a|s) \sum_{s',r} p(s',r|s,a) \left[r + \gamma v_{\pi}(s') \right]$$

The value of a state $v_{\pi}(s)$ is estimated based on the current estimate of the successor states $v_{\pi}(s')$. This is known as *bootstrapping*. The estimate is based on another set of estimates. The two sums are the ones that are represent branch-off nodes in the DP backup diagram in Figure 4-1. Compared to DP, MC is based on starting from a state and sampling the outcomes based on the current policy the agent is following. The state-value estimates are *averages over multiple runs*. In other words, the sum over model transition probabilities is replaced by averages, and hence the backup diagram for MC is a single long path from one state to the terminal state. The MC approach allowed you to build a scalable learning approach while removing the need to know the exact model dynamics. However, it created two issues: the MC approach works only for episodic environments, and the updates happen only at the end of the termination of an episode. DP had the advantage of using an estimate of the successor state to update the current state value without waiting for an episode to finish.

Temporal difference learning is an approach that combines the benefits of both DP and MC, using bootstrapping from DP and the sample-based approach from MC. The update equation for TD is as follows:

$$V(s) = V(s) + \alpha \left[R + \gamma \cdot V(s') - V(s) \right] \qquad (4\text{-}5)$$

The current estimate of the total return for state $S = s$, that is, G_t, is now given by bootstrapping from the current estimate of the successor state (s') shown in the sample run. In other words, G_t in Equation 4-2 is replaced by $R + \gamma \cdot V(s')$, an estimate. Compared to this, in the MC method, G_t was the discounted sum of rewards seen in the rollout of an episode.

The TD approach described is known as $TD(0)$, and it is a one-step estimate. The reason for calling it $TD(0)$ will become clearer toward the end of the chapter, when you learn about $TD(\lambda)$ and see that this one-step approach corresponds to $\lambda = 0$. For clarity's sake, look at the pseudocode in Figure 4-9 for the value estimation approach under $TD(0)$.

TD(0) FOR ESTIMATION

Input:
     π, the policy for which state values need to be estimated

Initialize:
     State values $V(s) = 0$ for all $s \in S$
     $\alpha \leftarrow$ step size

Loop for each episode:
     Choose a start state S
     Loop for each step in the episode:
         Take action A as per state S and policy π
         Observe reward R and next state S'
         $V(s) \leftarrow V(s) + \alpha[R + \gamma * V(s') - V(s)]$
         $S \leftarrow S'$

Figure 4-9. *TD(0) policy estimation*

Having seen all three approaches—DP, MC, and TD—consider the backup diagrams of all three approaches together, as shown in Figure 4-10. If you compare the starting and ending nodes of DP with that of TD(0), you will notice that both start at state s and the end at the next state s' after one transition. They both use *bootstrapping*, using the estimate of the next state values to estimate the current state values, which is an estimate formed by using another estimate. Next if you compare the branching factor of MC and TD(0), you will notice that both have no branches. MC starts from a state s and then rolls out a single trajectory until termination, while TD(0) also starts from s but rolls out just one step and hence stops at state s' at the next time step. TD(0) and MC both sample the episodes and use the observed rewards; MC is the sum total of rewards all the way to termination while TD(0) is the sum of immediate reward and estimate of value of the next state s'.

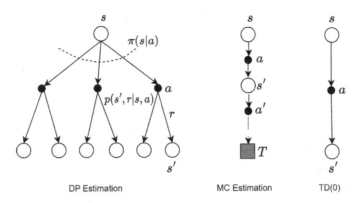

DP Estimation   MC Estimation   TD(0)

Figure 4-10. *Backup diagram of DP, MC, and TD(0) compared*

I now introduce another quantity that you are likely to see again and again in reinforcement learning literature. In Equation 4-5, the quantity $[R + \gamma \cdot V(s')]$ is the revised estimate of the total return from state $S = s$, and it is based on a one-step bootstrapped backup from successor state s'. Further, in Equation 4-5, $V(s)$ on the right side of assignment is the current estimate. With these insights, you can interpret Equation 4-5 as an update that is moving the estimate from $V(s)$ to $V(s) + \alpha(\text{Difference})$ where the difference δ_t is given by the following expression:

$$\delta_t = R_{t+1} + \gamma \cdot V(S_{t+1}) - V(S_t) \tag{4-6}$$

The difference δ_t is known as a *TD error*. It represents the error (difference) in the estimate of $V(s)$ based on reward R_{t+1} plus the discounted next time-step state value $V(S_{t+1})$ and the current estimate $V(S_t)$. The value of the state is moved by a proportion (a.k.a. learning rate or step rate α) of this error δ_t.

TD has an advantage over DP as it does not require the model knowledge (transition function). Further, TD method has an advantage over MC also. TD methods can update the state value at every step—that is, they can be used in an online setup through which you learn as the situation unfolds without waiting for the end of episode or they can also be used for continuing tasks which has no termination.

Temporal Difference Control

This section starts your journey into the realm of the real algorithms used in reinforcement learning. In the remaining sections of the chapter, you will learn about various methods used under TD learning. I will start with a simple one-step on-policy learning method called *SARSA*. This will be followed by a powerful off-policy technique called *Q-learning*. You will study some foundational aspects of Q-learning in this chapter, and in the next chapter, you will integrate deep learning with Q-learning, giving you a powerful approach called Deep Q Networks (DQN). Using DQN, you will be able to train game-playing agents on an Atari simulator.

In this chapter, you will also cover a variant of Q-learning called *expected SARSA*, another off-policy learning algorithm. I will then talk about the issue of maximization bias in Q-learning, taking you to *double Q-learning*. The advantages of *double Q-learning* is addressed in a subsequent section in this chapter. All the variants of Q-learning become very powerful when combined with deep learning to represent the state space, which will form the bulk of the next chapter. Toward the end of this chapter, I cover additional concepts such as *experience replay*, which make off-learning algorithms efficient with respect to the number of samples needed to learn an optimal policy. I then talk about a powerful and a bit involved approach called *TD(λ)* that combines MC and TD methods on a continuum controlled by *λ* between 0 to 1. Finally, you will look at an environment that has continuous state space and ways to discretize the state values for TD methods. The exercise will demonstrate the need for the approaches that you will take up in the next chapter, covering functional approximation and deep learning for state representation. After Chapters 5 and 6 on deep learning and DQN, I will explain another approach, called *policy optimization,* that revolves around directly learning the policy without needing to find the optimal state/action values.

You have been using the 4×4 grid world so far. You will now look at a few more environments that will be used in the rest of the chapter. You will write the agents in an encapsulated way so that the same agent/algorithm could be applied in various environments without any changes.

Cliff Walking

The first environment you will use is a variant of the grid world; it is part of the Gymnasium library called the cliff-walking environment.[2] In this environment, you have a 4×12 grid world, with the bottom-left cell with coordinates [3,0] being the start state *S* and the bottom-right state with coordinates [3,11] being the goal state *G*. The rest of the bottom row ([3,1] .... [3,10]) forms a cliff; stepping on it earns a reward of -100, and the agent is put back to the start state again.

---

[2] https://gymnasium.farama.org/environments/toy_text/cliff_walking/

The observation space represents each time a step earns a reward of -1 until the agent reaches the goal state. Similar to the 4×4 grid world, the agent can take a step in any direction [0:UP, 1:RIGHT, 2:DOWN, 3:LEFT] which is represented by a tuple of shape (1,) in the range {0,3}.

There are 3 x 12 + 1 possible states. The player cannot be at the cliff, nor at the goal, as the latter results in the end of the episode. What remains are all the positions of the first three rows plus the bottom-left cell. The observation is a value representing the player's current position as current_row * nrows + current_col (where both the row and col start at 0). For example, the stating position can be calculated as follows: 3 * 12 + 0 = 36. The observation is returned as an int(). The episode terminates when the agent reaches the goal state.

Figure 4-11 depicts the setup.

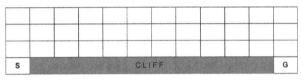

reward = -1 for all transitions. Stepping on cliff has reward of -100

Figure 4-11. *A 4×10 cliff world. There is a reward of -100 for stepping on the cliff and -1 for other transitions. The goal is to start from S and reach G with the best possible reward*

Taxi

The next environment you will look at is the "taxi problem."[3] There are four designated pick-up and drop-off locations (Red, Green, Yellow, and Blue) in the 5x5 grid world. The taxi starts off at a random square and the passenger at one of the designated locations. One of the four squares is chosen randomly as the destination location.

The goal is to move the taxi to the passenger's location, pick up the passenger, move to the passenger's desired destination, and drop off the passenger. Once the passenger is dropped off, the episode ends.

---

[3] https://gymnasium.farama.org/environments/toy_text/taxi/

The player receives positive rewards for successfully picking and dropping off the passenger at the correct start location and end destination. Negative rewards exist for incorrect attempts to pick up/drop off the passenger and for each step where another reward is not received. Specifically, the rewards are -1 per time step, +20 for successful pick/drop, and -10 for pick/drop from the wrong locations.

There are 500 discrete states since there are 25 taxi positions in the 5x5 grid, five possible locations of the passenger (including the case when the passenger is in the taxi), and four destination locations.

An observation is returned as an int() that encodes the corresponding state, calculated by ((taxi_row * 5 + taxi_col) * 5 + passenger_location) * 4 + destination.

Note that there are 400 states that can actually be reached during an episode. The missing states correspond to situations in which the passenger is at the same location as their destination, as this typically signals the end of an episode. Four additional states can be observed right after a successful episode, when both the passenger and the taxi are at the destination. This gives a total of 404 reachable discrete states.

There are six actions that can be taken deterministically. The taxi can move North, South, West, East, and then there are two actions of pick or drop passenger. The action shape is (1,) in the range {0, 5}. Values 0 to 3 represent the action of car moving in 4 directions, similar to the action values we saw in cliff walking environment. Value of 4 is the action of picking up the passenger and 5 represents the action of dropping the passenger.

The episode ends if the following happens: a) termination, whereby the taxi drops off the passenger, or b) truncation (when using the time_limit wrapper), which is when the length of the episode is 200.

Figure 4-12 shows the schematic of the environment.

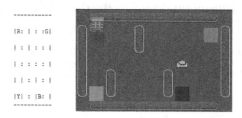

Figure 4-12. *5×5 taxi problem. There is a reward of -1 for all transitions. An episode ends when the passenger is dropped off at the destination. The left figure shows the schematic as used internally and the right figure shows the visual representation created using the PyGame library*

Cart Pole

The third and final environment you will look at is CartPole[4]. I covered this environment in Chapter 2. A pole is attached by an unactuated joint to a cart, which moves along a frictionless track. The pendulum is placed upright on the cart and the goal is to balance the pole by applying forces in the left and right directions on the cart.

Action space is a ndarray with shape (1,) which can take values {0, 1} indicating the direction of the fixed force the cart is pushed with. 0 is the action to push the cart to the left and 1 is the action to push the cart to the right.

Observation space is an ndarray with shape (4,) with the values corresponding to the following positions and velocities [Cart position, Car velocity, Pole angle, pole angular velocity]. You can refer to the accompanying documentation for the Gymnasium library to get more details on the specific ranges within which each of these values operates.

The reward is +1 at every time step, and the agent wants to maximize the reward by keeping the pole balanced for the longest time interval. The episode terminates once the pole angle is more than 12 degrees in either direction or the cart position is beyond 2.4 from the center—that is, either < -2.4 or > 2.4 or if the agent is able to balance the pole for 200 time or 500 steps. Figure 4-13 shows a snapshot of the environment in action.

Figure 4-13. *Cart pole problem. The objective is to balance the pole vertically for the maximum number of time steps with a reward of +1 for every time step*

Having explained the various environments, let's now jump into solving these with various TD algorithms, starting first with SARSA, an on-policy method.

On-Policy SARSA

Like the MC control methods, you will again leverage GPI. You will use a TD-driven approach for the policy value estimation/prediction step and will continue to use greedy maximization for the policy improvement. Just like with MC, you need to explore enough

---

[4]https://gymnasium.farama.org/environments/classic_control/cart_pole/

and visit all the states an infinite number of times to find an optimal policy. Therefore, similar to the MC method, you can use an ϵ-greedy policy and slowly reduce the ϵ value to zero—that is, in the limit as number of episodes increase, bring down the exploration to zero.

TD setup is model-free—that is, you have no prior comprehensive knowledge of transitions. There for a given state, to be able to maximize the return by choosing the right action in that state, you need to know the state-action values $Q(S, A)$ for all the actions in that state. And you need to do so for all the state. Therefore, the most standard way to find an optimal policy is to find all the Q values for all state/action pairs. You can reformulate TD estimation from Equation 4-5 to the one in Equation 4-7, essentially replacing $V(s)$ with $Q(s, a)$. Both the setups are Markov processes, with Equation 4-5 focusing on state-to-state transitions and Equation 4-7 shifting the focus to being state-action to state-action.

$$Q(S_t, A_t) \leftarrow Q(S_t, A_t) + \alpha \cdot \left[R_{t+1} + \gamma \cdot Q(S_{t+1}, A_{t+1}) - Q(S_t, A_t) \right] \quad (4\text{-}7)$$

Similar to Equation 4-6, the TD error is now given in context of q-values.

$$\delta_t = R_{t+1} + \gamma \cdot Q(S_{t+1}, A_{t+1}) - Q(S_t, A_t) \quad (4\text{-}8)$$

To carry out an update as per Equation 4-7, you need all the five values S_t, A_t, R_{t+1}, S_{t+1}, and A_{t+1}. This is the reason the approach is called SARSA. It is a short form to represent the tuple of five values—that is, state, action, reward, state, and action. You will follow an ϵ-greedy policy to generate the samples, update the q-values using Equation 4-7, and then based on the updated q-values create a new ϵ-greedy. The policy improvement theorem guarantees that the new policy will be better than the old policy unless the old policy was already optimal. Of course, for the guarantee to hold, you need to bring down the exploration probability ε to zero in the limit.

Note that for all episodic policies, the terminal states have $Q(S, A)$ equal to zero—that is, once in terminal state, the person cannot transition anywhere and will keep getting a reward of zero. This is another way of saying that the episode ends and $Q(S, A)$ is zero for all terminal states. Therefore, when S_{t+1} is a terminal state, Equation 4-7 will have $Q(S_{t+1}, A_{t+1}) = 0$, and the update equation will look like this:

$$Q(S_t, A_t) \leftarrow Q(S_t, A_t) + \alpha \cdot \left[R_{t+1} - Q(S_t, A_t) \right] \quad (4\text{-}9)$$

Let's now look at the pseudocode of the SARSA algorithm; see Figure 4-14.

SARSA on-policy TD control

Initialize:
    State-action values $Q(s, a) = 0$ for all $s \in S$ and $a \in A$.
    policy $\pi = \varepsilon$-greedy policy with some small $\varepsilon \in [0, 1]$
    learning rate (step size) $\alpha \in [0, 1]$
    discount factor $\gamma \in [0, 1]$

Loop for each episode:
    Start state S, choose action A based on ε-greedy policy
    Loop for each step till episode end:
        Take action A and observe reward R and next state S'
        Choose action A' using ε-greedy policy using current Q values
        If S' not terminal:

$$Q(S, A) \leftarrow Q(S, A) + \alpha \cdot [R + \gamma \cdot Q(S', A') - Q(S, A)]$$

        else:

$$Q(S, A) \leftarrow Q(S, A) + \alpha \cdot [R - Q(S, A)]$$

$$S \leftarrow S'; A \leftarrow A'$$

        [optionally reduce ε periodically towards zero]

Return policy π based on final Q values.

Figure 4-14. *SARSA, on-policy TD control*

The pseudocode starts with initializing all Q values with zero and an ϵ-greedy policy π. You iterate over multiple episodes and for each step in an episode, you carry out updates to state-action q-values using Equations 4-7 or 4-9. You also optionally reduce the exploration by driving $\epsilon \to 0$.

After enough iterations or once the change in q-values reduces below a threshold, you can return the final policy by finding the maximizing action for each state.

Let's now walk through the code in 4.c-sarsa-on-policy.ipynb, which implements SARSA. The code has a class named SARSAAgent, which implements the SARSA learning agent. It has two key functions—update(), which takes the tuple of state, action, reward, next_state, next_action, and done flag. It updates the q-value using the TD update Equations 4-7 and 4-9. The other function is get_action(state), which returns a random action with probability ϵ and argmax $Q(S,A)$ with probability $1 - \epsilon$. There is a generic function called train_agent(), outside the class, that trains the agent in a given environment. There are three helper functions: plot_rewards() to plot the reward per episode as the training progresses, print_policy() to print the optimal policy learned, and record_video() to record a video of a trained agent. Listing 4-3 shows the code for SARSAAgent, which implements the update inside the inner loop from Figure 4-14. The training code, which implements the training loop from Figure 4-14, is given in Listing 4-4. For the rest of the code, refer to the notebook.

Listing 4-3. SARSA On-Policy TD Control

```
class SARSAAgent:
    def __init__(self, alpha, epsilon, gamma, get_possible_actions):
        self.get_possible_actions = get_possible_actions
        self.alpha = alpha
        self.epsilon = epsilon
        self.gamma = gamma
        self._Q = defaultdict(lambda: defaultdict(lambda: 0))

    .....

    # carryout SARSA updated based on the sample (S, A, R, S', A')
    def update(self, state, action, reward, next_state, next_action, done):
        if not done:
            td_error = reward + \
                        self.gamma * self.get_Q(next_state, next_action) - \
                        self.get_Q(state, action)
        else:
            td_error = reward - self.get_Q(state, action)
        new_value = self.get_Q(state, action) + self.alpha * td_error
        self.set_Q(state, action, new_value)

    # get argmax for q(s,a)
    def max_action(self, state):
        ......
```

```
# choose action as per epsilon-greedy policy
def get_action(self, state):
    actions = self.get_possible_actions(state)
    if len(actions) == 0:
        return None
    if np.random.random() < self.epsilon:
        a = np.random.choice(actions)
        return a
    else:
        a = self.max_action(state)
        return a
```

Listing 4-4 contains the code for the training agent. It implements the complete SARSA algorithm. The outer loop goes over range(episode_cnt) and implements the outer loop from pseudocode to run agent over episode_cnt episodes. In the inside loop the agent starts from the initial state by calling env.reset() and then steps through the environment using ϵ-greedy policy, which is done via the action=agent. get_action(state) call. The agent.update(...) call carries out the SARSA update, as shown in Listing 4-3. You also reduce the epsilon exploration toward the end of inner loop with the help of agent.epsilon = agent.epsilon*0.99

Listing 4-4. SARSA Training Loop

```
# training algorithm
def train_agent(env, agent, episode_cnt=10000, tmax=10000, anneal_
eps=True):
    episode_rewards = []
    for i in range(episode_cnt):
        G = 0
        state, _ = env.reset()
        action = agent.get_action(state)
        for t in range(tmax):
            next_state, reward, termination, _, _ = env.step(action)
            next_action = agent.get_action(next_state)
            agent.update(state, action, reward, next_state, next_action,
            termination)
            G += reward
```

```
            if termination:
                episode_rewards.append(G)
                # to reduce the exploration probability epsilon over the
                # training period.
                if anneal_eps:
                    agent.epsilon = agent.epsilon * 0.99
                break
            state = next_state
            action = next_action
    return np.array(episode_rewards)
```

Figure 4-15 shows the graph of per-episode-reward as learning proceeds. You can see that the reward soon gets close to the optimal value. Figure 4-15 also shows the optimal policy learned. In the 4.c-sarsa-on-policy.ipynb notebook, there is a code to also replay the video of trained agent, which is shown in Listing 4-5. It is implemented by wrapping the environment with DummyVecEnv and the VecVideoRecorder environment class provided under the state-baseline-3 library.

Listing 4-5. Video Recording Code

```
# Helper function to record videos
def record_video(env_id, video_folder, video_length, agent):

    vec_env = DummyVecEnv([lambda: gym.make(env_id, render_mode="rgb_array")])
    # Record the video starting at the first step
    vec_env = VecVideoRecorder(vec_env, video_folder,
                               record_video_trigger=lambda x: x == 0, video_
                               length=video_length,
                               name_prefix=f"{type(agent).__name__}-{env_id}")

    obs = vec_env.reset()
    for _ in range(video_length + 1):
        action = agent.max_action(obs[0])
        obs, _, _, _ = vec_env.step([action])
    # video filename
    file_path = "./"+video_folder+vec_env.video_recorder.path.
    split("/")[-1]
```

```
# Save the video
vec_env.close()
return file_path
```

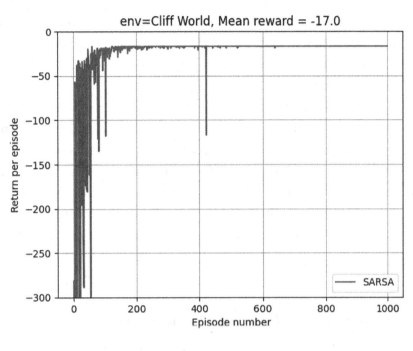

Figure 4-15. *Reward graph during learning under SARSA and policy learned by the agent*

The policy learned is to avoid the cliff by first going all the way up and then taking a right turn to walk toward the goal. This is surprising, as you would have expected the agent to learn the policy to skirt over the cliff and reach the goal, which would have been the shortest path by four steps as compared to the one learned by the agent.

This can be explained due to exploration in ϵ-greedy policy. As the policy continues to explore using ϵ-greedy, there is always a small chance that in any state next to the cliff, it could take a random action with probability of ϵ. And if this random action happens to be down, the agent will fall off into the cliff. Therefore, the optimal policy under any small random exploration is to completely avoid the cliff by climbing all the way up and then navigating to right to reach the goal. It demonstrates the issue of continued exploration even when enough has been learned about the environment—that is, when the same ε-greedy policy is used for sampling as well as for improvement. You will see how this issue is avoided in Q-learning in which off-policy learning is carried out using an exploratory behavior policy to generate training samples and a deterministic policy is learned as the optimal target policy.

In the 4.c-sarsa-on-policy.ipynb notebook, you also have code for running SARSA on Taxi environment. It also shows a video of trained agent. I think it is now time for you to share your own agent on HuggingFace with your friends, family, and colleagues. See if you train SARSA agent on some other simple environments from the Gymnasium library and upload the video and performance statistics of the trained agent to the HuggingFace hub. Refer to the 2.c-First-Agent.ipynb notebook for the relevant upload code. And if you do, please, tag me. My HuggingFace ID is nsanghi.

The next section studies the first off-policy TD algorithm, called *Q-learning*.

Q-Learning: An Off-Policy TD Control

In SARSA, you used the samples with the values S, A, R, S', and A' that were generated by the following policy. Action A' from state S' was produced using the ε-greedy policy, the same policy that was then improved in the "improvement" step of GPI. However, instead of generating A' from the policy, what if you looked at all the $Q(S', A')$ and chose the action A', which maximizes the value of $Q(S', A')$ across actions A' available in state S'?

You could continue to generate the samples (S, A, R, S') (notice no A' as the fifth value in this tuple that we saw in SARSA) using an exploratory policy like ε-greedy. You can improve the policy by choosing $A' = \arg\max_a Q(S', a)$. This small change in the approach creates a new way to learn the optimal policy, called *Q-learning*. It is no more an on-policy learning, rather an off-policy control method where the samples (S, A, R, S') are being generated by an exploratory policy, while you maximize $Q(S', A')$ to find a deterministic optimal target policy.

You are using exploration with the ε-greedy policy to generate the samples (S, A, R, S'). At the same time, you are exploiting the existing knowledge by finding the Q maximizing action $\arg\max_a Q(S', a)$ in state S'. This way the target for Q(S,A) the current state-action pair iterates to the optimal policy value as Q values converges to its fixed point. Using greedy policy for exploration and then exploiting the knowledge to find the deterministic policy is optimal.

The update rule for q-values is now defined as follows:

$$Q(S_t, A_t) \leftarrow Q(S_t, A_t) + \alpha \cdot \left[R_{t+1} + \gamma \cdot \max_{A_{t+1}} Q(S_{t+1}, A_{t+1}) - Q(S_t, A_t) \right] \quad (4\text{-}10)$$

Comparing the previous equation with Equation 4-7, you will notice the subtle difference between the two approaches and how that makes Q-learning an off-policy method. The off-policy behavior of Q-learning is handy, and it makes the approach sample efficient. I will touch upon this in a later section when I talk about *experience replay* or *replay buffer*. Figure 4-16 lists the pseudocode of Q-learning.

Q LEARNING OFF-POLICY TD CONTROL

Initialize:
    State-action values $Q(s, a) = 0$ for all $s \in S$ and $a \in A$.
    policy $\pi = \varepsilon$-greedy policy with some small $\varepsilon \in [0, 1]$
    learning rate (step size) $\alpha \in [0, 1]$
    discount factor $\gamma \in [0, 1]$

Loop for each episode:
    Start state S
    Loop for each step till episode end:
        Choose action A based on ε-greedy policy
        Take action A and observe reward R and next state S'
        If S' not terminal:
            $Q(S, A) \leftarrow Q(S, A) + \alpha \cdot [R + \gamma \cdot \max_{A'} Q(S', A') - Q(S, A)]$
        else:
            $Q(S, A) \leftarrow Q(S, A) + \alpha \cdot [R - Q(S, A)]$

        $S \leftarrow S'$

Return policy π based on final Q values.

Figure 4-16. *Q-learning, off-policy TD control*

If you compare the pseudocode of SARSA from Figure 4-14 with the pseudocode of Q-Learning from Figure 4-16, you will notice that there are two subtle differences. First, you generate action A using the greedy policy and do not worry that much about reducing ϵ, the exploration factor. Secondly, you do not choose the successor state action A', rather you take a max over all possible A' in the successor state S' to define the target in the update rule.

Next, let's look at the Python implementation of the Q-learning pseudocode and see how it performs on same environments as before—that is, Cliff world and Taxi. Head over to the code in 4.d-qlearning.ipynb, which shows the implementation of Q-learning.

Like SARSA, Q-learning agent code is very similar to SARSA agent code in Listings 4-3 and 4-4. The Q-learning agent code is given in Listing 4-6. Most of the code is similar to that of the SARSA agent in Listing 4-3, except the update rule. In the SARSA update, you are given a tuple of five values (S, A, R, S', A') and you use this to carry out the q-value update for $Q(S, A)$. While in Q-learning the update gets a tuple of only (S, A, R, S') and the update is carried out by finding the best A' from the current values of $Q(S, A')$ for A' applicable in state S'.

Listing 4-6. Q-Learning Update Rule

```
# Q learning update step
def update(self, state, action, reward, next_state, done):
    if not done:
        best_next_action = self.max_action(next_state)
        td_error = reward + \
                    self.gamma * self.get_Q(next_state, best_next_action) \
                    - self.get_Q(state, action)
    else:
        td_error = reward - self.get_Q(state, action)

    new_value = self.get_Q(state, action) + self.alpha * td_error
    self.set_Q(state, action, new_value)
```

The training code implementation for the rest of the Q-learning pseudocode from Figure 4-16 is given in Listing 4-7. Again, the training loop for Q-learning is very similar to the training loop for SARSA, as shown in Listing 4-4. Unlike SARSA, in Q-learning, you do not generate the next action A' using the code next_action = agent.get_ action(next_state). The update rule for Q-Learning just requires state, action, reward, next_state, and terminated, as discussed previously.

Listing 4-7. Q-Learning Training Loop

```
# training algorithm
def train_agent(env, agent, episode_cnt=10000, tmax=10000, anneal_
eps=True):
    episode_rewards = []
    for i in range(episode_cnt):
        G = 0
        state, _ = env.reset()
        action = agent.get_action(state)
        for t in range(tmax):
            next_state, reward, termination, _, _ = env.step(action)
            next_action = agent.get_action(next_state)
            agent.update(state, action, reward, next_state, next_action,
            termination)
            G += reward
            if termination:
                episode_rewards.append(G)
                # to reduce the exploration probability epsilon over the
                # training period.
                if anneal_eps:
                    agent.epsilon - agent.epsilon * 0.99
                break
            state = next_state
            action = next_action
    return np.array(episode_rewards)
```

Let's look at Q-learning being applied to the cliff world environment. The reward per episode improves with training and reaches the optimal value of -13 as compared to that of -17 under SARSA. As shown in Figure 4-17, a better policy under Q-learning is evident. Under Q-learning, the agent learns to navigate to the goal through the cells in the second row just above the cliff. Under Q-learning, the agent is learning a deterministic policy, and the environment is also deterministic. In other words, if the agent takes an action to move right, it will definitely move right and has zero chance of taking any random step in any other direction. There is no exploration component in the policy being learned by the agent. This is the key difference between Q-learning and SARSA. Under Q-learnin,g the exploration is used only in the behavior policy that is used to generate the samples but the final policy learned by the agent is a deterministic policy with zero exploration. Therefore, the agent learns the most optimal policy of going toward the goal by just grazing over the cliff.

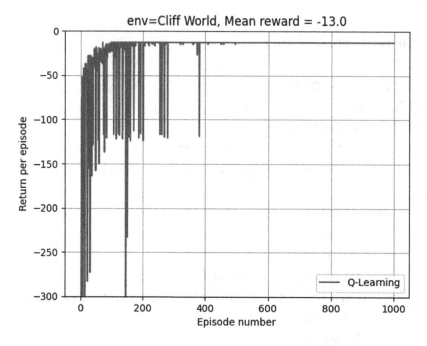

Figure 4-17. *Reward graph during learning under Q-learning and policy learned by agent*

The `4.d-qlearning.ipynb` notebook also has code for running Q-learning on the Taxi environment. Like before, it also shows a video of trained agent. I again encourage you to apply Q-learning on some other environment form Gymnasium and upload your trained agent video to HuggingFace.

I will be revisiting Q-learning in the next chapter on the deep learning approach to the state value approximation. Under that setup, Q-learning is called DQN. DQN with its variants will be applied to train a game-playing agent on some Atari games.

To conclude the discussion on Q-learning, let's look at a particular issue that Q-learning may introduce, that of maximization bias.

Maximization Bias and Double Learning

If you look back at Equation 4-10, you will notice that you are maximizing over A' to get the max value $Q(S', A')$. Similarly, in SARSA, you find a new ϵ-greedy policy that is also maximizing over Q values in a state to get the action in that state with highest q-value. Further, these q-values are estimates themselves of the true state-action values. In summary, you are using a max over the q-estimate as an "estimate" of the maximum value. Such an approach of "max of estimate" as an "estimate of max" introduces a +ve bias.

To see this, consider a scenario where the reward in a transition takes three values: 5, 0, and +5 with an equal probability of 1/3 for each value. The expected reward is 0, but the moment you see a +5, take that as part of the maximization, and then it never comes down. So, +5 becomes an estimate of the true reward that otherwise in expectation is 0. This is a positive bias introduced due to maximization step.

One of the ways to remove the +ve bias is to use a set of two q-values. One q-value is used to find the action that maximizes the q-value, and the other set of q-values is then used to find the q-value for that max action. Mathematically, it can be represented as follows:

$$\text{Replace } \max_A Q(S, A) \text{ with } Q_1\left(S, \arg\max_A Q_2(S, A)\right).$$

You are using Q_2 to find the maximizing action A, and then Q_1 is used to find the maximum q-value for that action A chosen earlier. It can be shown that such an approach removes the +ve or maximization bias. I will revisit this concept when I talk about DQN.

Expected SARSA Control

This section looks at another method that is kind of a hybrid between Q-learning and SARSA; it's called *expected SARSA*. It is similar to Q-learning except that "max" in 4-10 is replaced with an expectation, as shown here:

$$Q(S_t, A_t) \leftarrow Q(S_t, A_t) + \alpha \left[R_{t+1} + \gamma \cdot \sum_{a \in A_{t+1}} \pi(a|S_{t+1}) \cdot Q(S_{t+1}, a) - Q(S_t, A_t) \right] \quad (4\text{-}11)$$

Expected SARSA has a lower variance as compared to the variance seen in SARSA due to the random selection of A_{t+1}. In expected SARSA, instead of sampling, you take the expectation over all possible actions.

In the cliff world problem, you have deterministic actions, and hence you can set the learning rate $\alpha = 1$ without any major impact on the learning quality. The following pseudocode is for the algorithm. It mirrors Q-learning except for the update logic of taking *expectation* instead of *maximization*. You can run expected SARSA as on-policy, which is what you will do while testing it against the cliff world and taxi environments. See Figure 4-18.

EXPECTED SARSA TD CONTROL

Initialize:
    State-action values $Q(s, a) = 0$ for all $s \in S$ and $a \in A$.
    policy $\pi = \varepsilon$-greedy policy with some small $\varepsilon \in [0, 1]$
    learning rate (step size) $\alpha \in [0, 1]$
    discount factor $\gamma \in [0, 1]$

Loop for each episode:
    Start state S
    Loop for each step till episode end:
        Choose action A based on ε-greedy policy
        Take action A and observe reward R and next state S'
        If S' not terminal:
            $Q(S, A) \leftarrow Q(S, A) + \alpha \cdot [R + \gamma \cdot \sum_a \pi(a|S')Q(S', a) - Q(S, A)]$
        else:
            $Q(S, A) \leftarrow Q(S, A) + \alpha \cdot [R - Q(S, A)]$

        $S \leftarrow S'$

Return policy π based on final Q values.

Figure 4-18. *Expected SARSA TD control*

As you can see in Figure 4-18, the update rule uses the summation over all actions versus the max of actions for SARSA that you saw in Figure 4-14. The 4.e-expected_ sarsa.ipynb notebook contains the code for running the expected SARSA code. The code is similar to the SARSA or Q-learning except for the change in the update rule, as discussed previously. Listing 4-8 gives the code for the update rule implemented in the update(...) function inside the ExpectedSARSAAgent class. Refer to the notebook for the complete code, including a video of trained agents and the reward trajectories.

Listing 4-8. Expected SARSA TD Control

```
# Expected SARSA Update
def update(self, state, action, reward, next_state, done):
    if not done:
        best_next_action = self.max_action(next_state)
        actions = self.get_possible_actions(next_state)
        next_q = 0
        for next_action in actions:
            if next_action == best_next_action:
                next_q += (1-self.epsilon+self.epsilon/len(actions)) * \
                        self.get_Q(next_state, next_action)
            else:
                next_q += (self.epsilon/len(actions)) * \
                        self.get_Q(next_state, next_action)

        td_error = reward + self.gamma * next_q - self.get_Q(state, action)
    else:
        td_error = reward - self.get_Q(state, action)

    new_value = self.get_Q(state, action) + self.alpha * td_error
    self.set_Q(state, action, new_value)
```

Figure 4-19 shows the result of training the expected SARSA agent for the cliff world. The introduction of expectation—that is, summation over all actions in successor state—increases the computational complexity of the algorithm a bit but offers faster convergence than SARSA and Q-learning. This is evident from the episode level returns as training progresses. For the SARSA as seen in Figure 4-15, the return has downward spikes, indicating that training has not been completed. It shows no spike only after about 425 episodes are complete. For Q-Learning again the downward spikes die out around 400 episodes, as can be seen in Figure 4-17. While expected SARSA shows stabilization almost from 200 episodes, as visible in Figure 4-19. The convergence is clearly the fastest with expected SARSA. You can experiment and see that the changing learning rate α has no major impact on the convergence in the limit.

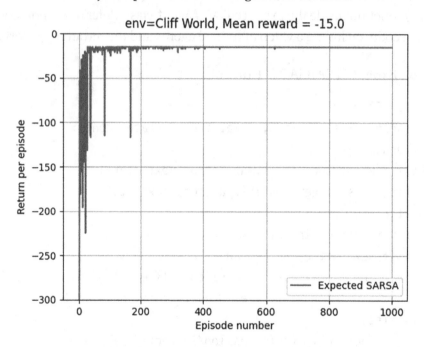

Policy learnt by Agent

```
>   >   >   >   >   >   >   >   >   V   >   V
>   >   >   >   >   >   >   >   >   >   >   V
^   ^   ^   ^   ^   ^   ^   ^   ^   ^   >   V
X   C   C   C   C   C   C   C   C   C   C   T
```

Figure 4-19. *Reward graph with expected SARSA in the cliff world and policy learned by the agent*

It is also interesting to note that the policy learned by the expected SARSA falls between Q-learning and SARSA. The agent under this policy goes through the middle row of the maze to reach the goal. In this case, you are using the expected SARSA as on-policy—that is, using the same ε-greedy policy to explore and improve. But possibly due to the expectation, it learns to improve from regular SARSA and finds it safe enough to go through the middle row. In the same notebook (`4.e-expected_sarsa.ipynb`), you can also see the result of running this algorithm for the taxi environment.

Replay Buffer and Off-Policy Learning

Off-policy learning involves two separate policies: behavior policy $b(a|s)$ to explore and generate examples, and $\pi(a|s)$, the target policy that the agent is trying to learn as the optimal policy. Accordingly, you could use the samples generated by the behavior policy again and again to train the agent. The approach makes the process sample efficient as a single transition observed by the agent can be used multiple times.

This is called *experience replay*. The agent collects experiences from the environment and replays those experiences multiple times as part of the learning process. In experience replay, you store the samples (`s, a, r, s', done`) in a buffer. The samples are generated using an exploratory behavior policy while you improve a deterministic target policy using q-values. Therefore, you can always use older samples from a behavior policy and apply them again and again. Keep the buffer size fixed to some predetermined size and keep deleting the older samples as you collect new ones. The process makes learning sample efficient by reusing a sample multiple time. The rest of the approach remains the same as an off-policy agent.

Let's apply this approach to the Q-learning agent. The code for the Q-learning agent remains the same. The `4.f-qlearning-exp-replay.ipynb` notebook contains a complete walkthrough for Q-learning agents with experience replay. There is no change in the update rule or any of the internal logic of Q-learning agent. The agent training code is the one that changes. You have two changes. First, you implement experience replay as shown in Listing 4-9 to store all transitions of the agent through the environment under the behavior policy. Each agent transition generates a tuple of state, action, reward, next state, done, which is stored in a list with fixed length as implemented by the `add(self,...)` method. Once the queue is full, the addition of a new item at the end of list leads to the earliest item in the queue being dropped out. Therefore, the queue holds the latest transitions up to a fixed look-back period. During training, you sample a batch of stored transitions to be used for Q-learning update. This is implemented by the `sample(self, batch_size)` metho, where `batch_size` defines the number of transitions to sample.

Listing 4-9. Replay Buffer

```
class ReplayBuffer:
    def __init__(self, size):
        self.size = size  # max number of items in buffer
        self.buffer = []   # array to hold buffer

    def __len__(self):
        return len(self.buffer)

    def add(self, state, action, reward, next_state, done):
        item = (state, action, reward, next_state, done)
        self.buffer = self.buffer[-self.size:] + [item]

    def sample(self, batch_size):
        idxs = np.random.choice(len(self.buffer), batch_size)
        samples = [self.buffer[i] for i in idxs]
        states, actions, rewards, next_states, done_flags =
        list(zip(*samples))
        return states, actions, rewards, next_states, done_flags
```

The other change from vanilla Q-learning is the training code. This is given in Listing 4-10. In vanilla Q-learning, the update of q-values is carried out online as you step through each individual transition. However, with experience replay, as an agent takes a step the transition is added to the experience replay buffer. This is followed by sampling batch_size transitions from the buffer, which are then used to update the q-values for target policy. The transition that is stored in experience replay is likely to be picked up by the sample method multiple times until this transition reaches the beginning of the queue and gets pushed out of the buffer.

Listing 4-10. Q-learning Training Code Using Experience Replay

```
def train_agent(env, agent, episode_cnt=10000, tmax=10000,
                anneal_eps=True, replay_buffer=None, batch_size=16):
    episode_rewards = []
    for i in range(episode_cnt):
        G = 0
        state, _ = env.reset()
```

```
    for t in range(tmax):
        action = agent.get_action(state)
        next_state, reward, done, _, _ = env.step(action)
        if replay_buffer:
            replay_buffer.add(state, action, reward, next_state, done)
            states, actions, rewards, next_states, done_flags = \
                    replay_buffer(batch_size)
            for i in range(batch_size):
                agent.update(states[i], actions[i], rewards[i],
                             next_states[i], done_flags[i])
        else:
            agent.update(state, action, reward, next_state, done)

        G += reward
        if done:
            episode_rewards.append(G)
            # to reduce the exploration probability epsilon over the
            # training period.
            if anneal_eps:
                agent.epsilon = agent.epsilon * 0.99
            break
        state = next_state
    return np.array(episode_rewards)
```

As discussed earlier, using experience replay leads to reuse of each transition tuple multiple times. Therefore, Q-learning with experience replay is sample-efficient. This is evident in the rewards trajectory—that is, the return per episode plotted in Figure 4-20. The convergence of return per episode is lot faster in Q-learning with experience replay. In Figure 4-20, you can see that the optimal value is reached almost by the time training code has seen 100 episodes. Compare this to Figure 4-17 for Q-learning, where the optimal value is reached around 200 episodes.

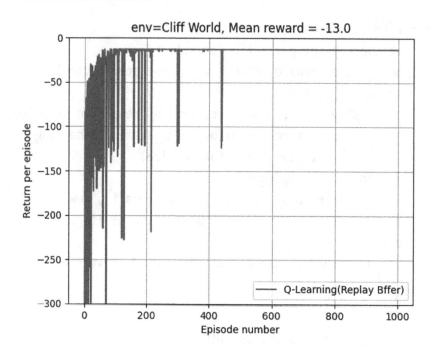

Figure 4-20. *Reward graph during learning under Q-learning with experience replay and policy learned by agent*

From Figure 4-20, it is also evident that, though episode return reaches optimal behavior faster, it shows more fluctuations. The more the exploration, the higher the transitions that are not optimal actions. In vanilla Q-learning, you do not see these exploratory actions too much as epsilon exploration drops to zero. However, in experience replay, you land up using these exploratory actions multiple times depending on the sampling logic of the replay buffer and the buffer length. This could lead to updates causing delayed convergence.

And just to point out the policy learned under experience replay is the same as the one learned in vanilla Q-learning.

As before, the `4.f-qlearning-exp-replay.ipynb` notebook contains the video replay of the trained agent as well as running of the same on Taxi environment.

In summary, the Q-agent with the replay buffer is supposed to improve the initial convergence by sampling repeatedly from the buffer. The sample efficiency will become more apparent when you look at DQN. Over the long run, there won't be any significant difference between the optimal values learned with or without the replay buffer. It has another advantage of breaking the correlation between samples. This aspect will also become apparent when you look at deep learning with Q-learning—that is, DQN.

Q-Learning for Continuous State Spaces

Until now all the examples you have looked at had discrete state spaces. All the methods studied so far could be categorized as tabular methods. The state action space was represented as a matrix with states along one dimension and actions along the cross-axis.

You will soon transition to continuous state spaces and make heavy use of deep learning to represent the state through a neural net. However, you can still solve many of the continuous state problems with some simple approaches. In preparation for the next chapter, let's look at the simplest approach of converting continuous values into discrete bins. The approach you will take is to round off continuous floating-point numbers with some precision, for example, for a continuous state space value between -1 to 1 being converted into -1, -0.9, -0.8, ... 0, 0.1, 0.2, ... 1.0.

4.g-qlearning_continuous_env.ipynb shows this approach in action. You will continue to use the Q-learning agent, experience reply, and learning algorithm from 4.f-qlearning-exp-replay.ipynb. However, this time you will be applying the learning on a continuous environment, that of CartPole, which was described in detail at the beginning of the chapter. The key change that you need is to receive the state values from environment, discretize the values, and then pass these along to the agent as observations. The agent only gets to see the discrete values and uses these discrete values to learn the optimal policy using Q-agent. Listing 4-11 shows the approach used for converting continuous state values into discrete ones. It is done by extending the Gymnasium-provided ObservationWrapper and implementing a function called observation, which takes in the continuous values from the CartPole environment and returns the discrete observation/state values.

Listing 4-11. Q-Learning (Off-Policy) in a Continuous State Environment

```
# We will use ObservationWrapper class from gymnasium to wrap our
environment.
# We need to implement observation() method which will receive the
# original state values from underlying environment
# In observation() we will discretize the state values
# which then will be passed to outside world by env
# the agent will use these discrete state values
# to learn an effective policy using q-learning
from gymnasium import ObservationWrapper

class Discretizer(ObservationWrapper):
    def observation(self, state):
        discrete_x_pos = round(state[0], 1)
        discrete_x_vel = round(state[1], 1)
        discrete_pole_angle = round(state[2], 1)
        discrete_pole_ang_vel = round(state[3], 1)

        return (discrete_x_pos, discrete_x_vel,
                discrete_pole_angle, discrete_pole_ang_vel)
```

Once the `Discretizer` class is defined to discretize the observations, you just need to wrap the Gymnasium environment while making the environment, as shown in Listing 4-12. The rest of the code sees no change between the two Python notebooks—`4.f-qlearning_continuous_env.ipynb` and `4.f-qlearning-exp-replay.ipynb`.

Listing 4-12. Using Discretizer on the Continuous State Environment

```
# create Cart Pole environment
env = gym.make('CartPole-v1')

# wrap our env with Discretizer
env = Discretizer(env)
```

The Q-agent with discrete discretization of states can get about 50 rewards compared to the maximum reward of 200. This is shown in Figure 4-21.

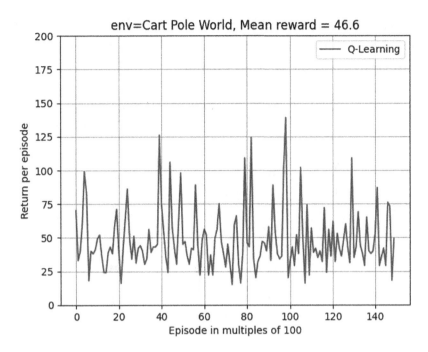

Figure 4-21. *Reward graph with Q-learning on the CartPole continuous state space environment*

In subsequent chapters, you will study other more powerful methods to obtain more rewards.

n-Step Returns

This section unifies the MC and TD approaches. MC methods sample the return from a state until the end of the episode, and they do not bootstrap. Accordingly, MC methods cannot be applied for continuing tasks. TD, on the other hand, uses one-step return to estimate the value of the remaining rewards. TD methods take a short view of the trajectory and bootstrap right after one step.

These methods are two extremes, and there are many situations when a middle-of-the-road approach could produce better results. The idea in *n-step* is to use the rewards from the next n steps and then bootstrap from $n + 1$ step to estimate the value of the remaining rewards. Figure 4-22 shows the backup diagrams for various values of n. On one extreme is one-step, which is the $TD(0)$ method that you just saw in the context of

SARSA, Q-learning, and other related approaches. At the other extreme is the ∞-step TD, which is nothing but an MC method. The broad idea is to see that the TD and MC methods are two extremes of the same continuum.

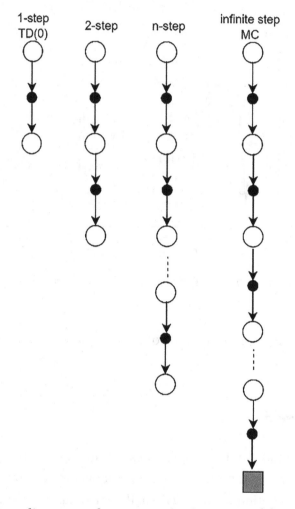

Figure 4-22. *Backup diagrams of n-step methods with TD(0) and MC at two extremes, n = 1 and n = ∞ (end of episode)*

The approach of n-step can be used in an on-policy setting. You can have n-step SARSA and n-step expected SARSA, and these are natural extensions of what you have learned so far. However, using an n-step approach in off-policy learning would require you to factor one more concept—the relative difference of observing a specific n-step transition across states under behavior policy versus that under target-policy. To use the data from behavior policy $b(a|s)$ for optimizing target policy $\pi(a|s)$, you need to multiply the n-step returns observed under the behavior policy with a ratio called *importance sampling ratio,* which was derived earlier in this chapter in Equation 4-4. The same is reproduced here:

$$\rho_{t:T-1} = \frac{\Pr_{\pi}\{\text{trajectory}\}}{\Pr_{b}\{\text{trajectory}\}} = \prod_{k=t}^{T-1} \frac{\pi\left(A_k|S_k\right)}{b\left(A_k|S_k\right)}$$

The importance sampling ratio ensures that the return of a trajectory observed under the behavior policy is adjusted up or down based on the relative chance of observing a trajectory under the target policy versus the chance of observing the same trajectory under the behavior policy.

Nothing comes for free, which is the case with importance sampling. The importance sampling ratio can cause wide variance. Plus, these are not computationally efficient. There are many advanced techniques like *discount-aware importance sampling* and *per-decision importance sampling* that I talked about in the earlier section on importance sampling. These approaches look at the importance sampling and rewards in various ways to reduce the variance and make these algorithms efficient.

I will not be going into the details of implementing these algorithms in this book. The key focus was to introduce these at a conceptual level and make you aware of the advanced techniques.

Eligibility Traces and TD(λ)

Eligibility traces unify the MC and TD methods in an algorithmically efficient way. TD methods, when combined with eligibility trace, produce $TD(\lambda)$ where $\lambda = 0$, making it equivalent to the one-step *TD* that you have studied so far. That's the reason that one-step TD is also known as TD(0). The value of $\lambda = 1$ makes it similar to the regular ∞-step TD or in other words an MC method. Eligibility trace makes it possible to apply MC methods to non-episodic tasks as well. I will cover only high-level concepts of eligibility trace and $TD(\lambda)$.

In the previous section, you looked at n-step returns with $n = 1$ taking you to the regular TD method and $n = \infty$ taking you to MC. I also touched upon the fact that neither extreme is good. MC or TD(0) has its own advantages and disadvantages. An algorithm performs best with some intermediate value of n. n-step offered a view on how to unify TD and MC. What eligibility offers is an efficient way to combine them without keeping track of the n-step transitions at each step. Until now you have looked at an approach of updating a state value based on the next n transitions in the future. This is called the *forward view*. However, you could also look backward—that is, at each time step t, and see the impact that the reward at time step t would have on the preceding n states in past. This is known as *backward view* and forms the core of $TD(\lambda)$. The approach allows an efficient implementation of integrating n-step returns in TD learning.

Look back at Figure 4-22. What if instead of choosing different values of n, you combined all the n-step returns with some weight? This is known as λ-return, and the equation is as follows:

$$G_t^\lambda = (1-\lambda) \sum_{n=1}^{T-t-1} \lambda^{n-1} G_{t:t+n} + \lambda^{T-t-1} G_t \qquad (4\text{-}12)$$

Here, $G_{t:t+n}$ is the n-step return, which uses the bootstrapped value of the remaining steps at the end of the nth step. It is defined as follows:

$$G_{t:t+n} = R_{t+1} + \gamma R_{t+2} + \ldots + \gamma^{n-1} R_{t+n} + \gamma^n V\left(S_{t+n}\right) \qquad (4\text{-}13)$$

If you put $\lambda = 0$ in Equation 4-12, you get the following:

$$G_t^0 = G_{t:t+1} = R_{t+1} + \gamma V\left(S_{t+1}\right)$$

The previous expression is similar to the target value for state-action updates of TD(0) in Equation 4-6.

On other hand, putting $\lambda = 1$ in Equation 4-12 makes $G_t^\lambda = G_t^0$ mimic MC and return G_t as follows:

$$G_t^1 = G_t = R_{t+1} + \gamma R_{t+2} + \ldots + \gamma^{T-t-1} R_T$$

The $TD(\lambda)$ algorithm uses the previous λ-return with a trace vector, known as *eligibility trace,* to have an efficient online TD method using the "backward" view. Eligibility trace keeps a record of how far in the past a state was observed and by how much would that state's estimate be impacted by the current return observed.

I will stop here with the basic introduction of λ-returns, eligibility trace, and $TD(\lambda)$. For a detailed review of the mathematical derivations as well as the various algorithms based on them, refer to the book *Reinforcement Learning: An Introduction* by Barto and Sutton, Second Edition.[5]

---

[5] http://incompleteideas.net/book/the-book-2nd.html

Relationships Between DP, MC, and TD

At the beginning of the chapter, you learned about how the DP, MC, and TD methods compare. I introduced the MC methods followed by the TD methods, and then I used n-step and TD(λ) to combine MC and TD as two extremes of the sample-based model-free learning.

To conclude the chapter, Table 4-1 summarizes the comparison of the DP and TD methods.

Table 4-1. *Comparison of DP and TD Methods in the Context of Bellman Equations*

	Full Backup (DP)	Sample Backup (TD)
Bellman expectation equation for $v_\pi(s)$	Iterative policy evaluation	TD prediction
Bellman expectation equation for $q_\pi(s, a)$	Q-policy iteration	SARSA
Bellman optimality equation for $q_\pi(s, a)$	Q-value iteration	Q-learning

Summary

This chapter looked at the model-free approach to reinforcement learning. It started by estimating the state value using the Monte Carlo approach. It explained the "first visit" and "every visit" approaches. It then looked at the bias and variance tradeoff in general and specifically in the context of the MC approaches.

With the foundation of MC estimation in place, the chapter looked at MC control methods connecting it with the GPI framework for policy improvement that was introduced in Chapter 3. You saw how GPI could be applied by swapping the estimation step of the approach from DP-based to an MC-based approach. The chapter looked in detail at the exploration-exploitation dilemma that needs to be balanced, especially in the model-free world where the transition probabilities are not known.

I then briefly talked about the off-policy approach in the context of the MC methods. And in this context, you learned about importance sampling where the policy to collect samples is called behavior policy and the policy being optimized is the target policy. Importance sampling allows the return from one only to be cast in the context of another policy.

TD was the next approach the chapter investigated with respect to model-free learning. It started off by establishing the basics of TD learning, starting with TD-based value estimation. This was followed by a deep dive into SARSA, an on-policy TD control method. The chapter then investigated Q-learning, a powerful off-policy TD learning approach, and some of its variants like expected SARSA.

In the context of TD learning, I also introduced the concept of state approximation to convert continuous state spaces into approximate discrete state values. The concept of state approximation will form the bulk of the next chapter and will allow you to combine deep learning with reinforcement learning.

Before concluding the chapter, you finally looked at n-step returns, eligibility traces, and TD(λ) as ways to combine TD and MC into a single framework.

CHAPTER 5

Function Approximation and Deep Learning

The previous three chapters looked at various approaches to planning and control—first at the dynamic programming (DP), then at the Monte Carlo approach (MC), and finally at the temporal difference (TD) approach. In all these approaches, you saw problems where the state space and actions were discrete. Only in the previous chapter, toward the end, did I talk about Q-learning in a continuous state space. You discretized the state values using an arbitrary approach and trained a learning model. This chapter extends that approach by talking about the theoretical foundations of approximation and how it impacts the setup for reinforcement learning. It will then look at the various approaches to approximating values, first with a linear approach that has a good theoretical foundation and then with a nonlinear approach with neural networks. This aspect of combining deep learning with reinforcement learning is the most exciting development and has moved reinforcement learning algorithms to scale.

As usual, the approach is to first look at things in the context of the *prediction/ estimation* setup, where the agent follows a given policy in order to learn the state value and/or state action values. This will be followed by talking about *control*—finding the optimal policy. The example will continue to be in a model-free world, where you do not know the transition dynamics. After the introduction, I will talk about the issues of convergence and stability in the world of function approximation. So far, convergence has not been a big issue in the context of the exact and discrete state spaces. However, function approximation brings about new issues that need to be considered for theoretical guarantees and practical best practices. The chapter also touches upon batch methods and compares them to the incremental learning approach discussed in the first part of this chapter.

The chapter closes with a quick overview of deep learning, basic theory, and the basics of building/training models using PyTorch, PyTorch Lightning, and TensorFlow.

© Nimish Sanghi 2024
N. Sanghi, *Deep Reinforcement Learning with Python*, https://doi.org/10.1007/979-8-8688-0273-7_5

Introduction

Reinforcement learning can be used to solve very big problems with many discrete state configurations or problems with continuous state space. Consider the game of backgammon, which has close to 10^{20} discrete states, or consider the game of Go, which has close to 10^{170} discrete states. Also consider environments like self-driving cars, drones, or robots—these all have a continuous state space.

Up to now, you have seen problems where the state space was discrete and also small in size, such as the grid world with ~100 states or the taxi world with 500 states. How do you scale the algorithms you have learned so far to bigger environments or environments with continuous state spaces? All along, the state values $V(s)$ or the action values $Q(s, a)$ have been represented with a table, with one entry for each value of state s or a combination of state s and action a. As the numbers increase, the table size is going to become huge, making it infeasible to be able to store state or state action values in a table. Further, there will be too many combinations, which can slow down the learning of a policy. The algorithm may spend too much time in states that have a very low probability in a real run of the environment.

Let's take a different approach now. Let's represent the state value (or state-action value) with the following function:

$$\hat{v}(s;w) \approx v_{\pi}(s)$$

$$\hat{q}(s, a;w) \approx q_{\pi}(s,a) \tag{5-1}$$

Instead of representing values in a table, they are now being represented by the function $\hat{v}(s;w)$ or $\hat{q}(s,a;w)$, where the parameter w is the parameter of the function that defines the policy $\pi(a|s)$ that the agent follows. And where s or (s, a) are the inputs to the state or state-value functions. The parameters w are the weights of a deep learning neural network or the tunable parameters of the function that defines the policy. An RL algorithm uses various learning approaches to tweak the values of w to obtain the desired behavior of the agent, first in terms of policy $\pi(a|s)$, which in turn leads to the state value function $\hat{v}(s;w)$ or the state action value function $\hat{q}(s,a;w)$. You choose the number of parameters $|w|$, which is lot smaller than the number of states $|s|$ or the number of state-action pairs ($|s| \times |a|$). The consequence of this approach is that there is a generalization of representation of the state of the state-action values. When you update the weight vector w based on some update equation for a specific state s or state action pair (s, a), it not only updates the value for that specific s or (s, a), but it also updates the values for many other states or state actions that are close to the original s or (s, a), the state or state action pair for which the update has been carried out. This depends on the geometry of the function. The other values of states near s will also be impacted by such an update, as shown previously. You are approximating the values with a function that is a lot more restricted than the number of states. Just to be specific, instead of updating $v(s)$ or $q(s, a)$ directly, you now update the parameter set w of the function, which in turn impacts the value estimates $\hat{v}(s;w)$ or $\hat{q}(s,a;w)$. Of course, like before, you carry out the w update using the MC or TD approach. There are various approaches to function approximation. You could feed the state vector (the values of all the variables that signify the state, such as position, speed, location, etc.) and get $\hat{v}(s;w)$, or you could feed the state and action vectors and get $\hat{q}(s,a;w)$ as an output. An alternative approach that is very dominant in the case of actions being discrete and coming from a small set is to feed state vector s and get $|A|$ number of $\hat{q}(s,a;w)$, one for each action possible ($|A|$ denotes the number of possible actions). Figure 5-1 shows the schematic of various approaches for using a function to represent $\hat{v}(s;w)$ or $\hat{q}(s,a;w)$.

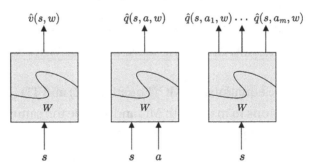

Figure 5-1. *Ways to represent $\hat{v}(s;w)$ or $\hat{q}(s,a;w)$ using the function approximation approach. The first and last ones are the most common ones used in this chapter*

There are various approaches to building such functional approximators, but this chapter explores two common approaches—linear approximators using tiling and nonlinear approximators using neural networks.

However, before getting into this, I need to revisit theoretical foundations to see what operations are required to make w move such that you can successively reduce the error between the target values and the current estimates of state or state-action values, $v(s)$ or $q(s, a)$.

Theory of Approximation

Function approximation is a topic studied extensively in the field of supervised learning, wherein based on training data you build a generalization of the underlying model. Most of the theory from supervised learning can be applied to reinforcement learning with functional approximation. However, RL with functional approximation brings to the fore new issues, such as how to bootstrap as well as its impact on nonstationarity. In supervised learning, while the algorithm is learning, the problem/model from which the training data was generated does not change. However, when it comes to RL with function approximation, the way the target (labeled *output* in supervised learning) is formed, it induces nonstationarity, and you need to come up with new ways to handle it. What I mean by *nonstationarity* is that you do not know the actual target values of $v(s)$ or $q(s, a)$. You use either the MC or TD approach to form estimates and then use these estimates as "targets." As you improve the estimates of the target values, you use the revised estimates as new targets. In supervised learning it is different—the targets are given and fixed during training. The learning algorithm has no impact on the targets. In reinforcement learning, you do not have actual targets, and you use the estimates of the target values. The agent is trying to find optimal actions, such as $v(s)$ or $q(s, a)$. The targets in the learning algorithm change. Therefore, the targets are not fixed or stationary during the learning.

Let's revisit the update equations for MC (Equation 4-2) and TD (Equation 4-5). I have modified the equations to make both MC and TD use the same notations of subscript t for the current time and $t + 1$ for the next instant. Both equations carry out the same update to move $V_t(s)$ closer to its target, which is $G_t(s)$ in the case of the MC update and $R_t + 1 + \gamma \cdot V_t(s')$ for the TD(0) update.

$$V_{t+1}(s) = V_t(s) + \alpha \left[G_t(s) - V_t(s) \right]$$ (5-2)

$$V_{t+1}(s) = V_t(s) + \alpha \left[R_{t+1} + \gamma \cdot V_t(s') - V_t(s) \right]$$ (5-3)

This is similar to what you do in supervised learning, especially in linear least square regression. You have the output values/targets $y(t)$ and you have the input features $x(t)$, together called *training data*. You can choose a model $Model_w[x(t)]$ like the polynomial linear model, decision tree, or support vectors, or even other nonlinear models like neural nets. The training data is used to minimize the error between what the model is predicting and the actual output values from the training set. This is done by minimizing a loss function as follows:

$$J(w) = \left[y(t) - \hat{y}(t;w) \right]^2 ; \text{where } \hat{y}(t;w) = Model_w \left[x(t) \right] \tag{5-4}$$

When $J(w)$ is a differentiable function, which is the case in this book, you can use gradient descent to tweak the weights/parameters w of the model to minimize the error/loss function $J(w)$. Usually, the update is carried out in batches multiple times using the same training data until the loss stops reducing further. The model with weights w has now learned the underlying mapping from input $x(t)$ to output $y(t)$. The way the delta update to the weights is carried out is given in the following equations:

$$\text{Gradient of } J(w) \text{ wrt } w \quad : = \nabla_w J(w)$$

For the loss function given in Equation 5-4, this becomes:

$$\nabla_w J(w) = -2 \cdot \left[y(t) - \hat{y}(t;w) \right] \cdot \nabla_w \hat{y}(t;w)$$

If you remember your calculus, you will remember that $\nabla_w J(w)$ is a vector in weight space—that is, it has $|w|$ components. The direction of this vector is more important than the actual values of the components. The direction of $\nabla_w J(w)$ is the direction in $|w|$ space, which results in a maximum rate of change of $J(w)$ as w changes. As $J(w)$ is the squared loss—that is, the difference between actual and predicted values, you want to reduce $J(w)$, the error, as much as possible. You can achieve this by moving w parameters in the negative direction of $\nabla_w J(w)$. This will give you a new set of w parameters such that the value of $J(w)$ is lowered.

$$w_{t+1} = w_t - \alpha \nabla_w J(w) \tag{5-5}$$

The weights move in the direction that minimize the loss—that is, the difference between the actual and predicted output values.

Moving on, let's spend some time talking about the various approaches to function approximation. The most common approaches are as follows:

- Linear combination of features. You combine the features (such as speed, velocity, position, etc.) weighted by vector w and use the computed value as the state value. Common approaches are as follows:

- Polynomials

- Fourier basis functions

- Radial basis functions

- Coarse coding

- Tile coding

- Nonlinear but differentiable approaches, with neural networks being the most popular and currently trending one.

- Nonparametric, memory-based approaches.

This book mostly talks about deep learning–based neural network approaches that suit unstructured inputs, such as images captured by the agent's vision system or freeform text using natural language processing (NLP). Later parts of this chapter and the next chapter are devoted to using deep learning–based function approximation, and you will see many variations with full code examples using PyTorch. But we are getting ahead of ourselves. Let's first examine a few common linear approximation approaches such as coarse coding and tile coding. As the focus of this book is on the use of deep learning in reinforcement learning, I do not devote a lot of time to various other linear approximation approaches. Also, just because I am not devoting time to all the linear approaches, it does not mean that they lack power of effectiveness. Depending on the problem at hand, the linear approximation approach may be the right way to go; it's effective, fast, and has convergence guarantees.

Coarse Coding

Let's revisit at the mountain car problem discussed in Figure 2-2. The car has a two-dimensional state, a position, and a velocity. Suppose you divide the two-dimensional state space into overlapping circles with each circle representing a feature. If state S lies inside a circle, that particular feature is present and has a value of 1; otherwise, the feature is absent and has a value of 0. The number of features is the number of circles. Let's say you have p circles; then you have converted a two-dimensional continuous state space into a p-dimensional state space, where each dimension can be 0 or 1. In other words, each dimension can belong to {0,1}.

Note   {0,1} represents the set of possible values—0 or 1. (0,1)—with regular brackets representing the range of values—that is, any value from 0 to 1 with 0 and 1 excluded. [0,1) represents the range of values between 0 to 1 plus the value on left—that is, 0 is included in the range and the value 1 on the right side of the range is excluded.

All those features represented by circles in which the state S lies will be "on" or equal to 1. Figure 5-2 shows an example. Two states are shown in the diagram, and depending on the circles that these points lie inside, the corresponding features will be turned on, while other features will be turned off. The generalization will depend on the size of the circles and how densely they are packed together. If the circles are replaced by ellipses, the generalization will be more in the direction of the elongation. You could also choose shapes other than circles to control the amount of generalization.

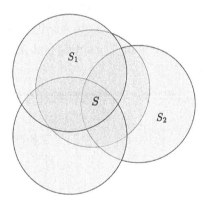

Figure 5-2. *Coarse coding in two dimensions using circles. The generalization depends on the size of circles as well as the density with which the circles are placed together*

Now consider the case with large, densely packed circles. A large circle makes the initial generalization wide where two faraway states are connected because they fall inside at least one common circle. However, the density (i.e., number of circles) allows you to control the fine-grained generalization. By having many circles, you ensure that even nearby states have at least one feature that is different between two states. This will hold even when each of the individual circles is big. With the help of experiments with varying configurations of the circle size and number of circles, you can fine-tune the size and number of circles to control the generalization appropriate for the problem/domain in question, controlling generalization with the size of the shape and granularity with the density of the shapes.

Tile Encoding

Tile encoding is a form of coarse coding that can be programmatically planned. It works well over multidimensional space, making it much more useful than generic coarse coding.

Let's consider a two-dimensional space like the mountain car I just talked about. You divide the space into nonoverlapping grids covering the whole space. Each such division is called *tiling*, as shown in the left diagram of Figure 5-3. The *tiles* are square here and, depending on the location of state S on this $2D$ space, only one tile is 1, while all other tiles are 0.

You then have a number of such *tiling* offset from each other. Suppose you use n tilings; then for a state, only one tile from each tiling will be on. In other words, if there are n tilings, then exactly n features will be 1, a single feature from each of the n tilings. Figure 5-3 shows an example.

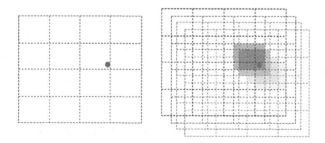

Figure 5-3. *Tile coding. You have 4×4=16 tiles in a single tiling, as shown in the left figure. And you have four tilings overlapping each other, as shown in four different colors in the right figure. A state (green color dot) lights up a single tile from each tiling. The generalization is controlled by the number of tiles in a single tiling as well as the total number of tilings*

Note that if the learning rate of step size were α(alpha) in Equations 5-1 and 5-2, you would now replace it with $\frac{\alpha}{n}$ where n is the number of tilings. This makes the algorithm scale free of the number of tilings. As both coarse coding and tile encoding use binary features, digital computers can speed up computations.

The nature of the generalization now depends on the following factors:

1. The number of tiles in a single tiling (the left figure in Figure 5-3).

2. The number of tilings (the right figure in Figure 5-3 showing four tilings in four different colors).

3. The nature of offset, whether uniform and symmetric or asymmetric.

There are some general strategies to determine the previous numbers. Consider the situation where each tile in a single tiling is a square of width w. For a continuous space of k dimensions, it will be a k-dimensional square with each side of width w. Suppose you have n tilings, and hence the tilings need to be offset from each other by a distance of $\frac{w}{n}$ in all dimensions. This is known as a *displacement vector*. The first heuristic is to choose n such that $n = 2^i \geq 4k$. The displacement in each direction is an odd multiple $(1, 3, 5, 7, ..., 2k - 1)$ of displacement vector $\frac{w}{n}$. The upcoming example uses a library to help you divide the two-dimensional mountain car state space into appropriate tilings. You will provide the 2D state vector, and the library will give you back the active tiles vector.

Challenges in Approximation

While I leverage the knowledge of supervised learning–based methods like gradient descent explained earlier, you have to keep two things in mind that make gradient-based methods harder to work in reinforcement learning as compared to the supervised learning.

First, in supervised learning, the training data is held constant. The data is generated from the model, which does not change during the data generation process. It is a ground truth that is given to you. You are building a function whose parameters are identified by using this training data to learn input to output mapping. The data provided to the training algorithm is external to the learning process/algorithm. It does not depend on the algorithm in any way. It is given as constant and independent of the learning algorithm. Unfortunately, in RL, especially in a model-free setup, this is not the case. The data used to generate training samples is based on the policy the agent is following, and it is not a complete picture of the underlying model. As you explore the environment, you learn more and adjust the policy. This change in policy leads to exploration of news states and actions, which in turns generates a new set of training data for the next step of iteration. You either use the MC-based approach of observing an actual trajectory or bootstrap under TD to form an estimate of the target value, the $y(t)$. As you explore, learn more, and adjust the policy, the target $y(t)$ for a given specific state changes, which is not the case in supervised learning. This is known as the problem of *nonstationary targets*.

Second, supervised learning is based on the theoretical premise of samples being uncorrelated to each other, mathematically known as i.i.d. (for "independent identically distributed") data. All the algorithms for regression, classification, trees, and support vector machines are based on this i.i.d. assumption. However, in RL, the data you see depends on the policy that the agent followed to generate the data. In a given episode, the states you see are dependent on the policy the agent is following at that instant. States that come in later time steps depend on the action (decisions) the agent took earlier. In other words, the data is *correlated*. The next state s_{t+1} you see depends on the current state s_t and the action a_t agent takes in that state. Note that this assumes that the Markov property (the future state s_{t+1}) depends only on the current state s_t and action a_t you take. There is no dependence on the past—that is, either actions or states from time step $t-1$ or before. The relationship can be expressed by way of a dependency graph, as shown in Figure 5-4.

This temporal dependence on data in RL means that you cannot shuffle the data in the case of RL and that each data point, unlike with supervised learning, cannot be used in isolation. The whole trajectory of data points from the beginning of an episode to the end must be used together.

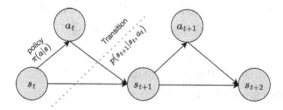

Figure 5-4. *Policy and transition dynamics under Markov assumptions*

These two issues make function approximation harder in an RL setup. As you go along, you will see the various approaches that have been taken to address these challenges. These approaches do not always come with theoretical guarantees—rather with practical hacks and ways that have proven useful and do carry some intuitive reasoning.

With a broad understanding of the approach, it is time now to start with the usual course of first looking at value *prediction/estimate* to learn a function that can represent the value functions $v(s)$ or state-action value functions $q(s, a)$. You will then look at the *control* aspect—that is, the process of optimizing the policy $\pi(a|s)$ that the agent should follow. It will follow the usual pattern of using Generalized Policy Iteration (GPI), just like the approach in the previous chapter—alternating between predicting/estimating the state-action values and using a greedy policy to control/identify the best action. And you keep iterating in this loop until some convergence is observed.

Incremental Prediction: MC, TD, TD(λ)

This section looks at the prediction problem,—that is, how to estimate the state values using function approximation.

Following along, you will now attempt to extend the supervised training process of tuning a model using training data consisting of inputs and targets. This is done to function approximation under RL using the loss function in Equation 5-4 and weight update, as shown in Equation 5-5. If you compare the loss function in Equation 5-4 and the MC/TD updates in Equations 5-2 and 5-3, you can draw a parallel by thinking of MC and TD updates as operations, which are trying to minimize the error between the actual target $V_\pi(s)$ and the current estimate $V_t(s)$. You can represent the loss function as follows:

$$J(w) = E_\pi \left[V_\pi(s) - V_t(s) \right]^2 \qquad (5\text{-}6)$$

This is similar to the loss function for supervised learning, as shown in Equation 5-4. Following the same derivation as in Equation 5-5 and using stochastic gradient descent (i.e., replacing expectation with update in each sample), you can write the update equation for weight vector w as follows:

$$w_{t+1} = w_t - \alpha \cdot \nabla_w J(w)$$

$$w_{t+1} = w_t + \alpha \cdot \left[V_\pi(s) - V_t(s;w) \right] . \nabla_w V_t(s;w) \qquad (5\text{-}7)$$

Here, w represents the weight vector of the model—that is, the tunable parameters of the model that are progressively improved to fit the model to the state values observed for a given policy. However, unlike supervised learning, you do not have the actual/target output values $V_\pi(s)$; rather, you use estimates of these targets. With MC, the estimate/target of $V_\pi(s)$ is $G_t(s)$, while the estimate/target under TD(0) is $R_{t+1} + \gamma \cdot V_t(s')$. Accordingly, the updates under MC and TD(0) with functional approximation can be written as follows.

MC update:

$$w_{t+1} = w_t + \alpha \cdot \left[G_t(s) - V_t(s;w) \right] . \nabla_w V_t(s;w) \tag{5-8}$$

TD(0) update:

$$w_{t+1} = w_t + \alpha \cdot \left[R_{t+1} + \gamma \cdot V_t(s';w) - V_t(s;w) \right] . \nabla_w V_t(s;w) \tag{5-9}$$

A similar set of equations can be written for q-values. You will see that in the next section. This is along the same lines of what you did for the MC and TD control sections in the previous chapter; first you use state action values in a model-free world to build accurate estimates of the state action values under the current policy and then use greedy updates to find values, maximizing actions in each individual state. Moving back and continuing with the derivation of weight update equation, let's first consider the setup of linear approximation where the state value $\hat{v}(s;w)$ can be expressed as a dot product of state vector $x(s)$ and weight vector w, each with vector component dimension index represented as i:

$$\hat{v}(s;w) = x(s)^T \cdot w = \sum_i x_i(s) \cdot w_i \tag{5-10}$$

In the previous expression, the gradient of $\hat{v}(s;w)$ with respect to w will be the state vector $x(s)$—that is:

$$\nabla_w V_t(s;w) = x(s) \tag{5-11}$$

Combining Equation 5-11 with Equation 5-7 gives the following:

$$w_{t+1} = w_t + \alpha \cdot \left[V_\pi(s) - V_t(s;w) \right] . x(s) \tag{5-12}$$

As discussed earlier, you do not know the true state value $V_\pi(s)$, and hence you use an estimate $G_t(s)$ in the MC approach and an estimate $R_{t+1} + \gamma \cdot V_t(s')$ in the TD(0) approach. This gives you the weight update rules for MC and TD in the linear approximation case as follows.

MC update:

$$w_{t+1} = w_t + \alpha . \left[G_t(s) - V_t(s;w) \right] . x(s) \qquad (5\text{-}13)$$

TD(0) update:

$$w_{t+1} = w_t + \alpha . \left[R_{t+1} + \gamma . V_t(s';w) - V_t(s;w) \right] . x(s) \qquad \bullet \qquad (5\text{-}14)$$

In simple terms, the update to weights—that is, the second term on the right side of Equation 5-14 can be expressed as follows:

Update = learning rate x prediction error x feature value

Where learning rate is α, prediction error is denoted as $R_{t+1} + \gamma \cdot V_t(s';w) - V_t(s;w)$ and feature value is $x(s)$.

Let's tie this back to the table-based approach for the discrete states that you saw in the previous chapter. The table lookup—that is, using state or state-action values as an index to look into a table or a dictionary to get the values—is a special case of the linear approach. Consider that each component of $x(s)$ is either 1 or 0 and only one of them can have a value of 1, with the rest of the features being 0. $x^{table}(s)$ is a column vector of size p in which only one element can have a value of 1 at any point, with the rest of the elements equal to 0. Depending on the state the agent is in, the corresponding element would be 1. Therefore, the number of elements in the column vector (p) is equal to the number of states the agent can exit in. This is as shown here:

$$x^{table}(s) = \begin{pmatrix} 1(s = s_1) \\ .. \\ .. \\ 1(s = s_p) \end{pmatrix}$$

The weight vector comprises the values of state v(s) for each $s = s_1, s_2, ...s_p$.

$$w = \begin{pmatrix} w_1 \\ .. \\ .. \\ w_p \end{pmatrix} = \begin{pmatrix} v(s_1) \\ .. \\ .. \\ v(s_1) \end{pmatrix}$$

Using these expressions in Equation 5-10, you get the following:

$$\hat{v}(s = s_k; w) = x(s)^T \cdot w = v(s_k)$$

You can see that the value of state $s = s_k$ can be expressed as a dot product of the state vector $x(s)$ and the weight vector w. If you substitute this expression back into Equation 5-13 for the MC update or into Equation 5-14 for the TD(0) update, you get the familiar update rules derived directly for discrete state situation in Chapter 4.

Here is the MC update:

$$V_{t+1}(s) = V_t(s) + \alpha \cdot \left[G_t(s) - V_t(s) \right], s \in s_1, s_1, \ldots, s_p \qquad (5\text{-}15)$$

Here is the TD(0) update:

$$V_{t+1}(s) = V_t(s) + \alpha \cdot \left[R_{t+1} + \gamma \cdot V_t(s') - V_t(s) \right], s \in s_1, s_1, \ldots, s_p \qquad (5\text{-}16)$$

The previous derivation was to tie back the table lookup back as a special case of a more general linear function approximation.

One more point to note that I glossed over while deriving the update equations were the details of the target estimates $G_t(s)$ in MC and $R_{t+1} + \gamma \cdot V_t(s')$ in TD(0). These estimates are not some constant fixed value. These are also dependent on the policy—for MC, directly on the sample episode return $G_t(s)$ and on the value of next state $V_t(s')$ in the case of TD(0). As an example, let's revisit Equation 5-6 and replace $V_\pi(s)$ with the TD target and then take the gradient.

$$J(w) = \left[V_\pi(s) - V_t(s) \right]^2$$

Replacing $V_\pi(s)$ with the value of target based on the model , you have:

$$J(w) = \left[R_{t+1} + \gamma \cdot V_t(s'; w) - V_t(s; w) \right]^2$$

If you take the derivative of $J(w)$ with respect to w, you will actually get two terms, one due to the derivative of $V_t(s'; w)$, the next state value, and another term due to the derivative of $V_t(s; w)$, the current state value. Such an approach of taking both gradient contributions $\nabla V_t(s'; w)$ and $\nabla V_t(s; w)$ introduces shifting of target values and has shown to worsen the speed of learning. First, the reason is that you want the targets to stay constant, and hence you need to ignore the contribution due to $\nabla V_t(s'; w)$. Second, conceptually with gradient descent, you are trying to pull the value of the current state

$V_t(s; w)$ toward its target. Taking the second contribution term $\nabla V_t(s'; w)$ means that you are trying to move the value of the next state $S = s'$ toward the value of the current state $S = s$.

In summary, you take only the derivative of the current state value $V_t(s; w)$ and ignore the derivative of the next state value $V_t(s'; w)$. The approach makes the value estimation look similar to the approach used with supervised learning with stationary targets. This is also the reason that sometimes the gradient descent method used in Equations 5-8 and 5-9 are also called *semi-gradient* methods. It is because you are taking gradient for the current state and ignoring any update in weights due to the gradient of the next state.

As I touched on earlier, the convergence of algorithms is not guaranteed anymore, unlike the guarantee you had in the tabular setup due to contraction theorem. However, most of the algorithms with some careful considerations do converge in practice. Table 5-1 shows the convergence of various prediction/estimation algorithms. I will not be going into the detailed explanation of these convergence properties. Such a discussion is more suited for a book with a focus on the theoretical aspects of learning. This book is a practical one, with just enough theory to understand the background and appreciate the nuances of algorithms. The core focus is on coding these algorithms in deep learning libraries.

Table 5-1. *Convergence of Prediction/Estimation Algorithms*

Policy Type	Algorithm	Table Lookup	Linear	Nonlinear
On-policy	MC	Y	Y	Y
	TD(0)	Y	Y	N
	TD(λ)	Y	Y	N
Off-policy	MC	Y	Y	Y
	TD(0)	Y	N	N
	TD(λ)	Y	N	N

In a later section, you will see how the combination of bootstrapping (e.g., TD), function approximation and off-policy, all present together, can impact stability adversely unless careful consideration is given to the learning process.

The next section looks at control problems, which define how to optimize a policy with a function approximation.

Incremental Control

This section moves from state value prediction to that of control—that is, finding the optimal policy. This chapter follows a similar approach as in the previous chapter. You start with function approximation to estimate the q-values. The q-value is represented by a model with weight/parameters w.

$$\hat{q}(s,a;w) \approx q_\pi(s,a) \tag{5-17}$$

As before, you form a loss function between the target and current value. This is being done so that, based on the sample of q-values, you can tweak the parameters for the model to estimate the q-values.

$$J(w) = E_\pi\left[\left(q_\pi(s,a) - \hat{q}(s,a;w)\right)^2\right] \tag{5-18}$$

In order to make fit better, you need to reduce the loss function in Equation 5-18. You can achieve that by doing gradient descent on weight w so that loss is reduced iteratively.

$$w_{t+1} = w_t - \alpha \cdot \nabla_w J(w)$$

where,

$$\nabla_w J(w) = \left[q_\pi(s,a) - \hat{q}(s,a;w)\right] \cdot \nabla_w \hat{q}(s,a;w) \tag{5-19}$$

Like before, in the case of a linear model where $\hat{q}(s,a;w)$ is represented as a linear equation $\hat{q}(s,a;w) = x(s,a)^T \cdot w$, you can simplify the equation. The derivative $\nabla_w \hat{q}(s,a;w)$, in a linear case, as shown previously, will become $\nabla_w \hat{q}(s,a;w) = x(s,a)$.

There is one more change you need to make to Equation 5-19 like you did in the previous section. You do not know the true q-value $q_\pi(s, a)$. Like before, you replace it with the estimates using either MC or TD, giving you a set of weight update equations.

MC update:

$$w_{t+1} = w_t + \alpha \cdot \left[G_t(s) - q_t(s,a;w)\right] \cdot \nabla_w \hat{q}(s,a;w) \tag{5-20}$$

TD(0) update:

$$w_{t+1} = w_t + \alpha \cdot \left[R_{t+1} + \gamma \cdot q_t(s',a';w) - q_t(s,a;w)\right] \cdot \nabla_w \hat{q}(s,a;w) \tag{5-21}$$

These equations allow you to carry out q-value estimation/prediction. This is the *evaluation* step of generalized policy iteration, where you carry out multiple rounds of gradient descents to improve on the q-value estimates for a given policy and get them close to the actual target values.

Evaluation is followed by greedy policy maximization to improve the policy. Figure 5-5 shows the process of iteration under GPI with function approximation. You can observe that the arrows are not touching the bottom line $q_w = q_\pi$, which indicates that while converging you do not need to go all the way. You could carry out a few gradient steps and then do a policy update with greedy maximization, followed again with another cycle of gradient steps and a greedy policy update.

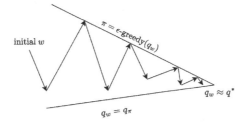

Figure 5-5. *Generalized policy iteration with function approximation*

Semi-gradient *n*-step SARSA Control

Having looked at the theoretical derivation, it is now time to apply all the learning of this chapter to train an agent. Just a bit more theory—an actual algorithm pseudocode that implements SARSA control. If you remember from the previous chapter, SARSA is an on-policy algorithm. You will use n-step returns to improve the variance of estimates as you saw in the previous chapter. As you will be implementing Equation 5-20, this algorithm will be called "Semi-gradient N-step SARSA control". Figure 5-6 shows the pseudocode of the algorithm. As with the standard SARSA setup, you use an ϵ-greedy policy to sample the environment and collect tuples of state, action, reward, next state, and next action $(s_t, a_t, r_{t+1}, s_{t+1}, a_{t+1})$ and then use these observations to form n-step returns. This is followed by a gradient step using Equation 5-20 to take a step and improve the estimate of q-values. As q-values improve and you use these q-values to run the next episode using ϵ-greedy policy, you are implicitly carrying out the policy update. Note that you use ϵ-greedy policy, as you are in an on-policy setup and you need exploration to make sure that you do enough exploration of the state space.

Semi Gradient n-Step SARSA Control (Episodic)

Input:

A differentiable function $\hat{q}(s, a; w) : |S| \times |A| \times \mathbb{R}^d \to \mathbb{R}$

Other paramters: steps size α, exploration prob. ϵ, number of steps: n

Initialize:

Initialize weight vector $\mathbf{w} : \mathbb{R}^d$ abitrarily like $\mathbf{w} = \mathbf{0}$

Three arrays to store (S_t, A_t, R_t) which access using "mod $n + 1$"

Loop for each episode:

Start episode with S_0 (non-termial)

Select and store A_0, ϵ-greedy action using $\hat{q}(S_0, \bullet; w)$

$T \leftarrow \infty$

Loop for $t = 1, 2, 3, \ldots$

| if $t < T$, then:

| Take action A_t

| Observe and store R_{t+1} and S_{t+1}

| if S_{t+1} is termial then:

| $T \leftarrow t + 1$

| else:

| select and store A_{t+1} using ϵ-greedy $\hat{q}(S_t, \bullet; w)$

| $\tau \leftarrow t - n + 1$ (τ is inex whose estimate is being updated)

| If $\tau \geq 0$:

| $G \leftarrow \displaystyle\sum_{i=\tau+1}^{\min(\tau+n, T)} \gamma^{i-\tau-1} \cdot R_i$

| if $t + n < T$, then $G \leftarrow G + \gamma^n \hat{q}(S_{\tau+n}, A_{\tau+n}; w)$

| $w \leftarrow w + \alpha \cdot [G - \hat{q}(S_\tau, A_\tau; w)] \cdot \nabla_w \hat{q}(S_\tau, A_\tau; w)$

Until $\tau = T - 1$

Figure 5-6. *n-step semi-gradient SARSA for episodic control*

The w update uses Equation 5-20 with the target being G, the n-step return.

You will now see how to apply this to mountain car environment, which has continuous state space. You will be using tile encoding studied earlier in this chapter to build a function approximator to the state values. Tile encoding is a binary feature approximator. In the setup:

$$\hat{q}(S, A; w) = x(S, A)^T \cdot w$$

where $x(S, A)$ is the tile encoded feature vector.

Accordingly, $\nabla\hat{q}(S_\tau, A_\tau; w) = x(S, A)$

In this example of tile encoding, you use the implementation of tile coding by Rich Sutton, which you can explore in the Python script file `tiles3.py`. This is an implementation of grid-style tile codings, based originally on the UNH CMAC code,[1] but by now highly changed. You provide a function, called "`tiles`", that maps floating and integer variables to a list of tiles, and a second function "`tiles-wrap`" that does the same while wrapping some floats to provided widths (the lower wrap value is always 0). The float variables are gridded at unit intervals, so generalization will be by approximately one in each direction, and any scaling will have to be done externally before calling tiles.

`Num-tilings` should be a power of 2, for example, 16. To make the offsetting work properly, it should also be greater than or equal to four times the number of floats. The first argument is either an index hash table of a given size (created by `make-iht(size)`), an integer "`size`" (range of the indices from 0), or nil (for testing, indicating that the tile coordinates are to be returned without being converted to indices).

You use the `tiles3.py` to build the `QEstimator` class, as shown in Listing 5-1. The `QEstimator` class holds the weights and tiles the observation space. The complete code is given in the Python notebook called `5.a-n-step-SARSA.ipynb`.

Listing 5-1. n-Step SARSA Control: Mountain Car. Q-Estimator initialization

```
class QEstimator:

    def __init__(self, step_size, num_of_tilings=8, tiles_per_dim=8, max_
    size=2048, epsilon=0.0):
        self.max_size = max_size
        self.num_of_tilings = num_of_tilings
        self.tiles_per_dim = tiles_per_dim
        self.epsilon = epsilon
        self.step_size = step_size / num_of_tilings

        self.table = IHT(max_size)
        self.w = np.zeros(max_size)
```

---

[1] http://www.ece.unh.edu/robots/cmac.htm

```
        self.pos_scale = self.tiles_per_dim / (env.observation_space.
        high[0] \

                                                    - env.observation_
                                                    space.low[0])
        self.vel_scale = self.tiles_per_dim / (env.observation_space.
        high[1] \

                                                    - env.observation_
                                                    space.low[1])
```

As shown in Listing 5-2, the get_active_features function takes as input continuous two-dimensional values of S and the discrete input action A to return the tile-encoded active_feature $x(S, A)$—that is, binary tiling features that are active for a given (S, A). The q_predict function also takes as input (S, A) and returns the estimate $\hat{q}(S, A;w) = x(S, A)^T \cdot w$. It internally calls get_active_features to first get features and carries out a dot product with the weight vector.

Listing 5-2. n-Step SARSA Control: q-predict implementation in 5.a-n-step-SARSA.ipynb

```
    def get_active_features(self, state, action):
        pos, vel = state
        active_features = tiles(self.table, self.num_of_tilings,
                            [self.pos_scale * (pos - env.observation_space.
                            low[0]),
                             self.vel_scale * (vel- env.observation_space.
                             low[1])],
                            [action])
        return active_features

    def q_predict(self, state, action):
        pos, vel = state
        if pos == env.observation_space.high[0]:  # reached goal
            return 0.0
        else:
            active_features = self.get_active_features(state, action)
            return np.sum(self.w[active_features])
```

The weight update equation shown toward the end of the algorithm in Figure 5-6 is what the q_update function carries out. The get_eps_greedy_action functions carries out action selection using ε-greedy $\hat{q}(S_0, \bullet; w)$. Listing 5-3 shows the code.

Listing 5-3. n-Step SARSA Control: q-update implementation in 5.a-n-step-SARSA.ipynb

```
# learn with given state, action and target
def q_update(self, state, action, target):
    active_features = self.get_active_features(state, action)
    q_s_a = np.sum(self.w[active_features])
    delta = (target - q_s_a)
    self.w[active_features] += self.step_size * delta

def get_eps_greedy_action(self, state):
    pos, vel = state
    if np.random.rand() < self.epsilon:
        return np.random.choice(env.action_space.n)
    else:
        qvals = np.array([self.q_predict(state, action) for action in
        range(env.action_space.n)])
        return np.argmax(qvals)

def get_action(self, state):
    pos, vel = state
    qvals = np.array([self.q_predict(state, action) for action in
    range(env.action_space.n)])
    return np.argmax(qvals)
```

The main training algorithm of the SARSA agent, as shown in Figure 5-6, is implemented in the function sarsa_n, calling functions from QEstimator as required. Listing 5-4 contains the code. The implementation is straightforward Python code following the pseudocode of Figure 5-6.

Listing 5-4. n-Step SARSA Control: SARSA Agent Implementation in 5.a-n-step-SARSA.ipynb

```python
def sarsa_n(qhat, step_size=0.5, epsilon=0.0, n=1, gamma=1.0, episode_cnt
= 10000):
    episode_rewards = []
    for _ in range(episode_cnt):
        state,_ = env.reset()
        action = qhat.get_eps_greedy_action(state)
        T = float('inf')
        t = 0
        states = [state]
        actions = [action]
        rewards = [0.0]
        while True:
            if t < T:
                next_state, reward, terminated, _, _ = env.step(action)
                states.append(next_state)
                rewards.append(reward)

                if terminated:
                    T = t+1
                else:
                    next_action = qhat.get_eps_greedy_action(next_state)
                    actions.append(next_action)

            tau = t - n + 1

            if tau >= 0:
                G = 0
                for i in range(tau+1, min(tau+n, T)+1):
                    G += gamma ** (i-tau-1) * rewards[i]
                if tau+n < T:
                    G += gamma**n * qhat.q_predict(states[tau+n],
                    actions[tau+n])
                qhat.q_update(states[tau], actions[tau], G)
```

```
    if tau == T - 1:
        episode_rewards.append(np.sum(rewards))
        break
    else:
        t += 1
        state = next_state
        action = next_action

return np.array(episode_rewards)
```

Similar to many examples in the previous chapter, there is also a helper function called `plot_rewards` that plots the per-episode returns and records a video of the trained agent.

Figure 5-7 shows the result of running this algorithm to train the mountain car. Within 50 episodes, the agent reaches a steady state, and it reaches the goal of hitting the flag on the right side of the valley in about 110 time steps.

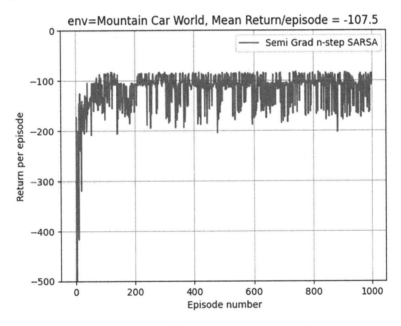

Figure 5-7. *n-step semi-gradient SARSA with MountainCar*

This completes the n-step semi-gradient SARSA example.

Semi-gradient SARSA(λ) Control

This section looks at the semi-gradient SARSA(λ) algorithm with eligibility traces. As you saw in the previous chapter, in the MC update using the full episode return—that is $\lambda = 1$, TD(0) uses one step next state estimate to form the target—that is, $\lambda = 0$. n-step uses the returns from next n-steps followed by the q-value estimate for the state after n-step to form the target estimate, which is midway through the MC of the full episode and TD(0) of one step. How do you decide the value of n? Eligibility traces uses the full spectrum of one step update to full episode return by adding all the rewards for all possible values of n. The weight of these returns is controlled by the value of λ, which can vary from 0 to 1. In other words, SARSA(λ) further generalizes the n-step SARSA. When the state or state-action values are represented by binary features with linear function approximation (like in the case of the mountain car with tile encoding), you get the algorithm in Figure 5-8. This algorithm introduces a concept of eligibility trace that has the same number of components as the weight vector. The *weight vector* is the long-term memory over many episodes to generalize from all the examples shown. The *eligibility trace* is short-term memory lasting less than the length of episode. It helps during the learning process by having an effect on weight. Consider the case of a wave pattern that slops upward in the graph. The weight vector is like the slop of the upward overall trend and eligibility trace is like the local neighborhood fluctuations from one peak of the wave to another, which is going up and down in that local vicinity. I will not go into the detailed derivations of the update rules. You can refer to the book titled *Reinforcement Learning: An Introduction*[2] for a detailed explanation of the concepts and mathematical derivations.

---

[2] http://incompleteideas.net/book/the-book.html

Semi Gradient SARSA(λ) Control (Episodic)

Input:

A Tile Function $F(s, a)$ giving the active tile indices (i) for a given state and action s, a

A differentiable function $\hat{q}(s, a; w) : |S| \times |A| \times \mathbb{R}^d \rightarrow \mathbb{R}$

Other paramters: steps size α, exploration prob. ϵ, trace decay $\lambda \in [0, 1]$

Initialize:

Initialize weight vector $\mathbf{w} : \mathbb{R}^d$ abitrarily like $\mathbf{w} = \mathbf{0}$

Initialize eligibility trace $\mathbf{z} : \mathbb{R}^d$ abitrarily like $\mathbf{z} = \mathbf{0}$

Three arrays to store (S_t, A_t, R_t) which access using "mod $n + 1$"

Loop for each episode:

Start episode with S_0 (non-termial)

Select and store A_0, ϵ-greedy action using $\hat{q}(S_0, \bullet; w)$

$\mathbf{z} \leftarrow \mathbf{0}$

$T \leftarrow \infty$

Loop for each step

Take action A, observe R and S'

$\delta \leftarrow R$

Loop for active indices i in $F(S, A)$:

$\delta \leftarrow \delta - w_i$

$z_i \leftarrow z_i + 1$ (accumulating trace)

or, $z_i \leftarrow 1$ (relacing trace)

if S' termial:

$\mathbf{w} \leftarrow \mathbf{w} + \alpha \cdot \delta \cdot \mathbf{z}$

goto next episode

select A' using ϵ-greedy $\hat{q}(S', \bullet; w)$

loop for active indices i in $F(S', A')$:

$\delta \leftarrow \delta + \gamma \cdot w_i$

$\mathbf{w} \leftarrow \mathbf{w} + \alpha \cdot \delta \cdot \mathbf{z}$

$\mathbf{z} \leftarrow \gamma \cdot \lambda \cdot \mathbf{z}$

$S \leftarrow S'; A \leftarrow A'$

Figure 5-8. *Semi-gradient SARSA(λ) for episodic control when features are binary and the value function is a linear combination of the feature vector and the weight vector*

As in the previous subsection, you will run this algorithm on the mountain car environment. The 5.b-lambda-sarsa.ipynb notebook contains the full code. Listing 5-5 gives the code for the implementation of eligibility trace update as shown in Figure 5-8. Accumulating trace adds one to the previous trace for active tiles, while replacing trace makes the trace for index *i* as one for active tile index *i*.

Listing 5-5. Accumulating Trace and Replacing Trace Implementation from 5.b-lambda-sarsa.ipynb

```python
def accumulating_trace(trace, active_features, gamma, lambd):
    trace *= gamma * lambd
    trace[active_features] += 1
    return trace

def replacing_trace(trace, active_features, gamma, lambd):
    trace *= gamma * lambd
    trace[active_features] = 1
    return trace
```

The QEstimator class from n-step SARSA is modified to store the trace values and to use the trace in the weight q_update update function, as shown in Listing 5-6.

Listing 5-6. q-update Implementation Using Trace from 5.b-lambda-sarsa.ipynb

```python
# learn with given state, action and target
def q_update(self, state, action, reward, next_state, next_action):

    active_features = self.get_active_features(state, action)

    q_s_a = self.q_predict(state, action)
    target = reward + self.gamma * self.q_predict(next_state, next_action)
    delta = (target - q_s_a)

    if self.trace_fn == accumulating_trace or self.trace_fn ==
    replacing_trace:
        self.trace = self.trace_fn(self.trace, active_features, self.gamma,
        self.lambd)
```

```
else:
    self.trace = self.trace_fn(self.trace, active_features, self.
    gamma, 0)

self.w += self.step_size * delta * self.trace
```

The `sarsa_lambda` function implements the overall learning algorithm given in Figure 5-8. It is a straight implementation of the pseudocode from that figure. Listing 5-7 shows the code.

Listing 5-7. sarsa-lambda from `5.b-lambda-sarsa.ipynb`

```
def sarsa_lambda(qhat, episode_cnt = 10000, max_size=2048, gamma=1.0):
    episode_rewards = []
    for i in range(episode_cnt):
        state, _ = env.reset()
        action = qhat.get_eps_greedy_action(state)
        qhat.trace = np.zeros(max_size)
        episode_reward = 0
        while True:
            next_state, reward, terminated, _, _ = env.step(action)
            next_action = qhat.get_eps_greedy_action(next_state)
            episode_reward += reward
            qhat.q_update(state, action, reward, next_state, next_action)
            if terminated:
                episode_rewards.append(episode_reward)
                break
            state = next_state
            action = next_action
    return np.array(episode_rewards)
```

You also have a function to run the trained agent through some episodes and record the behavior. Once you have trained the agent and generated the animation, you can run the MP4 file and see the strategy that the agent follows to reach the goal.

Figure 5-9 shows the result of running the SARSA(λ) algorithm to train the mountain car. You can see that the results are similar to those in Figure 5-7. This is too small a problem, but for larger problems, eligibility trace–driven algorithms will show better and faster convergence.

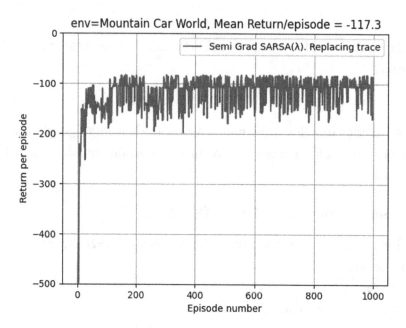

Figure 5-9. *Semi-gradient SARSA(λ) with the mountain car*

This concludes the example of running SARSA(λ) on mountain car environment.

Convergence in Functional Approximation

Let's start exploring convergence by looking at an example. As shown in Figure 5-10, let's consider a two-state transition as part of some MDP. Let's assume you will use function approximation with the value of the first state being w and the value of the second state being $2w$. Note that w is a single number and not a vector.

Figure 5-10. *Two-step transition under functional approximation*

Assume that $w = 10$ and the agent transitions from the first state to the second state—that is, from the state with the value 10 to the state with the value 20. This example also assumes that the transition from the first state to the second state is the only possible transition in the first state, and the reward for this transition is zero every time. Let the learning rate be $\alpha = 0.1$.

Let's now apply Equation 5-14 to the first state in Figure 5-10.

$$w_{t+1} = w_t + \alpha \cdot \left[R_{t+1} + \gamma \cdot V_t(s';w) - V_t(s;w) \right] \cdot x(s)$$

Now substitute the values. Substitute $\alpha = 0.1$, which is what you have assumed, $R_{t+1} = 0$ which is also the assumption. The state values are given in Figure 5-10— next state $V_t(s';w) = 2w_t$, current state value $V_t(s;w) = w_t$. The state feature vector is one dimensional and has a value of 1—that is, $x(s) = \nabla_w V_t(s;w) = 1$. With all these substitutions, you get:

$$w_{t+1} = w_t + 0.1 \cdot \left[0 + \gamma \cdot 2w_t - w_t \right] \cdot 1$$

For example:

$$w_{t+1} = w_t + 0.1 \cdot w_t \cdot (2\gamma - 1)$$

Now assume that γ is close to 1, and the current weight w_t is 10. The updated weight will look like this:

$$w_{t+1} = 10 + 0.1 \cdot 10 \cdot (2 - 1) = 11$$

Actually, as long as $(2\gamma - 1) > 0$, every update will lead to the divergence of weight w. It goes to show that function approximation can cause divergence. This is because of the value generalization where updating the value of a given state also updates the value of nearby or related states. There are three angles to the instability issue as listed.

> *Function approximation*: A way to generalize for a very large state space using a model with the number of parameters smaller than the total number of possible states.

> *Bootstrapping*: Forming target values using estimates of state values, for example, in TD(0) the target being the estimate $R_{t+1} + \gamma \cdot V_t(s';w)$

> *Off-policy learning*: Training the agent using a behavior policy but learning a different optimal policy.

The presence of these three components together significantly increases the chances of divergence even in a simple prediction/estimation scenario. The control and optimization problems are even more complex to analyze. It has also been shown that instability can be avoided as long as all three are not present together. That begs the question, could you drop any of the three and assess the impact of such a drop?

Function approximation, especially using neural networks, has made RL practical for large real-world problems. Other alternatives are not practical. Bootstrapping has made the process sample efficient. The alternative of forming targets by watching the whole episode, while feasible, is not very practical. Off-policy learning can be replaced with on-policy, but again for RL to become close to the way humans learn, you need off-policy to learn about some problem/situation by exploring another, similar problem. Therefore, there are no easy answers to this. You cannot drop any of the three requirements with low impact. That's the theoretical side. In practice, most of the time, algorithms converge with some careful monitoring and tweaking—such as use of replay buffers, learning rates, and ideas like double Q-learning. Chapters 6 and 7 discuss some of these practical tweaks in the context of DQN.

While training, it is important to periodically evaluate the agent and plot the curve of the return per episode. If it shows wild fluctuations, it is an indication that training is unstable. Remember that it will almost never be a smooth curve, but it should also not be wildly fluctuating.

Gradient Temporal Difference Learning

The semi-gradient TD learning with the update equation shown in Equation 5-9 does not follow a true gradient. While taking the gradient of a loss function, you kept the estimate of the target—that is, $R_{t+1} + \gamma \cdot V_t(s'; w)$, constant. It did not appear in the derivative with respect to the weight w. The real Bellman error is $R_{t+1} + \gamma \cdot V_t(s'; w) - V_t(s; w)$, and the derivative of it ideally should have had gradient terms for both $V_t(s'; w)$ and $V_t(s; w)$.

There is a variation of this known as *gradient temporal difference learning* that follows the true gradient and offers convergence in all the cases of table lookup, linear and nonlinear functional approximation, and both the under on-policy and off-policy methods. Adding this to the mix of algorithms, Table 5-1 can be modified as shown in Table 5-2. I do not go into the mathematical proof of it in this book because the focus of the book is on the practical implementations of the algorithms.

Table 5-2. *Convergence of Prediction/Estimation Algorithms*

Policy Type	Algorithm	Table Lookup	Linear	Nonlinear
On-policy	MC	Y	Y	Y
	TD(0)	Y	Y	N
	TD(λ)	Y	Y	N
	Gradient TD	Y	Y	Y
Off-policy	MC	Y	Y	Y
	TD(0)	Y	N	N
	TD(λ)	Y	N	N
	Gradient TD	Y	Y	Y

Continuing further, Table 5-3 lists the convergence of control algorithms.

Table 5-3. *Convergence of Control Algorithms*

Algorithm	Table Lookup	Linear	Nonlinear
MC control	Y	(Y)	N
On Policy TD (SARSA)	Y	(Y)	N
Off policy Q-Learning	Y	N	N
Gradient Q-Learning	Y	Y	N

(Y) fluctuates around the near-optimal value function. The guarantee of convergence is off in all nonlinear cases.

Batch Methods (DQN)

Up to now, the chapter has been focusing on algorithms that were incremental—that is, you sampled transition and then using the values you updated the weight vector w with the help of stochastic gradient descent. But this approach is not sample efficient. You discard a sample after using it only once. However, with nonlinear function approximation, especially with neural networks, you need multiple passes over the network to make the network weights converge to true values. Further, in many real-life scenarios like robotics, you need sample efficiency on two counts: neural networks with the slow convergence and hence the need for multiple passes, and generating samples in real life being very slow. In this section on batch reinforcement methods, I take you through the use of batch methods specifically with Deep Q Networks, which are the deep neural network version of the off-policy Q-learning.

Like before, you estimate the state value using function approximation as shown in Equation 5-1: $\hat{v}(s,w) \approx v_\pi(s)$.

Consider that you somehow know the actual state value $v_\pi(s)$, and you are trying to learn the weight vector w to arrive at a good estimate $\hat{v}(s,w) \approx v_\pi(s)$. You collect a batch of experiences.

$$D = \left\{ <s_1, v_1^\pi>, <s_2, v_2^\pi>, \ldots, <s_T, v_T^\pi> \right\}$$

You then form a least squares loss as the average of the difference between the true value and the estimate and then carry out gradient descent to minimize error. You use a mini-batch gradient descent to take a sample of past experiences and use that to move the weight vector using a learning rate α.

$$LS(w) = E_D \left[\left(v_\pi(s) - \hat{v}(s;w) \right)^2 \right]$$

Then approximate the expectation as the mean of the samples:

$$LS(w) = \frac{1}{N} \sum_{i=1}^{N} \left(v_\pi(s_i) - \hat{v}(s_i;w) \right)^2 \tag{5-22}$$

Taking gradient of $LS(w)$ with respect to w and using the negative gradient to adjust w, you get the weight update Equation 5-23. Note that it is similar to Equation 5-7

$$w_{t+1} = w_t + \alpha \cdot \frac{1}{N} \sum_{i=1}^{N} \left[v_\pi(s_i) - \hat{v}(s_i;w) \right] \cdot \nabla_w \hat{v}(s_i;w) \tag{5-23}$$

As before, you can move from state values $v(s; w)$ to state-action values $q(s, a; w)$ and carry out a similar update with q-values. To remind the readers once more, the purpose of moving from state values to state-action values is to be able to find the optimal policy for a given state. As you have q-values for all possible actions in a state, you can choose the optimal action in that state, do so for all states and then again carry out the estimation/prediction of q-values followed by optimal action selection. Keep repeating these until some kind of convergence of policy and q-values.

$$w_{t+1} = w_t + \alpha \cdot \frac{1}{N} \sum_{i=1}^{N} \left[q_\pi(s_i, a_i) - \hat{q}(s_i, a_i; w) \right] \cdot \nabla_w \hat{q}(s_i, a_i; w) \tag{5-24}$$

If you remember from past derivations, you do not know the true value functions, $v_\pi(s_i)$ or $q_\pi(s_i, a_i)$. As before, you replace the true values with the estimates using either the MC or TD approach. Let's now look at a flavor of this called DQN, the deep learning version of Q-learning shown in Chapter 4. In DQN, an off-policy algorithm, you sample the current state s, take a step a as per the current behavior policy, and an ϵ-greedy policy using current q-values. You observe the reward r and next state s'. You use $\max_{a'} \hat{q}(s', \cdot; w)$ for all actions a' possible in the state s' to form the target. This is shown here:

$$q_\pi(s_i, a_i) = r_i + \max_{a'} \hat{q}\left(s_i', a_i'; w_t^-\right)$$

Here you have used a different weight vector w_t^- to calculate the estimate of the target. Essentially, you have two networks, one called *online* with weights w, which is being updated as per Equation 5-24, and a second similar network called *target network* but with a copy of the weight called w^-. Weight vector w^- is updated less frequently, say after every 100 updates of online network weight w. This approach keeps the target network constant and allows you to use the machinery of supervised learning. Also note that the subscript i denotes the sample in the mini-batch and t denotes the index at which weights are updated. The final update equation with all this put together can be written as follows:

$$w_{t+1} = w_t + \alpha \cdot \frac{1}{N} \sum_{i=1}^{N} \left[r_i + \gamma \max_{a_i'} \tilde{q}\left(s_i', a_i'; w_t^-\right) - \hat{q}(s_i, a_i; w_t) \right] \cdot \nabla_{w_t} \hat{q}(s_i, a_i; w_t) \tag{5-25}$$

This equation looks very daunting. Let's unpack it. The expression $r_i + \gamma \max_{a_i'} \tilde{q}\left(s_i', a_i'; w_t^-\right)$ is the target value for $q(s_i, a_i)$ and is formed using a different neural network with weights w_t^-. The current q-value estimate is given by $\hat{q}(s_i, a_i; w_t)$, which is done using a network with weights w_t. You calculate the error by subtracting current estimate from the target and then multiply this with the gradient of current estimate—that is, $\nabla_{w_t} \hat{q}(s_i, a_i; w_t)$. This is then averaged over all the N samples and added to current weight w_t using learning rate α. To collect the samples, you run the agent through the environment using the ε-greedy policy and collect experiences in a buffer called replay buffer D.

You also update target network weights w_t^- once in a while (say after every 100 batch updates of w_t). You use the updated q-values with ε-exploration to add more experiences to the replay buffer and carry out the whole cycle once again. This in essence is the DQN approach. There is a lot more to say about this in the next chapter, which is completely devoted to DQN and its variants. For now, I leave the topic here and move on.

Linear Least Squares Method

Experience replay as used in the batch method finds the least square solution, minimizing the error between the target as estimated using TD or MC and the current value function estimate. However, it takes many iterations to converge. However, if you use linear function approximation for value function $\hat{v}(s;w) = x(s)^T w$ for prediction and $\hat{q}(s,a;w) = x(s,a)^T w$ for control, you can find the least square solution directly. Let's look at the prediction first.

Start with Equation 5-22 and substituting $\hat{v}(s;w) = x(s)^T w$, you get this:

$$LS(w) = \frac{1}{N} \sum_{i=1}^{N} \left(v_\pi(s_i) - x(s_i)^T . w \right)^2$$

Taking gradient of $LS(w)$ with respect to w and setting it to zero, you get the following:

$$\sum_{i=1}^{N} x(s_i) v_\pi(s_i) = \sum_{i=1}^{N} x(s_i) . x(s_i)^T w$$

Solving for w gives the following:

$$w = \left(\sum_{i=1}^{N} x(s_i) . x(s_i)^T \right)^{-1} \sum_{i=1}^{N} x(s_i) v_\pi(s_i) \qquad (5\text{-}26)$$

The previous solution involves the inversion of a NxN matrix that requires $O(N^3)$ computations. However, using Shermann-Morrison, you can solve this in $O(N^2)$ time. As before, you do not know the true value $v_\pi(s_i)$. You replace the true value with its estimate using MC, TD(0), or TD(λ) estimates giving you linear least square MC (LSMC), LSTD, or LSTD(λ) prediction algorithms.

LSMC: $v_\pi(s_i) \approx G_i$

LSTD: $v_\pi(s_i) \approx R + \gamma \hat{v}(s_i'; w)$

LSTD(λ): $v_\pi(s_i) \approx G_i^\lambda$

All these prediction algorithms have good convergence both for off-policy and on-policy.

Moving ahead, you extend the analysis to control using the q-value linear function approximation and GPI where the previous approach is used for the q-value prediction, followed by the greedy q-value maximization in the policy improvement step. This is known as *linear least square policy iteration* (LSPI). You iterate through these cycles of prediction followed by improvement until the policy converges—that is, until the weights converge. The final result of linear least square Q-learning (LSPI) is shown here, without going through the derivation.

Here is the prediction step:

$$w = \left(\sum_{i=1}^{N} x(s_i, a_i) \cdot \left(x(s_i, a_i) + \gamma \, x(s_i', \pi(s_i')) \right)^T \right)^{-1} \sum_{i=1}^{N} x(s_i, a_i) r_i$$

where (i) is the ith sample (s_i, a_i, r_i, s_i') from experience replay D and $\pi(s_i') = \arg\max_{a'} \hat{q}(s_i', a_i'; w)$ —that is, the action with maximum q-value in state s'.

And here is the control step:

$$\pi(s_i') = \arg\max_{a'} \hat{q}\left(s_i', a_i'; w^{\text{updated}} \right)$$

In the control step, for each state s, you change the policy that maximizes the q-value after the weight update w carried out in the previous prediction step.

As you move farther into the book, the function approximation of either state values $v(s)$, state-action values $q(s, a)$ and policy $\pi(s|a)$ will be based on some kind of neural network model. The training of these is governed by use of deep learning frameworks like PyTorch, TensorFlow, or JAX. In this edition of the book, I largely focus on PyTorch, which has gained significant traction since the first edition of this book. PyTorch and its related ecosystem of libraries is dominating at the moment. For completeness' sake, I do cover TensorFlow variants occasionally. To prepare you, the next section is a quick introduction to these underlying deep learning frameworks.

Deep Learning Libraries

Previous sections in the chapter showed that, to use the function approximation approach, you need to have an efficient way to calculate the derivatives of state-value function $\nabla_w \hat{v}(s;w)$ or the action-value function $\nabla_w \hat{q}(s,a;w)$. If you are using neural networks, you use back-propagation to calculate these derivatives at each layer of the network. This is where libraries like PyTorch and TensorFlow come into the picture. Similar to the NumPy library, they also carry out vector/matrix computations in an efficient manner. Further, these are highly optimized to handle tensors (arrays with more than two dimensions), both using CPU and GPUs.

In neural networks, you need to be able to back-propagate to calculate the gradients of error with respect to the weights of layers at all levels. Both these libraries are highly abstracted and optimized to handle this for you behind the scenes. You just need to build the forward computation, taking input through all the computations to give the final output. The libraries keep track of the computation graph and allow you to carry out the gradient update to weights with just one function call.

This section takes you through the popular frameworks such as PyTorch, PyTorch Lightning, and TensorFlow. It is introductory level coverage. If you are familiar with these libraries, you can skip this section. Also, if you are interested in diving deep into the topic, you'd do well to take up further studies after getting the basic introduction. You can refer to the resources on the Internet for more details.

PyTorch

PyTorch is a framework that allows manipulation of tensors using constructs like NumPy. In addition, it has modules to allow auto differentiation to carry out back-propagation, which forms the backbone of training a neural network.

What Are Neural Networks

Deep learning is based on artificial neural networks, which are made up of neurons. A neuron takes inputs, calculates the weighted sum, and then passes the sum through some kind of nonlinear function $f(h)$ (called the activation function), as shown in Figure 5-11.

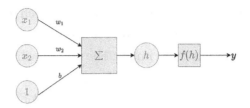

Figure 5-11. *A neuron in a neural network*

The training in deep learning refers to finding the weight vector $[w_1, w_2, b]$ using back-propagation. Of course, the neural network model is composed of billions of such neurons grouped into layers and many such layers are stacked together. The arrangement of the neural network layers has many varieties, giving rise to architectures such as CNN (Convolutional Neural Networks), RNN (Recurrent Neural Networks), and in last few years to the transformers architecture. The world of large language models (LLMs) is dominated by transformers at its core and comprises tens and hundreds of billions of weight parameters.

A basic level architecture is comprised of simple layers, with all neurons from the previous layers connected to all the neurons of the current layer. This formation is called a *linear dense layer architecture*, as shown in Figure 5-12.

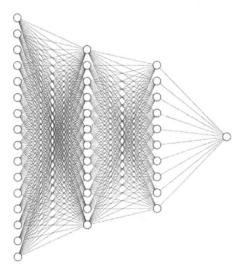

Figure 5-12. *A simple, fully connected neural network*

You can take the MNIST image dataset and train a PyTorch model to predict the digit based on the grayscale image that's 24x24 pixels. Accordingly, the input layer will have 768 inputs translating to flattened 24x24 image. The output will have ten units, which will produce the probability that input image is a "1" or "2" and so on. The complete code of end-to-end PyTorch training is given in the Python notebook called `5.c-Intro-to-pytorch.ipynb`.

Several hidden layers take 768 dimensional inputs to the first 192 units and then to the 128 units, with the final output layer taking 128 units to 10 units. You create the model in PyTorch by extending the PyTorch `Model` class and implementing the forward method to take input image vector and output a ten-dimensional output vector. Listing 5-8 shows the code that creates this network in PyTorch. The `__init__` function defines the layers of the network using the PyTorch supplied `Linear` layer module. The `forward` function implements the flow of input tensor through the network, outputting the ten-dimensional output vector. It returns the raw score, which is converted into probability in the `predict` function with the help of a `softmax` function. The `predict` function is just for convenience. It is not required. The only two things that are required are defining the layers of the network in `init` and implementing the `forward` function.

Listing 5-8. Custom Network for MNIST Digit Prediction from `5.c-Intro-to-pytorch.ipynb`

```
class NN(nn.Module):
    def __init__(self):
        super().__init__()
        self.fc1 = nn.Linear(784, 192)
        self.fc2 = nn.Linear(192, 128)
        self.fc3 = nn.Linear(128, 10)

    def forward(self, x):
        ''' Forward pass through the network, returns the output logits '''
        x = self.fc1(x)
        x = F.relu(x)
        x = self.fc2(x)
        x = F.relu(x)
        x = self.fc3(x)

        return x
```

```
    def predict(self, x):
        ''' To predict classes by calculating the softmax '''
        logits = self.forward(x)
        return F.softmax(logits, dim=1)

model = NN()
model
```

You load the MNIST dataset and prepare dataloaders to supply images in batches to the training code, as shown in Listing 5-9. It consists of use of three high-level functions—transforms transforms the images and does things like normalization and conversion of image to PyTorch tensors, datasets downloads the dataset locally and then wraps the downloaded dataset into a DataLoader, which supplies images to the rest of the code in batches. These are highly optimized functions with many parameters to tweak the behavior based on the specific need. You can read more about these in the PyTorch documentation.

Listing 5-9. Data Loading in PyTorch from 5.c-Intro-to-pytorch.ipynb

```
# Define a transform to normalize the data
transform = transforms.Compose([transforms.ToTensor(),
                                transforms.Normalize(0.5, 0.5),
                               ])
# Download and load the training data
trainset = datasets.MNIST('MNIST_data/', download=True, train=True,
transform=transform)
trainloader = torch.utils.data.DataLoader(trainset, batch_size=128,
shuffle=True)

# Download and load the test data
testset = datasets.MNIST('MNIST_data/', download=True, train=False,
transform=transform)
testloader = torch.utils.data.DataLoader(testset, batch_size=128,
shuffle=True)
```

Training with Back-Propagation

After having defined the model and prepared the data, the next step is to train the model. This involves first defining the optimizer that will be used to keep track of gradients during back-propagation and the code to adjust the weights on the neural network. Adam is a very popular optimizer with advanced and adaptive capabilities to adjust the training step α as training progresses. There are many more optimizers. Optimizer design is a big field of research on its own. You also define a loss function, which must be minimized as part of training the network. Listing 5-10 shows the code for these steps. This is then followed by the training code, in which you go over the training data in batches, calculate the loss function value, and use the optimizer to do a back-propagation, adjusting the weights.

Listing 5-10. Model Training in PyTorch from `5.c-Intro-to-pytorch.ipynb`

```
# Create an optimizer to train the network by carrying out back propagation
model = NN()
optimizer = optim.Adam(model.parameters(), lr=0.001)
loss_fn = nn.CrossEntropyLoss()

# Train network
epochs = 1
steps = 0
running_loss = 0
eval_freq = 10
for e in range(epochs):
    for images, labels in iter(trainloader):
        steps += 1
        images.resize_(images.size()[0], 784)

        optimizer.zero_grad()

        output = model.forward(images)
        loss = loss_fn(output, labels)
        loss.backward()
        optimizer.step()

        running_loss += loss.item()
```

```
if steps % eval_freq == 0:
    # Test accuracy
    accuracy = 0
    for ii, (images, labels) in enumerate(testloader):

        images = images.resize_(images.size()[0], 784)
        predicted = model.predict(images).data
        equality = (labels == predicted.max(1)[1])
        accuracy += equality.type_as(torch.FloatTensor()).mean()

    print("Epoch: {}/{}".format(e+1, epochs),
          "Loss: {:.4f}".format(running_loss/eval_freq),
          "Test accuracy: {:.4f}".format(accuracy/(ii+1)))
    running_loss = 0
```

Once the model is trained, you can pass a new image and observe the probability distribution of the model's prediction. I created a custom function called view_classification for the same. To pass images through the network and observe the outcome, you need a short three-line code, as shown in Listing 5-11.

Listing 5-11. Prediction Using Trained Model from 5.c-Intro-to-pytorch.ipynb

```
logits = model.forward(img[None,])
```

```
# Predict the class from the network output
prediction = F.softmax(logits, dim=1)
```

```
view_classification(img.reshape(1, 28, 28), prediction[0])
```

The output of passing an image of 4 is shown in Figure 5-13.

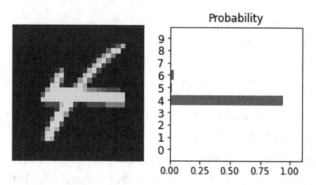

Figure 5-13. *Prediction of trained model on input image of the "4" digit*

In summary, the steps involved are 1) designing the model based on the problem and keeping specific input size and output size in mind; 2) preparing the training data; 3) defining the loss function and optimizer and finally; 4) the training code to train the model. You will be following these steps in increasing complex ways depending on the need of the agent learning algorithm in question.

PyTorch Lightning

PyTorch Lightning is a library written to speed up model development and training in PyTorch. PyTorch Lightning is the deep learning framework with "batteries included" for professional AI researchers and machine learning engineers who need maximal flexibility while super-charging performance at scale.

Lightning organizes PyTorch code to remove boilerplate code and unlock scalability. While doing so, it still offers full flexibility to try any idea without the boilerplate. It helps you focus on the core parts of your model training and provides you with a standard way to organize your code.

Most of the code will remain the same as you saw in the raw PyTorch notebook. The code to prepare data loaders is the same as before. You first create a network model in PyTorch like before. Just for variety, I have used another shorter version to create the neural network, this time using the nn.Sequential class from PyTorch, as shown in Listing 5-12. The complete code for end-to-end training using PyTorch Lightning is given in the 5.d-Intro-to-pytorch-lighning.ipynb notebook.

Listing 5-12. PyTorch Model from `5.d-Intro-to-pytorch-lightning.ipynb`

```
model = nn.Sequential(
           nn.Linear(28 * 28, 192),
           nn.ReLU(),
           nn.Linear(192, 128),
           nn.ReLU(),
           nn.Linear(128,10)
)

# print model summary
model

#### OUTPUT ####
Sequential(
  (0): Linear(in_features=784, out_features=192, bias=True)
  (1): ReLU()
  (2): Linear(in_features=192, out_features=128, bias=True)
  (3): ReLU()
  (4): Linear(in_features=128, out_features=10, bias=True)
)
```

When using PyTorch Lightning, you create a lightning model by extending the `pl.LightningModule` and implementing specific functions. This removes the need for the long boilerplate code you have for training when using pure PyTorch without Lightning. In this case, as shown in Listing 5-13, you create a `LightningModule` called `MNISTClassifier`. In the `__init__` function, you initialize the model defined in Listing 5-12. Next, you define the `training_step` function, which takes a batch of input training data, passes it through the model, and calculates the loss, which needs to be minimized. This function must return the loss which will be used by the optimizer to calculate the gradients during back-propagation. You then use `self.log` to log the results, which under the hood uses TensorBoard to log. You also have the choice to forward logging data to "weights and biases" or such other logging and experimentation services. Next, you implement a `test_step` which largely mimics the `training_step` producing metrics that you want to track during model performance evaluation. The last function you define is `configure_optimizers` to define the optimizer that will be used during back-propagation of gradients during the training phase.

Listing 5-13. PyTorch Lightning Module from 5.d-Intro-to-pytorch-
lightning.ipynb

```python
# define the LightningModule
class MNISTClassifier(pl.LightningModule):
    def __init__(self, model):
        super().__init__()
        self.model = model

    def training_step(self, batch, batch_idx):
        # training_step defines the train loop.
        # it is independent of forward
        x, y = batch
        x = x.view(x.size(0), -1)
        logits = self.model(x)
        loss = F.cross_entropy(logits,y)
        # Logging to TensorBoard (if installed) by default
        self.log("train_loss", loss)
        return loss

    def test_step(self, batch, batch_idx):
        # training_step defines the train loop.
        # it is independent of forward
        x, y = batch
        x = x.view(x.size(0), -1)
        logits = self.model(x)
        loss = F.cross_entropy(logits, y)
        self.log("test_loss", loss)

    def configure_optimizers(self):
        optimizer = optim.Adam(self.parameters(), lr=0.001)
        return optimizer

# init the MNIST Classifier
classifier = MNISTClassifier(model)
```

With PyTorch Lightning, once the Lightning module is defined, it requires one line of code to run the training code. There is no need to load data in a loop and run the backward step manually in the code. Listing 5-14 shows the training code you need to execute to train the model. You just need to define a `trainer` and call the `fit` method on it by passing the `PyTorchLightning` module defined earlier and the `dataloaders` to it.

Listing 5-14. PyTorch Lightning Training from `5.d-Intro-to-pytorch-lightning.ipynb`

```
# train the model (hint: here are some helpful Trainer arguments for rapid
idea iteration)
trainer = pl.Trainer(max_epochs=2)
trainer.fit(model=classifier, train_dataloaders=trainloader)
```

The prediction code using the trained model remains the same as in PyTorch. Figure 5-14 shows the performance of the trained model for one sample of input.

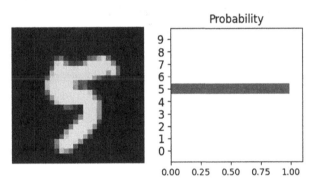

Figure 5-14. *Prediction of a trained model on input image of a "5" digit*

As you can see, the use of PyTorch Lightning helps reduce the boilerplate code overhead.

TensorFlow

TensorFlow also largely follows the same pattern as PyTorch. It has its own set of data loaders, model building patterns, loss and optimizer definitions, and training loops. I'll do a quick dive. The end-to-end training using TensorFlow is given in the `5.e-Intro-to-tensorflow.ipynb` notebook. Listing 5-15 shows the code for downloading and creating the MNIST dataset under TensorFlow.

Listing 5-15. TensorFlow Data Loading from `5.e-Intro-to-tensorflow.ipynb`

```
mnist = tf.keras.datasets.mnist

(x_train, y_train), (x_test, y_test) = mnist.load_data()
x_train, x_test = x_train / 255.0, x_test / 255.0
```

As with PyTorch, you define the neural network model using the `tf.keras.model.Sequential` class, as shown in Listing 5-16. It also gives the code for remaining steps— that is, defining the loss function as well as tying the model with the optimizer and loss function using the `model.compile` function call. Finally, like PyTorch Lightning, you use `model.fit` to train the model using the training data.

Listing 5-16. TensorFlow Model Training from `5.e-Intro-to-tensorflow.ipynb`

```
model = tf.keras.models.Sequential([
  tf.keras.layers.Flatten(input_shape=(28, 28)),
  tf.keras.layers.Dense(192, activation='relu'),
  tf.keras.layers.Dense(128, activation='relu'),
  tf.keras.layers.Dense(10)
])

# Create loss_fn
loss_fn = tf.keras.losses.SparseCategoricalCrossentropy(from_logits=True)

# Train model
model.fit(x_train, y_train, epochs=2)

# evaluate model
model.evaluate(x_test,  y_test, verbose=2)
```

The code to evaluate the output of the trained model on a sample image is shown in Listing 5-17. It is largely the same as what you saw in the case of PyTorch.

Listing 5-17. TensorFlow Model Prediction from `5.e-Intro-to-tensorflow.ipynb`

```
# check the prediction on same image as the one used before training
logits = model(img)

# Predict the class from the network output
prediction = tf.nn.softmax(logits).numpy()
view_classification(img[0], prediction[0])
```

Like PyTorch or PyTorch Lightning, the model's accuracy is very high after training. The probability distribution of the model's output in a sample image mimics the kind of values you see in Figures 5-13 and 5-14.

This completes the introduction of the popular three frameworks. There are many more that are emerging, such as JAX, FastAI, Glueon from AWS, Microsoft, and so on. I will stick to PyTorch most of the time and may occasionally show some sample code using TensorFlow.

Summary

In this chapter, the primary focus was to look at the use of function approximation for very large or continuous state spaces that cannot be handled using the approach of table-based learning that you saw in previous chapters.

The chapter talked about what it means to carry out optimization with function approximation. I also showed how the concepts of training in supervised learning, training a model to produce values close to targets, can be applied in reinforcement learning. The chapter emphasized the need for the proper handling due to moving targets and the sample interdependence that is present in RL.

You then looked at various strategies of functional approximation, including linear and nonlinear ones. I also demonstrated that table-based methods are just special cases of linear approximation. This was followed by a detailed discussion of incremental methods for prediction and control. You saw these being applied to the mountain car for build training agents using n-step SARSA and SARSA(λ).

Next, the chapter talked about batch methods in general and explored the complete derivation of the update rule for DQN, a popular algorithm from the family of batch methods. You then looked at linear least squares methods for prediction and control. Along the way, I kept highlighting the convergence issues in general as well as convergence in the specific methods being discussed.

I concluded the chapter with a brief introduction to deep learning frameworks like PyTorch, PyTorch Lightning, and TensorFlow.

CHAPTER 6

Deep Q-Learning (DQN)

This chapter takes a deep dive into Q-learning combined with function approximation using neural networks. Q-learning in the context of deep learning using neural networks is also known as *Deep Q Networks* (DQN). The chapter first summarizes what I have talked about so far with respect to Q-learning. You will then look at code implementations of DQN on simple problems. Following this, you will take stock of Gymnasium and learn how it differs from OpenAI Gym. The next focus is on the Stable Baselines 3 (SB3) library and its associated ecosystem. You will also explore hyperparameter optimization using Optuna and plotting, both in the context of SB3. With the basic machinery in place, I will extend DQN to cover Atari game playing agent. The topics of DQN are followed by a quick exploration of various other RL environments and associated libraries covering the use of RL in financial market trading and in the field of robotics. However, be aware that these are still research environments useful to gain more insights. As some of these have multi-dimensional continuous value actions, there are better algorithms than DQN for training agents. This chapter focuses on learning about these environments and how to set them up. You will apply DQN where appropriate and detailed coverage of those algorithms appears in future chapters.

Finally, most of these algorithms will be covered using PyTorch, which in recent years has become the go-to deep learning framework displacing TensorFlow. TensorFlow has seen a steady decline over the last four to five years. The chapter includes an example in TensorFlow and, through that, you will realize that that big part of expertise in RL lies in problem framing and having a good understanding of the training algorithms. The role of deep learning framework is rather limited. Once you master the concepts using one framework, it is very easy to switch to another framework like TensorFlow or JAX.

225

© Nimish Sanghi 2024

N. Sanghi, *Deep Reinforcement Learning with Python*, https://doi.org/10.1007/979-8-8688-0273-7_6

Deep Q Networks

To introduce DQN, I first recap what you have learned so far. Chapter 4 talked about Q-learning as a model-free off-policy TD control method. You first looked at the online version, where you used an exploratory behavior policy (ϵ-greedy) to take a step (i.e., take action A) while in state S. The reward R and next state S' were then used to update the q-value $Q(S, A)$. Figure 4-16 explained the approach in pseudocode and Listing 4-6 showed the code implementing the same. The update rule, as shown in Equation 4-10, is reproduced here. You are encouraged to review the section "Q-Learning: An Off-Policy TD Control" in Chapter 4 before you proceed.

$$Q(S_t, A_t) \leftarrow Q(S_t, A_t) + \alpha \cdot \left[R_{t+1} + \gamma \cdot \max_{A_{t+1}} Q(S_{t+1}, A_{t+1}) - Q(S_t, A_t) \right] \qquad (6\text{-}1)$$

I briefly talked about maximization bias and the approach of double Q-learning, wherein you use two tables of q-values. Following this, you looked at the approach of using a sample multiple times to convert online TD updates to batch TD updates, making it more sample efficient. This introduced the concept of a replay buffer. While it was only about sample efficiency in the context of discrete state and state-action spaces, with function approximation using neural networks, it becomes pretty much a must-have to make deep learning neural networks converge. You will revisit this again, and when I talk about prioritized replay, you will look at other options to sample from the replay buffer.

Moving along, in Chapter 5, you looked at various approaches of function approximation. You looked at tile encoding as a way to achieve linear function approximation. I then talked about DQN, that is, batch Q-learning with neural networks as function approximators. I went through a long derivation to arrive at a weight (i.e., neural network parameters) update equation, as given in Equation 5-25. That equation is reproduced here:

$$w_{t+1} = w_t + \alpha \cdot \frac{1}{N} \sum_{i=1}^{N} \left[r_i + \gamma \max_{a_i'} \tilde{q}(s_i', a_i'; w_t^-) - \hat{q}(s_i, a_i; w_t) \right] \cdot \nabla_{w_t} \hat{q}(s_i, a_i; w_t) \qquad (6\text{-}2)$$

Note that I use the subscript i to denote the sample in the mini-batch and t to denote the time/loop index at which the weights are updated. Equation 6-2 is the one you will be using extensively in this chapter and the next. You will make various adjustments to this equation as you read about different modifications and study their impact.

I also talked about not having a theoretical guarantee of convergence under nonlinear function approximation with a gradient update. I have more to say about that in this chapter. The Q-learning approach was for discrete states and actions, where the update to the q-value was using Equation 6-1 as compared to adjusting the weight parameters for the deep learning–based approach in DQN. The Q-learning case had guarantees of convergence, while with DQN you had no such guarantee. DQN is computationally intensive as well. However, despite these shortcomings, DQN makes it possible to train agents using even raw images and not just specific observation vectors like position, velocity, angle, grid number, and so on. This is not at all conceivable in the case of plain Q-learning. Let's now put Equation 6-2 into practice to train the DQN agents in various environments.

Let's revisit the `CartPole` problem, which has a four-dimensional continuous state with values for the current cart position, velocity, angle of the pole, and angular velocity of the pole. The actions are of two types: push the cart to the left or push the cart to the right, with the aim of keeping the pole balanced as long as possible. This environment is part of OpenAI Gym library, which now has been forked to Gymnasium, as discussed earlier. The following are the details of the environment:

```
Observation:
    Type: Box(4)
    Num     Observation              Min          Max
    0       Cart Position            -4.8         4.8
    1       Cart Velocity            -Inf         Inf
    2       Pole Angle               -0.418 rad   0.418 rad
    3       Pole Angular Velocity    -Inf         Inf

Actions:
    Type: Discrete(2)
    Num        Action
    0          Push cart to the left
    1          Push cart to the right
```

You will build a small neural network with four as the input dimension since the state/observation vector is four dimensional, two hidden layers, followed by an output layer of dimension two, as the number of possible actions is two. There are two output q-values, that is $Q(S, A)$, one for each action. Figure 6-1 shows the network diagram.

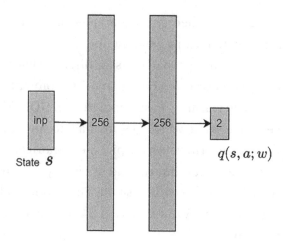

Figure 6-1. *The simple neural network*

There are many ways to build neural networks in PyTorch. The most flexible way is to sub-class the torch.nn module. In the init method of subclass, you define the building blocks like the linear layers, activation functions, and so on. Then you implement the forward method, the parent abstract method, which takes the input, passes it through the layers of the network defined in the init method, and returns the final output. The second method defines a list of all the layers in the sequence, all the way from input to output and passes that list to nn.Sequential. The second approach is a quicker one with minimal code, although it does not give you flexibility especially when it comes to multiple inputs, outputs, and skip connections.

This chapter uses the first approach, with the second approach embedded inside the init function, as shown in Listing 6-1. You also define a few additional convenience methods as part of the network. You will define three such methods. The first one is get_ qvalues and it takes state/observation as a numpy array instead of a torch tensor that the forward expects. It then converts the numpy array to torch tensor, passes it through the forward method, and then again converts the returned PyTorch tensor from forward back to numpy. Such a method makes it easier to use the network inside the environment step function code, which otherwise would require you to convert the tensor to a NumPy or Python array while implementing the logic to step through the environment. Ideally these kinds of manipulations are best abstracted away and kept close to the network PyTorch code. If you use GPU, such a structure of the code makes it easier to handle the movement of batch of observations from CPU to GPU, process it through the network, and then shift it back to CPU making the back and forth move from CPU to GPU transparent to rest of the code. You can see this pattern in the first and third line of code inside the get_qvalues function shown in Listing 6-1. While moving from NumPy to tensor you use the device argument to move the observations to the appropriate device that's hosting the neural network. In line three, you use the Tensor.cpu() chained call to move the q-values output by network back to CPU. The get_qvalues function takes a batch of states/observations as input, that is, a tensor of dimension $(N \times 4)$ where N is the number of samples. It passes the state values through the network to produce the q-values. The output vector has size $(N \times 2)$; one row for each input. Each row has two q-values, one for the left-push action and another for the right-push action.

The second function you'll implement is sample_actions. It implements the ϵ-greedy policy. It uses the q-values returned by the get_qvalues function as a NumPy array to choose the best action $(1 - \epsilon)$ fraction times and choose a random action ϵ fraction of times. It is what you use during learning to keep exploring the unvisited states. As discussed earlier you start with a high ϵ value like 0.6-0.8 and gradually reduce it down close to 0, with 0.05 being a popular choice. The function takes in a batch of q-values $(N \times 2)$. It uses the ε-greedy policy (equation 4-3) to choose an action. The output is the $(N \times 1)$ vector.

The third function, called get_action, is a truncated version of sample_actions. You use a greedy policy with no exploration. This is implemented in the get_action function. There is no random selection. In this function you deterministically and with zero exploration always return the best action with maximum q-value. As with sample_actions, the get_action function also takes a batch of q-values $(N \times 2)$ and produces $(N \times 1)$ action vector, one action per N observations in the batch. This function is used post-training to get the best action for a given state.

Listing 6-1 shows the code in PyTorch with the full code available in the 6.a-dqn-pytorch.ipynb notebook.

Listing 6-1. DQN Network from 6.a-dqn-pytorch.ipynb

```
class DQNAgent(nn.Module):
    def __init__(self, state_shape, n_actions, epsilon=0):

        #... boiler plate code omitted
        # ...
        # a simple NN
         self.network = nn.Sequential()
         self.network.add_module('layer1', nn.Linear(state_dim, 256))
```

```python
        # code omitted ---- all other layers of network come here
        self.network.add_module('layer3', nn.Linear(256, n_actions))
        self.parameters = self.network.parameters

    def forward(self, state_t):
        # pass the state at time t through the network to get Q(s,a)
        qvalues = self.network(state_t)
        return qvalues

    def get_qvalues(self, states):
        # input is an array of states in numpy and output is Qvals as
        numpy array
        states = torch.tensor(np.array(states), device=device, dtype=torch.
        float32)
        qvalues = self.forward(states)
        return qvalues.data.cpu().numpy()

    def get_action(self, states):
        states = torch.tensor(np.array(states), device=device, dtype=torch.
        float32)
        qvalues = self.forward(states)
        best_actions = qvalues.argmax(axis=-1)
        return best_actions

    def sample_actions(self, qvalues):
        # sample actions from a batch of q_values using epsilon
        greedy policy
        epsilon = self.epsilon
        batch_size, n_actions = qvalues.shape
        random_actions = np.random.choice(n_actions, size=batch_size)
        best_actions = qvalues.argmax(axis=-1)
        should_explore = np.random.choice(
            [0, 1], batch_size, p=[1-epsilon, epsilon])
        return np.where(should_explore, random_actions, best_actions)
```

Note   While not essential, you will gain more from the code discussions if you have some prior knowledge of PyTorch or TensorFlow. You should be able to create basic networks, define loss functions, and carry out basic training steps for optimization.

Next, you look at the way a similar structure is created in TensorFlow. TensorFlow's eager mode is very much like PyTorch. In TensorFlow, you have a similar sequential model that's available under `tf.keras.models.Sequential`. In TensorFlow, you have to implement a `__call__` method instead of the `forward` method that you used in PyTorch. The rest of the code looks very similar and therefore I do not list the code here. Interested readers can refer to the `6.b-dqn-tensorflow.ipynb` notebook for full details.

The code for the replay buffer is simple. You have an array called `self.buffer` to hold the previous examples. The `add` function takes in `state, action, reward, next_state, and done`, the values from a single step/transition by the agent, and adds it as a tuple to the buffer. If the buffer has reached its full length, it discards the oldest transition/tuple to make space for the new addition. The `sample` function takes an integer `batch_size` and returns `batch_size` samples/transitions from the buffer. In this vanilla implementation, each transition stored in the buffer has equal probability of getting sampled. Listing 6-2 shows the code for the replay buffer from the `6.a-dqn-pytorch.ipynb` notebook.

Listing 6-2. Replay Buffer (Same in PyTorch or TensorFlow) from `6.a-dqn-pytorch.ipynb`

```
class ReplayBuffer:
    def __init__(self, size):
        self.size = size #max number of items in buffer
        self.buffer =[] #array to hold buffer
        self.next_id = 0

    #... boiler plate code omitted

    def add(self, state, action, reward, next_state, done):
        item = (state, action, reward, next_state, done)
        if len(self.buffer) < self.size:
            self.buffer.append(item)
```

```
        else:
            self.buffer[self.next_id] = item
        self.next_id = (self.next_id + 1) % self.size

    def sample(self, batch_size):
        idxs = np.random.choice(len(self.buffer), batch_size)
        samples = [self.buffer[i] for i in idxs]
        states, actions, rewards, next_states, done_flags =
        list(zip(*samples))
        return np.array(states), np.array(actions),
        np.array(rewards), np.array(next_states), np.array(done_flags)
```

Moving ahead, you have a utility function called play_and_record that takes in an env (e.g., CartPole), an agent (e.g., DQNAgent), an exp_replay (ReplayBuffer), the agent's start_state, and n_steps (i.e., number of steps/actions to take in the environment). The function makes the agent take n_steps number of steps, starting from the initial state start_state. The steps are taken based on the current ε-greedy policy that the agent is following using agent.sample_actions and records these n_steps transitions in the buffer. Refer to the notebook from Listing 6-1 (6.a-dqn-pytorch.ipynb) for the detailed code showing how this is implemented.

Next, you look at the learning process. You first build the expression for loss, L, that you want to minimize. It is the averaged squared error between the target value of the current state action using a one-step TD value and the current state value. As discussed in Chapter 5, you use a copy of the original neural network that has weights w^- (w with superscript $-$). You use the loss to calculate the gradient of the agent (online/original) network's weight w and take a step in the negative direction of the gradient to reduce the loss. Note that as discussed in Chapter 5, you keep the target network with weights w^- frozen and update these weights on a less frequent basis. To quote from the section on batch methods of DQN in Chapter 5:

Here you have used a different weight vector w_t^- to calculate the estimate of the target. Essentially, you have two networks, one called online with weights w, which is being updated as per Equation 5-24, and a second similar network called target network but with a copy of the weight called w^-. Weight vector w^- is updated less frequently, say after every 100 updates of online network weight w. This approach keeps the target network constant and allows you to use the machinery of supervised learning.

The loss function L is as follows:

$$L = \frac{1}{N}\sum_{i=1}^{N}\left[r_i + \left(1 - \text{done}_i\right).\gamma.\max_{a'}\hat{q}\left(s_i', a_i'; w_t^-\right) - \hat{q}\left(s_i, a_i; w_t\right)\right]^2 \tag{6-3}$$

You take a gradient (derivative) of loss L with respect to w and then use this gradient to update the weights w of the online network. You update the weights in the negative direction of gradient, as you want to reduce the loss by adjusting the values of weight w. This concept is from calculus that the gradient (also known as the derivative or the slope) of a function with respect to its variable gives the direction of maximum increase and therefore by adjusting the w in the negative direction of the gradient, you walk toward a new w that offers the fastest decrease in the function value.

This concept translates to the following equations:

$$\nabla_{w_t} L = -\frac{1}{N}\sum_{i=1}^{N}\left[r_i + \left(1 - \text{done}_i\right).\gamma.\max_{a'}\hat{q}(s_i', a_i'; w_t^-) - \hat{q}(s_i, a_i; w_t)\right] \cdot \nabla_{w_t}\hat{q}\left(s_i, a_i; w_t\right) \tag{6-4}$$

$\hat{q}\left(s', A; w^-\right)$ is calculated using target network whose weights are held constant and refreshed periodically from the agent learning network.

The target is given by following:

- Nonterminal state: $r_i + \gamma.\max_{a'}\hat{q}(s_i', a_i'; w_t^-)$

- Terminal state: r_i

The value of $done_i$ is 1 when the environment terminated as a result of the action a_i. In Equation 6-4, by multiplying with $(1 - done_i)$ you get the two values of the target depending on the state of $done_i$ flag.

The gradient $\nabla\hat{q}\left(s_i, a_i; w_t\right)$ is calculated by back-propagation of the loss outputted by neural network and this is one of the primary reasons for using deep learning frameworks like PyTorch and TensorFlow, which automate the calculation of these gradients and have made these frameworks so powerful.

The basic flow is to design the network to output $\hat{q}\left(s_i, a_i; w_t\right)$ and then write a loss function to calculate loss L as given in Equation 6-3, which uses the network output $\hat{q}\left(s_i, a_i; w_t\right)$ and the target values to calculate the average squared loss as per Equation 6-3. The implementation of Equation 6-4 is automatically done by deep learning frameworks with a simple call of something like loss.backward(). Once the gradient has been calculated you have the value as given in Equation 6-4. Next, you adjust the current weights of the network with a scaled amount of this gradient. The scaling factor α is the learning rate and it controls the amount of adjustment you make in each update. Usually, you start with higher α early in the training phase and gradually reduce it down as training of network progresses.

$$w_{t+1} \leftarrow w_t - \alpha \nabla_w L \qquad (6\text{-}5)$$

Listing 6-3 shows the code to calculate L, as discussed previously. This is done by the compute_td_loss function. It takes in a batch of states, actions, rewards, next_states, and done_flags. It also takes in the discount parameter γ as well as the agent/online and target networks. The function first converts the NumPy arrays to PyTorch tensors and moves them to a CPU or GPU device. Next the states tensor batch is passed through the agent network to output the predicted_qvalues for all actions. You choose the specific q-value for the action tensor batch a_i to produce $\hat{q}(s_i, a_i; w_t)$, shown in the code as tensor predicted_qvalues_for_actions. Next, you compute the predicted_next_qvalues for all actions by passing the batch of next_states through the target network. You then take a max over the actions to produce $\max_{a_i} \hat{q}(s_i', a_i'; w_t^-)$, shown in the code as next_state_values. Next, you calculate the target values $r_i + (1 - \text{done}_i) \cdot \gamma \cdot \max_{a'} \hat{q}(s_i', a_i'; w_t^-)$ shown in the code as target_qvalues_for_actions. The last line of the code calculates the loss by taking a difference of target_qvalues and predicted q_values, squaring, adding, and then averaging over the batch to return the computed loss L.

Listing 6-3. Compute TD Loss in PyTorch from 6.a-dqn-pytorch.ipynb

```
def compute_td_loss(agent, target_network, states, actions, rewards,
next_states, done_flags, gamma=0.99, device=device):

    # convert numpy array to torch tensors
    states = torch.tensor(states, device=device, dtype=torch.float)
    actions = torch.tensor(actions, device=device, dtype=torch.long)
    rewards = torch.tensor(rewards, device=device, dtype=torch.float)
    next_states = torch.tensor(next_states, device=device,
    dtype=torch.float)
    done_flags = torch.tensor(done_flags.astype('float32'),device=device,
    dtype=torch.float)
```

```
# get q-values for all actions in current states
# use agent network
predicted_qvalues = agent(states)

# compute q-values for all actions in next states
# use target network
predicted_next_qvalues = target_network(next_states)

# select q-values for chosen actions
predicted_qvalues_for_actions = predicted_qvalues[range(
    len(actions)), actions]

# compute Qmax(next_states, actions) using predicted next q-values
next_state_values,_ = torch.max(predicted_next_qvalues, dim=1)

# compute "target q-values"
target_qvalues_for_actions = rewards + gamma * next_state_values *
(1-done_flags)

# mean squared error loss to minimize
loss = torch.mean((predicted_qvalues_for_actions -
                    target_qvalues_for_actions.detach()) ** 2)

return loss
```

At this point, you have all the machinery needed to train the agent to balance the pole. But first you need to define some hyperparameters like batch_size, total training steps total_steps, and the rate at which exploration ϵ will decay. It starts at 1.0 and slowly reduces to 0.05 as the agent learns the optimal policy. You also define an optimizer, which takes in the parameters of the agent network, that is, weight vector w. This is provided by passing agent.parameters() to the constructor function of the optimizer. There are many optimizers and Adam is a popular choice. The name is derived from adaptive moment estimation. The optimizer is called Adam because it uses estimations of the first and second moments of the gradient to adapt the learning rate for each weight of the neural network. You can read more about it in the PyTorch and TensorFlow documentation. The other important training parameter that you set is the frequency the target network is updated-that is, network with weights w^-. In this case you set it to 100 using refresh_target_network_freq = 100, which means that for every 100 updates of weight w of agent network, you will update the target network. This way the target network provides a stable target for weight w to converge as well as it is not too infrequent leading to target network weights becoming stale. It is a hyperparameter and in real life you will probably be required to tune this. You also define the frequency of logging the training loss for plotting purposes and the frequency with which the trained agent will be evaluated to assess the quality of the policy agent is learning. Finally, you define `max_grad_norm` to clip the gradient's values. This is again a standard practice in deep learning, and it is a hyperparameter. All of these parameters are shown in Listing 6-4.

Listing 6-4. Training Parameters from `6.a-dqn-pytorch.ipynb`

```
#setup some parameters for training
timesteps_per_epoch = 1
batch_size = 32
total_steps = 50000

#init Optimizer
opt = torch.optim.Adam(agent.parameters(), lr=1e-4)

# set exploration epsilon
start_epsilon = 1
end_epsilon = 0.05
eps_decay_final_step = 2 * 10**4

# setup some frequency for logging and updating target network
loss_freq = 20
refresh_target_network_freq = 100
eval_freq = 1000

# to clip the gradients
max_grad_norm = 5000
```

Now you are all set to start training the agent. Listing 6-5 shows the code. You first reset the environment to obtain the initial state. After this point you will run a loop for a number of times equal to `total_steps`—that is, the number of training steps that the agent will be trained for. This is one of the parameters defined in Listing 6-4. In the loop you first set the exploration ϵ as per the settings defined in Listing 6-4. You use a one-line helper function called `epsilon_schedule` to do this. The implementation of this function is shown in Listing 6-5. You use the `play_and_record` function discussed earlier to fill the

Replay Buffer shown in Listing 6-1. You use a hyper-parameter called `timesteps_per_epoch` to control the number of steps you will take in each step of training. Usually this is set to 1 so that each time you run a training step you add one more tuple of training data to the buffer. This way, as the agent learns and explores new parts of the *state*, *action* space, the replay buffer is updated. Next, you sample a batch of stored transitions from the replay buffer with the size of buffer controlled by the `batch_size` parameter. The batch of training data so collected is passed to the `compute_td_loss` discussed in Listing 6-3. The return value stored in variable `loss` is explained in Equation 6-3. Next, you call `loss.backward()` which almost magically calculates the gradient $\nabla_w L$, as defined in Equation 6-4. This is the power of deep learning frameworks, which implement automatic differentiation. As discussed, you clip the gradients to ensure that each individual gradient value in the vector of gradients $\nabla_w L$ is within your defined bounds. Next, you call `opt.step()` to run the weight update as defined in Equation 6-5. This completes the core code in the training loop. Further in the training loop, you have some minor housekeeping code. You need to zero the vector of gradient $\nabla_w L$ after the update of the weight vector w. You also need to refresh the target network as per the `refresh_target_network_freq` parameter. Finally, you have to log the training loss, determine the mean return per episode as part of validation, and plot all these values to show the progress of training. Listing 6-5 shows all the code discussed. Some of the housekeeping code has been omitted from the listing to keep the focus sharp on what you need to understand. Readers interested in exploring the full code can refer to the `6.a-dqn-pytorch.ipynb` notebook.

Listing 6-5. Main Training Loop from `6.a-dqn-pytorch.ipynb`

```
def epsilon_schedule(start_eps, end_eps, step, final_step):
    return start_eps + (end_eps-start_eps)*min(step, final_step)/final_step

state,_ = env.reset()
for step in trange(total_steps + 1):

    # reduce exploration as we progress
    agent.epsilon = epsilon_schedule(start_epsilon, end_epsilon, step,
    eps_decay_final_step)

    # take timesteps_per_epoch and update experience replay buffer
    _, state = play_and_record(state, agent, env, exp_replay, timesteps_
    per_epoch)
```

```
# train by sampling batch_size of data from experience replay
states, actions, rewards, next_states, done_flags = exp_replay.
sample(batch_size)

# loss = <compute TD loss>
loss = compute_td_loss(agent, target_network,
                       states, actions, rewards, next_states,
                       done_flags,
                       gamma=0.99,
                       device=device)

loss.backward()
grad_norm = nn.utils.clip_grad_norm_(agent.parameters(), max_grad_norm)
opt.step()
opt.zero_grad()

if step % loss_freq == 0:
    td_loss_history.append(loss.data.cpu().item())

if step % refresh_target_network_freq == 0:
    # Load agent weights into target_network
    target_network.load_state_dict(agent.state_dict())

if step % eval_freq == 0:
    # eval the agent
    mean_rw_history.append(evaluate(
        make_env(env_name), agent, n_games=3, greedy=True, t_max=1000)
    )

# plotting code omitted
```

You now have a fully trained agent. You train the agent and plot the mean reward per episode periodically as you train the agent for 50,000 steps. In the left plot of mean return per episode in Figure 6-2, the x-axis value of 10 corresponds to the 10,000[th] step. You also plot the TD loss every 20 steps, and that's why you have the x-axis on the right plot going from 0 to 2500, that is, 0 to 2500x20=50,000 steps. Unlike supervised learning, the targets are not fixed. Keep the target network fixed for a short duration and update it periodically by refreshing the target network weights with the online network. Also, as discussed, nonlinear function approximation (neural networks) with off-policy learning

(Q-learning) and bootstrapped targets (the target network being just an estimate of the actual value and the estimate formed using the current estimates of other q-values) has no convergence guarantee. The training may see losses going up and exploding or fluctuating. This loss graph is counterintuitive as compared to the loss graphs in the usual supervised learning. Figure 6-2 shows the graphs from the training DQN.

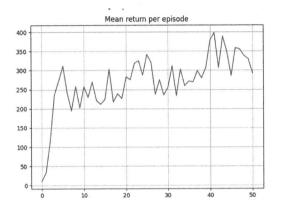
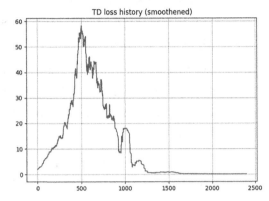

Figure 6-2. *Training curves for DQN*

I have more to say about other options for DQN training and logging using frameworks like Stable Baselines3 (SB3), RL Baselines3 Zoo (SB3 Zoo), weights and biases and others. However, before covering those issues, I take a quick detour to explain the origin of environment API from OpenAI and its forked version under Gymnasium. This understanding will be helpful, as some of the environment libraries and frameworks (like SB3, SB3 Zoo, and a few others) still follow an older version of API.

OpenAI Gym vs Farma Gymnasium

OpenAI made some breaking changes to the core API. It impacted workflows like how to create environments, what value is returned when you reset the environment, how to set the random seed, what values are returned by step function, and so on. These changes were introduced initially in OpenAI Gym v0.25.0 in July 2022 as opt-in and breaking change in v0.26.0 in Sep 2022.[1] The details are as follows:

---

[1] https://github.com/openai/gym/releases

- Step: The step function was modified to return five values: `obs`, `reward`, `termination`, `truncation`, and `info`. Original return values were `obs`, `reward`, `done`, and `info`. The reason for this change was to make a distinction between termination and truncation. `done=True` did not distinguish between the environment terminating and the episode truncating. In most cases where time limit is not a factor, you can set `done = truncation or termination`. You can read more about it here.[2]

- Render: The Render API was changed such that the mode had to be specified during `gym.make` with the `render_mode` keyword, after which the render mode is fixed. You can refer to release comments of Gym v0.25.0.

- Reset info: The `Env.reset` function was changed to return two values (`obs` and `info`) versus the older API returning only `obs`.

- No Seed function: While `Env.seed` was a helpful function, this was almost solely used for the beginning of the episode and was hence added to `gym.reset(seed=...)`.

In October 2022, the role of maintenance and updates to Gym was moved over to a new non-profit organization called the Farma Foundation. The code was moved over to a new Python package named `Gymnasium`, which carried over the new version of the APIs introduced in OpenAI Gym 0.26.0.

However, the story does not end here. Some of the third-party environments and libraries like SB3 and RL SB3 Zoo use parts or full older version of the API. SB3 and SB3 Zoo use the older API for `step` and `reset`. They retain the new `render` method, requiring `render_mode` to be specified right in the `gymnasium.make` call.

There are other environment libraries that are still using the pre 0.25.0—the popular `gym v0.21` version of API. Depending on the environment frameworks you plan to use, you should read the documentation and make sure you understand the version the library supports and if the library is compatible with OpenAI Gym or Frama Gymnasium.

Gymnasium also provides some environment wrappers to wrap environments following the old API to become compatible with newer post v0.26 API. Key wrappers that serve this compatibility need are as follows:

---

[2] `https://github.com/Farama-Foundation/gym-docs/pull/115`

- The wrapper `gymnasium.wrappers.EnvCompatibility` wraps an environment following the older version to become compatible with the newer API.

- The wrapper `gymnasium.wrappers.StepAPICompatibility` can transform an environment from new the step API to the old and vice versa.

- The wrapper `gymnasium.wrappers.PassiveEnvChecker` is a passive environment checker wrapper that surrounds the `step`, `reset`, and `render` functions to determine whether they follow the Gymnasium API.

Recording Videos of Trained Agents

Now that you understand the API changes between Gym and Gymnasium, let's continue with the topic of DQN training. Having trained the agent and reviewed the performance by way of plots, next you want to generate the video of the trained agent in action. I talked about it briefly in Chapter 2. However, this time I dive a bit deeper, which will help you gain a better understanding of the Stable-Baselines3 (SB3) library. SB3 gives you a unified structure for all the popular RL algorithms. It has optimized implementations of most of the popular RL algorithms used for training agents. It has integration with various experiment sharing, tracking, and logging services, such as weights and biases, HuggingFace, and MLFlow. SB3 also provides vectorized environments. Vectorized environments stack multiple independent environments into a single environment. Instead of training an RL agent on one environment per step, vectorized environments allow you to train it on n environments per step. You will use two such vector environment classes—DummyVecEnv and VecVideoRecorder—to help you record the video of trained agent.

DummyVecEnv creates a simple vectorized wrapper for multiple environments, calling each environment in sequence on the current Python process. This can also be used for RL methods that require a vectorized environment, but that you want a single environment to train with. You will use the latter purpose of wrapping a single environment with DummyVecEnv. The reason is that the recording API expects a vectorized environment. Listing 6-6 shows the code for the record_video function. The first line of code creates an array with a single CartPole environment. You'll use Gymnasium (imported as gym) to call its make function and pass the array to DummyVecEnv's init function.

Next, you wrap the vectorized environment with VecVideoRecorder, providing arguments such as the vectorized environment for which you want to record the video, the folder path where the video will be recorded, the trigger callback which starts the recording, and the name prefix to specify the prefix of the video file.

Once vec_env has been wrapped in VecVideoRecorder, you run the standard rollout of episode, starting from the reset of the environment with obs = vec_env.reset(). Note that the SB3 vectorized environment's reset function returns a single value—the observation that conforms to the older version of the Gym API. Next, you run a loop to step through the environment with the policy being the trained agent.

You extract the generated video filename so that it can be returned by the record_ video function and consumed by the play_video function. You can refer to the accompanying notebook for the implementation of the play_video function. Finally, you call the vec_env.close() function to close the environment, stop the video, and free up all the resources.

Listing 6-6. Video Recording Code from 6.a-dqn-pytorch.ipynb

```python
def record_video(env_id, video_folder, video_length, agent):

    vec_env = DummyVecEnv([lambda: gym.make(env_id, render_mode="rgb_
    array")])
    # Record the video starting at the first step
    vec_env = VecVideoRecorder(vec_env, video_folder,
                        record_video_trigger=lambda x: x == 0, video_
                        length=video_length,
                        name_prefix=f"{type(agent).__name__}-{env_id}")

    obs = vec_env.reset()
    for _ in range(video_length + 1):
        action = agent.get_action(obs).detach().cpu().numpy()
        obs, _, _, _ = vec_env.step(action)
    # video filename
    file_path = "./"+video_folder+vec_env.video_recorder.path.
    split("/")[-1]
    # Save the video
    vec_env.close()
    return file_path
```

Confusingly, Gymnasium also has a set of vectorized environments that follow the new API with a `step` function returning a batch of `observations`, `rewards`, `terminations`, `truncations`, and `infos`. Even the Gymnasium vectorized environment's `reset` function adheres to the newer version, returning a batch of observations and info from the vectorized environment. You could have implemented the same video recording code using the Gymnasium API. Interested readers can refer to the Gymnasium documentation for more details.

Next, you leverage the same code you saw in Chapter 2 with which you can push the trained agent and the network weights along with the video to HuggingFace. The code is similar to what you saw in Chapter 2. Inside the `6.a-dqn-pytorch.ipynb` notebook, you can look at the section named "Share the agent using HuggingFace" and explore the arguments passed to the `package_to_hub()` function call.

End-to-End Training with SB3

Up to this point, you handcrafted the agent network, the loss calculation, and the training code for the DQN algorithm. While this is good for learning, eventually in projects you want to leverage well implemented scalable DQN algorithms with all the bells and whistles. SB3 provides you with most of the popular algorithms with parameters to control the behavior. Let's look at the complete code to create and train the agent using SB3.

In terms of actual code, you first create the `CartPole` environment, and this code is similar to what you used earlier while handcrafting the DQN algorithm in earlier sections. Next, you create `policy_kwargs` to define the network layers and activation function to be used inside the neural network. You create a similar network like before with two hidden layers of 256 nodes each and ReLU as the activation function. There are multiple ways to define the network. The most basic is to go with the default and not create `policy_kwargs`, the policy configuration dictionary. This approach creates a default network of two hidden layers with 64 nodes each and ReLU activations. The most complex and flexible way is to create your own custom network similar to Listing 6-1. For details, refer to the documentation in SB3 under the "Custom Policy" topic.

Once `policy_kwargs` is defined, you instantiate the `model` with a call to the SB3 provided DQN class. It takes in `MlpPolicy` as the type of policy, `env` as the variable holding the environment, and `policy_kwargs` as the custom configuration of that policy. There are many other parameters you might need to tweak for your specific use case.

This pretty much completes the creation code. Next, you call `model.learn()` with parameters like training time steps, logging intervals, and flags to control the display of training progress. After the training is over, you save the model with a call to `model.save()`, providing the path where you want the model to be saved. The six-line code given in Listing 6-7 is sufficient to instantiate and train a DQN on the `CartPole` environment.

Listing 6-7. End-to-End DQN with SB3 from `6.a-dqn-pytorch.ipynb`

```
# create the DQN agent
from stable_baselines3 import DQN

# create the environment
env_name = env_name = 'CartPole-v1'
env = gym.make(env_name, render_mode="rgb_array")

#define a policy matching ours
#define the activation function and the network layers size
policy_kwargs = dict(activation_fn=torch.nn.ReLU,
                     net_arch=[256, 256])
model = DQN("MlpPolicy", env, policy_kwargs=policy_kwargs, verbose=1)

model.learn(total_timesteps=1e5, log_interval=500, progress_bar=True)
model.save("logs/6_a/sb3/dqn_cartpole")
```

End to End Training with SB3 Zoo

As discussed earlier, RL Baselines3 Zoo is a training framework for reinforcement learning (RL). It provides scripts for training, evaluating agents, tuning hyperparameters, plotting results, and recording videos. In addition, it includes a collection of tuned hyperparameters for common environments and RL algorithms, and agents trained with those settings.

To expose you to various options, this example uses weights and biases (wandb) integration to log these experiments. With wandb, you can capture the logging and share it with other team members. It also helps you track all your experiments, the specific parameters used in a given run, the training logs, and various other system level details. You first need to create a wandb account by going to www.wandb.ai. Once you have created an account and logged in, click your name/photo on the top-right of the page and then navigate to "User Settings". Once on the Settings page, navigate to the end of the page and look for a sub-section named "API Keys" under the "Danger Zone" section. Click the "New Key" button to create the key and copy the key generated somewhere, as you will need it in your Python code to allow it to access your wandb account.

Similarly, you also need to create an account on HuggingFace and generate an API key/token, which you can do in Chapter 2. You can reuse the same token here and for all the following chapters. Listing 6-8 shows the code for using RL SB3 Zoo for training an agent, playing/uploading trained agent performance.

You first log in to your wandb account so that you can log the experiment results there. Next, you run rl_zoo3.train on the command line with various parameters to train and track the agents. Most of the command-line parameters are self-explanatory based on their names. The command-line parameters are over and above what can be seen in Listing 6-8. You can refer to RL SB3 Zoo documentation for more details.[3] Sometimes, we have found it easier to refer to the script code to understand all the parameters. For example, the Train script code can be browsed in GitHub at the location given in the reference.[4]

Once you have trained the agent, you evaluate it using an rl_zoo3.enjoy call and record the video with the help of rl_zoo3.recrod_video. Finally, rl_zoo3.push_to_hub can be used to push the trained agent weights, video, and other relevant details to HuggingFace.

Listing 6-8. End-to-end DQN with SB3 Zoo from 6.a-dqn-pytorch.ipynb

```
# login to wandb
import wandb
wandb.login()

# Train the agent
!python -m rl_zoo3.train --algo dqn --env CartPole-v1 --save-freq 10000 \
--eval-freq 10000 --eval-episodes 10 --log-interval 400 --progress \
--track --wandb-project-name dqn-cartpole -f logs/6_a/rlzoo3/

# Evaluate trained agent
```

---

[3] https://rl-baselines3-zoo.readthedocs.io/en/master/

[4] https://github.com/DLR-RM/rl-baselines3-zoo/blob/master/rl_zoo3/train.py

```
!python -m rl_zoo3.enjoy --algo dqn --env CartPole-v1 --no-render
--n-timesteps 5000 --folder logs/6_a/rlzoo3

# Record a video
!python -m rl_zoo3.record_video --algo dqn --env CartPole-v1 --exp-id 0 -f
logs/6_a/rlzoo3/ -n 1000

# Login into Hugging face if not done so earlier
# To log to our Hugging Face account to be able to upload models to
the Hub.
from huggingface_sb3 import load_from_hub, package_to_hub, push_to_hub
from huggingface_hub import notebook_login

notebook_login()
!git config --global credential.helper store

# Share on HuggingFace
!python -m rl_zoo3.push_to_hub --algo dqn --env CartPole-v1 --exp-id 0 \
--folder logs/6_a/rlzoo3 --n-timesteps 1000 --verbose 1 --load-best  \
--organization nsanghi --repo-name dqn-cart-pole-rlzoo -m "Push to Hub"
```

Hyperparameter Optimization**

You have seen in the previous examples that there are a lot of hyperparameters that can impact the performance of the model. On a real-life RL training project, you are likely to spend a lot of time fine-tuning many of these parameters. First-time readers can skip this section and come back later as and when the need arises. This is not an essential read for understanding the RL algorithms.

RL SB3 Zoo is integrated with Optuna.[5] Optuna offers three key features 1) automated search for optimal hyperparameters using Python conditionals, loops, and syntax, 2) efficiently search large spaces and prune unpromising trials for faster results, and 3) parallelize hyperparameter searches over multiple threads or processes without modifying code. I cover high-level concepts of how Optuna works with RL SB3 Zoo. For more details, refer to the documentation of Optuna.

---

[5]https://optuna.readthedocs.io/en/stable/tutorial/index.html

To optimize, you first need to define the range over which you want to tune hyperparameters. You also define how these values will change over the range—for example, will they change by an integer value, or will the change be a float number, and will the change follow a log domain where the parameter values are sampled from log domain. A hyperparameter changing values over 0.1, 1, 10, 100, 1000 are equally spaced when you take the log. The previous range in base-10 log works out to -1, 0, 1, 2, 3, which is equally spaced. Log space sampling is the preferred choice when varying the learning rate α. The change could also be a categorical, meaning there is a defined set of values that it can take—for example, batch_size = [8, 16, 32, 64]. In RL SB3 Zoo, there is a file named rl_zoo3/hyperparams_opt.py[6] that provides a suggested list of tunable hyperparameters along with the ranges for each of the hyperparameters. The list and range are defined for each RL algorithm separately. Listing 6-9 shows the setting for DQN.

The hyperparameter sampler takes an argument trial of the type optuna.Trial, which defines a single instance of the run. For a given instance of the experiment run, you choose a specific set of hyperparameter values and return the same back for consumption by the rest of the optimization pipeline. The key part is that hyperparameter value sampling is done by the calls to trial.suggest() and the algorithm used for sampling is determined by the --sampler command-line argument while calling rl_zoo3.train. There are three possible values: random, tpe, and skopt. Random is a random search and can be a fairly good option in most cases. TPE stands for Tree-structured Parzen estimator algorithm. This sampler is based on independent sampling. On each trial, for each parameter, TPE fits one Gaussian Mixture Model (GMM) l(x) to the set of parameter values associated with the best objective values, and another GMM g(x) to the remaining parameter values. It chooses the parameter value x that maximizes the ratio l(x)/g(x).[7] Scikit-Optimize, or skopt, is a simple and efficient library that minimizes (very) expensive and noisy black-box functions. It implements several methods for sequential model-based optimization. skopt aims to be accessible and easy to use in many contexts.[8]

---

[6] https://github.com/DLR-RM/rl-baselines3-zoo/blob/master/rl_zoo3/hyperparams_opt.py
[7] https://proceedings.neurips.cc/paper_files/paper/2011/file/86e8f7ab32cfd12577bc2619bc635690-Paper.pdf
[8] https://scikit-optimize.github.io/stable/

Listing 6-9. Hyperparameter Search Space for DQN from hyperparams_opt.py

```python
def sample_dqn_params(trial: optuna.Trial, n_actions: int, n_envs: int,
additional_args: dict) -> Dict[str, Any]:
    """

    Sampler for DQN hyperparams.

    :param trial:
    :return:
    """
    gamma = trial.suggest_categorical("gamma", [0.9, 0.95, 0.98, 0.99,
    0.995, 0.999, 0.9999])
    learning_rate = trial.suggest_float("learning_rate", 1e-5, 1, log=True)
    batch_size = trial.suggest_categorical("batch_size", [16, 32, 64, 100,
    128, 256, 512])
    buffer_size = trial.suggest_categorical("buffer_size", [int(1e4),
    int(5e4), int(1e5), int(1e6)])
    exploration_final_eps = trial.suggest_float("exploration_final_
    eps", 0, 0.2)
    exploration_fraction = trial.suggest_float("exploration_
    fraction", 0, 0.5)
    target_update_interval = trial.suggest_categorical("target_update_
    interval", [1, 1000, 5000, 10000, 15000, 20000])
    learning_starts = trial.suggest_categorical("learning_starts", [0,
    1000, 5000, 10000, 20000])

    train_freq = trial.suggest_categorical("train_freq", [1, 4, 8, 16, 128,
    256, 1000])
    subsample_steps = trial.suggest_categorical("subsample_steps", [1,
    2, 4, 8])
    gradient_steps = max(train_freq // subsample_steps, 1)

    net_arch_type = trial.suggest_categorical("net_arch", ["tiny", "small",
    "medium"])

    net_arch = {"tiny": [64], "small": [64, 64], "medium": [256, 256]}[net_
    arch_type]
```

```python
hyperparams = {
    "gamma": gamma,
    "learning_rate": learning_rate,
    "batch_size": batch_size,
    "buffer_size": buffer_size,
    "train_freq": train_freq,
    "gradient_steps": gradient_steps,
    "exploration_fraction": exploration_fraction,
    "exploration_final_eps": exploration_final_eps,
    "target_update_interval": target_update_interval,
    "learning_starts": learning_starts,
    "policy_kwargs": dict(net_arch=net_arch),
}

if additional_args["using_her_replay_buffer"]:
    hyperparams = sample_her_params(trial, hyperparams, additional_
    args["her_kwargs"])

return hyperparams
```

The other important part of the hyperparameter optimization is the concept of pruning, which refers to the strategy for the termination of unpromising trials. In some literature, it is also known as early stopping. In SB3 Zoo, pruning is controlled via the `--prunner` command-line argument, which can take one of the three possible values: `halving,` `median,` or `none`. Successive halving, or halving, has a simple idea behind the algorithm. Given an input budget, uniformly allocate the training budget to a set of arms (number of parameter combinations) for a predefined number of iterations, evaluate their performance, throw out the worst half, and repeat until just one hyperparameter configuration remains. You can read more about the concept of pruning in the paper titled "Non-stochastic Best Arm Identification and Hyperparameter Optimization" by Kevin Jamieson and Ameet Talwalkar.[9] Median pruning refers to the strategy that a trial is pruned if the trial's best intermediate result is worse than the median of the intermediate results of previous trials at the same step.

---

[9] https://arxiv.org/abs/1502.07943

rl_zoo.train provides a few other controls on the optimization. The --n-startup-trials command-line argument defines the number of trials before the Optuna sampler starts. The --n-trials command-line argument defines the number of trials for optimizing hyperparameters. The --max-total-trials command-line argument defines the number of (potentially pruned) trials for optimizing hyperparameters. This last argument applies to the entire optimization process and takes precedence over --n-trials. There are a few more and you can read about them in the rl_zoo3.train script.[10]

Not to forget—the most important command-line argument is --optimize. It launches rl_zoo3.train in hyperparameter optimization mode.

Lastly, there are many ways to specify hyperparams_opt.py. Listing 6-10 shows a sample call of rl_zoo.train with DQN and the CartPole environment.

Listing 6-10. Hyperparameter Optimization with rl_zoo3.train from 6.a-dqn-pytorch.ipynb

```
!python -m rl_zoo3.train --algo dqn --env CartPole-v1 --save-freq 10000 \
--eval-episodes 10 --log-interval 100 --progress --track \
--wandb-project-name dqn-cartpole \
--optimization-log-path logs/optimization/ --log-folder logs --optimize-
hyperparameters --n-jobs 2 \
--study-name dqn-test --max-total-trials 25
```

Note that, based on the settings, it may take a long time to complete the optimization run. Moving on to plotting the optimization results, the rl_zoo3/plots and scripts folders in the rl_zoo3 repository provide a few scripts to plot the results. In the scripts folder, parse_study.py helps parse an optimization run and saves the *n* best hyperparameter configuration. Listing 6-11 shows an example of running this script after the optimization run in Listing 6-10 completes.

Listing 6-11. Saving "n" Best Configurations After an Optimization Run from the 6.a-dqn-pytorch.ipynb notebook

```
!python -m rl_zoo3.scripts.parse_study -i ./logs/dqn/report_CartPole-
v1_500-trials-50000-tpe-median_1698290229.pkl --print-n-best-trials 10 \
--save-n-best-hyperparameters 10 --folder ./logs/dqn/6.a/best-params/
```

---

[10] https://github.com/DLR-RM/rl-baselines3-zoo/blob/master/rl_zoo3/train.py

Integration with Rliable library(**)

In 2021, a paper called "Deep Reinforcement Learning at the Edge of the Statistical Precipice" by Agarwal et al.,[11] received the Outstanding award at NeurIPS 2021. To quote from the paper:

> *"Deep reinforcement learning (RL) algorithms are predominantly evaluated by comparing their relative performance on a large suite of tasks. Most published results on deep RL benchmarks compare point estimates of aggregate performance such as mean and median scores across tasks, ignoring the statistical uncertainty implied by the use of a finite number of training runs...."*

The authors go on to say that the shift toward computationally demanding benchmarks, especially with benchmarks based on environments like ALE (Arcade Learning Environments), has increasingly led to a practice of evaluating only on a small number of tasks, which is worsening the statistical uncertainty of the point estimates. The authors illustrate this point using ALE and then go on to recommend use of

> *"...interval estimates of aggregate performance and propose performance profiles to account for the variability in results, as well as present more robust and efficient aggregate metrics, such as interquartile mean scores, to achieve small uncertainty in results."*

You can read more about it in the link to the blog accompanying the paper. The authors also released a Python library called `rliable`, which has been integrated into RL SB3 Zoo. There are three key scripts provided by SB3 Zoo for plotting. `scripts/all_plots.py`, `scripts/plot_from_file.py` for plotting evaluations, and `scripts/plot_train.py` for plotting training reward/success. The `6.a-dqn-pytorch.ipynb` notebook contains some samples.

First, you need to do multiple training runs to collect diverse data points for meaningful plotting. For this, you can rerun the training script from Listing 6-8 multiple times, with some random variations of the hyperparameters. You'll create various training trajectories and have multiple validation results. Once you have had enough runs, the next step is to prepare the results file. Listing 6-12 contains the code for running these and other plotting scripts. The first command line in Listing 6-12 is about preparing the file for all plots. It requires parameters like the algorithm, the environment,

---

[11] https://agarwl.github.io/rliable/

the location where the data from the training runs is stored, and the output location where you want to store file generated by this script. After the results have been saved in a file, you feed the same into the second command, as shown in Listing 6-12. `plot_from_file` takes as input the file produced from the previous `all_plots` command and, based on the command-line parameters, it uses `rliable` or other graphing libraries to create the graphs. Some of the command-line arguments pertain to the `rliable` library. You may need to explore the `rliable` documentation for a better understanding of what they control and what is the significance of each of them. You can look into the Plotting section of SB3 Zoo documentation to understand the various command-line options these scripts have. Like `all_plots`, there is another script called `plot_train` that plots the training graphs.

Listing 6-12. Plotting from Notebook `6.a-dqn-pytorch.ipynb`

```
!python -m rl_zoo3.plots.all_plots --algo dqn --env CartPole-v1 \
-f logs/6_a/rlzoo3/ -o logs/6_a/rlzoo3/dqn_results

!python -m rl_zoo3.plots.plot_from_file -i logs/6_a/rlzoo3/dqn_
results.pkl \
-latex -l DQN --output logs/6_a/rlzoo3/dqn_test_plot.svg

!python -m rl_zoo3.plots.plot_from_file -i logs/6_a/rlzoo3/dqn_
results.pkl \
--rliable --versus --iqm --boxplot \
-latex -l DQN --output logs/6_a/rlzoo3/dqn_test1_plot.svg
```

This completes the implementation of a full DQN to train an agent. The core idea was to focus on the algorithm and teach you how to write a DQN learning agent. You will now use the same implementation with minor tweaks so that the agent can play Atari games using the game image pixel values as the state. Note that you will be building a more powerful neural network and will need to train agents over many more time steps running into a million steps. Therefore, unless you have access to a GPU locally, you may consider running the Atari DQN notebook on Google Colab. The notebook contains commands for all the package installations required to run the notebook on Google Colab and is thus self-contained.

Atari Game-Playing Agent Using DQN

In a 2013 seminal paper titled "Playing Atari with Deep Reinforcement Learning,"[12] the authors used a deep learning model to create a neural network–based Q-learning algorithm. They christened it Deep Q Networks. This is exactly what you implemented in the previous section. You will now briefly learn about the additional steps the authors took to train the agent to play Atari games. It is mostly the same as the previous section with two key differences: 1) uses game image pixel values as state inputs that necessitate some preprocessing, and b) uses convolutional networks inside an agent instead of linear layers that you saw in the previous section. The rest of the approach to calculate the loss L and carry out the training remains the same. Note that training with convolution networks takes a lot of time, especially on regular PCs/laptops. Get ready to watch the training code run for hours, even on moderately powerful GPU-based machines.

You can find the complete code to train the agent in PyTorch in the Python notebook called `6.c-dqn_atari_pytorch.ipynb`. The Gymnasium library has implemented many of the transformations needed for an Atari game image, and wherever possible, these examples use the same.

Atari Environment in Gymnasium

Let's first talk about Atari games environment as provided within Gymnasium and also the image preprocessing done to get the image pixel values ready for feeding into the deep learning network. Gymnasium provides a set of Atari 2600 environments simulated through Stella and the Arcade Learning Environment.

The Atari 2600 Video Computer System (VCS), introduced in 1977, was the most popular home video game system of the early 1980s. You can enjoy all the popular Atari 2600 games on your PC thanks to Stella. Stella is a multi-platform Atari 2600 VCS emulator released under the GNU General Public License (GPL).

The Arcade Learning Environment (ALE) is a simple framework that allows researchers and hobbyists to develop AI agents for Atari 2600 games. It is built on top of the Atari 2600 emulator called Stella and separates the details of emulation from agent design. ALE provides Python bindings and supports OpenAI Gym and Farma Gymnasium.

---

[12] https://arxiv.org/pdf/1312.5602.pdf

ALE-py doesn't include the Atari ROMs (`pip install gymnasium[atari]`), which are necessary to make Atari environments in the Gymnasium make function. The Python installation instructions in the beginning of the book take care of this.

The action space of Atari games is comprised of 18 actions numbered from 0 to 17. By default, most environments use a smaller subset of the legal actions, excluding those actions that don't have an effect in that specific game. If users are interested in using all possible actions, pass the keyword argument `full_action_space=True` to `gymnasium. make`. I will not be using this as the neural networks will be used to output only relevant actions for the specific game in question.

Atari games have three types of observation spaces: the most common one is 1) RGB image of size 210x160 with type `Box(0, 255, (210, 160, 3), np.uint8)`, 2) the grayscale version of the image, and 3) the RAM state of 128 bytes. You can control this using the `obs_type` argument to the `gym.make` function call.

As the Atari games are entirely deterministic, agents could achieve state-of-the-art performance by simply memorizing an optimal sequence of actions while completely ignoring observations from the environment. There are several methods to avoid this. Quoting from the Gymnasium documentation:

- *Sticky actions:* Instead of always simulating the action passed to the environment, there is a small probability that the previously executed action is used instead. In the v0 and v5 environments, the probability of repeating an action is 25% while in v4 environments, the probability is 0%. Users can specify the repeat action probability using `repeat_action_probability to gymnasium.make`.

- *Frameskipping:* On each environment step, the action can be repeated for a random number of frames. This behavior may be altered by setting the `argument  frameskip` keyword to either a positive integer or to a tuple of two positive integers. If `frameskip` is an integer, frame skipping is deterministic, and in each step the action is repeated `frameskip` many times. Otherwise, if `frameskip` is a tuple, the number of skipped frames is chosen uniformly at random between `frameskip[0]` (inclusive) and `frameskip[1]` (exclusive) in each environment step.

Some games allow users to set a difficulty level and a game mode. Different modes/difficulties may have different game dynamics and different action spaces. This example does not vary either of these. It uses the default of 0.

You will be implementing an agent for the game of Breakout. In Breakout, there is a paddle at the bottom, and the idea is to move the paddle to ensure the ball does not fall below it. The player needs to use the paddle to hit the ball and take out as many bricks as possible. Each time the ball misses the paddle, the player loses a life. The player has five lives to start with. Breakout has an action space of 0-3 ['NOOP', 'FIRE', 'RIGHT', 'LEFT'] with 0: "No Operation (NOOP), 1: FIRE to start the game, 2: move paddle right and 3: move paddle left. You can refer to the Gymnasium documentation for more details.[13] Figure 6-3 shows three frames of the game.

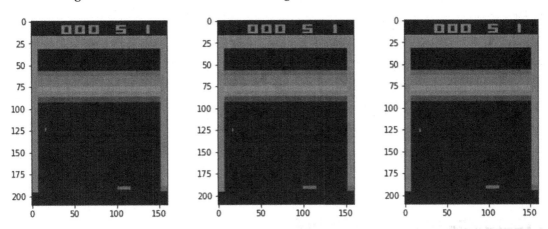

Figure 6-3. *Atari Breakout game images*

The Atari game images are 210×160-pixel images with a 128-color palette. Listing 6-13 shows the code used to create the environment and print a few frames of the game.

Listing 6-13. Creating an Atari Breakout Game Instance from notebook 6.c-dqn_atari_pytorch.ipynb

```
def make_env(env_name, frameskip=5, repeat_action_probability=0.25, render_
mode='rgb_array', mode=0, difficulty=0):
    # remove time limit wrapper from environment
    env = gym.make(env_name,
                    render_mode=render_mode,
                    frameskip=frameskip,
                    repeat_action_probability=repeat_action_probability,
                    mode=mode,
```

---

[13] https://gymnasium.farama.org/environments/atari/breakout/

```
                      difficulty=difficulty
                  ).unwrapped
    return env
env_name = "ALE/Breakout-v5"

env = make_env(env_name)
env.reset(seed=127)

n_cols = 4
n_rows = 2
fig = plt.figure(figsize=(16, 9))

for row in range(n_rows):
    for col in range(n_cols):
        ax = fig.add_subplot(n_rows, n_cols, row * n_cols + col + 1)
        ax.imshow(env.render())
        env.step(env.action_space.sample())
plt.show()
```

Preprocessing and Training

You will do preprocessing to prune down the image size and make the convolutional network run faster. You also scale down the images. You can convert the images to grayscale again to reduce the size of the input vector, with one channel of grayscale instead of three channels of colors for RGB (the red, green, and blue channels). A preprocessed single frame of the image of size (1×84×84—channel, width, height) in PyTorch just gives the static state. The position of the ball or paddle does not tell you about the direction in which both are moving. Accordingly, you will stack a few consecutive frames of the game images together to train the agent. You will stack four reduced-size grayscale images that will the state S feeding into the neural network. The input (i.e., state S) will be of size 4×84×84 in PyTorch and 84×84×4 in TensorFlow, where 4 refers to the four frames of the game images and 84×84 is the grayscale image size of each frame. Stacking four frames together will allow the agent network to infer the direction of movement of the ball and paddle.

Listing 6-14 shows the code to create the environment, including the preprocessing steps. When you plot a single observation of the environment after the preprocessors, you see that state/observation has a shape of 4x84x84. In the make_env function, you first call gym.make with frameskip=1 which is the default behavior, meaning that the next call to step will produce the next game frame. The reason you do this is because you will be wrapping this with additional wrappers, which will take care of stacking multiple frames together. Therefore, you do not want any frame skipping in the call to gym.make. Note that all along you import Gymnasium as gym. Accordingly, in this code I refer to the package as Gymnasium or Gym interchangeably, for example, gymnasium.make or gym.make. Both refer to the Gymnasium library and calling the make function.

Next, you wrap the Gymnasium Breakout environment with AtariPreprocessing to set screen_size to 84 pixels, and the pixel values are mapped to 0-1 range instead of 0-255 range with the help of scale_obs=True. The complete list of the arguments to AtariPreProcessing can be seen in the Gymnasium documentation.[14] The same documentation page also provides details with respect to many other wrappers that can be used to wrap the vanilla environments to augment or alter the default behaviors. You will use the default value of frame_skip=4, as you need to stack four frames together to provide the context of ball and pad direction before feeding the observations to the neural network. Therefore, the next wrapper you use is FrameStack with num_stack=4 to stack four frames of game images together. Finally, based on the clip_rewards flag, you further wrap the environment with the TransformReward wrapper. This wrapper takes the input environment and a function to transform the reward. In this case, you use a simple lambda function lambda r: np.sign(r) to return just the sign of the reward, dropping the magnitude. This is in line with the process followed in the original paper.

Listing 6-14. Creating an Atari Breakout Game Instance with Required Wrappers to Train the Agent from notebook 6.c-dqn_atari_pytorch.ipynb

```
from gymnasium.wrappers import AtariPreprocessing
from gymnasium.wrappers import FrameStack
from gymnasium.wrappers import TransformReward

def make_env(env_name,
            clip_rewards=True):

    env = gym.make(env_name,
                   render_mode='rgb_array',
                   frameskip=1
```

---

[14] https://gymnasium.farama.org/api/experimental/wrappers/

```
            )
    env = AtariPreprocessing(env, screen_size=84, scale_obs=True)
    env = FrameStack(env, num_stack=4)
    if clip_rewards:
        env = TransformReward(env, lambda r: np.sign(r))
    return env

env = make_env(env_name)
obs, _ = env.reset()
n_actions = env.action_space.n
state_shape = env.observation_space.shape

print("Observation Shape:", state_shape) #prints 4x84x84

obs = obs[:] #unpack lazyframe
obs = np.transpose(obs,[1,0,2]) #move axes
obs = obs.reshape((obs.shape[0], -1))

plt.figure(figsize=[15,15])
plt.title("Agent observation (4 frames left to right)")
plt.imshow(obs, cmap='gray')
plt.show()
```

The previous preprocessing step produces the final state that you will be feeding into the network. This is 4×84×84 in PyTorch, where 4 refers to the four frames of the game images and 84×84 is the grayscale image size of each frame. Figure 6-4 shows the observation as produced by the code in Listing 6-14 and which is used as input to the neural network.

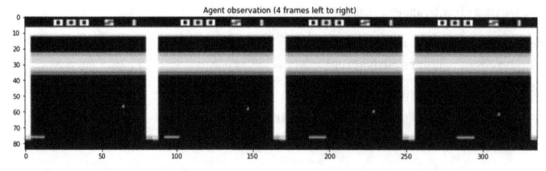

Figure 6-4. *Processed image to be used as state input into the neural network*

Next, you build the neural network that will take in the previous image, that is, the state/observation S, and produce q-values for all four actions in this case. The shape of the network you will build is given here:

- Input: PyTorch tensor of shape [batch_size, 4, 84, 84]

- First hidden convolutional layer: 16 nos of 8x8 filters with stride 4 and ReLU activation producing output of shape [batch_size, 16, 20, 20]

- Second hidden convolution layer: 32 nos of 4x4 filters with stride of 2 and ReLU activation producing output of shape [batch-size, 32, 9, 9]

- Flatten the observation to shape [batch, 32x9x9] = [batch_size, 2592]

- Third hidden fully connected linear layer with 256 outputs and ReLU activation producing output of shape [batch_size, 256]

- Output layer is a linear layer with "n_actions" units with no activation producing output of shape [batch_size, 4]

You implement the usual functions—forward, get_qvalues, and sample_actions. These are similar to what you did in the 6.a-dqn_pytorch.ipynb notebook for the CartPole environment. Listing 6-15 shows the code implementing what you just read about when creating a DQNAgent specific to the Atari game in question.

Listing 6-15. DQNAgent Implementation from Notebook 6.c-dqn_atari_ pytorch.ipynb

```
class DQNAgent(nn.Module):
    def __init__(self, state_shape, n_actions, epsilon=0):

        super().__init__()
        self.epsilon = epsilon
        self.n_actions = n_actions
        self.state_shape = state_shape

        state_dim = state_shape[0]
        # a simple NN with state_dim as input vector (input is state s)
        # and self.n_actions as output vector of logits of q(s, a)
        self.network = nn.Sequential()
```

```python
        self.network.add_module('conv1', nn.Conv2d(4,16,kernel_size=8,
        stride=4))
        self.network.add_module('relu1', nn.ReLU())
        self.network.add_module('conv2', nn.Conv2d(16,32,kernel_size=4,
        stride=2))
        self.network.add_module('relu2', nn.ReLU())
        self.network.add_module('flatten', nn.Flatten())
        self.network.add_module('linear3', nn.Linear(2592, 256)) #2592
        calculated above
        self.network.add_module('relu3', nn.ReLU())
        self.network.add_module('linear4', nn.Linear(256, n_actions))

        self.parameters = self.network.parameters

    def forward(self, state_t):
        # pass the state at time t through the network to get Q(s,a)
        qvalues = self.network(state_t)
        return qvalues

    def get_qvalues(self, states):
        # input is an array of states in numpy and outout is Qvals as
        numpy array
        states = torch.tensor(states, device=device, dtype=torch.float32)
        qvalues = self.forward(states)
        return qvalues.data.cpu().numpy()

    def sample_actions(self, qvalues):
        # sample actions from a batch of q_values using epsilon
        greedy policy
        epsilon = self.epsilon
        batch_size, n_actions = qvalues.shape
        random_actions = np.random.choice(n_actions, size=batch_size)
        best_actions = qvalues.argmax(axis=-1)
        should_explore = np.random.choice(
            [0, 1], batch_size, p=[1-epsilon, epsilon])
        return np.where(should_explore, random_actions, best_actions)
```

This is then followed by the usual implementation of experience replay, functions to play and record the game, creating an initial copy of target network as a deep copy of agent network, the `compute_td_loss` function to compute TD loss, instantiating all the machinery of agent, target network, and replay buffer.

Once these are done, you fill the replay buffer with an initial set of actions by having the agent randomly choose actions. You then go on to create the hyperparameters, as shown in Listing 6-16.

At this point, the agent is ready to be trained. Note that to get meaningful behavior, you need to train the agent close to one million steps or so. The network is also substantially more complex than the one used in the `CartPole` environment. Therefore, it takes a long time to train the Atari game playing agent for one million steps. You are advised to run the code on Google Colab or run it on a local computer that has GPU computing capabilities.

Listing 6-16. Hyperparameters from Notebook 6.c-dqn_atari_pytorch.ipynb

```
#set up some parameters for training
timesteps_per_epoch = 1
batch_size = 32

total_steps = 100
# total_steps = 3 * 10**6
# We will train only for a sample of 100 steps
# To train the full network on a CPU will take hours.
# in fact even GPU training will be fairly long
# Those who have access to powerful machines with GPU could
# try training it over 3-5 million steps or so

#init Optimizer
opt = torch.optim.Adam(agent.parameters(), lr=1e-4)

# set exploration epsilon
start_epsilon = 1
end_epsilon = 0.05
eps_decay_final_step = 1 * 10**6

# setup some frequency for logging and updating target network
loss_freq = 20
```

```
refresh_target_network_freq = 100
eval_freq = 1000

# to clip the gradients
max_grad_norm = 5000
```

The code for training the agent is similar to what you saw for `CartPole` in the `6.a-dqn_pytorch.ipynb` notebook. The code for recording and playing the video of trained agent, pushing the trained agent parameters and video to HuggingFace and so on, are pretty similar as before. Therefore, I do not explicitly show the listings of the code here. Interested readers may refer to the `6.c-dqn_atari_pytorch.ipynb` notebook. You can also train the same agent with Stable Baselines3 (SB3) or RL SB3 Zoo. You are encouraged to write that code using `CartPole` notebook code as a starting point. The key difference you will have to watch out for is the preprocessing of Breakout environment with specific Atari wrappers, as discussed earlier in this section. The default policy also must be switched from `MLPPolicy` to `CnnPolicy`, either with default values or with custom network as per your needs.

This completes the implementation and training of DQN for the Atari game. Now that you know how to train a DQN agent, in Chapter 7, an optional chapter, you will investigate some issues and various approaches you could take to modify the DQN. I close the chapter by reviewing the various other environments that are integrated with Gymnasium and/or Stable Baselines3, including the kind of problems they solve and the technical considerations for using these environments.

Overview of Various RL Environments and Libraries

Until now in this book, you have looked at Gymnasium and its earlier version OpenAI Gym. They both, barring some minor differences as discussed in the beginning of this chapter, specify a standard interface for defining RL environments—the need for such environments to have functions like `reset, step, render,` and so on.

You also looked at specific environments like Cart Pole, Mountain Car, Lunar Lander, and Atari Breakout. These environments follow the specifications of Gymnasium/Gym and are made available inside the Gymnasium library. There are many other third-party, open-source environments for different use cases which are either integrated within Gymnasium by the group maintaining the Gymnasium codebase or by other open-source libraries. To help you appreciate that RL is used not just for playing games but for very large and varied scenarios, the next section look at some of the other environments. What I discuss next is by no means exhaustive, rather just a sample. Readers, based on their interest, may decide to dive into the details of libraries/environments.

PyGame

PyGame is a Python-based library. This library includes several modules for playing sound, drawing graphics, handling mouse inputs, and so on. Many RL environments in gymnasium use PyGame to implement rendering logic in environments for human and/ or visual space representation of the game state.

PyGame is not just for RL. It is a wider library that's used for implementing stand-alone games and interactive environments using the Python programing language. You can possibly take a full course on PyGame to understand the details; it has an extensive set of features.

As an RL engineer, you will not be not needing to dive into the details. Even if you design your customer environments, you will use other higher-level libraries that may internally use PyGame to implement the graphics.

MuJoCo

MuJoCo stands for Multi-Joint dynamics with Contact. It is a physics engine for facilitating research and development in robotics, biomechanics, graphics and animation, and other areas where fast and accurate simulation is needed. The unique dependencies for this set of environments can be installed via `pip install gymnasium[mujoco]`.

Gymnasium has eleven MuJoCo environments: Ant, HalfCheetah, Hopper, Humanoid, HumanoidStandup, InvertedDoublePendulum, InvertedPendulum, Pusher, Reacher, Swimmer, and Walker2d. A sample screenshot of these environments is shown in Figure 6-5.

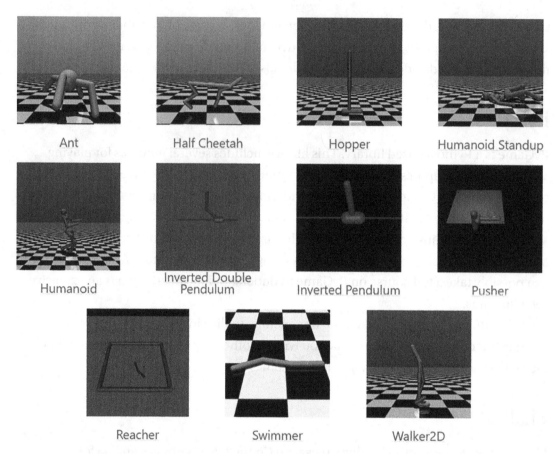

Figure 6-5. *MuJoCo environments in Gymnasium*

All of these environments are stochastic in terms of their initial state, with Gaussian noise added to a fixed initial state in order to add stochasticity. The state spaces for MuJoCo environments in Gymnasium consist of two parts that are flattened and concatenated together: a position of a body part (`mujoco-py.mjsim.qpos`) or joint and its corresponding velocity (`mujoco-py.mjsim.qvel`). Often, some of the first positional elements are omitted from the state space since the reward is calculated based on their values, leaving it up to the RL algorithm to infer those hidden values indirectly.

If you want to use these environments from gymnasium, you also need to install the MuJoCo engine. DeepMind acquired MuJoCo in late 2021 and open-sourced the code, making it free for everyone. Instructions on installing the MuJoCo engine can be found on the website[15] and on the GitHub repository.[16]

---

[15] https://mujoco.org/

[16] https://github.com/google-deepmind/mujoco

Using MuJoCo with Gymnasium also requires the `mujoco` framework be installed (this dependency is installed with the previous `pip` command). Environments can be configured by changing the XML files or by tweaking the parameters of their classes.

Similar to Atari, you can use RL Zoo to train the agent. Listing 6-17 shows the terminal command to train the agent. The complete code with the required installation steps can be found in the `6.d-robotic-env.ipynb` notebook.

Listing 6-17. Training the MuJoCo Environment 6.d-robotic-env.ipynb

```
!python -m rl_zoo3.train --algo ppo --env Reacher-v4 --progress \
--log-interval 400 --save-freq 10000 --eval-freq 10000 --eval-episodes 10 \
--n-timesteps 100000 --conf-file ppo_config_6e.py \
--track --wandb-project-name ppo-reacher-v4 -f logs/6_e/rlzoo3/
```

Unity ML Agents

The Unity Machine Learning Agents Toolkit (ML-Agents[17]) is an open-source project that enables games and simulations to serve as environments for training intelligent agents. ML-Agents provides PyTorch-based implementations of state-of-the-art algorithms to enable game developers and hobbyists to easily train intelligent agents for 2D, 3D and VR/AR games. Researchers can also use the provided simple-to-use Python API to train agents using reinforcement learning, imitation learning, neuro evolution, or any other methods. These trained agents can be used for multiple purposes, including controlling non-playing character (NPC) behavior (in a variety of settings such as multi-agent and adversarial), automated testing of game builds, and evaluating different game design decisions pre-release.

The `ML-agents` toolkit comes with 17+ sample Unity environments. It has support for training single-agent, multi-agent cooperative, and multi-agent competitive scenarios via several deep reinforcement learning algorithms. You can very easily add your own customer environment or custom training algorithm.

It also provides a way to wrap Unity Learning environments as a OpenAI Gym environment for you to use it in your code as a straight drop-in in place of any other environment. You can read about it in the documentation for `ml-agents` in GitHub, the link for the same is referenced at the beginning of this sub-section. Unfortunately, as of

---

[17] https://github.com/Unity-Technologies/ml-agents

writing this book, `ml-agents` supports the OpenAI gym API version and not the latest API version from Gymnasium and there is no clear direction from the maintainers of the code on the migration from Gym to Gymnasium.

You can refer to the `ml-agent` tutorials or go to the `https://huggingface.co/docs/hub/ml-agents` link, which walks through the way `ml-agents` can be used in HuggingFace ecosystem.

PettingZoo

PettingZoo is a simple, Pythonic interface capable of representing general *multi-agent reinforcement learning* (MARL) problems. PettingZoo includes a wide variety of reference environments, helpful utilities, and tools for creating your own custom environments. While I mention this library here for the sake of completeness and for putting together all environment related discussions in a single place in the book, I will be covering PettingZoo in greater detail in a subsequent chapter on multi-agent RL. Some common multi-agent environments include:

- Atari multiplayer games

- Board games like Chess and Go

- Card games like rummy

- Multi-Particle Environments (MPEs), where particle agents can move, communicate, see each other, push each other around, and interact with fixed landmarks

- The SISL environments, which are a set of three cooperative multi-agent benchmark environments, created at SISL (Stanford Intelligent Systems Laboratory) and released as part of "Cooperative multi-agent control using deep reinforcement learning"

Bullet Physics Engine and Related Environments

Bullet is a physics engine which simulates collision detection as well as soft and rigid body dynamics. It has been used in video games and for visual effects in movies. Erwin Coumans, its main author, won a Scientific and Technical Academy Award[4] for his work on Bullet.[18]

---

[18] `https://en.wikipedia.org/wiki/Bullet_%28software%29`

PyBullet is a fast and easy to use the Python module for robotics simulation and machine learning, with a focus on sim-to-real transfer. PyBullet wraps the Bullet API. PyBullet can be easily used with TensorFlow and OpenAI Gym. Researchers from Google Brain, Stanford AI Lab, OpenAI, and many other labs use PyBullet.

Then come the various RL environments, which further wrap PyBullet to create custom RL environments that follow the OpenAI gym API signatures. Some of the common ones are as follows:

- `Pandas-gym`: PyBullet based simulations of a robotic arm moving objects.

- `Parallel Game engine (PGE)`: Intended for fast and pretty 3D artificial intelligence experiments. However, this open-source project has not seen any update since 2018 and may be considered stale.

- `PyFlyt`: UAV Flight Simulator Environments for Reinforcement Learning Research. It is a library for testing reinforcement learning algorithms on various UAVs. Built on the Bullet physics engine, it offers flexible rendering options, time-discrete steppable physics, Python bindings, and support for custom drones of any configuration, be it biplanes, quadcopters, rockets, and anything you can think of. It is still under development and supports the newer Gymnasium API signatures.

There are a few more that are mentioned in the Gymnasium documentation under section on third-party environments.

CleanRL

CleanRL is a deep reinforcement learning library that provides high-quality, single-file implementation with research-friendly features. The implementation is clean and simple. With single file implementation, every detail about an algorithm variant is put into a single stand-alone file. For example, `ppo_atari.py` has 340 lines of code and contains all the implementation details on how PPO works with Atari games, so it is a great reference implementation to read.

CleanRL only contains implementations of online deep reinforcement learning algorithms. Refer to `https://github.com/corl-team/CORL`, which shares a similar design philosophy as CleanRL but for offline RL algorithms.

CleanRL uses "Poetry" which is yet another Python packaging and dependency management tool and achieves a similar purpose as that of venv or conda. You can explore Poetry's dependency management and its pros and cons by visiting the Poetry documentation.[19]

Listing 6-18 shows the set of commands you can run to install Poetry, followed by CleanRL and then training an agent DQN. The code in Listing 6-18 has been tried on WSL2 Ubuntu on Windows 11 and should also work out-of-the-box on Ubuntu and macOS. For macOS, you may want to use the Homebrew package manager for installing Python and the other libraries.

Listing 6-18. Running DQN Using CleanRL

```
# to install python and pip followed by poetry

sudo apt install python3-pip
python3 -m pip install --user -U pipx
pipx install poetry

# install CleanRL and associated python packages
git clone https://github.com/vwxyzjn/cleanrl.git && cd cleanrl
poetry install

# if you get some error during the execution, you may want to install
specific version of SB3
poetry run pip install "stable_baselines3==2.0.0a1"

# to enter poetry shell which is like conda activate
poetry shell

# to login into wandb and huggingface
# You will need api-key/token from your wandb and huggingface accounts
wandb login
huggingface-cli login

# train CartPole with DQN implementation by CleanRL
# please use your HuggingFace name instead of "nsanghi" for parameter
"--hf-entity"
```

---

[19] https://python-poetry.org/

```
python -m cleanrl.dqn  --seed 1  --env-id CartPole-v1     --total-
timesteps 50000 --track --capture-video --save-model --upload-model --hf-
entity nsanghi

# See your results
# replace "nsanghi" with your hf-account name
https://wandb.ai/nsanghi/cleanRL/runs/99rex9ov
https://huggingface.co/nsanghi/CartPole-v1-dqn-seed1

# please look at clearn_rl/dqn.py in the github repo
# for more command line options and default values
# see https://docs.cleanrl.dev/get-started/installation/#prerequisitesmore
# for additional environments and different algos and backends

# to leave poetry shell
exit
```

MineRL

MineRL is a Python 3 library that provides a OpenAI Gym interface for interacting with the video game Minecraft, accompanied with datasets of human gameplay. Started as a research project at Carnegie Mellon University, MineRL aims to assist in the development of various aspects of artificial intelligence within Minecraft.

Refer to the documentation[20] for more details on how to install and use this. You can also refer to the Tutorial section of the documentation for examples on how to run MineRL on Colab.

FinRL

FinRL is an open-source framework for financial reinforcement learning. FinRL provides a framework that supports various markets, State of the Art (SOTA) deep learning RL (DRL) algorithms, benchmarks of many quant finance tasks, live trading, and so on.

The DRL framework is powerful in solving dynamic decision-making problems by learning through interactions with an unknown environment, thus exhibiting two major advantages: portfolio scalability and market model independence. Automated trading

---

[20] https://minerl.readthedocs.io/en/latest/index.html

is essentially making dynamic decisions, namely, to decide where to trade, at what price, and what quantity, over a highly stochastic and complex stock market. Taking many complex financial factors into account, DRL trading agents build a multi-factor model and provide algorithmic trading strategies, which are difficult for human traders. Interested readers can read more about it on GitHub.[21] The documentation also contains some notebooks showcasing on its use.

FlappyBird Environment

The GitHub repository markub3327/flappy-bird-gymnasium[22] contains the implementation of Gymnasium environment for the FlappyBird game.

FlappyBird, FlappyBird-v0, has 12 dimensional observation spaces with the following meanings:

1. The last pipe's horizontal position

2. The last top pipe's vertical position

3. The last bottom pipe's vertical position

4. The next pipe's horizontal position

5. The next top pipe's vertical position

6. The next bottom pipe's vertical position

7. The next next pipe's horizontal position

8. The next next top pipe's vertical position

9. The next next bottom pipe's vertical position

10. The player's vertical position

11. The player's vertical velocity

12. The player's rotation

The action space is discrete, with two values:

1. Do nothing - "0"

2. Flap - "1"

---

[21] https://github.com/AI4Finance-Foundation/FinRL

[22] https://github.com/markub3327/flappy-bird-gymnasium

And reward is comprised of:

+0.1 - Every frame it stays alive

+1.0 - Successfully passing a pipe

-1.0 - Dying

Interested readers can refer to the `6.e-dqn_flappy_pytorch.ipynb` notebook for training and evaluating the FlappyBird environment using DQN. It is very similar to what you saw for Atari DQN. This brings you to the end of this chapter.

Summary

This was a long chapter, where you looked at DQN in depth and the various RL environments. It started with a quick recap of Q-learning and the derivation of the update equation for DQN. You then looked at the implementation of DQN in both PyTorch and TensorFlow for a simple `CartPole` environment. You saw the complete code for the DQN algorithm.

You followed it with a review of the Gymnasium API and compared the changes it has undergone from its original version under OpenAI Gym. You looked at how Stable Baselines3 (SB3) and RL SB3 Zoo (SB3 Zoo) can be used for training agents, where you leverage the pre-implemented DQN and various other algorithms. A quick review of the process to record the video of a trained agent in action was also discussed.

The next two sections were on hyperparameter optimization using Optuna and SB3 and plotting using the Rliable library.

This completed the walkthrough of DQN and its associated concepts. Next, I talked about applying DQN to Atari games, which resulted in discussing preprocessing steps commonly required for games like these.

The final section was devoted to the various RL environments ,ranging from MuJoCo for simple joint robots, to Unity-based ML agents, and areas like finance under the FinRL library.

CHAPTER 7

Improvements to DQN**

This chapter looks at various enhancements and variations to DQN. Specifically, it looks at Prioritized Replay, DDQN (Double Q-Learning), Dueling DQN, NoisyNets DQN, C-51 (Categorical 51-Atom DQN), Quantile Regression DQN, and Hindsight Experience Replay. All the examples in this chapter are coded using PyTorch. This is an optional chapter with each variant of DQN as a standalone topic. You can skip this chapter in the first pass and come back to it when you want to explore specific variants of DQN.

The first variant of DQN covered in this chapter is prioritized replay.

Prioritized Replay

In the previous chapter, you saw how to use a batch version of updates in DQN that addresses some key issues in the online version, such as updates being done with each transition and a transition being discarded right after that one step of learning. The following are the key issues in the online versions:

- The training samples (transitions) are correlated, breaking i.i.d. (independent identically distributed) assumptions. With online learning, you have correlated transitions in a sequence. Each transition is linked to the previous one. This breaks the i.i.d. assumption that is required to apply a gradient descent.

- As the agent learns and discards, it might never get to visit the initial exploratory transitions. If the agent goes down a wrong path, it will keep seeing examples from that part of the state space. It may settle on a suboptimal solution.

- With neural networks, learning on a single transition basis is hard and inefficient. There will be too much variance for the neural networks to learn anything effective. Neural networks work best when they learn in batches of training samples.

273

N. Sanghi, *Deep Reinforcement Learning with Python*, https://doi.org/10.1007/979-8-8688-0273-7_7

These were addressed in DQN by using an experience replay where all transitions are stored. Each transition is a tuple of state, action, reward, next_state and done. As the buffer gets full, you discard old samples to add new ones. You then sample a batch from the current buffer with each transition in the buffer having equal probability of being selected in a batch. It allows rare and more exploratory transitions to be picked multiple times from the buffer. However, a plain-vanilla *experience replay* does not have any way to choose the important transitions with some priority. Would it help to somehow assign an importance score to each transition stored in the replay buffer and sample the batch from the buffer using these importance scores as the probability of selection, assigning a higher probability of selection to important transitions as signified by their respective importance score?

This is what the authors of the paper "Prioritized Experience Replay"[1] from DeepMind explored in 2016. This chapter follows the main concepts of that paper to create your own implementation of experience replay and apply it to the DQN agent for the CartPole environment. I start by talking a little bit about how these importance scores are assigned and how the loss L is modified.

The key approach of this paper is to assign importance scores to training samples in the buffer using their TD errors. When a batch of samples is picked from the buffer, you calculate the TD error as part of the loss L calculation. The TD error is given by this equation:

$$\delta_i = r_i + \left((1 - done_i) \cdot \gamma \cdot \max_{a'} \hat{q}(s_i', a'; w_t^-) \right) - \hat{q}(s_i, a_i; w_t)$$

(7-1)

This appears inside Equation 6-3 where you calculate the loss. The error is squared and averaged over all samples to calculate the magnitude of updates to weight vectors, as shown in Equations 6-4 and 6-5. The magnitude of TD error δ_i denotes the contribution that the sample transition *(i)* would have on the update. The authors used this reasoning to assign an importance score p_i to each sample, where p_i is given by this equation:

$$p_i = |\delta_i| + \varepsilon$$

(7-2)

[1] https://arxiv.org/pdf/1511.05952.pdf

A small constant ϵ is added to avoid the edge case of p_i being 0 when TD error δ_i is 0. When a new transition is added to the buffer, you assign it the max of p_i across all current transitions in the buffer. The reason for doing that is to increase the chance of this transition getting picked when you sample a batch of transitions for TD learning. When a batch is picked for training, you calculate the TD error δ_i of each sample as part of the loss/gradient calculation. The TD error as calculated is then used to update the importance score of these samples in the buffer.

There is another approach of rank-based prioritization that the paper talks about. Using that approach, $p_i = \dfrac{1}{rank(i)}$, where $rank(i)$ is the rank of transition (i) when the replay buffer transitions are sorted based on $|\delta_i|$.

In the sample code, you will be using the first approach, which is called *proportional prioritization*. Next, at the time of sampling, you convert p_i to probabilities by using the following equation:

$$P(i) = \frac{p_i^\alpha}{\sum_i p_i^\alpha} \tag{7-3}$$

Here, $P(i)$ denotes the probability of transition (i) in the buffer getting sampled while picking the training batch. This assigns a higher sampling probability to the transitions that have a higher TD error. Here, α is a hyperparameter, which was tuned using grid search, and the authors found $\alpha = 0.6$ to be the best for the proportional variant that you will be implementing.

The previous approach of breaking the uniform sampling with some kind of sampling based on importance introduces bias. You need to correct the bias while calculating the loss L. In the paper, this was corrected using *importance sampling* by weighing each sample with weight w_i and then summing it up to get the revised loss function L. The equation for calculating weights is as follows:

$$w_i = \left(\frac{1}{N} \cdot \frac{1}{P(i)} \right)^\beta \tag{7-4}$$

Here, N is the number of samples in the training batch, and $P(i)$ is the probability of selecting a sample as calculated in the previous expression. β is another hyperparameter. $\beta = 1$ fully compensates for the non-uniform sampling introduced by $P(i)$ importance sampling in the previous step. You can also exploit the flexibility of annealing the amount of importance-sampling correction over time, by defining a schedule on the exponent β that reaches 1 only at the end of learning. In practice, you could linearly anneal β from its initial value β_0 to 1. Note that the choice of this hyperparameter interacts with choice of prioritization exponent α; increasing both simultaneously prioritizes sampling more aggressively at the same time as correcting for it more strongly. In this example, you use a constant value of 0.4 which is taken from some of the experiments in the paper.

The weights are further normalized by $\dfrac{1}{max_i w_i}$ to ensure that the weights stay within bounds:

$$w_i^{IS} = \frac{1}{max_i w_i} w_i \qquad (7\text{-}5)$$

With these changes in place, the loss L equation is also updated to weigh each transition in the batch with w_i as follows:

$$L = \frac{1}{N} \sum_{i=1}^{N} \left[r_i + \left((1 - \text{done}_i) \cdot \gamma \cdot \max_{a'} \hat{q}\left(s_i', a'; w_t^-\right) \right) - \hat{q}\left(s_i, a_i; w_t\right) \right]^2 \cdot w_i^{IS} \qquad (7\text{-}6)$$

Notice the w_i^{IS} in the equation. After L is calculated, you follow the usual gradient step using back-propagation of the loss gradient with respect to the online neural network weights w (denoted as w_t, with a subscript t to highlight that weights change with as you iterate over batches for training).

Remember, the TD error in the previous equation is used to update the importance score back in the replay buffer for these transitions in the current training batch. This completes the theoretical discussion on the prioritized replay.

You will now look at the implementation. The complete code of training a DQN agent with prioritized replay is given in 7.a-dqn_prioritized_replay.ipynb. You are advised to study the code in detail along with the referenced paper after going through the explanations given here. The ability to follow the academic papers and match the details in the paper to working code is an important part of becoming a good practitioner. The explanation in this section is to just get you started. For a firm grasp of the material, you should follow the accompanying code in detail. It will be even better if you try to code it yourself after examining how the code works.

Getting back to the explanation, you first look at the prioritized replay implementation, which is the major change in the code from the previous DQN training notebook. Listing 7-1 shows the code for prioritized replay. Most of the code is similar to that of the plain ReplayBuffer you saw earlier. You now have an additional array called self.priorities to hold the importance/priority score p_i for each sample. The add function is modified to assign p_i to the new sample being added. It is just the max of values in the array self.priorities. The sample function is the one that has undergone maximum change. The first probabilities are calculated using Equation 7-3, and then the weights are calculated using Equations 7-4 and 7-5. The function now returns two additional arrays: the array of weights np.array(weights) and the array of index np.array(idxs). The array of indexes contains the index of the samples in the buffer that were sampled in the batch. This is required so that after the calculation of TD error in the loss step, you can update the priority/importance back in the buffer. The update_priorities(idxs, new_priorities) function is exactly for that purpose.

Listing 7-1. Prioritized Replay from 7.a-dqn-prioritized-replay.ipynb

```python
class PrioritizedReplayBuffer:
    def __init__(self, size, alpha=0.6, beta=0.4):
        ### init code omitted from this listing

    def add(self, state, action, reward, next_state, done):
        item = (state, action, reward, next_state, done)
        max_priority = self.priorities.max()
        if len(self.buffer) < self.size:
            self.buffer.append(item)
        else:
            self.buffer[self.next_id] = item
        self.priorities[self.next_id] = max_priority
        self.next_id = (self.next_id + 1) % self.size

    def sample(self, batch_size):
        N = len(self.buffer)
        priorities = self.priorities[:N]
        probabilities = priorities ** self.alpha
        probabilities /= probabilities.sum()
        weights = (N * probabilities) ** (-self.beta)
        weights /= weights.max()

        idxs = np.random.choice(len(self.buffer), batch_size,
        p=probabilities)
```

```
        samples = [self.buffer[i] for i in idxs]
        states, actions, rewards, next_states, done_flags =
        list(zip(*samples))
        weights = weights[idxs]

        return  (np.array(states), np.array(actions), np.array(rewards),
                 np.array(next_states), np.array(done_flags),
                 np.array(weights), np.array(idxs))

    def update_priorities(self, idxs, new_priorities):
        self.priorities[idxs] = new_priorities+self.epsilon
```

Next, let's look at the loss calculation. The code is almost similar to the TD loss computation you saw in Listing 6-3. There are two changes. The first is multiplying the TD error with weights, in line with Equation 7-6. The second change is calling update_priorities from inside the function to update the priorities back in the buffer. The function now takes two new arguments. First one is weights, which is denoted as w_i^{IS} in Equation 7-6. The second argument is buffer_idxs and it is used to update the priorities of the transitions forming the training batch. Listing 7-2 highlights the code changes of the revised compute_td_loss_priority_replay function.

Listing 7-2. TD Loss with Prioritized Replay from 7.a-dqn-prioritized-replay.ipynb

```
def compute_td_loss_priority_replay(agent, target_network, replay_buffer,
                                    tates, actions, rewards, next_states,
                                    done_flags, weights, buffer_idxs,
                                    gamma=0.99, device=device):

    # portions of the code similar to plain DQN omitted

    #compute each sample TD error. Notice multiplication by weights
    loss = ((predicted_qvalues_for_actions - target_qvalues_for_actions.
    detach()) ** 2) * weights

    # code omitted

    # new code to update back the priorities of the batch data
    with torch.no_grad():
        new_priorities = (predicted_qvalues_for_actions.detach()
```

```
            - target_qvalues_for_actions.detach())
    new_priorities = np.absolute(new_priorities.detach().numpy())
    replay_buffer.update_priorities(buffer_idxs, new_priorities)

return loss
```

The training code remains the same as before. You can look at the `7.a-dqn-prioritized-replay.ipynb` notebook to see the details. As before, you train the agent, and you can see that the agent learns to balance the pole really well with this approach. Figure 7-1 shows the training curves.

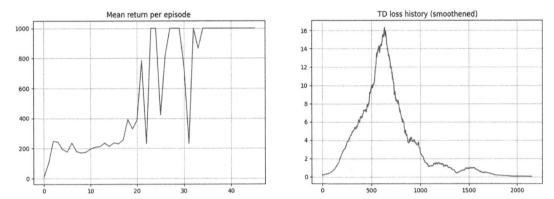

Figure 7-1. *Training curve for DQN agent with prioritized experience replay on CartPole*

The paper has lot more to say about the theoretical background of prioritized replay and why it can help. Plus the paper also talks about the impact it has on Atari games as well as how such an approach could be used for supervised learning with class imbalances. This completes the section on prioritized replay. You are advised to refer to the original paper and code notebooks for further details.

Double DQN (DDQN)

The authors of the paper "Deep Reinforcement Learning with Double Q-Learning"[2] explore the possibility and show that DQN algorithm, which combines Q-learning with a deep neural network, suffers from substantial overestimations, especially in some

---

[2] https://arxiv.org/pdf/1509.06461v3.pdf

Atari games. The authors then propose the idea of applying double Q-learning to DQN calling it *double DQN*. The idea of double Q-learning is to reduce overestimations by decomposing the max operation in the target into two separate steps of selection of action and evaluation of the selected action. Although not fully decoupled, the target network in the DQN architecture provides a natural candidate for the second value function, without having to introduce additional networks. The authors therefore propose to evaluate the greedy policy according to the online network, but using the target network to estimate its value.

Let's look at the max operation in the regular DQN. You calculate the TD target as follows:

$$Y^{DQN} = r + \gamma \cdot \max_{a'} \hat{q}(s', a'; w_t^-) \tag{7-7}$$

I have simplified the equation a bit by dropping the subscript (i) as well as removing the ($1-$done) multiplier, which drops the second term for the terminal state. I have done so to keep the explanation uncluttered. Now, let's unwrap this equation by moving the *max* inside. The previous update can be equivalently written as follows:

$$r + \gamma \cdot \hat{q}\left(s', \arg\max_{a'} \hat{q}\left(s', a'; w_t^-\right); w_t^-\right)$$

I have moved the max inside by first taking the max action and then taking the q-value for that max action. This is similar to directly taking the max q-value. In the previous unwrapped equation, you can clearly see that you are using the same network weight w_t^-, first for selecting the best action and then for getting the q-value for that action. This is what causes maximization bias. The authors of the paper suggested an approach that they called double DQN (DDQN), where the weight for selecting the best action, $\arg\max_{a'} \hat{q}(s', a')$, comes from the online network with weight w_t and then the target network with weight w_t^- is used to select the q-value for that best action. This change results in the updated TD target as follows:

$$r + \gamma \cdot \hat{q}\left(s', \arg\max_{a'} \hat{q}(s', a'; w_t); w_t^-\right) \tag{7-8}$$

Note that the inner network for selecting the best action is now using online network weights w_t. Everything else remains the same. You calculate the loss as before and then use the gradient step to update the weights of the online network. You also periodically update the target network weights with the weights from the online network. The updated loss function that you use is as follows (bringing back the *done* flag and index i):

$$Y_i^{DDQN} = r_i + \left((1 - \text{done}_i) \cdot \gamma \cdot \hat{q}\left(s_i', \arg\max_{a'} \hat{q}(s_i', a'; w_t); w_t^- \right) \right) \tag{7-9}$$

$$L = \frac{1}{N} \sum_{i=1}^{N} \left[Y_i^{DDQN} - \hat{q}(s_i, a_i; w_t) \right]^2 \tag{7-10}$$

The authors show that the double DQN approach leads to significant reduction in overestimation bias, which in turn leads to better policies. Let's now look at the implementation details. The only thing that will change as compared to the DQN implementation is the way loss is calculated. You use Equations 7-9 and 7-10 to calculate the loss. Everything else—including the DQN policy agent code, the replay buffer, and the way training is carried out to step through back-propagation of gradients—remains the same. Listing 7-3 shows the revised loss function calculation. You calculate the current q-value with q_s = agent(states) and then, for each row, pick the q-value corresponding to the action a_i. These action-values are stored in q_s_a. You then use the agent network with weight w (w_t with iteration time index t) to calculate the q-values for the next states: q_s1 = agent(next_states). The q_s1 tensor is used to find the best action for each row with a1_max = torch.max(q_s1, dim=1), and then you use the target network with the best action to find the target q-value: q_s1=agent(next_states) and then q_s1_a1max = q_s1_target[range(len(a1max)), a1max]. The rest of the code for TD loss calculations remains the same.

Listing 7-3. TD Loss with Double DQN (DDQN) from 7.b-ddqn.ipynb

```
def td_loss_ddqn(agent, target_network, states, actions, rewards, next_
states, done_flags,
                gamma=0.99, device=device):

    # convert numpy array to torch tensors
    # code omitted - same as DQN loss

    # get q-values for all actions in current states
    # use agent network
    q_s = agent(states)

    # select q-values for chosen actions
    q_s_a = q_s[range(
        len(actions)), actions]
```

```
# compute q-values for all actions in next states
# use agent network (online network)
q_s1 = agent(next_states).detach()

# compute Q argmax(next_states, actions) using predicted next q-values
_,a1max = torch.max(q_s1, dim=1)

#use target network to calculate the q value for best action
chosen above
q_s1_target = target_network(next_states)

q_s1_a1max = q_s1_target[range(len(a1max)), a1max]

# compute "target q-values"
target_q = rewards + gamma * q_s1_a1max * (1-done_flags)

# mean squared error loss to minimize
loss = torch.mean((q_s_a - target_q).pow(2))

return loss
```

Running DDQN on CartPole produces the training graph shown in Figure 7-2. You may not notice a big difference, as CartPole is too simple a problem to show the benefits. In addition, the training algorithm is running for a small number of episodes to demonstrate the algorithms. For quantified benefits of the approach, you should look at the referenced paper.

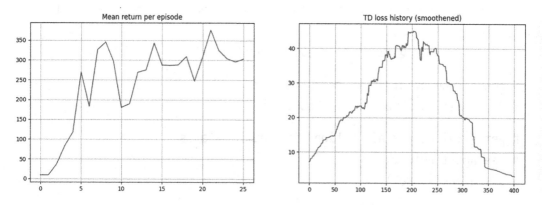

Figure 7-2. *Training curve for DDQN on CartPole from* `7.b-ddqn.ipynb`

This completes the discussion on DDQN. The next section covers dueling DQN.

Dueling DQN

Up until now, all of your networks took in state S and produced the q-values $Q(S, A)$ for all actions A in the state S. Figure 6-1 shows a sample of such a network. However, many times in a particular state, there is no impact from taking any specific action. Consider the case that a car is driving in the middle of road and there are no cars around it. In such a scenario, the action of taking a slight left or right, or speeding a bit or breaking a bit, has no impact; these actions all produce similar q-values. Is there a way to separate the average value in a state and the advantage of taking a specific action over that average? That's the approach that the authors of the paper titled "Dueling Network Architectures for Deep Reinforcement Learning"[3] took in 2016. They showed that it led to significant improvements, and the improvement was higher as the number of possible actions in a state grew.

Let's derive the computation that the dueling DQN network performs. You saw the definition of state-value and action-value functions in Chapter 2 in Equations 2-9 and 2-10, which are reproduced here:

$$v_\pi(s) = E_\pi \left[G_t | S_t = s \right]$$

$$q_\pi(s, a) = E_\pi \left[G_t | S_t = s, A_t = a \right] \tag{7-11}$$

Then, in Chapter 5 on function approximation, you saw these equations change a bit with the introduction of parameters w when you switched to representing the state/action values as parameterized functions:

$$\hat{v}(s; w) \approx v_\pi(s)$$

$$\hat{q}(s, a; w) \approx q_\pi(s, a) \tag{7-12}$$

Both sets of Equations 7-11 and 7-12 show that v_π measures the value of being in a state in general, and q_π shows the value of taking a specific action from the state S.

---

[3] https://arxiv.org/pdf/1511.06581.pdf

If you subtract Q from V, you get something that is called *advantage A*. Note that there is a bit of overload of notations. A inside $Q(S, A)$ represents action, and A_π on the left side of equation represents the advantage of taking an action a.

$$A_\pi(s,a) = Q_\pi(s,a) - V_\pi(s)$$

The authors created a network that takes in a state S as input and, after a few layers of network, produces two streams—one giving state value V and another giving advantage A with part of the network being individual sets of layers, one for V and one for A. Finally, the last layer combines advantage A and state value V to recover Q by rearranging the Equation 7-13:

$$Q_\pi(s,a) = V_\pi(s) + A_\pi(s,a) \tag{7-13}$$

Figure 7-3 shows a representative network architecture for the same.

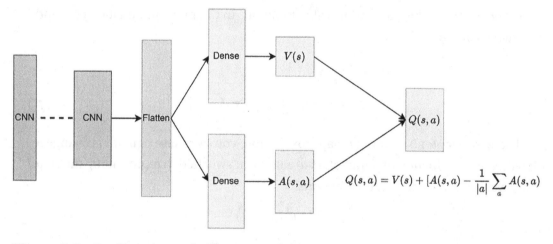

Figure 7-3. *Dueling network. The network has a common set of weights in the initial layers, denoted as w_1, and then it branches off to have one set of weights w_2, producing value V and another set of weights w_3, producing advantage A*

Before you proceed, let's rewrite Equation 7-13 with weights w_1, w_2, w_3 explicitly shown in expression. The revised equation looks like this:

$$\hat{Q}(s, a; w_1, w_2, w_3) = \hat{V}(s; w_1, w_2) + \hat{A}(s, a; w_1, w_3) \tag{7-14}$$

Do you notice a problem in learning this kind of expression where you estimate V and A independently? There is no unique V and A that represent the Q. If you add a constant term to V and subtract the same term from A, the expression for Q_π remains unchaned. Therefore, this network will not learn a unique V, the state value. The combined network could learn correct Q for all the states but the state values V for each state could be shifted by some random amount individually, making it hard to use V as a true representative of the state values.

One way to force this issue of identifiability/uniqueness the authors suggested to force the advantage to have zero value for the chosen action, that is,. an action with maximum advantage.

$$\hat{Q}\left(s,a;w_1,w_2,w_3\right)=\hat{V}\left(s;w_1,w_2\right)+\left(\hat{A}\left(s,a;w_1,w_3\right)-\max_{a'}\hat{A}\left(s,a';w_1,w_3\right)\right) \qquad (7\text{-}15)$$

The authors in the study found an alternative formulation to offer better stability where the max operation was replaced by an average.

$$\hat{Q}(s,a;w_1,w_2,w_3)=\hat{V}\left(s;w_1,w_2\right)+\left(\hat{A}\left(s,a;w_1,w_3\right)-\frac{1}{|a'|}\sum_{a'}\hat{A}\left(s,a';w_1,w_3\right)\right) \qquad (7\text{-}16)$$

On the one hand, this loses the original semantics of V and A because they are now off-target by a constant, but on the other hand it increases the stability of the optimization. With Equation 7-16 the advantages only needs to change as fast as the mean, instead of having to compensate any change to the optimal action's advantage in 7-15.

In the previous equation, weight w_1 corresponds to the initial common part of the network, w_2 corresponds to the part of the network that predicts state value $\hat{V}$, and finally w_3 corresponds to the part of network that predicts advantage $\hat{A}$.

The authors named this architecture *dueling networks* as it has two networks fused together with an initial common part. Since the dueling network is at the agent network level, it is independent of the other components like the type of replay buffer or the way the weights are learned (i.e., simple DQN or double DQN). Accordingly, you can use the dueling network independent of the type of replay buffer or the type of learning. In this walk-through, you will be using a simple replay buffer with uniform probability of selection for each transition in the buffer. Further, you will be using a DQN agent.

The authors studied the improvement of this approach over a baseline single network, as shown Figure 7-4. It clearly shows the benefit of a dueling architecture.

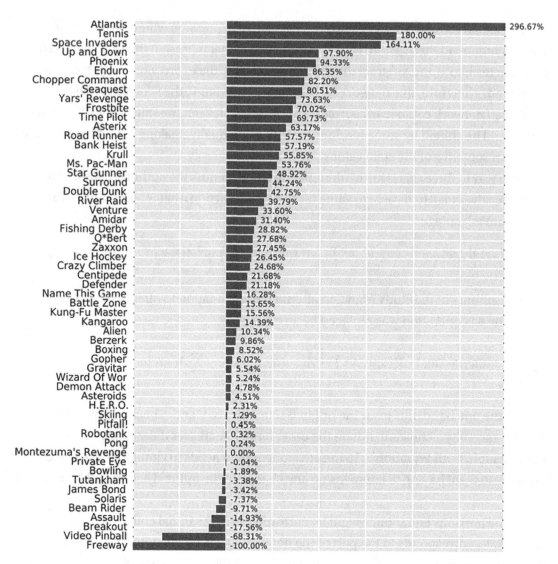

Figure 7-4. *Improvements of a dueling architecture over a baseline single network*
Source: Fig 4 of the paper[4]

Compared to the notebook for vanilla DQN—6.a-dqn_pytorch.ipynb—the only change is the way the network is constructed. Listing 7-4 shows the code for the dueling agent network.

---

[4]https://arxiv.org/pdf/1511.06581.pdf

Listing 7-4. Dueling Network from 7.c-dueling_dqn.ipynb

```
class DuelingDQNAgent(nn.Module):
    def __init__(self, state_shape, n_actions, epsilon=0):

        super().__init__()
        self.epsilon = epsilon
        self.n_actions = n_actions
        self.state_shape = state_shape

        state_dim = state_shape[0]
        # a simple NN with state_dim as input vector (input is state s)
        # and self.n_actions as output vector of logits of q(s, a)
        self.fc1 = nn.Linear(state_dim, 64)
        self.fc2 = nn.Linear(64, 128)
        self.fc_value = nn.Linear(128, 32)
        self.fc_adv = nn.Linear(128, 32)
        self.value = nn.Linear(32, 1)
        self.adv = nn.Linear(32, n_actions)

    def forward(self, state_t):
        # pass the state at time t through the network to get Q(s,a)
        x = F.relu(self.fc1(state_t))
        x = F.relu(self.fc2(x))
        v = F.relu(self.fc_value(x))
        v = self.value(v)
        adv = F.relu(self.fc_adv(x))
        adv = self.adv(adv)
        adv_avg = torch.mean(adv, dim=1, keepdim=True)
        qvalues = v + adv - adv_avg
        return qvalues

    # rest of the code is similar to that of DQN
```

You have two layers in a common network (self.fc1 and self.fc2). For *V* prediction, you have another two layers (self.fc_value and self.value) on top of fc1 and fc2. Similarly, for advantage estimation, you again have a separate set of two layers (self.fc_adv and self.adv) on top of fc1 and fc2. These outputs are combined to give the modified q-value as per Equation 7-16. The rest of the code, that is, the calculation of the TD loss and the gradient descent for the weight update, remains the same as that for DQN. Figure 7-5 shows the result of training the previous network on `CartPole`.

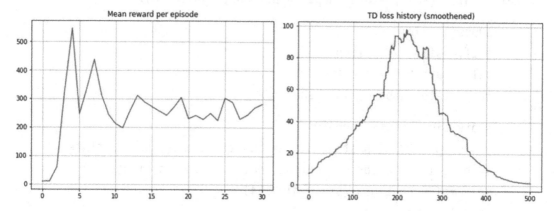

Figure 7-5. *Training curves for a dueling network from .c-dueling_dqn.ipynb*

As stated, you can try to substitute `ReplayBuffer` with `PrioritizedReplayBuffer`. You could also use DDQN instead of DQN as a learning agent. This concludes the discussion on dueling DQN. The next section discusses a very different variant.

NoisyNets DQN

Recall that you need to explore parts of state space. You have been doing this using a ε-greedy policy. Under this exploration, you take the max q-value action with probability (1- ε), and a random action with probability ε. The authors of a recent 2018 paper titled "Noisy Networks for Exploration,"[5] used a different approach of adding random perturbations to linear layers as parameters, and like network weights, these are also learned.

The usual linear layers are affine transformations, as given by the following:

$$y = wx + b$$

[5] https://arxiv.org/pdf/1706.10295.pdf

In noisy linear versions, you introduce random perturbations in the weights as given by the following equation:

$$y = \left(\mu^w + \sigma^w \odot \epsilon^w\right)x + \left(\mu^b + \sigma^b \odot \epsilon^b\right) \tag{7-17}$$

In the previous equation, μ^w, σ^w, μ^b, and σ^b are the weights of the network that are learned. ϵ^w and ϵ^b are random noises that introduce randomness leading to exploration. Figure 7-6 shows a schematic diagram of the noisy version of the linear layer, which explains Equation 7-17.

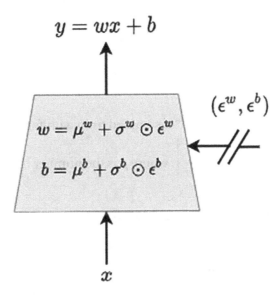

Figure 7-6. *The noisy linear layer. The weights and biases are a linear combination of the mean and standard deviation, which are learned just like the weights and biases in a regular linear layer (Figure 4 of the referenced paper* `https://arxiv.org/pdf/1706.10295.pdf`)

In terms of specific noise distributions in linear layers of a noisy network, the authors sampled two types: Independent Gaussian noise, where each weight has its own Gaussian noise, and Factorized Gaussian noise, which introduces separate noise for each output and input. The primary advantage of Factorized Gaussian noise lies in its ability to lower the time required for random number generation in various algorithms. This reduction is particularly crucial for single-thread agents like DQN and dueling, where computational overhead can be significant. Consequently, Factorized noise is employed in DQN and dueling, while Independent noise is used in the distributed A3C, where computation time is less of a concern.

In line with the authors' recommendations, you will implement the factorized version discussed in the paper, where each element $\epsilon_{i,j}^{w}$ of the matrix ϵ^{w} is factored. Suppose you have p units of inputs and q units of outputs. Accordingly, you generate a p-size vector of Gaussian noise ϵ_i and a q-size vector of Gaussian noise ϵ_j. Each $\epsilon_{i,j}^{w}$ and ϵ_{j}^{b} can now be written as follows:

$$\epsilon_{i,j}^{w} = f(\epsilon_i) \cdot f(\epsilon_j),$$

$$\epsilon_{j}^{b} = f(\epsilon_j),$$

$$f(x) = \text{sgn}(x) \cdot \sqrt{|x|} \tag{7-18}$$

For factorized networks like the one used here, the authors suggest initializing the weights as follows:

- Each element $\mu_{i,j}$ of μ^{w} and μ^{b} is sampled from a uniform distribution in the range of $U\left[-\dfrac{1}{\sqrt{p}}, \dfrac{1}{\sqrt{p}} \right]$, where p is the number of input units.

- Similarly, each element $\sigma_{i,j}$ of σ^{w} and σ^{b} is initialized to a constant $\dfrac{\sigma_0}{\sqrt{p}}$ with the hyperparameter σ_0 set to 0.5.

At a high level, the following modifications are applied to the previous DQN algorithm: a) you will not be using ϵ-greedy exploration. Instead the policy greedily optimizes the (randomized) action-value function; 2) the fully connected layers are parameterized as a noisy network, where the parameters are drawn from the noisy network parameter distribution after every replay step, which means the same noisy network is fixed used for a batch of transitions, the gradient is calculated, and the weights are adjusted. The network weights are sampled again from this modified weights before taking the next step.

Section C of the paper provides the pseudocode for NoisyNet-DQN and NoisyNet-Dueling, that is, the NoisyNet version of DQN and Dueling DQN.

Let's now look at the code required to implement NoisyNet-DQN. You create a noisy layer along the lines of the linear layer provided by PyTorch. You do so by extending nn.Module from PyTorch. It is a simple and standard implementation where you create weight vectors and noise vectors in the init function. You create the linear layer parameters μ^{w}, σ^{w}, μ^{b}, and σ^{b} from Equation 7-17. These are created using four individual calls to nn.Paramater. You also create a buffer in memory to store the noise tensors ϵ^w and ϵ^b. They are not created using nn.Parameter as they are not parameters of the model. They are sampled from standard Gaussian and stored for use in the forward function. Therefore, you use the PyTorch register_buffer function to create memory space for these tensors. Then you write a forward function to take an input and transform it through a set of noisy linear and regular linear layers, as per Equation 7-17. During training, you use the noisy version 7-17 and during inference, you use the usual version of linear layer without adding any noise.

You also need some additional functions. In this case, I wrote a function called reset_noise to generate the noise ϵ^w and ϵ^b. This function internally uses a helper function called _noise to generate standard Gaussian noise. The reset_parameters function resets the parameters following the strategy outlined in Equation 7-18. Listing 7-5 shows the code for the noisy linear layer.

Listing 7-5. Noisy Linear Layer from 7.d-noisynet_dqn.ipynb

```python
class NoisyLinear(nn.Module):
    def __init__(self, in_features, out_features, sigma_0 = 0.4):

        #boilerplate code omitted

        self.mu_w = nn.Parameter(torch.FloatTensor(out_features, in_
        features))
        self.sigma_w = nn.Parameter(torch.FloatTensor(out_features, in_
        features))
        self.mu_b = nn.Parameter(torch.FloatTensor(out_features))
        self.sigma_b = nn.Parameter(torch.FloatTensor(out_features))

        self.register_buffer('epsilon_w', torch.FloatTensor(out_features,
        in_features))
        self.register_buffer('epsilon_b', torch.FloatTensor(out_features))

        self.reset_noise()
        self.reset_params()

    def forward(self, x, training=False):
        if training:
            w = self.mu_w + self.sigma_w * self.epsilon_w
            b = self.mu_b + self.sigma_b * self.epsilon_b
```

```
    else:
        w = self.mu_w
        b = self.mu_b
    return F.linear(x, w, b)

def reset_params(self):
    k = 1/self.in_features
    k_sqrt = math.sqrt(k)
    self.mu_w.data.uniform_(-k_sqrt, k_sqrt)
    self.sigma_w.data.fill_(k_sqrt*self.sigma_0)
    self.mu_b.data.uniform_(-k_sqrt, k_sqrt)
    self.sigma_b.data.fill_(k_sqrt*self.sigma_0)

def reset_noise(self):
    eps_in = self._noise(self.in_features)
    eps_out = self._noise(self.out_features)
    self.epsilon_w.copy_(eps_out.ger(eps_in))
    self.epsilon_b.copy_(self._noise(self.out_features))

def _noise(self, size):
    x = torch.randn(size)
    x = torch.sign(x)*torch.sqrt(torch.abs(x))
    return x
```

You could use a noisy net with DQN, DDQN, dueling DQN, and prioritized replay in various combinations. However, for the purpose of the walk-through, I focus on using a regular replay buffer with DQN. You can also train using the regular DQN approach and not DDQN. The implementation of NoisyDQN mirrors that of DQN from the previous chapter. The only difference is that you now need to use a noisy linear layer, as shown in Listing 7-5. You also pass as an additional parameter to the forward function of NoisyDQN called `training`, which is `true` during training and made `false` during inference. Listing 7-6 shows the init and forward functions of NoisyDQN. You have no ε-greedy selection in the function `sample_actions` of the DQN agent. The `reset_noise` function also resets the noise after each batch. This is in line with the recommendations of the paper to decorrelate. Listing 7-6 contains the NoisyDQN version with these modifications. The rest of the implementation is similar to the vanilla DQN agent.

Listing 7-6. NoisyDQN Agent from 7.d-noisynet_dqn.ipynb

```python
class NoisyDQN(nn.Module):
    def __init__(self, state_shape, n_actions):
        super(NoisyDQN, self).__init__()
        self.n_actions = n_actions
        self.state_shape = state_shape
        state_dim = state_shape[0]
        # a simple NN with state_dim as input vector (input is state s)
        # and self.n_actions as output vector of logits of q(s, a)
        self.fc1 = NoisyLinear(state_dim, 64)
        self.fc2 = NoisyLinear(64, 128)
        self.fc3 = NoisyLinear(128, 32)
        self.q = NoisyLinear(32, n_actions)
        self.training=False

    def forward(self, state_t):
        # pass the state at time t through the network to get Q(s,a)
        x = F.relu(self.fc1(state_t, training=self.training))
        x = F.relu(self.fc2(x, training=self.training))
        x = F.relu(self.fc3(x, training=self.training))
        qvalues = self.q(x)
        return qvalues

    def get_qvalues(self, states):
        # input is an array of states in numpy and output is Qvals as
        numpy array
        states = torch.tensor(states, device=device, dtype=torch.float32)
        qvalues = self.forward(states)
        return qvalues.data.cpu().numpy()

    def get_action(self, states):
        states = torch.tensor(np.array(states), device=device, dtype=torch.
        float32)
        qvalues = self.forward(states)
        best_actions = qvalues.argmax(axis=-1)
        return best_actions
```

```python
def sample_actions(self, qvalues):
    # sample actions from a batch of q_values using greedy policy
    batch_size, n_actions = qvalues.shape
    best_actions = qvalues.argmax(axis=-1)
    return best_actions

def reset_noise(self):
    self.fc1.reset_noise()
    self.fc2.reset_noise()
    self.fc3.reset_noise()
    self.q.reset_noise()
```

The training code also has a minor modification. You reset the noise by calling reset_noise after each batch of samples has been processed. Listing 7-7 shows the code with this change. Refer to the 7.d-noisynet_dqn.ipynb notebook for the complete code.

Listing 7-7. NoisyDQN Batch Training Step from 7.d-noisynet_dqn.ipynb

```python
def train_agent_noisy(env, agent, target_network, optimizer, td_loss_fn):

    # code omitted
    # Fill experience replay using full random policy
    # reset noise at each action
    # also initiaize housekeeping array

    state, _ = env.reset(seed=seed)

    #enable training flag
    agent.training = True
    target_network.training = True

    for step in trange(total_steps + 1):

        # take timesteps_per_epoch and update experience replay buffer

        # train by sampling batch_size of data from experience replay
        states, actions, rewards, next_states, done_flags = exp_replay.
        sample(batch_size)

        # loss = <compute TD loss>
        optimizer.zero_grad()
```

```
agent.reset_noise()
target_network.reset_noise()
loss = td_loss_fn(agent, target_network,
                    states, actions, rewards, next_states, done_flags,
                    gamma=0.99, device=device)
loss.backward()
grad_norm = nn.utils.clip_grad_norm_(agent.parameters(), max_
grad_norm)
optimizer.step()

if step % refresh_target_network_freq == 0:
    # Load agent weights into target_network
    target_network.load_state_dict(agent.state_dict())

 # logging and evaluation code
 # log using noise equal to zero i.e. training=False

state, _ = env.reset(seed=seed)
#disable training flag
agent.training = False
target_network.training = False
```

Training a NoisyDQN in the `CartPole` environment produces the training curves, as given in Figure 7-7. You may not see any significant difference between this variant and DQN (or for that matter, all the variants). The reason is that you are using a simple problem and training it for a short number of episodes. You are also not optimizing the parameters. The idea in the book is to teach you the inner details of a specific variant.

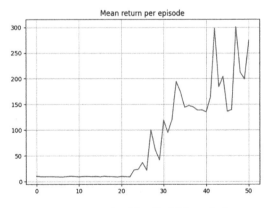
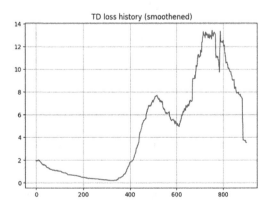

Figure 7-7. *NoisyNet DQN training graph*

For a thorough study of the improvements and other observations, you are advised to refer to the original papers referenced in this chapter. You should also go through the accompanying Python notebooks in detail, and after you have grasped the details, you should try to code the example anew. Figure 7-8 shows the training curve from the paper where the NoisyNet DQN and the NoisyNet dueling learning progress is compared to the baseline version.

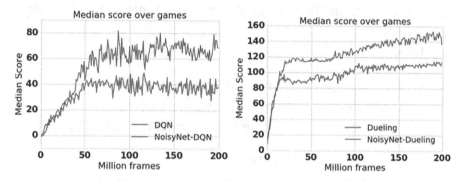

Figure 7-8. *NoisyNet comparison to the baseline from Figure 2 of the paper* `https://arxiv.org/pdf/1706.10295.pdf`

You are also encouraged to try to code a noisy version of dueling DQN. Further, you could also try the DDQN variant of learning. In other words, with what you have learned so far, you could try the following combinations:

- DQN

- DDQN (impacts how you learn)

- Dueling DQN (impacts the architecture of training)

- Dueling DDQN

- Replace the vanilla `ReplayBuffer` with a prioritized replay buffer

- Replace ε-exploration with NoisyNets in any of the previous approaches

- Code all combinations

- Try many other Gymnasium environments, making appropriate changes, if any, to the network

- Run some of them on Atari, especially if you have access to a GPU machine

Categorical 51-Atom DQN (C51)

In a 2017 paper titled "A Distributional Perspective on Reinforcement Learning,"[6] the authors argued in favor of the distributional nature of RL. Instead of looking at the expected values like the q-value, they looked at Z, a random distribution whose expectation is Q.

Up to now you have been outputting $Q(s, a)$ values for an input state s. The number of units in the output were of size n_action. In a way, the output values were the expected $Q(s, a)$ as shown here:

$$Q(s,a) = E\big[R(s,a)\big] + \gamma E\big[Q(s',a')\big]$$

In practice you have been estimating it using the Monte Carlo technique of averaging over a single sample or number of samples to form an estimate $\hat{Q}(s,a)$ of the actual expected value $E[Q(s, a)]$ as discussed earlier in chapter on TD learning.

In categorial 51-Atom DQN, for each $Q(s, a)$ (i.e. one per action and in total n_action of them), you produce an estimate of the distribution of $Q(s, a)$ values: n_atom (51 to be precise) values for each $Q(s, a)$. The network is now predicting the entire distribution modeled as a categorical probability distribution instead of just estimating the mean value of $Q(s, a)$.

$$Q(s,a) = \sum_i z_i p_i(s,a)$$

$p_i(s, a)$ is the probability that the action value at (s, a) will be z_i.

You now have n_action * n_atom outputs, that is, n_atom outputs for each value of n_action. Further, these outputs are probabilities. For one action, you have n_atom probabilities, and these are the probability of the q-value being in any one of the n_atom discrete values in the range V_min to V_max. Refer to refer to the paper for more details.

In the C51 version of distributional RL, the authors took i to be 51 atoms (support points) over the values -10 to 10. You will use the same setup in the code. As these values are parameterized in the code, you are welcome to change them and explore the impact.

After the Bellman updates are applied, the values shift and may not fall on the 51 support points. There is a step of projection to bring back the probability distribution to the support points of 51 atoms, that is, to the immediate value of z_i.

---

[6] https://arxiv.org/pdf/1707.06887.pdf

The loss is also replaced from the mean squared error to cross-entropy loss. The agent is trained using an ε-greedy policy, similar to DQN. The math is fairly involved, and it will be a good exercise for you to go through the paper together with the code to link each code line to the specific detail in the paper. This is an important skill a practitioner of RL needs to have.

Let's now look at code. Similar to the DQN approach, you have the CategoricalDQN class, as shown in Listing 7-8, which is the neural network that takes the state *s* as input to produce the distribution Z of Q(*s*, *a*). There is a function to calculate the TD loss: td_loss_categorical_dqn. As discussed earlier, you need a projection step to bring the values back to the n_atom support points, which is carried out in the compute_projection function. The compute_projection function is used inside td_loss_categorical_dqn while calculating the loss. The rest of the training remains the same as before.

Listing 7-8. CategoricalDQN from 7.e-c51_ dqn.ipynb

```python
class CategoricalDQN(nn.Module):
    def __init__(self, state_shape, n_actions, n_atoms=51, Vmin=-10,
    Vmax=10, epsilon=0):

        # code omitted
        # a simple NN with state_dim as input vector (input is state s)
        # and self.n_actions as output vector of logits of q(s, a)
        self.fc1 = nn.Linear(state_dim, 64)
        self.fc2 = nn.Linear(64, 128)
        self.fc3 = nn.Linear(128, 32)
        self.probs = nn.Linear(32, n_actions * n_atoms)

    def forward(self, state_t):
        # pass the state at time t through the network to get Q(s,a)
        x = F.relu(self.fc1(state_t))
        x = F.relu(self.fc2(x))
        x = F.relu(self.fc3(x))
        probs = F.softmax(self.probs(x).view(-1, self.n_actions, self.n_
        atoms), dim=-1)
        return probs
```

```
def get_probs(self, states):
    # input is an array of states in numpy and output is Qvals as
    numpy array
    # code omitted

def get_qvalues(self, states):
    # input is an array of states in numpy and output is Qvals as
    numpy array
    states = torch.tensor(states, device=device, dtype=torch.float32)
    probs = self.forward(states)
    support = torch.linspace(self.Vmin, self.Vmax, self.n_atoms)
    qvals = support * probs
    qvals = qvals.sum(-1)
    return qvals.data.cpu().numpy()

def get_action(self, states):
    # similar to get_qvalues just that it returns the best action
    return best_actions

def sample_actions(self, qvalues):
    # similar to get_action with added e-greedy exploration
    should_explore = np.random.choice(
        [0, 1], batch_size, p=[1-epsilon, epsilon])
    return np.where(should_explore, random_actions, best_actions)
```

Figure 7-9 shows the training curves for running this through the `CartPole` environment.

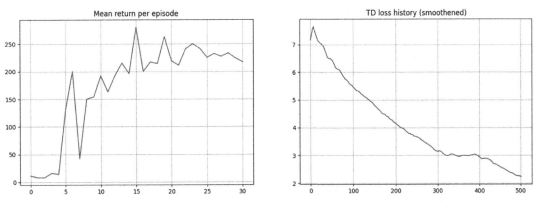

Figure 7-9. *Categorical 51 Atom DQN (C51) training graph*

Quantile Regression DQN

A little after the paper on C51 algorithm was published in mid-2017, some of the original authors with a few additional ones, all from DeepMind, came up with a variant that they called *Quantile Regression DQN*. In a paper titled "Distributional Reinforcement Learning with Quantile Regression,"[7] the authors used a slightly different approach than the original C51, however, in the same focus area of distribution RL.

Similar to the C51 approach of distributional RL, the QR-DQN approach also depended on using quantiles to predict the distribution of $Q(s, a)$ instead of predicting an estimate of the average of $Q(s, a)$. Both C51 and QR-DQN are variants of distributional RL and were produced by scientists from DeepMind.

The C51 approach modeled the distribution of $Q^\pi(s, a)$ named $Z^\pi(s, a)$ as a categorical distribution of probability over fixed points in the range of V_{min} to V_{max}. The probability over these points was learned by the network. Such an approach resulted in use of a *projection* step after the Bellman update to bring the new probabilities back to the fixed support points of n_atoms, spread uniformly over V_{min} to V_{max}. While the result worked, there was a bit of disconnect with the theoretical basis on which the algorithm was derived.

In QR-DQN, the approach is slightly different. The support points are still N but now the probabilities were fixed to 1/N with the location of these points being learned by the network. To quote the authors:

> We "transpose" the parametrization from C51: whereas the former uses N fixed locations for its approximation distribution and adjusts their probabilities, we assign fixed, uniform probabilities to N adjustable locations

The DQN network for QR DQN is similar to C51 DQN, as shown in Listing 7-9. The network takes a batch of states to produce the distribution Z of $Q(s, a)$.

Listing 7-9. QRDQN from 7.f-qr_ dqn.ipynb

```
class QRDQN(nn.Module):
    def __init__(self, state_shape, n_actions, N=51, epsilon=0):

        # code omitted
```

---

[7] https://arxiv.org/pdf/1710.10044.pdf

```
    # a simple NN with state_dim as input vector (input is state s)
    # and self.n_actions as output vector of logits of q(s, a)
    self.fc1 = nn.Linear(state_dim, 64)
    self.fc2 = nn.Linear(64, 128)
    self.fc3 = nn.Linear(128, 32)
    self.q = nn.Linear(32, n_actions*N)

def forward(self, state_t):
    # pass the state at time t through the network to get Q(s,a)
    x = F.relu(self.fc1(state_t))
    x = F.relu(self.fc2(x))
    x = F.relu(self.fc3(x))
    qvalues = self.q(x)
    qvalues= qvalues.view(-1, self.n_actions, self.N)
    return qvalues

def get_qvalues(self, states):
    # input is an array of states in numpy and output is Qvals as
    numpy array

def get_action(self, states):
    # return best action from a batch of states

def sample_actions(self, qvalues):
    # sample actions from a batch of q_values using epsilon
    greedy policy
```

The loss used in QR DQN is that of quantile regression loss mixed with Huber loss and called *Quantile Huber loss*. Equations 9 and 10 in the linked paper give the details. I do not provide the code listings as I want you to read the paper and match the equations in the paper with the code in the 7.f- qr_dqn.ipynb notebook. The paper is dense with mathematics, so unless you are very comfortable with advanced math, you should try to focus on the higher-level details of the approach.

Figure 7-10 shows the training reward and loss curves.

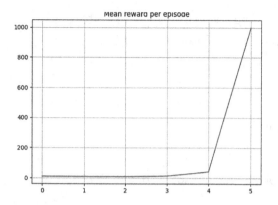
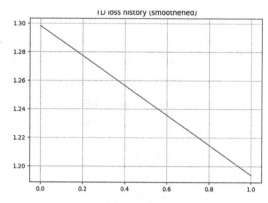

Figure 7-10. *Quantile regression DQN*

Hindsight Experience Replay

In the 2018 paper by OpenAI titled "Hindsight Experience Replay,"[8] the authors presented a sample efficient approach to learn in an environment where the rewards are sparse. The common approach is to shape the reward function in a way to guide the agents toward optimization. This is not generalizable.

Compared to RL agents, which learn from a successful outcome, humans seem to learn not just from that but also from unsuccessful outcomes. This is the basis of the idea proposed in the hindsight replay approach known as *hindsight experience replay* (HER). While HER can be combined with various RL approaches, in this code walk-through, you will use HER with dueling DQN, called HER-DQN.

In the HER approach, after an episode is played out, say an unsuccessful one, you form a secondary objective where the original goal is replaced with the last state before termination as a goal for this trajectory.

Say an episode has been played out $s_0, s_1, ... s_T$. Normally you store a tuple of $(s_t, a_t, r, s_{t+1}, done)$ in the replay buffer. Let's say the goal for this episode was g which could not be achieved in this run. In the HER approach, you will store the following in the replay buffer:

---

[8] https://arxiv.org/pdf/1707.01495.pdf

- $(s_t || g, a_t, r, s_{t+1} || g, done)$

- $(s_t || g', a_t, r(s_t, a_t, g'), s_{t+1} || g', done)$: other state transitions based on synthetical goals like last state of the episode as a sub-goal say g'. The reward is modified to show how state transition $s_t \rightarrow s_{t+1}$ was good or bad for the sub-goal of g'.

The original paper discusses various strategies for forming these subgoals. You will use the one called future:

- future replays with k random states which come from the same episode as the transition being replayed and were observed after it.

You will also use a different kind of environment from past notebooks. You will use an environment of bit-flipping experiment. Say you have a vector with n-bits, each being binary in the range {0,1}. Therefore, there are 2^n combinations possible. At reset, the environment starts in a n-bit configuration chosen randomly and the goal is also randomly picked to be some different n-bit configuration. Each action is to flip a bit. The bit to be flipped is the policy $\pi(a|s)$ that agent is trying to learn. An episode ends if the agent finds the right configuration matching the goal or when agent has exhausted n actions in an episode.

Listing 7-10 contains the code for the environment. The complete code is in the 7.g-her_dqn.ipynb notebook.

Listing 7-10. Bit-Flipping Environment from 7.g-her_dqn.ipynb

```
class BitFlipEnvironment:

    def __init__(self, bits):
        self.bits = bits
        self.state = np.zeros((self.bits, ))
        self.goal = np.zeros((self.bits, ))
        self.reset()

    def reset(self):
        self.state = np.random.randint(2, size=self.bits).astype(np.
        float32)
        self.goal = np.random.randint(2, size=self.bits).astype(np.float32)
        if np.allclose(self.state, self.goal):
            self.reset()
        return self.state.copy(), self.goal.copy()
```

```
def step(self, action):
    self.state[action] = 1 - self.state[action]  # Flip the bit on
    position of the action
    reward, done = self.compute_reward(self.state, self.goal)
    return self.state.copy(), reward, done

def render(self):
    print("State: {}".format(self.state.tolist()))
    print("Goal : {}\n".format(self.goal.tolist()))

@staticmethod
def compute_reward(state, goal):
    done = np.allclose(state, goal)
    return 0.0 if done else -1.0, done
```

You have implemented your own render and step functions so that the interface of this environment remains similar to the ones in Gymnasium and so that you can use the previously developed machinery. The compute_reward custom function returns the reward and done flags when given the input of a state and a goal.

The authors show that, with a regular DQN, where the state (configuration of n-bits) is represented as a deep network, it is almost impossible for a regular DQN agent to learn beyond 15-digit combinations. However, coupled with the HER-DQN approach, the agent is able to learn easily, even for large-digit combinations like 50 or so. Figure 7-11 shows the complete pseudocode from the paper with certain modifications to make it match with the notations.

HINDSIGHT EXPERIENCE REPLAY (HER)

Input:

    An off policy algorithm (e.g. DQN or its variants)

    A strategy of sampling goals for replay *(future, episode, random)*

    A reward function (e.g. 0 if goal not met and 1 if goal met)

Initialize:

    Initialize neural network $\mathbf{A}$

    Initialize Replay Buffer $\mathbf{R}$

loop for each episode in $(1, M)$:

    start episode with s_0 (non terminal) and goal g

    for $t = 0, T - 1$:

        select a_t using a behavior policy (e.g. ε-greedy)

            $a_t \leftarrow \pi_b\left(s_t\|g\right)$

        take action a_t and observe reward r_t next state s_{t+1}

        record transition $\left(s_t, a_t, r_t, s_{t+1}, done_t\right)$ in a temporary array $\mathbf{ET}$

    for $t = 0, T - 1$:

        store $\left(s_t\|g, a_t, r_t, s_{t+1}\|g, done_t\right)$ in $\mathbf{R}$

        sample additional goals using sampling strategy e.g. (future)

        for _ in $0, k$:

            select a transition k from trajectory in future

            $g' \leftarrow s'_k$, next state from k^{th} transition in trajectory array

            calculate new reward r' and $done'$ flag using $\left(s_t, a_t, g'\right)$

            store transition $\left(s_t\|g', a_t, r', s_{t+1}\|g', done'\right)$ in $\mathbf{R}$

    for $t = 1, N$:

        sample a batch from replay buffer

        perform one step of gradient descent

Figure 7-11. *HER using a future strategy*

You use dueling DQN, which is similar to the one used for `CartPole` earlier in this chapter. For the BitFlipping environment, you use a smaller version of the same network. The most interesting part of the code is the implementation of the HER algorithm, as per the pseudocode given in Figure 7-11.

Let's look at the way replay buffer is filled. There are two ways as discussed previously. The first is to fill the replay buffer usual way just with the difference that you augment the states with the goal as depicted by $(s_t||g, a_t, r, s_{t+1} | g, done)$. The more interesting part is the additional transitions that are filled into the buffer using synthetic goals, that is, $(s_t||g', a_t, r(s_t, a_t, g'), s_{t+1} | g', done)$. You use the *future* sampling strategy among the many options that are discussed in the paper. The future strategy involves replay with k random states as synthetic goals g' which come from the same episode as the transition being replayed and were observed after it. The reward $r(s_t, a_t, g')$ is 0 if the bits of s_t and g'. The reward is -1 if these two do not match. Listing 7-11 shows the way this is accomplished. This is part of the overall `train_her` agent training function in the 7.g-her_dqn.ipynb notebook.

Listing 7-11. HER Replay Buffer Filling Using a Future Strategy from 7.g-her_dqn.ipynb

```
for t in range(steps_taken):

    # Usual experience replay
    state, action, reward, next_state, done = episode_trajectory[t]
    state_, next_state_ = np.concatenate((state, goal)),
    np.concatenate((next_state, goal))
    exp_replay.add(state_, action, reward, next_state_, done)

    # Hindsight experience replay
    for _ in range(future_k):
        future = random.randint(t, steps_taken)  # index of future
        time step

        # take future next_state from (s,a,r,s',d) and set as goal
        new_goal = episode_trajectory[future][3]
        new_reward, new_done = env.compute_reward(next_state, new_goal)
        state_, next_state_ = np.concatenate((state, new_goal)), \
                np.concatenate((next_state, new_goal))
        exp_replay.add(state_, action, new_reward, next_state_, new_done)
```

Figure 7-12 shows the training curve. For a 50-bit `BitFlipping` environment, the agent with HER can successfully solve the environment 100 percent of the time. Remember, the environment starts with a random combination of 50 bits and has another random combination as the goal. The agent has a maximum of 50 flipping moves to reach the goal combination. An exhaustive search would require the agent to try each of the 2^{50} combinations except the one it started with.

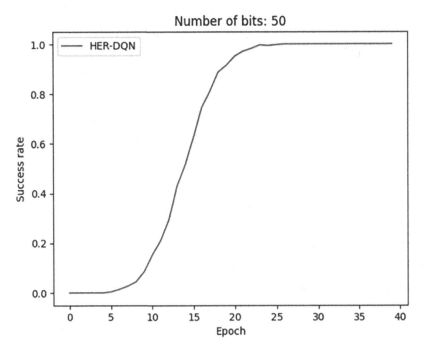

Figure 7-12. *Success rate graph: bit flipping environment with HER using a future strategy*

This brings you to the end of discussion on HER as well as to the end of chapter.

Summary

This was a short but dense chapter, where you looked at various modifications to the vanilla DQN.

The chapter began by looking at prioritized replay, where the samples from the buffer are picked based on a certain importance score assigned to them proportional to the magnitude of the TD error.

Following this, you revisited double Q-learning in the context of DQN, known as double DQN. This is an approach that impacts the way learning takes place and attempts to reduce the maximization bias.

You then looked at dueling DQN in which two networks with an initial shared network were used. This was followed by NoisyNets in which ε-greedy exploration was replaced by noisy layers.

Next, you looked at two flavors of distributional RL under which the network produced Z, a distribution of q-values. Instead of producing the expected action-value $Q(S, A)$, in Distributional RL, the whole distribution is produced, specifically a categorical distribution. You also saw the use of the projection step and losses like *cross entropy* and *quantile Huber loss*.

The last section was on hindsight experience replay, which addresses learning in environments with sparse rewards. Previous learning approaches were centered on learning only from successful outcomes, but hindsight replay allows you to learn from unsuccessful outcomes as well.

Many of the algorithms and approaches you saw in this chapter are state-of-the-art research. You will gain a lot by looking at the original papers as well as going through the code line by line. I have also suggested the various combinations that you could try to code to further cement the concepts in your mind.

This chapter concludes the exploration of value-based methods, where you learn a policy by first learning with V or Q functions and then use these to find an optimal policy. The next chapter switches to policy-based methods, where you find the optimal policy without the involvement of the intermediary step of learning V/Q functions.

CHAPTER 8

Policy Gradient Algorithms

Up to now, the book has focused on model-based and model-free methods. All the algorithms using these methods estimate the state or state-action values for a given current policy as the first step. In the second step, these estimated values are used to find a better policy by choosing the best action in a given state. These two steps are carried out in a loop until no further improvement in values is observed. In this chapter, you look at a different approach for learning optimal policies, by directly operating in the policy space. You will learn to improve the policies without explicitly learning or using state or state-action values.

You will also learn that the policy-based approach and the value-based approach are not two disjoint approaches. There are approaches that combine the value-based and policy-based approaches such as the actor-critic methods.

The core of the chapter establishes the definition and mathematically derives the key parts of policy-based optimization. The policy-based approach is currently one of the most popular family of approaches for solving large and continuous space problems in reinforcement learning.

Introduction

You started your journey by first looking at simple model-based approaches wherein you solved small, discrete state space problems by iterating over Bellman equations. Following this, the book discussed a model-free setup using the Monte Carlo and temporal difference approaches. I then extended the analysis to cover large or continuous state spaces using function approximation. In particular, you learned about DQN and many of its variants as the way to go for policy learning.

309

© Nimish Sanghi 2024
N. Sanghi, *Deep Reinforcement Learning with Python*, https://doi.org/10.1007/979-8-8688-0273-7_8

The core idea of all these approaches is to first learn the value of the current policy and then make iterative improvements to the policy to get better rewards. This is done using the general framework of *Generalized Policy Iteration* (GPI). If you think for a minute, you will realize that the real objective is to learn a good policy. You use value functions as an intermediary step to finding a good policy.

This approach of learning value functions to improve a policy is an indirect one. It is not always easy to learn values as compared to learning good policies directly. Consider the case of encountering a bear on a jogging trail. What comes to your mind first? Does your brain try to evaluate the state (the bear in front of you) and the possible actions ("freeze," "pet the bear," "run for your life," or "attack the bear")? Or do you "run" with almost certainty, that is, follow a policy of `action="run"` with a probability of 1.0? I am sure the answer is the latter one. Consider another example of playing a game of Atari Breakout, the one used in the DQN examples in the previous chapter. Consider the situation where the ball is near the right edge of your paddle and moving away from the paddle (the "state"). As a human player, what do you do? Do you try to evaluate $Q(s, a)$, the state-action values for the two actions, to decide whether the paddle needs to move right or left? Or do you just look at the state and learn to move the paddle toward the right to avoid having the ball fall? Again, I am sure the answer is the second one. In both of these examples, the latter and easier alternative is learning to act directly instead of learning the value first, followed by use of the state value to find the best action among the possible choices.

Pros and Cons of Policy-Based Methods

The previous examples show that in many cases, it is easier to learn the policy (which action to take in a given state) as compared to learning the value functions and then using them to learn a policy. Why then did I go through the approach of value methods like SARSA, Q-learning, DQN, and so on, at all? Because policy learning, while easier, is not a bed of roses. It has its own set of challenges, especially based on your current knowledge and available algorithms, which are listed here.

Advantages:

- **Better convergence:** The convergence during training toward an intended behavior is better for policy-based methods, as they directly optimize the policy instead of value-based algorithms that go through two step process of value function learning to policy optimization.

- **Effective in high-dimensional and continuous action spaces:** Policy-based methods can handle high dimensional and continuous action spaces.

- **Adaptability to complex environments:** Policy-based methods are adaptive to complex and dynamic environments, allowing for effective decision-making in situations where the optimal policy may change over time.

- **Learn stochastic policies:** Policy-based methods that are learning stochastic policies inherently incorporate exploration, thereby offering a better balance between exploration and exploitation.

Disadvantages:

- **Sample inefficiency and computational overload:** Policy-based methods require many more episodes to learn optimal behavior. Therefore, these methods may need longer runtime and extra computational load.

- **Convergence issues:** Vanilla policy-based methods at times converge to local maximums instead of global ones. However, to a large extent these can be addressed with methods like PPO.

- **Policy evaluation inefficient and with high variance:** Vanilla policy-based methods do not maintain value functions, causing policy evaluation to be inefficient. Therefore, evaluation must be carried out over high number of episodes to keep the variance of policy estimate within a reasonable variance.

Elaborating on these points, remember the DQN learning curves? You saw how the value of policy varied a lot during training. In the "cart pole" problem, you saw that the score (as shown in the left graph of all the training progress diagrams in the previous chapter) fluctuated a lot. There was no steady convergence toward a better policy. Policy-based methods, especially with some additional controls that I cover toward the end of the chapter, ensure a smooth and steady progress toward better policies as you move through the learning process.

The action space has always been a small set of possible actions. Even in the case of function approximation coupled with deep learning like DQN, the action spaces were limited to single digits. The actions were unidimensional. Imagine trying to control various joins of a walking robot together. You would need to make decisions about each of the joints of the robot, and these individual choices together would make a complete action in a given state of the robot. Further, the individual actions of each joint are not going to be discrete ones. Most likely, the actions, such as speed of motors or angles by which an arm or a leg needs to move, are going to be in a continuous range. Policy-based methods are more suited to handle these actions.

In all the value-based methods you have seen so far, you always learned an optimal policy—a deterministic one where you knew with certainty the best action to take in a given state. I had to introduce the concept of exploration to try different actions using ε-greedy policies, where you started with a non-zero exploration probability, which steadily reduced to a very small value as training progressed and the agent learned to act with better actions. The final output was always a deterministic policy.

However, deterministic policies are not always *optimal*. There are situations where the optimal policy is to take multiple actions with some probability distribution, especially in multi-agent environments. If you have some experience with game theory, you will realize this immediately from setups like prisoner's dilemma and the corresponding Nash equilibrium. Anyway, let's look at a simple situation.

Have you ever played the game of rock-paper-scissors? It is a two-player game. In a single turn, each player must choose one of the three options—scissors, rock, or paper. Both players do so at the same time and reveal their choice at the same time. The rules state that scissors beat paper because scissors can cut paper, rock beats scissors because a rock can be used to smash the scissors, and paper beats rock because the paper can cover the rock.

What is the best policy? There are no clear winners. If you always choose, let's say, rock, then I as your opponent will exploit that knowledge and always choose paper. Can you think of any other deterministic policy (i.e., of choosing one of the three all the time)? To prevent the opponent from exploiting your strategy, you must be completely random in making the choice. You must choose scissors or rock or paper randomly with equal probability, that is, use a stochastic policy. Deterministic policies are a specialized form of stochastic policies in which one choice has a probability of 1.0 with all other actions having a 0 probability. Stochastic policies are more general, and that's what the policy-based approach learns.

Here is the deterministic policy:

$$a = \pi_\theta(s)$$

In other words, this is the specific action a to be taken when in state s.

Here is the stochastic policy:

$$a \sim \pi_\theta\left(a|s\right)$$

In other words, this is the distribution of action probability in a given state s. To choose an action, you sample from this probability the distribution of the actions in a given state.

There are disadvantages as well. Policy-based approaches, while having good convergence, can converge into local maxima. The second big disadvantage is that policy-based methods do not learn any direct representation of value functions, making evaluating the value of a given policy inefficient. Evaluating the policy usually requires rolling out multiple episodes of an agent using a policy and summing up the reward sequence of each rollout to form an estimate of the state or state-action value, essentially the MC approach. The MC approach comes with the issue of high variance in estimates that require multiple rollouts to take an average and get an estimate with lower variance. I discuss ways to address this concern by combining the best of both worlds of value-based methods and policy-based methods. This is known as the *actor-critic* family of algorithms.

Policy Representation

In Chapter 5, which covered model-free setups with function approximation, I represented value functions in Equation 5-1 as follows:

$$\hat{v}\left(s;w\right) \approx v_\pi\left(s\right)$$

$$\hat{q}\left(s,a;w\right) \approx q_\pi\left(s,a\right)$$

There was a model (a linear model or a neural network) with weights w. It represented the state value v and state-action value q with functions parameterized by weights w. Instead, you will now parameterize the policy directly as follows:

$$\pi\left(a|s;\theta\right) \approx \pi_\theta\left(a|s\right)$$

θ is the parameter of the function that defines the policy function, and what the learning algorithm will tune to make the agent learn to act in a meaningful way, generating desired rewards. Next you'll see how this parameterization is done for cases when actions are discrete and when actions form a continuous range of values.

Discrete Cases

For not-too-large a discrete action space, you can parameterize another function $h(s, a; \theta)$ for the state-action pairs. The probability distribution is formed using a soft-max of h.

$$\pi(a|s;\theta) = \frac{e^{h(s,a;\theta)}}{\sum_b e^{h(s,b;\theta)}}$$

The value $h(s, a; \theta)$ is known as *logits* or *action preferences*. It is like the approach used in the supervised classification case. In supervised learning, you input X, the observations. Here in RL, you input the state S to the model. The outputs of the model in supervised cases are the logits of input X belonging to different classes. And in RL, the output of the model is *action preference h* for taking that specific action, *a*.

Continuous Cases

In continuous action spaces, the Gaussian representation of a policy is a natural choice. Suppose the action space is continuous and multidimensional with a dimension of, say, d. The model will take state S as input and produce the multidimensional mean vector $\mu \in R^d$. Variance $\sigma^2 I_d$ can also be parameterized or can be kept constant. The policy the agent will follow is a Gaussian policy with mean μ and a variance $\sigma^2 I_d$. The agent will poll actions from this probability distribution.

$$\pi(a|s;\theta) \sim N(\mu, \sigma^2 I_d)$$

Here, action a & $\mu \in R^d$.

Having formulated the notations of how policy is represented as a parameterized function, you will now shift your focus to the way the policy is learned by the agent directly, without resorting to learning state and state-action values.

Policy Gradient Derivation

The approach to deriving a policy-based algorithm is similar to what you do in supervised learning. Here is the outline of steps to follow to come up with the algorithm:

1. Form an objective that you want to maximize, just like in supervised learning. It will be the total reward received by following a policy. That is the objective you want to maximize.

2. Derive the gradient update rule to carry out the gradient ascent. You are doing gradient ascent and not gradient descent, as the objective is to maximize the total average return per episode/rollout.

3. You need to recast the gradient update formula into an expectation so that the gradient update can be approximated using samples, such as an Monte Carlo estimate of an expectation.

4. Now you formally convert the update rule into an algorithm that can be used with auto-differentiation libraries like PyTorch, JAX, or TensorFlow.

Objective Function

Let's start with the objective you want to maximize. As mentioned, it will be the value of the policy, that is, the return an agent can get by following a policy. There are many variations to the expected return representation. You will look at some of these and learn briefly about the context of when to use which representation. However, the detailed derivation of the algorithm is done using one of the variants, as the derivations for other reward formulations are similar. The reward function and its variants are as follows:

- Episodic undiscounted: $J(\theta) = \sum_{t=0}^{T-1} r_r$

- Episodic discounted: $J(\theta) = \sum_{t=0}^{T-1} \gamma^t r_t$

- Infinite horizon discounted: $J(\theta) = \sum_{t=0}^{\infty} \gamma^t r_t$

- Average reward: $J(\theta) = \lim_{T \to \infty} \frac{1}{T} \sum r_t$

Most of this derivation will follow an episodic undiscounted structure of reward, just to keep the math simple and focus on the key aspects of the derivation.

WHAT ROLE DOES γ PLAY?

I also want to give you a feel for the discount factor, γ. The discount is used in infinite formulation to keep the sum bounded. Usually, you use a discount value of 0.99 or something similar to keep the sum theoretically bounded. In some formulations, the discount factor also determines what role the rate of interest plays in financial markets—that is, a reward today is worth more than the same reward tomorrow. Using a discount factor brings in the concept of favoring the reward today.

The discount factor is also used to reduce the variance in the estimation by providing a soft cutoff of the time horizon. Suppose you get a reward of 1 at every time step and you use the discount factor of γ. The sum of this infinite series is $\dfrac{1}{1-\gamma}$. Say you have $\gamma = 0.99$. The infinite series sum is then equal to 100. Therefore, you can think of a discount of 0.99 as limiting your horizon to 100 steps, in which you collected a reward of 1 in each of the 100 steps, to give you a total return of 100.

To summarize, a discount of γ implies a time horizon of $\dfrac{1}{1-\gamma}$ steps.

Using γ also ensures that the impact of changing policy actions in the initial part of the trajectory has a higher impact on the overall quality of the policy versus the impact of decisions in the later part of the trajectory.

Moving back to derivation, you can now calculate the gradient update for improving a policy. The agent follows a policy as parameterized by θ. This involves a bit of math and if you want to skip ahead, skip to Equations 8-16 and 8-17, and follow along from that point.

$$\pi_\theta\left(a_t | s_t\right) \tag{8-1}$$

The agent follows the policy and generates the trajectory τ as follows:

$$s_1 \rightarrow a_1 \rightarrow s_2 \rightarrow a_2 \rightarrow \ldots. \rightarrow s_{T-1} \rightarrow a_{T-1} \rightarrow s_T \rightarrow a_T$$

Here, s_T is not necessarily the terminal state but some time horizon T up to which you are considering the trajectory.

The probability of observing trajectory τ during a rollout depends on the transition probabilities $p(s_{t+1}|s_t, a_t)$ and the policy $\pi_\theta(a_t|s_t)$. It is given by the following expression:

$$p_\theta(\tau) = p_\theta(s_1, a_1, s_2, a_2, \cdots, s_T, a_T) = p(s_1)\prod_{t=1}^{T}\pi_\theta(a_t|s_t)p(s_{t+1}|s_t, a_t) \tag{8-2}$$

The expected return from following the policy is given by the following:

$$J(\theta) = E_{\tau \sim p_\theta(\tau)}\left[\sum_t r(s_t, a_t)\right] \tag{8-3}$$

You want to find the θ that maximizes the expected return $J(\theta)$. In other words, the optimal $\theta = \theta^*$ is given by the following expression:

$$\theta^* = \operatorname*{argmax}_\theta E_{\tau \sim p_\theta(\tau)}\left[\sum_t r(s_t, a_t)\right] \tag{8-4}$$

Before moving forward, consider how you will evaluate the objective $J(\theta)$. You convert the expectation in Equation 8-3 to an average over samples, that is, you run the agent through the policy multiple times, collecting N trajectories. You calculate the total reward (total reward in a trajectory is also called the return of an episode or a single rollout) in each trajectory and take an average of the return across N trajectories. This is a Monte Carlo (MC) estimation of the expectation. This is what is meant when I talked about evaluating a policy. The expression you obtain is the following:

$$J(\theta) \approx \frac{1}{N}\sum_{i=1}^{N}\sum_{t=1}^{T}r(s_t^i, a_t^i) \tag{8-5}$$

In the notation in Equation 8-5, s_t^i represents the state s in step t of rollout instance i. There are T steps in a rollout and there are N rollouts in total.

Derivative Update Rule

Moving on, let's try to find the optimal θ. To keep the notations easier to understand, I replace $\sum_t r(s_t, a_t)$ with $r(\tau)$. Rewriting Equation 8-3, you get the following:

$$J(\theta) = E_{\tau \sim p_\theta(\tau)}\left[r(\tau)\right] = \int p_\theta(\tau)r(\tau)d\tau \tag{8-6}$$

Take the gradient/derivative of Equation 8-6 with respect to θ.

$$\nabla_\theta J(\theta) = \nabla_\theta \int p_\theta(\tau) r(\tau) d\tau \tag{8-7}$$

Using linearity, you can move the gradient inside the integral.

$$\nabla_\theta J(\theta) = \int \nabla_\theta p_\theta(\tau) r(\tau) d\tau \tag{8-8}$$

With the log derivative trick, you know that $\nabla_x f(x) = f(x) \nabla_x \log f(x)$. Using this, you can write the previous Equation 8-8 as follows:

$$\nabla_\theta J(\theta) = \int p_\theta(\tau) \left[\nabla_\theta \log p_\theta(\tau) r(\tau) \right] d\tau \tag{8-9}$$

THE LOG DERIVATIVE TRICK

Start with a probability distribution $p_\theta(x)$, where x is the random variable and θ represents the parameters of the probability distribution. Say you have this expression:

$$G(\theta) = E_{x \sim p_\theta(x)} \left[f(x) \right]$$

You now want to calculate the derivate of $G(\theta)$ w.r.t \theta. For simplicity's sake, assume that x and θ are both single dimensional variables. You write $G(\theta)$ by converting expectation to an integral. You can rewrite $G(\theta)$ as:

$$G(\theta) = \int p_\theta(x) f(x) dx$$

Taking a derivative w.r.t θ on both sides, you get:

$$\frac{d}{d\theta} G(\theta) = \frac{d}{d\theta} \int p_\theta(x) f(x) dx$$

Taking a derivative inside due to linearity property, you get:

$$\frac{d}{d\theta} G(\theta) = \int \left(\frac{d}{d\theta} p_\theta(x) \right) f(x) dx$$

Now multiply and divide the right side by $p_\theta(x)$:

$$\frac{d}{d\theta} G(\theta) = \int \left(\frac{p_\theta(x)}{p_\theta(x)} \frac{d}{d\theta} p_\theta(x) \right) f(x) dx$$

This gives:

$$\frac{d}{d\theta}G(\theta) = \int p_\theta(x)\left(\frac{1}{p_\theta(x)}\frac{d}{d\theta}p_\theta(x)\right)f(x)dx$$

The expression inside the brackets on the left side is the derivative of $\log p_\theta(x)$:

$$\frac{d}{d\theta}G(\theta) = \int p_\theta(x)\left(\frac{d}{d\theta}\log p_\theta(x)\right)f(x)dx$$

Now convert the integral back to expectation over $p_\theta(x)$:

$$\frac{d}{d\theta}G(\theta) = E_{\theta \sim p_\theta(x)}\left[\frac{d}{d\theta}\log p_\theta(x)f(x)\right]$$

This is log derivative trick, where the derivative of a function is the function in an expectation and you want to take the derivate of the function with respect to the parameters of the probability distribution over which the expectation has been taken. The log derivative trick allows you to represent the derivative, which in turn allows you to take Monte Carlo estimates of the derivative expression.

You can now write the integral back as the expectation, which gives the following expression:

$$\nabla_\theta J(\theta) = E_{\tau \sim p_\theta(\tau)}\left[\nabla_\theta \log p_\theta(\tau)r(\tau)\right] \tag{8-10}$$

Now expand the term $\nabla_\theta \log p_\theta(\tau)$ by writing out the full expression of $p_\theta(\tau)$ from Equation 8-2.

$$\nabla_\theta \log p_\theta(\tau) = \nabla_\theta \log\left[p(s_1)\prod_{t=1}^{T}\pi_\theta(a_t|s_t)p(s_{t+1}|s_t,a_t)\right] \tag{8-11}$$

You know that the log of the product of terms can be written as a sum of the log of terms. In other words:

$$\log\prod_i f_i(x) = \sum_i \log f_i(x) \tag{8-12}$$

Substituting Equation 8-12 in Equation 8-11, you get the following:

$$\nabla_\theta \log p_\theta(\tau) = \nabla_\theta\left[\log p(s_1) + \sum_{t=1}^{T}\left(\log \pi_\theta(a_t|s_t) + \log p(s_{t+1}|s_t,a_t)\right)\right] \tag{8-13}$$

The only term in Equation 8-13 dependent on θ is $\pi_\theta(a_t|s_t)$. The other two terms—$\log p(s_1)$ and $\log p(s_{t+1}|s_t, a_t)$—do not depend on θ. Accordingly, you can simplify Equation 8-13 as follows:

$$\nabla_\theta \log p_\theta(\tau) = \sum_{t=1}^{T} \nabla_\theta \log \pi_\theta(a_t|s_t) \tag{8-14}$$

By substituting Equation 8-14 into the expression for $\nabla_\theta J(\theta)$ in Equation 8-10, as well as expanding $r(\tau)$ as $\sum_t r(s_t, a_t)$, you get the following:

$$\nabla_\theta J(\theta) = E_{\tau \sim p_\theta(\tau)} \left[\left(\sum_{t=1}^{T} \nabla_\theta \log \pi_\theta(a_t|s_t) \right) \left(\sum_{t=1}^{T} r(s_t, a_t) \right) \right] \tag{8-15}$$

You can now replace the outer expectation with an estimate/average over multiple trajectories to get the following expression for the *gradient of policy objective*:

$$\nabla_\theta J(\theta) \approx \frac{1}{N} \sum_{i=1}^{N} \left[\left(\sum_{t=1}^{T} \nabla_\theta \log \pi_\theta(a_t^i|s_t^i) \right) \left(\sum_{t=1}^{T} r(s_t^i, a_t^i) \right) \right] \tag{8-16}$$

where superscript index i denotes the ith trajectory.

To improve the policy, take a +ve step in the direction of $\nabla_\theta J(\theta)$ with a learning rate α.

$$\theta = \theta + \alpha \cdot \nabla_\theta J(\theta) \tag{8-17}$$

Summarizing, you have designed a model that takes state s as the input and produces the policy distribution $\pi_\theta(a|s)$ as the output of the model. You use a policy as determined by cthe urrent model parameters θ to generate trajectories and calculate the total return of each trajectory. You then calculate $\nabla_\theta J(\theta)$ using Equation 8-16 and change the model parameters θ using the expression $\theta = \theta + \alpha \nabla_\theta J(\theta)$ in Equation 8-17.

Intuition Behind the Update Rule

This section develops some intuition behind Equation 8-16 by explaining the equation in words. You carry out an average over N trajectories and that is the outermost sum. What is it that you average over the trajectories? For each trajectory, you look at the total reward obtained in that trajectory, that is, the second term inside the summation. This return is multiplied with the first term, which is the sum of log probabilities of all the actions along that trajectory.

Suppose the total reward $r(\tau^i)$ of a trajectory was +ve. Each gradient inside the first inner sum, $\nabla_\theta \log \pi_\theta\left(a_t^i | s_t^i\right)$, which is the gradient of the log probability of that action—gets multiplied by the total reward $r(\tau^i)$. It results in an individual *gradient-log* term being amplified by the total reward of the trajectory, and in Equation 8-17, its contribution is to move the model parameters θ in the +ve direction of $\nabla_\theta \log \pi_\theta\left(a_t^i | s_t^i\right)$, that is, increase the probability of taking action a_t^i when the system was in state s_t^i. However, for the case when $r(\tau^i)$ was a -ve quantity, Equations 8-16 and 8-17 lead to moving θ in the -ve direction, resulting in a decrease of the probability of taking action a_t^i when the system is in state s_t^i.

You could summarize the whole explanation by saying that policy optimization is all about trial and error. You roll out multiple trajectories. The probability of all the actions along the trajectory is increased for those trajectories that are good. For the trajectories that are bad, the probability of all the actions along those bad trajectories are reduced, as depicted in Figure 8-1.

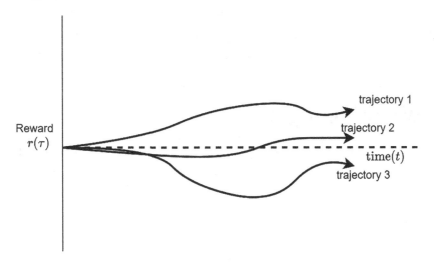

Figure 8-1. *Trajectory rollouts. Trajectory 1 is good, so you want the model to produce more of these. Trajectory 2 is neither good nor bad, and the model should not worry too much about it. Trajectory 3 is bad, so you want the model to reduce the probability of it*

Consider this interpretation by comparing the expression in Equation 8-17 with that of maximum likelihood. If you just wanted to model the probability of seeing the trajectories you saw, you would get the maximum likelihood estimation—you observe some data (trajectories), and you want to build a model with the highest probability of producing the observed data/trajectory. This is the maximum likelihood model building. Under that, you would get an equation as follows:

$$\nabla_\theta J_{MLE}(\theta) \approx \frac{1}{N} \sum_{i=1}^{N} \left[\sum_{t=1}^{T} \nabla_\theta \log \pi_\theta \left(a_t^i | s_t^i \right) \right] \qquad (8\text{-}18)$$

In Equation 8-18, you are just increasing the probability of the actions to increase the overall probability of trajectories that were observed. You are not making any differentiation for the good vs bad trajectories. In the case of policy gradients in Equation 8-16, you are doing something similar to the maximum likelihood estimation (MLE), but with the addition that you are weighing the log probability gradients with the return of the trajectory so that the probability of good trajectories is increased, and the probability of bad trajectories is decreased.

Before concluding this section, one observation I want to make is regarding the Markov property and partial observability. You have not really used the Markov assumption in the derivation. In the end, Equation 8-16 just increases the probability of good stuff and reduces the probability of bad stuff. You have not used Bellman equations so far. Therefore, a policy gradient approach can also work for non-Markovian setups.

The REINFORCE Algorithm

You'll now convert Equation 8-16 into an algorithm for policy optimization. The basic algorithm is shown in Figure 8-2. It is known as REINFORCE.

REINFORCE

Input:
A model with parameters θ taking state s as input and producing $\pi_\theta\left(a|s\right)$
Other parameters: step size α

Initialize:
Initialize weights $\boldsymbol{\theta}$

Loop:
Sample $\{\tau^i\}$, a set of N trajectories from current policy $\pi_\theta\left(a_t|s_t\right)$
Update model parameters θ:

$$\nabla_\theta J\left(\theta\right) \approx \frac{1}{N}\sum_{i=1}^{N}\left[\left(\sum_{t=1}^{T}\nabla_\theta \log \pi_\theta\left(a_t^i|s_t^i\right)\right)\left(\sum_{t=1}^{T}r\left(s_t^i, a_t^i\right)\right)\right]$$
$$\theta = \theta + \alpha\nabla_\theta J\left(\theta\right)$$

Figure 8-2. *The REINFORCE algorithm*

Consider some implementation-level details. Suppose you use a neural network as a model that takes a state value as input and produces the *logit* (log probability) of taking all possible actions in that state. Figure 8-3 shows a diagram of such a model.

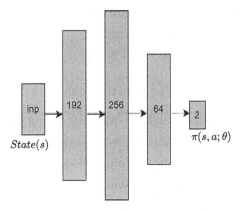

Figure 8-3. *Neural network model for predicting policy*

You can use auto-differentiation libraries such as PyTorch, JAX, or TensorFlow. You do not explicitly calculate the differentiation. Equation 8-16 gives you the expression for $\nabla_\theta J(\theta)$. With PyTorch or such other libraries, you need an expression $J(\theta)$ which, during backward propagation, will use auto differentiation to calculate $\nabla_\theta J(\theta)$ and then update the weights θ of the neural network using Equation 8-17. The neural network model will take state s as input and produce $\pi_\theta(a_t|s_t)$. Therefore, you need to use this output and carry out further computation to arrive at an expression for $J(\theta)$. The auto-differential packages like PyTorch or TensorFlow will automatically calculate the gradient $\nabla_\theta J(\theta)$ from the expression for $J(\theta)$. The correct expression for $J(\theta)$ is as follows:

$$J(\theta) \approx \frac{1}{N}\sum_{i=1}^{N}\left[\left(\sum_{t=1}^{T}\log\pi_\theta\left(a_t^i|s_t^i\right)\right)\left(\sum_{t=1}^{T}r\left(s_t^i,a_t^i\right)\right)\right] \tag{8-19}$$

You can check and confirm that the gradient of this expression will give you the correct value for $\nabla_\theta J(\theta)$, as shown in Equation 8-16. Start with Equation 8-19 for $J(\theta)$ and take a derivative with respect to θ. It will give you an expression for $\nabla_\theta J(\theta)$ that matches the Equation 8-16.

Note that the expression $J(\theta)$ in Equation 8-19 is not same as the one in Equation 8-6. While moving from expression of $J(\theta)$ in Equation 8-6 to the derivative $\nabla_\theta J(\theta)$ in Equation 8-16 and then back to an expression of $J(\theta)$ in Equation 8-19, you introduced the log trick. You removed the terms from probability that were not dependent on θ and rather were the properties of the environment, the state transition probabilities in form of $p(s_{t+1}|s_t, a_t)$. The expression $J(\theta)$ is known as a *pseudo-objective*. It is an expression whose derivate matches the correct derivative expression $\nabla_\theta J(\theta)$, shown in Equation 8-16. This expression implements in auto-diff libraries like PyTorch. It is just being used to tune the agent and, as such, has no physical significance.

You then calculate the log probabilities $\log\pi_\theta\left(a_t^i|s_t^i\right)$, weigh the probabilities with the total reward of the trajectory $\sum_{t=1}^{T}r\left(s_t^i,a_t^i\right)$, and then calculate the negative log likelihood (NLL, or cross-entropy loss) of the weighted quantity, giving you the expression in Equation 8-20. It is similar to the approach for training a multiclass classification model in a supervised learning setup. The only difference is weighing the log probability by the trajectory reward. This is the approach you will take when actions are discrete. The loss you implement in PyTorch is as follows:

$$L_{\text{cross-entropy}}(\theta) = -1 \cdot \frac{1}{N}\sum_{i=1}^{N}\left[\left(\sum_{t=1}^{T}\log\pi_\theta\left(a_t^i|s_t^i\right)\right)\left(\sum_{t=1}^{T}r\left(s_t^i,a_t^i\right)\right)\right] \tag{8-20}$$

Note that PyTorch (and other auto-diff libraries) minimize the loss by taking a step in a negative direction of the loss. And note that the -ve gradient of Equation 8-20 is a +ve gradient of Equation 8-19 due to the presence of a factor of -1 in Equation 8-20.

Therefore, a step in the negative direction of the gradient of $L_{\text{cross-entropy}}$ is a step in the direction that increases the value of $J_{(\theta)}$ in Equation 8-19. By extension, this increases the value of the actual objective $J(\theta)$, as shown in Equation 8-6, which is the average of the return per episode.

Having looked at the formulation when action space is discrete, you can now turn your attention to the formulation under continuous action space. As discussed earlier in this chapter, the model parameterized by θ will take state s as input and produce the mean μ of a multivariate normal distribution. This considers the case where the variance of the normal distribution is known and fixed to some small value, say $\sigma^2 I_d$.

$$\pi(a|s;\theta) \sim N\left(\mu, \sigma^2 I_d\right)$$

Say for state s_t^i the mean that the model produces is $\mu_\theta\left(s_t^i\right)$. The value of $\log \pi_\theta\left(a_t^i|s_t^i\right)$ is then given by the following:

$$\log \pi_\theta\left(a_t^i|s_t^i\right) = \log \frac{1}{\sqrt{2\pi}\sigma} e^{-\frac{1}{2\sigma^2}\left(a_t^i - \mu_\theta\right)^2}$$

$$= -\frac{1}{2}\log 2\pi - \log \sigma - \frac{1}{2\sigma^2}\left(a_t^i - \mu_\theta\right)^2 \tag{8-21}$$

The only value in the previous expression that depends on model parameters θ is $\mu_\theta\left(s_t^i\right)$. Now take a gradient of Equation 8-21 with regard to θ. You obtain the following:

$$\nabla_\theta \log \pi_\theta\left(a_t^i|s_t^i\right) = \text{const} \cdot \left(a_t^i - \mu_\theta\right) \nabla_\theta \mu_\theta\left(s_t^i\right) \tag{8-22}$$

To implement this in PyTorch, JAX or TensorFlow, you will form a modified mean squared error, just like the approach you took with the previous case of discrete actions. You find an expression whose derivative will yield the above expression of $\nabla_\theta \log \pi_\theta\left(a_t^i|s_t^i\right)$. By observation of Equation 8-22, you can find an expression, as shown in Equation 8-23, whose derivative will be equal to the one in Equation 8-22.

$$\left(a_t^i - \mu_\theta\right)^2 \tag{8-23}$$

Now you weigh the mean squared error with the trajectory return and that will form the loss equation you implement in PyTorch, which is given in Equation 8-24.

$$L_{MSE}(\theta) = \frac{1}{N}\sum_{i=1}^{N}\left[\left(\sum_{t=1}^{T}(a_t^i - \mu_\theta)^2\right)\left(\sum_{t=1}^{T}r(s_t^i, a_t^i)\right)\right]$$

(8-24)

Again, note that using gradient of $L_{MSE}(\theta)$ and then taking a step in the -ve direction of the gradient will produce the following expression:

$$-\nabla_\theta L_{MSE}(\theta) = \frac{1}{N}\sum_{i=1}^{N}\left[\left(\sum_{t=1}^{T}(a_t^i - \mu_\theta)\nabla_\theta\,\mu_\theta(s_t^i)\right)\left(\sum_{t=1}^{T}r(s_t^i, a_t^i)\right)\right]$$

(8-25)

A step in the direction of $-\nabla_\theta L_{MSE}(\theta)$ is a step in the direction of $\nabla_\theta J(\theta)$, as given in Equation 8-19. This step increases the value of $J(\theta)$ from Equation 8-6, that is, maximizing the return from the policy.

To summarize, implementations in auto-diff libraries like PyTorch require you to form cross-entropy loss in discrete action space and mean squared loss in the case of continuous action space, where the loss term is weighted by the total return of that trajectory to which a specific pair (s_t^i, a_t^i) belongs. This is similar to the approach you take in supervised learning, except for the extra step of weighing by the trajectory return $r(\tau^i) = \sum_{t=1}^{T}r(s_t^i, a_t^i)$.

To reiterate, remember that *weighted cross-entropy loss* or *weighted mean-squared loss* has no physical meaning or significance. It is just a convenient expression that allows you to use auto-diff capabilities of PyTorch to calculate the gradients using back propagation and then take a step to improve the policy. Compared to this, in supervised learning, the losses do signify the quality of prediction. In supervised learning, lower loss means that the model can make accurate predictions. However, there is no such inference or meaning that can be attached to the weighted loss in the case of the policy gradient scenario in reinforcement learning. That's why they are called pseudo-loss/objectives.

Variance Reduction with Rewards-to-Go

The expression derived in Equation 8-16, if used in its current form, has an issue. It has high variance. You will now leverage the temporal nature of the problem to do some variance reduction.

When you roll out the policy (i.e., take actions as per the policy) to produce a trajectory, you calculate the total reward of the trajectory $r(\tau)$. As shown in Equation 8-16, each of the action probability terms for the actions in the trajectory are weighted by this trajectory reward.

However, the world is causal when the agent interacts with the environment. An action taken in a time step, say t, can only impact the reward you see after that action. The reward you see prior to time step t is not impacted by the action you take at time step t or any subsequent actions. Future actions cannot impact past rewards. You will use this property to drop certain terms in Equation 8-16. This will help you reduce the variance.

You may ask why? Remember that you are using the sample trajectories as the Monte Carlo estimate of the expected trajectory reward. Each action in the trajectory comes from a probability distribution. The first sum inside Equation 8-16 is adding all these samples of the random variables. From basic probability, you know that as you add more random variables to an existing sequence of random variables, the total variance of the sum will grow linearly. Therefore, by dropping irrelevant terms, you help reduce the length of the sequence and hence the variance.

The steps that derive the revised formula under the causal world assumption are shown here. Note that it is not a rigorous mathematical proof.

Start with Equation 8-15.

$$\nabla_{\theta} J(\theta) = E_{\tau \sim p_{\theta}(\tau)} \left[\left(\sum_{t=1}^{T} \nabla_{\theta} \log \pi_{\theta}(a_t | s_t) \right) \left(\sum_{t=1}^{T} r(s_t, a_t) \right) \right]$$

Now change the index of summation for the reward term from t to t' and move the reward sum term inside the first summation over actions from policy π_{θ}. This gives you the following expression:

$$\nabla_{\theta} J(\theta) = E_{\tau \sim p_{\theta}(\tau)} \left[\left(\sum_{t=1}^{T} \left(\nabla_{\theta} \log \pi_{\theta}(a_t | s_t) \sum_{t'=1}^{T} r(s_{t'}, a_{t'}) \right) \right) \right]$$

Inside the summation over index t, you drop the reward terms that came before time t. At time t, the action you take can only impact the reward that comes at time t as *well as those that come later*. This leads to changing the second inner sum starting from $t' = t$ instead of $t' = 1$. The revised expression is as follows:

$$\nabla_{\theta} J(\theta) = E_{\tau \sim p_{\theta}(\tau)} \left[\left(\sum_{t=1}^{T} \left(\nabla_{\theta} \log \pi_{\theta}(a_t | s_t) \sum_{t'=t}^{T} r(s_{t'}, a_{t'}) \right) \right) \right]$$

The inner sum $\sum_{t'=t}^{T} r(s_{t'}, a_{t'})$ is no longer the total reward of the trajectory. Rather, it is the reward of the remaining trajectory that you see from *time* = t to T. As you may recall, this is nothing but a sample of the q-value for the current policy when in state-action pair $(s_{t'}, a_{t'})$. The q-value for a policy is the expected reward you get from time t until the end, after you take a step/action a_t at time t while in state s_t. You can also call it *reward-to-go*. Since the expression $\sum_{t'=t}^{T} r(s_{t'}, a_{t'})$ is for only one trajectory, you denote it is a single sample estimate of the expected reward-to-go. The updated gradient equations are as follows:

$$\hat{Q}_t^i = \sum_{t'=t}^{T} r(s_{t'}, a_{t'}) \tag{8-26}$$

$$\nabla_\theta J(\theta) = \frac{1}{N} \sum_{i=1}^{N} \sum_{t=1}^{T} \nabla_\theta \log \pi_\theta(a_t^i | s_t^i) \hat{Q}_t^i \tag{8-27}$$

To use this equation in PyTorch, you just need to make a minor modification to the naive REINFORCE algorithm from Figure 8-2. Instead of weighing each log probability term with the total trajectory reward, you will now weigh it with the reward-to-go from that time step. Figure 8-4 shows the revised REINFORCE algorithm using reward-to-go.

REINFORCE with *Reward to go*

Input:

A model with parameters θ taking state s as input and producing $\pi_\theta\left(a|s\right)$

Other parameters: step size α

Initialize:

Initialize weights $\boldsymbol{\theta}$

Loop:

Sample $\{\tau^i\}$, a set of N trajectories from current policy $\pi_\theta\left(a_t|s_t\right)$

Calculate the reward to go $\hat{Q}_t^i = \sum_{t'=t}^{T} r\left(s_{t'}^i, a_{t'}^i\right)$

Update model parameters θ:

$$\nabla_\theta J\left(\theta\right) \approx \frac{1}{N}\sum_{i=1}^{N}\left[\left(\sum_{t=1}^{T}\nabla_\theta\log\pi_\theta\left(a_t^i|s_t^i\right)\right)\cdot\hat{Q}_t^i\right]$$
$$\theta = \theta + \alpha\nabla_\theta J\left(\theta\right)$$

Figure 8-4. *The REINFORCE algorithm using reward-to-go*

You have been doing a lot of theory so far in this chapter, and there is a bit of overload of mathematical formula. I have tried to keep it minimal, and if there is one takeaway from the chapter thus far, it is that of the REINFORCE algorithm in Figure 8-4. Let's now put this equation into practice. You will implement REINFORCE from Figure 8-4 to the usual `CartPole` problem with continuous state space and discrete actions.

Before you do, I want to introduce one last mathematical term. The exploration of state-action space in policy gradient algorithms comes from the fact that you learn a stochastic policy that assigns some probability to all the actions for a given state instead of choosing the best possible action using DQN. To ensure that exploration is maintained and to ensure that $\pi_\theta(a|s)$ does not collapse into a single action with high probability, you introduce a regularization term known as *entropy*.

Entropy of a distribution $X \sim p(X)$ is defined as follows:

$$H(X) = \sum_{x} -p(x).\log p(x) \tag{8-28}$$

To keep enough exploration, you will want the probability to have a spread-out distribution and not let the probability distribution peak around a single value or a small region too soon. The bigger the spread of a distribution, the higher the entropy H(x) of a distribution. Accordingly, the term fed into the PyTorch minimizer is as follows:

$$Loss(\theta) = -J(\theta) - H\left(\pi_\theta\left(a_t^i|s_t^i\right)\right)$$

$$= -\frac{1}{N}\sum_{i=1}^{N}\left[\sum_{t=1}^{T}\left(\log\pi_\theta\left(a_t^i|s_t^i\right)\sum_{t'=t}^{T}\gamma^{t'-t}r\left(s_{t'}^i,a_{t'}^i\right)\right) - \beta\sum_{a_i}\pi_\theta\left(a_t^i|s_t^i\right).\log\pi_\theta\left(a_t^i|s_t^i\right)\right] \qquad (8\text{-}29)$$

In this code example, you are taking only one trajectory, that is, $N = 1$. You will, however, average it over the number of actions to get the average loss. The function you will implement is as follows:

$$Loss(\theta) = -J(\theta) - H\left(\pi_\theta\left(a_t|s_t\right)\right)$$

$$= -\frac{1}{T}\sum_{t=1}^{T}\left(\log\pi_\theta\left(a_t|s_t\right)G(s_t) - \beta\cdot\sum_{a_i}\pi_\theta\left(a_t|s_t\right).\log\pi_\theta\left(a_t|s_t\right)\right) \qquad (8\text{-}30)$$

Where,

$$G(s_t) = \sum_{t'=t}^{T}\gamma^{t-t'}r\left(s_{t'}^i, a_{t'}^i\right)$$

Note that the discount factor γ was reintroduced in the previous expressions. You had dropped γ to reduce overload of terms in the derivations and now have brought it back for actual implementation. The purpose that γ serves was explained in the beginning of the chapter.

With this complete all the machinery required to implement this first policy gradient algorithm in the form of REINFORCE with reward-to-go. Now you'll look at an implementation and how it can be used to train an agent in cart-world gymnasium environment.

Let's now walk through the implementation. You can find the complete code in the 8.a-reinforce.ipynb notebook. I explained the environment CartPole in earlier chapters. It has a four-dimensional continuous state space and a discrete action space of two actions:—move left and move right. You first define a simple policy network with one hidden layer of 192 units and ReLU activation. The final output has no activation. Listing 8-1 shows the code.

Listing 8-1. Policy Network in PyTorch

```
model = nn.Sequential(
            nn.Linear(state_dim,192),
            nn.ReLU(),
            nn.Linear(192,n_actions),
)
```

Next, you define a `generate_trajectory` function that takes the current policy to generate a trajectory of (`states, actions, rewards`) for one episode. It uses a helper function called `predict_probs` to do so. You'll use `torch.no_grad` in the `predict_probs` function to avoid extra accumulation of gradients in the PyTorch training step. Listing 8-2 shows the code. It starts with initializing the environment and then successively takes steps following the current policy until there are 1,000 steps or until the current episode is terminated, whichever is earlier. It returns the list of (`states, actions, rewards`) tuples observed during the episode.

Listing 8-2. generate_trajectory and predict_probs from 8.a-reinforce.ipynb

```
def generate_trajectory(env, n_steps=1000):
    """

    Play a session and generate a trajectory
    returns: arrays of states, actions, rewards
    """

    states, actions, rewards = [], [], []
    # initialize the environment
    s, _ = env.reset()

    #generate n_steps of trajectory:
    for t in range(n_steps):
        action_probs = predict_probs(np.array([s]))[0]
        #sample action based on action_probs
        a = np.random.choice(n_actions, p=action_probs)
        next_state, r, done, _, _ = env.step(a)

        #update arrays
        states.append(s)
```

```
        actions.append(a)
        rewards.append(r)

        s = next_state
        if done:
            break
    return np.array(states), np.array(actions), np.array(rewards)

def predict_probs(states):
    """

    params: states: [batch, state_dim]
    returns: probs: [batch, n_actions]
    """

    states = torch.tensor(np.array(states), device=device, dtype=torch.
    float32)
    with torch.no_grad():
        logits = model(states)
    probs = nn.functional.softmax(logits, -1).detach().numpy()
    return probs
```

You also have another helper function to convert the individual step returns

$r(s_{t'},a_{t'})$ to reward-to-go as per the expression: $G(s_t) = \sum_{t'=t}^{T} \gamma^{t-t'} r(s_{t'}^i, a_{t'}^i)$. You use a loop that starts from the end and calculates the cumulative sum with a discount factor at each step, while going from the end of the list to the beginning. There are many other efficient ways to do this. See if you can think of other vectorized ways to do this. As a hint, check the documentation of the cumsum function in PyTorch. Listing 8-3 contains the implementation of this function.

Listing 8-3. get_rewards_to_go from 8.a-reinforce.ipynb

```
def get_rewards_to_go(rewards, gamma=0.99):

    T = len(rewards) # total number of individual rewards
    # empty array to return the rewards to go
    rewards_to_go = [0]*T
    rewards_to_go[T-1] = rewards[T-1]
```

```
for i in range(T-2, -1, -1): #go from T-2 to 0
    rewards_to_go[i] = gamma * rewards_to_go[i+1] + rewards[i]

return rewards_to_go
```

You are now ready to implement the training. You build the loss function that you are going to feed into the PyTorch optimizer. As explained, you are going to implement the expression given in Equation 8-29. Listing 8-4 contains the code used for a single step of training. As discussed earlier, I have added an additional term representing the entropy of the action probability distribution. This prevents the policy from collapsing into a single action. Entropy is calculated using the expression $H(X) = \sum_x - p(x). \log p(x)$. The function takes a tuple of (states, actions, rewards) observed during a trajectory, calculates the rewards-to-go followed, then calculates the training loss, which is comprised of weighted cross-entropy loss and entropy. The optimizer uses the loss to inprove the parameters of the neural network.

Listing 8-4. Training for One Trajectory from 8.a-reinforce.ipynb

```
#init Optimizer
optimizer = torch.optim.Adam(model.parameters(), lr=1e-3)

def train_one_episode(states, actions, rewards, gamma=0.99, entropy_
coef=1e-2):

    # get rewards to go
    rewards_to_go = get_rewards_to_go(rewards, gamma)

    # convert numpy array to torch tensors
    states = torch.tensor(states, device=device, dtype=torch.float)
    actions = torch.tensor(actions, device=device, dtype=torch.long)
    rewards_to_go = torch.tensor(rewards_to_go, device=device,
    dtype=torch.float)

    # get action probabilities from states
    logits = model(states)
    probs = nn.functional.softmax(logits, -1)
    log_probs = nn.functional.log_softmax(logits, -1)

    log_probs_for_actions = log_probs[range(len(actions)), actions]
```

```
#Compute loss to be minimized
J = torch.mean(log_probs_for_actions*rewards_to_go)
H = -(probs*log_probs).sum(-1).mean()

loss = -(J+entropy_coef*H)

optimizer.zero_grad()
loss.backward()
optimizer.step()

return loss.detach().cpu() #to show progress on training
```

You are now ready to carry out the training. Listing 8-5 shows how to train the agent for 10,000 steps, plotting the progress as training progresses. You can stop the training once you achieve a mean reward of 500. The code initializes two arrays to keep track of the per-episode return and loss. Inside the loop, it follows a simple process, as explained in the algorithm for REINFORCE with rewards-to-go shown in Figure 8-4. Toward the end of the loop, the code periodically checks the performance of the agent and stops the training if the average score per episode has crossed 500. There is also code that plots the performance by showing graphs of return per episode and the training loss during the course of the training.

Listing 8-5. Training the Agent from 8.a-reinforce.ipynb

```
loss_history = []
return_history = []

for i in range(5000):
    states, actions, rewards = generate_trajectory(env)
    loss = train_one_episode(states, actions, rewards)
    # return_history.append(np.sum(rewards))
    loss_history.append(loss)

    if i != 0 and i % eval_freq == 0:
        mean_return = np.mean(return_history[-eval_freq:])
        if mean_return > 500:
            break
```

```
if i != 0 and i % eval_freq == 0:
    # eval the agent
    eval_env = make_env(env_name)
    return_history.append(
        evaluate(eval_env, model)
    )
    eval_env.close()
    clear_output(True)

    plt.figure(figsize=[16, 5])
    plt.subplot(1, 2, 1)
    plt.title("Mean return per episode")
    plt.plot(return_history)
    plt.grid()

    assert not np.isnan(loss_history[-1])
    plt.subplot(1, 2, 2)
    plt.title("Loss history (smoothened)")
    plt.plot(smoothen(loss_history))
    plt.grid()
    plt.show()

env.close()
```

By the end of training, the agent has learned to balance the pole well. You will also notice that the program needed fewer iterations and less time to achieve this result as compared to the DQN-based approach. Figure 8-5 shows the graphs observed during training.

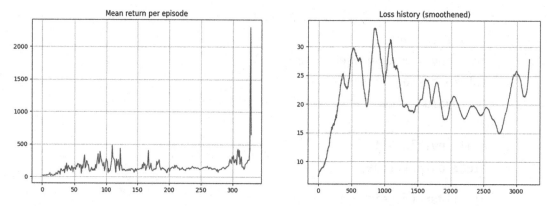

Figure 8-5. *Training progess from 8.a-reinforce.ipynb*

As you can observe from Figure 8-5, the training is very unstable and the return per episode fluctuates a lot. You could try to tune the hyperparameters, such as relative weight given entropy of the action, probability distribution in function via `entropy_coef`, number of training steps, network, and so on. You could tune these manually or using the Optuna as explained Chapter 6.

This brings you to the end of section on the REINFORCE algorithm. Note that REINFORCE is an *on-policy* algorithm. After every step of training, you regenerate a new trajectory, throwing away the observations from the previous trajectory. For an environment like `CartPole`, it is inconsequential. However, for environments where it is expensive to generate samples, throwing away observations at the end of each trajectory is very inefficient. You'll need to look further and define off-policy approaches.

Further Variance Reduction with Baselines

Strating with the original policy gradient update expression in Equation 8-15, you converted the expectation to an estimate using the average in Equation 8-16. Following this, you learned how the variance could be reduced by considering rewards-to-go instead of full trajectory rewards. The expression with the rewards-to-go was given in Equation 8-27.

In this section, you look at yet another change that makes the policy gradient even more stable. Consider the motivation. Assume you have performed three rollouts of the trajectory following a policy. And let's say the trajectory returns were 300, 200, and 100.

To keep the explanation simple, consider the case of total rewards and total trajectory probability of the gradient update equation, as given in Equation 8-10, reproduced here:

$$\nabla_\theta J(\theta) = E_{\tau \sim p_\theta(\tau)}\left[\nabla_\theta \log p_\theta(\tau) r(\tau)\right]$$

What will the gradient update do? It will weigh the gradient of the log probability of the first trajectory with 300, the second trajectory with 200, and the third trajectory with 100. This means that the probability of each of these three trajectories is being increased by some different amount. Figure 8-6 shows a pictorial representation.

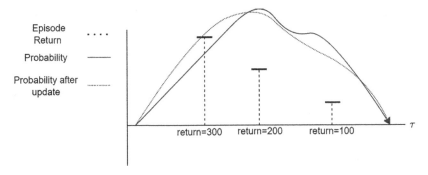

Figure 8-6. *The gradient update for a policy with actual trajectory rewards*

As you can see in Figure 8-6, you are increasing the probability of all three trajectories with different weighting factors, all +ve weights, making the probability of all trajectories go up. Ideally, you would increase the probability of the trajectory with a return of 300 and reduce the probability of the trajectory with a return of 100, as it is not a very good trajectory. You would want the policy to change such that it does not generate the trajectory with a return of 100 too often. However, using the current approach, the revised probability curve is becoming flatter, as it is trying to increase the probability of all three trajectories and the total area under the probability curve must be 1 given that it is a probability curve.

Let's consider a scenario in which you subtract the average return of three trajectories, (300 + 200 + 100)/3 = 200 from each of the three rewards. You get the revised trajectory returns of 100, 0, and -100 (300-200; 200-200; 100-200). Let's use the revised trajectory returns as the weights to carry out the gradient update. Figure 8-7 shows the outcome of such an update. You can see that the probability curve is becoming narrower and sharper as its spread reduces along the x-axis.

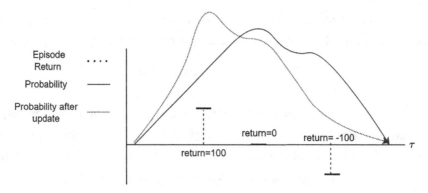

Figure 8-7. *The gradient update for a policy with trajectory rewards reduced by a baseline*

Reducing the rewards with a baseline reduces the variance of the update. In the limit, whether you use or do not use a baseline, the outcome will be same. The introduction of a baseline does not alter the optimal solution. It just reduces the variance and speeds up the learning. I will show mathematically that the introduction of a baseline does not alter the expected value of the gradient update. The baseline can be a fixed baseline across all trajectories and all steps in the trajectory, or it can be a changing quantity that depends on the state. *However, it cannot depend on the action.* Let's first go through the derivation in which the baseline is a function of the state s_t^i.

Update Equation 8-15 to introduce a baseline.

$$\nabla_\theta J(\theta) = E_{\tau \sim p_\theta(\tau)}\left[\left(\sum_{t=1}^{T}\nabla_\theta \log \pi_\theta\left(a_t|s_t\right)\right)\left(r(\tau) - b(s_t)\right)\right]$$

Separate out the term of $b(s_t)$ and evaluate what the expectation would be.

$$E_{\tau \sim p_\theta(\tau)}\left[\left(\sum_{t=1}^{T}\nabla_\theta \log \pi_\theta\left(a_t|s_t\right)\right)b(s_t)\right]$$

Now move the first inner sum out due to the property of linearity of expectation to obtain the expression.

$$\sum_{t=1}^{T}E_{a_t \sim \pi_\theta(a_t|s_t)}\left[\nabla_\theta \log \pi_\theta\left(a_t|s_t\right)b(s_t)\right]$$

You have switched the expectation from $\tau \sim p_\theta(\tau)$ to $a_t \sim \pi_\theta(a_t|s_t)$. This is because you moved the first inner sum outside the expectation, and after that, the only term that depends on a probability distribution is the action a_t with probability $\pi_\theta(a_t|s_t)$.

Now just focus on the inner expectation: $E_{\tau \sim p_\theta(\tau)}\left[\nabla_\theta \log \pi_\theta(a_t|s_t)b(s_t)\right]$. You can write this as an integral, as shown here:

$$E_{a_t \sim \pi_\theta(a_t|s_t)}\left[\nabla_\theta \log \pi_\theta(a_t|s_t)b(s_t)\right]$$

$$= \int \pi_\theta(a_t|s_t)\left(\nabla_\theta \log \pi_\theta(a_t|s_t)\right)b(s_t)\,da_t$$

$$= \int \pi_\theta(a_t|s_t)\frac{\nabla_\theta \pi_\theta(a_t|s_t)}{\pi_\theta(a_t|s_t)}b(s_t)\,da_t$$

$$= \int \nabla_\theta \pi_\theta(a_t|s_t)\,b(s_t)\,da_t$$

$$= b(s_t)\nabla_\theta \int \pi_\theta(a_t|s_t)\,da_t$$

As $b(s_t)$ does not depend on a_t, you can take it out. Again, due to the linearity of the integral, you can swap the gradient and the integral. Now the integral will evaluate to 1, as that is the total probability using the curve for $\pi_\theta(a_t|s_t)$. Accordingly, you get the following:

$$E_{a_t \sim \pi_\theta(a_t|s_t)}\left[\nabla_\theta \log \pi_\theta(a_t|s_t)b(s_{t'})\right]$$

$$= b(s_t)\nabla_\theta(1) = b(s_t).0$$

$$= 0$$

The previous derivation shows that subtracting a baseline that depends on the state or on a constant will not change the expectation. *The condition is that it should not depend on the action a_t.*

Therefore, REINFORCE with the baseline will go through the updates as follows:

$$\nabla_\theta J(\theta) = E_{\tau \sim p_\theta(\tau)}\left[\left(\sum_{t=1}^{T}\nabla_\theta \log \pi_\theta(a_t|s_t)\right)\left(r(\tau)-b(s_t)\right)\right] \tag{8-31}$$

You can further modify and combine the rewards-to-go using the same rationale of causal world where actions today do not influence the rewards in the past, rather only the current and future rewards. You can combine Equations 8-31 and 8-27 to get the following set of equations:

$$\hat{Q}\left(s_t^i, a_t^i\right) = \sum_{t'=t}^{T} \gamma^{t'-t} r\left(s_{t'}^i, a_{t'}^i\right)$$

$$\nabla_\theta J(\theta) = \frac{1}{N} \sum_{i=1}^{N} \sum_{t=1}^{T} \nabla_\theta \log \pi_\theta \left(a_t^i | s_t^i\right) \left[\hat{Q}\left(s_t^i, a_t^i\right) - b\left(s_t^i\right)\right] \qquad (8\text{-}32)$$

Equation 8-32 is REINFORCE with a baseline and rewards-to-go. You have used two tricks to reduce the variance of the vanilla REINFORCE. You used a temporal structure to remove the rewards that are in the past and not impacted by the actions in the present. Then you used the baseline to have the bad policies get -ve rewards and the good policies +ve rewards. Both these tricks lead to lower variance as you go through the learning.

Note that REINFORCE and all its variations are *on-policy algorithms*. After the weights of the policy are updated, you need to roll out new trajectories. The old trajectories are no longer representative of the old policy. This is one of the reasons that, like the value-based on-policy methods, REINFORCE is also sample inefficient. You cannot use transitions from earlier policies. You must discard them and generate new transitions after each weight update.

Actor-Critic Methods

In this section, you'll further refine the algorithm by combining the policy gradient with the value functions to get something called the *actor-critic* family of algorithms (A2C/A3C). You'll first define a term called *advantage A(s, a)*.

Defining Advantage

I first talk about the expression $\hat{Q}\left(s_t^i, a_t^i\right)$ in Equation 8-32. It is the reward-to-go in a given trajectory (i) and in a given state s_t:

$$\hat{Q}\left(s_t^i, a_t^i\right) = \sum_{t'=t}^{T} r\left(s_{t'}^i, a_{t'}^i\right)$$

To evaluate the $\hat{Q}$ value in this expression, you use the Monte Carlo simulation. In other words, you are adding all the rewards from that time step t until the end, that is, until T. It will again have high variance, as it is only one trajectory estimate of the expectation. In a previous chapter on model-free policy learning, you saw that MC methods have zero bias but high variance. Comparatively, TD methods have some bias but low variance and can lead to faster convergence due to lower variance. Could you do something similar here also? What is this reward-to-go? What is the expectation of the expression $\hat{Q}\left(s_t^i, a_t^i\right)$? It is nothing but the q-value of the state-action pair (s_t, a_t). If you had access to the q-value, you could replace the summation of individual rewards with the q-estimate.

$$\hat{Q}\left(s_t^i, a_t^i\right) = q\left(s_t^i, a_t^i; \phi\right) \tag{8-33}$$

Let's roll the value of $q(s_t, a_t)$ by one time step. This is similar to the $TD(0)$ approach in Chapter 5. You can write $\hat{Q}^i\left(s_t, a_t\right)$ as follows:

$$\hat{Q}\left(s_t^i, a_t^i\right) = r\left(s_t^i, a_t^i\right) + V\left(s_{t+1}^i\right) \tag{8-34}$$

This is the undiscounted rollout. As discussed at the beginning of the chapter, you will do all the derivations in the context of the finite horizon undiscounted setting. The analysis can be easily extended to other settings. You will switch to a more general case in the final pseudocode for the algorithm while restricting the analysis to the undiscounted case.

Looking at Equation 8-33 again, can you think of a good baseline $b\left(s_t^i\right)$ that could be used? How about using the state value $V\left(s_t^i\right)$? As explained, you can use any value, as the baseline provided it does not depend on the action a_t^i. $V\left(s_t^i\right)$ is one such quantity that depends on the state s_t^i but does not depend on the action a_t^i since the action a_t^i is taken once the agent reaches the agent in state s_t^i. In other words, $V\left(s_t^i\right)$ depends on the policy you follow, which impacts the actions in general you take in that state, but it does not depend on the specific action a_t^i you want to take.

$$b\left(s_t^i\right) = V\left(s_t^i\right) \tag{8-35}$$

Using the previous expression:

$$\hat{Q}\left(s_t^i, a_t^i\right) - b\left(s_t^i\right) = \hat{Q}\left(s_t^i, a_t^i\right) - V\left(s_t^i\right) \tag{8-36}$$

The right side is known as the advantage $\hat{A}\left(s_t^i, a_t^i\right)$. It is the extra benefit/reward you get by following a policy at state s_t^i to take step a_t^i, which gives a reward of $\hat{Q}\left(s_t^i, a_t^i\right)$ versus the average reward you get in state s_t^i under a given policy as denoted by $V\left(s_t^i\right)$. You can now substitute Equation 8-34 in 8-36 to get the following:

$$\hat{A}\left(s_t^i, a_t^i\right) = r\left(s_t^i, a_t^i\right) + V\left(s_{t+1}^i\right) - V\left(s_t^i\right) \tag{8-37}$$

The changes you have made to the basic reinforce algorithm equation will bring significant benefits to the learning process. These changes form the basis for the set of algorithms known as *Advantage Actor-Critic* algorithms, which you will study next.

Advantage Actor-Critic (A2C)

Continuing from the previous section, you can rewrite the gradient update given in Equation 8-32. Here is the original gradient update from Equation 8-32:

$$\nabla_\theta J(\theta) = \frac{1}{N} \sum_{i=1}^{N} \sum_{t=1}^{T} \nabla_\theta \log \pi_\theta\left(a_t^i | s_t^i\right) \left[\hat{Q}\left(s_t^i, a_t^i\right) - b\left(s_t^i\right)\right]$$

Substituting, $b\left(s_t^i\right) = V\left(s_t^i\right)$ from Equation 8-35, you get the following:

$$\nabla_\theta J(\theta) = \frac{1}{N} \sum_{i=1}^{N} \sum_{t=1}^{T} \nabla_\theta \log \pi_\theta\left(a_t^i | s_t^i\right) \left[\hat{Q}\left(s_t^i, a_t^i\right) - V\left(s_t^i\right)\right] \tag{8-38}$$

Look at the inner expression $\hat{Q}\left(s_t^i, a_t^i\right) - V\left(s_t^i\right)$. Q is the value of following a specific step a_t using the current policy. In other words, the actor and V is the average value of the following current policy, that is, the critic. The actor is trying to maximize the reward, and the critic is telling the algorithm how good or bad that specific step was, compared to the current average. The *actor-critic* approach is a family of algorithms with an *actor* using policy gradients to improve actions, and a *critic* informing the algorithm about the goodness of current policy. You have $\hat{A}(s_t^i, a_t^i) = \hat{Q}(s_t^i, a_t^i) - V(s_t^i)$ as the advantage of a specific action a_t^i over the general state value under current policy.

There are two ways to handle $\hat{Q}(s_t^i, a_t^i)$. You can use the original definition of:

$$\hat{Q}(s_t^i, a_t^i) = \sum_{t'=t}^{T} r(s_{t'}^i, a_{t'}^i)$$

This is the MC approach, where you evaluate the rewards-to-go $\hat{Q}(s_t^i, a_t^i)$ as a sample estimate of one rollout. This has high variance, as you are making an estimate based on a single rollout. Under the MC approach the gradient update for the Actor-Critic algorithm takes the following form:

$$\nabla_\theta J(\theta) = \frac{1}{N}\sum_{i=1}^{N}\sum_{t=1}^{T}\nabla_\theta \log \pi_\theta\left(a_t^i | s_t^i\right)\left[\sum_{t'=t}^{T} r(s_{t'}^i, a_{t'}^i) - V(s_t^i)\right] \tag{8-39}$$

Or you could take the *TD*(0) approach, as shown in Equation 8-34, where $\hat{Q}(s_t^i, a_t^i) = r(s_{t'}^i, a_{t'}^i) + V(s_{t+1}^i)$. The *TD*(0) approach will take you to Equation 8-37, as shown here:

$$\hat{Q}(s_t^i, a_t^i) - b(s_t^i) - r(s_t^i, a_t^i) + V(s_{t+1}^i) - V(s_t^i)$$

You substitute this into the gradient update Equation 8-32 to get the gradient update for Actor-Critic, as the following update equation:

$$\nabla_\theta J(\theta) = \frac{1}{N}\sum_{i=1}^{N}\sum_{t=1}^{T}\nabla_\theta \log \pi_\theta\left(a_t^i | s_t^i\right)\left[r(s_t^i, a_t^i) + V(s_{t+1}^i) - V(s_t^i)\right] \tag{8-40}$$

This expression is the update reason you will implement and it's known as *advantage actor-critic* (A2C). Note that the actor-critic is a family of approaches, with A2C and A3C being two specific instances of it. Sometimes in the literature, actor-critic is also interchangeably referred to as A2C. At the same time, some papers refer to A2C as a synchronous version of A3C, which I will talk about in the next section.

You must have two estimators. One is a policy network $\pi_\theta(a|s)$ with parameters θ and another network with parameters ϕ to estimate the $V_\phi(s_t)$. The general approach to design a single network with two heads having a small subnet—one head for outputting the policy π and another to output the state-value V. Figure 8-8 shows a sample of such a network. This is the architecture you will be implementing in the code walkthrough.

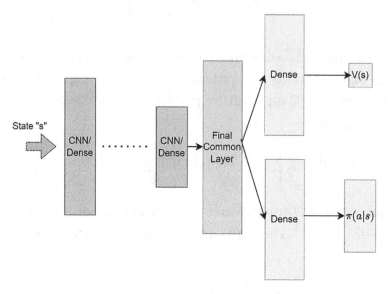

Figure 8-8. *Actor-critic network with common weights in the initial layers*

Now look at the pseudocode. Figure 8-9 shows the complete pseudocode for the actor-critic (also known as A2C for advantage actor-critic) of the Actor-Critic algorithm. In a loop, you sample a trajectory. You use this trajectory to calculate the rewards-to-go, which will act as a target value for the $V(s)$ head of the neural network shown in Figure 8-8. The second head producing the policy $\pi(a|s)$ is tuned using the cross-entropy loss weighted by the advantage, as per Equation 8-40. Note that while forming the cross-entropy loss for policy, you also use the output of the $V(s)$ produced by the other head.

Advantage Actor Critic algorithm

Input:

A model with parameters θ taking state s as input and producing $\pi_\theta\left(a|s\right)$

A model with parameters ϕ taking state s as input and producing $V_\phi(s)$

Other parameters: step sizes α, β

Initialize:
Initialize weights $\boldsymbol{\theta}$ and $\boldsymbol{\phi}$

Loop:

Sample τ^i, a set of N trajectories from current policy $\pi_\theta\left(a_t|s_t\right)$

Calculate the reward to go $\hat{Q}_t^i = \sum_{t'=t}^T r\left(s_{t'}^i, a_{t'}^i\right)$

Fit Value function $V_\emptyset(s)$ to $\hat{Q}_t^i$ by forming mean square error:

$$L = \left(V_\emptyset(s) - \hat{Q}_t^i\right)^2$$

Carry out stochastic gradient step on L to adjust ϕ:

$$\phi = \phi - \beta * \nabla_\emptyset L$$

Update policy model parameters \theta:

Calculate pseudo Cross-Entropy-Loss: $L_{CE}\left(\theta\right) = -J\left(\theta\right)$:

$$J\left(\theta\right) = \frac{1}{N}\sum_{i=1}^N\sum_{t=1}^T \log \pi_\theta\left(a_t^i|s_t^i\right).\ \left(r\left(s_t^i, a_t^i\right) + V_\emptyset(s_{t+1}) - V_\emptyset(s_t)\right)$$

Carry out gradient step on θ:

$$\theta = \theta + \alpha\nabla_\theta J\left(\theta\right)$$

Figure 8-9. *Advantage actor-critic algorithm*

The pseudocode in Figure 8-9 uses one-step undiscounted return for advantage $\hat{A}\left(s_t^i, a_t^i\right)$:

$$\hat{A}\left(s_t^i, a_t^i\right) = r\left(s_t^i, a_t^i\right) + V_\phi\left(s_{t+1}\right) - V_\phi\left(s_t\right)$$

There are many other ways to formulate the advantage $\hat{A}\left(s_t^i, a_t^i\right)$. The discounted one-step version is as follows:

$$r\left(s_t^i, a_t^i\right) + \gamma V_\phi\left(s_{t+1}\right) - V_\phi\left(s_t\right)$$

Similarly, the discounted n-step return version is as follows:

$$\left(\sum_{t'=t}^{t+n-1} \gamma^{t'-t} r\left(s_{t'}^i, a_{t'}^i\right)\right) + \gamma^n V_\phi\left(s_{t+n}\right) - V_\phi\left(s_t\right)$$

There is also one more way, that is, using the MC approach. In that approach, you directly use rewards-to-go. The advantage is as follows:

$$\hat{A}\left(s_t^i, a_t^i\right) = \sum_{t'=t}^{T} \gamma^{t'-t} r\left(s_{t'}^i, a_{t'}^i\right) - V\left(s_t\right)$$

So, you have many ways to formulate the advantage function—you can choose to use discount γ or use it undiscounted, which is the version you will implement in this code. You will be implementing the version given in Equation 8-40 with a discount and the one explained in the pseudocode in Figure 8-9. That brings you to the next section, which covers the code walkthrough of the A2C implementation in PyTorch.

Implementation of the A2C Algorithm

In this code walkthrough, you will make the following changes to the actor-critic algorithm given in Figure 8-9:

- Like you did with REINFORCE, you will introduce the entropy regularizer.

- You will use the reward-to-go MC approach, that is,
  $\hat{A}\left(s_t^i, a_t^i\right) = \hat{Q}\left(s_t^i, a_t^i\right) - V_\phi\left(s_t\right) = \sum_{t'=t}^{T} r\left(s_{t'}^i, a_{t'}^i\right) - V_\phi\left(s_t\right)$ instead of the
  TD(0) approach of $\hat{A}\left(s_t^i, a_t^i\right) = r\left(s_t^i, a_t^i\right) + V_\phi\left(s_{t+1}\right) - V_\phi\left(s_t\right)$

- Instead of training two separate loss training steps of the first fitting $V(s)$ and then doing the policy gradient, you will form a single loss objective that will carry out the $V(s)$ fitting as well as the policy gradient step together with the entropy regularizer.

The loss using actor-critic with the previous modifications is as follows:

$$Loss(\theta,\phi) = -J(\theta,\phi) - H\left(\pi_\theta\left(a_t^i | s_t^i\right)\right)$$

$$= -\frac{1}{N}\sum_{i=1}^{N}\left[\sum_{t=1}^{T}\left(\log\pi_\theta\left(a_t^i | s_t^i\right)\left[\hat{Q}\left(s_t^i, a_t^i\right) - V_\phi\left(s_t^i\right)\right]\right) - \beta\sum_a \pi_\theta\left(a | s_t^i\right).\log\pi_\theta\left(a | s_t^i\right)\right]$$

As with REINFORCE, you will carry out weight updates after each trajectory. Therefore, N will be equal to 1 in the previous equation. Instead, you will average it over the number of actions to get the average loss. The function you will implement is as follows:

$$Loss(\theta,\phi) = -\frac{1}{T}\left[\sum_{t=1}^{T}\left(\log\pi_\theta\left(a_t | s_t\right)\left[\hat{Q}\left(s_t, a_t\right) - V_\phi\left(s_t\right)\right]\right) - \beta\sum_a \pi_\theta\left(a | s_t\right).\log\pi_\theta\left(a | s_t\right)\right]$$

Now turn your attention to the actual PyTorch code, which you can find in 8.b-action_critic.ipynb. Let's first talk about the network. You will have a joint network with shared weights, one producing the policy action probabilities and another the value of state. For CartPole, it is a simple network and is shown in Figure 8-10.

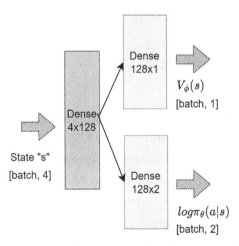

Figure 8-10. *Actor-critic network for the CartPole enviroment from 8.b-action_critic.ipynb*

Listing 8-6 shows the implementation of this network in PyTorch. It is a straight implementation of the network, as shown in Figure 8-10. CartPole has a state dimension of 4. This input is passed through a common layer of 128 units in output, as represented by self.fc1 in the code. The output of this layer is then passed through two separate networks. One is a linear layer with output dimension matching the number of actions as denoted by the self.actor dense layer in the code. The other output produces state value V(s) and has a dimension of 1 since state value is a scaler quantity. This layer is denoted as self.critic in the listing.

Listing 8-6. Actor-Critic Network in PyTorch from 8.b-actor_critic.ipynb

```
class ActorCritic(nn.Module):
    def __init__(self):
        super(ActorCritic, self).__init__()
        self.fc1 = nn.Linear(state_dim, 128)
        self.actor = nn.Linear(128,n_actions)
        self.critic = nn.Linear(128,1)
    def forward(self, s):
        x = F.relu(self.fc1(s))
        logits = self.actor(x)
        state_value = self.critic(x)
        return logits, state_value
model = ActorCritic()
```

The other change big change is the way you implement the training code for one episode. It is similar to the code in Listing 8-4 for REINFORCE; however, with a change to introduce $V(s_t)$ as a baseline value. Listing 8-7 gives the complete code for train_one_episode. You first calculate the reward-to-go. Next, you convert the list of tuples (state, action, rewards) to the appropriate tensor formats. The state is then passed through the network, producing the state-value $V(s)$ and $log\pi_\theta(a|s)$. These values are used to calculate the loss, as explained previously. You also add the entropy of the action distribution as a penalty to ensure that the action probabilities do not collapse into a single action.

Listing 8-7. train_one_episode for Actor-Critic Using MC Rewards-To-Go
in PyTorch

```
def train_one_episode(states, actions, rewards, gamma=0.99, entropy_
coef=1e-2):

    # get rewards to go
    rewards_to_go = get_rewards_to_go(rewards, gamma)

    # convert numpy array to torch tensors
    states = torch.tensor(states, device=device, dtype=torch.float)
    actions = torch.tensor(actions, device=device, dtype=torch.long)
    rewards_to_go = torch.tensor(rewards_to_go, device=device,
    dtype=torch.float)

    # get action probabilities from states
    logits, state_values = model(states)
    probs = nn.functional.softmax(logits, -1)
    log_probs = nn.functional.log_softmax(logits, -1)

    log_probs_for_actions = log_probs[range(len(actions)), actions]

    advantage = rewards_to_go - state_values.squeeze(-1)

    #Compute loss to be minimized
    J = torch.mean(log_probs_for_actions*(advantage))
    H = -(probs*log_probs).sum(-1).mean()

    loss = -(J+entropy_coef*H)

    optimizer.zero_grad()
    loss.backward()
    optimizer.step()

    return loss.detach().cpu() #to show progress on training
```

Note that the code to train over multiple trajectories is the same as before. The
curves of the episode return and loss during training are shown in Figure 8-11.
Comparing the graphs in this figure with the previous ones for REINFORCE in Figure 8-5,
you can see that the training using A2C happens faster with steady progress toward a
better policy.

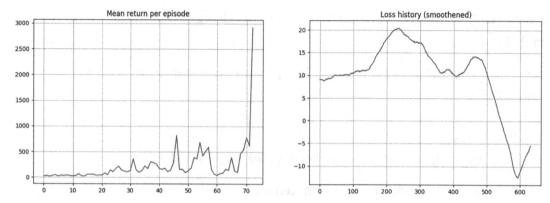

Figure 8-11. *Training progess A2C from 8.a-action_critic.ipynb*

Note that actor-critic is also an *on-policy* approach, just like with REINFORCE.

Asynchronous Advantage Actor-Critic

In 2016, the authors of the paper "Asynchronous Methods for Deep Reinforcement Learning,"[1] introduced the asynchronous version of A2C. The basic idea is simple. You have a global server that is the "parameter" server providing the network parameters, θ and ϕ. There are multiple actor-critic agents running in parallel. Each actor-critic agent obtains the parameter from the server, carries out trajectory rollouts, and does the gradient update on θ, ϕ. The agent updates the parameter back to the server. It allows for faster learning, especially in an environment that uses simulators, such as robotic environments. You can first train an algorithm using A3C on multiple instances of the simulator. A subsequent learning would be to further fine-tune/train the algorithm on a physical robot in the real environment.

Figure 8-12 shows a high-level schematic of A3C. Note that it is a simplistic explanation of the approach. For actual implementation details, you are advised to refer to the referenced paper in detail.

---

[1] https://arxiv.org/pdf/1602.01783.pdf

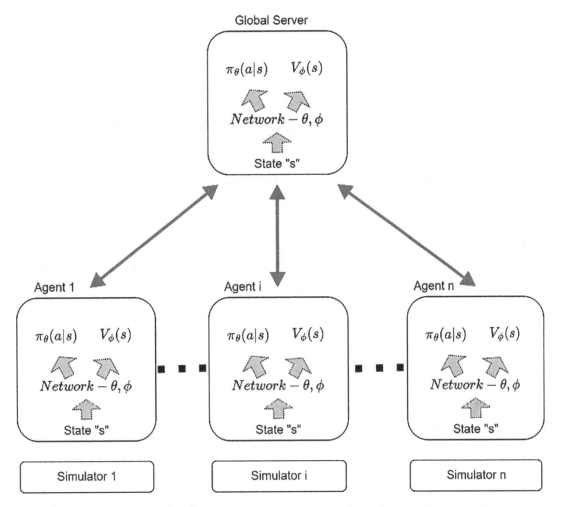

Figure 8-12. *Asynchronous advantage actor-critic (A3C)*

As explained, some papers refer to the synchronous version with multiple agents training together as the A2C version of A3C, that is, A3C without the asynchronous part. However, at times, the actor-critic with one agent is also referred to as *advantage actor-critic* (A2C). In the end, the actor-critic is a family of algorithms where you use two networks together: a value network to estimate the $V(s)$ and a policy network to estimate the policy $\pi_\theta(a|s)$. You are leveraging the best of both worlds: value-based methods and policy gradient methods.

Next, you will look at some other approaches to the policy algorithm. These approaches, with their variants, are considered state of the art, especially the PPO algorithm. PPO has been used in Large Language Models to fine-tune the models for safety, instructions, and for mimicking human preferences. This approach is called RLHF (Reinforcement Learning with Human Feedback). I cover this in detail in a future chapter.

Trust Region Policy Optimization Algorithm

The methods detailed so far in this chapter are also known as *vanilla policy gradients* (VPG). The policy trained using VPG is a stochastic one that offers exploration on its own without explicitly using an ε-greedy exploration. As the training progresses, the policy distribution becomes sharper, centering on optimal actions. This could happen despite the use of entropy penalty term, since there is only so much you add it as penalty. If you add too much weight to the entropy penalty term, the model will be significantly constrained and will start producing uniform distribution probability across all possible actions.

The peaking of policy starts to reduce the exploration, making the algorithm exploit more and more what it has learned. It can cause the policy to get stuck at a local maxima. Using entropy as a regularizer/penalty is not the only approach.

As you saw in the policy gradient approach in previous sections, specifically in Equation 8-17, you update the policy parameters by a small amount given by the following equation:

$$\theta_{\text{new}} = \theta_{\text{old}} + \alpha \cdot \nabla_\theta J(\theta)\big|_{\theta = \theta_{\text{old}}}$$

In other words, the gradient is evaluated at the old policy parameter $\theta = \theta_{\text{old}}$ and then updated by taking a small step determined by the step size α. VPG tries to keep the new and old policies near to each other in the parameter space by restricting the change in policy parameters from θ_{old} to θ_{new} using a learning rate α. However, just because policy parameters are nearby, that does not guarantee that old and new policies (i.e., the action probability distribution) are close to each other. A small change in parameter θ could lead to significant divergence in the policy probabilities. Ideally, you should keep the old and new policies close to each other in the probability space and not in the parameter space. This is the key insight as detailed by authors of the 2015 paper titled "Trust Region Policy Optimization."[2] Before diving into the details, I spend a few minutes talking about a metric called the Kullback-Liebler divergence (KL divergence).

---

[2] https://arxiv.org/pdf/1502.05477.pdf

KULLBACK-LIEBLER DIVERGENCE (KL DIVERGENCE)

KL divergence is a measure of how different the two probabilities are. It comes from the field of information theory, and diving deep into it would require a book of its own. I only attempt to explain the formula and some intuition behind it without getting into the mathematical proof.

Suppose you have two discrete probability distributions, P and Q defined over some range of values (called *support*). Let the support be "x" going from 1 to 6—just like the case of a rolling die. $P_X(X = x)$ defines the probability of $X = x$ using the probability distribution P. Similarly, you have another probability distribution Q defined over the same support. As an example, consider a die with six faces with a probability distribution.

x	1	2	3	4	5	6
$P(x)$	1/6	1/6	1/6	1/6	1/6	1/6
$Q(x)$	2/9	1/6	1/6	1/6	1/6	1/9

The die Q is loaded to show less of 6 and more of 1, while P is a fair die with equal probabilities of showing any face of the dice.

The KL divergence between P and Q is expressed as follows:

$$D_{KL}(P\|Q) = \sum_x P(x)\log\frac{P(x)}{Q(x)} \tag{8-41}$$

Calculate $D_{KL}(P\|Q)$ for the previous table.

$$D_{KL}(P\|Q) = \frac{1}{6}\log\frac{1/6}{2/9} + \frac{1}{6}\log\frac{1/6}{1/6} + \frac{1}{6}\log\frac{1/6}{1/6} + \frac{1}{6}\log\frac{1/6}{1/6} + \frac{1}{6}\log\frac{1/6}{1/6} + \frac{1}{6}\log\frac{1/6}{1/9}$$

$$= \frac{1}{6}\log\frac{3}{4} + \frac{1}{6}\log 1 + \frac{1}{6}\log 1 + \frac{1}{6}\log 1 + \frac{1}{6}\log 1 + \frac{1}{6}\log\frac{3}{2}$$

$$= \frac{1}{6}\log\frac{3}{4} + \frac{1}{6}x0 + \frac{1}{6}x0 + \frac{1}{6}x0 + \frac{1}{6}x0 + \frac{1}{6}\log\frac{3}{2}$$

$$= \frac{1}{6}\left(\log\frac{3}{4} + \log\frac{3}{2}\right)$$

$$= \frac{1}{6}\log\left(\frac{3}{4}x\frac{3}{2}\right)$$

$$= \frac{1}{6}\log\left(\frac{9}{8}\right) = 0.0196$$

You can satisfy yourself by putting P = Q to obtain $D_{KL}(P\|Q)$ = 0. When two probabilities are equal, the KL divergence is 0. For any other two unequal probability distributions, you will always get a +ve KL divergence. The farther apart the distribution is, the higher the KL divergence value. There is a rigorous mathematical proof to show that KL divergence is always +ve and 0 only when the two distributions are equal.

Also note that KL divergence is not symmetric.

$$D_{KL}\left(P\|Q\right) \neq D_{KL}\left(Q\|P\right)$$

KL divergence is a kind of pseudo-measure of distance between two probability distributions in probability space. The KL divergence formula for continuous probability distributions is given as follows:

$$D_{KL}\left(P\|Q\right) = \int P\left(x\right)\log\frac{P\left(x\right)}{Q\left(x\right)}dx \tag{8-42}$$

With the basics of KL divergence defined, it's time to move back to the discussion of TRPO algorithm.

Coming back to TRPO, you need to keep the new and old policies close to each other, not in parameter space but in probability space. This is the same as saying that you want the KL divergence to be bounded in each update step to ensure that the new and old policies do not diverge too far.

$$D_{KL}\left(\theta\|\theta_{k}\right) \leq \delta$$

Here, θ_k is the current policy parameter, and θ is the parameter for the updated policy.

Now turn your attention to the objective that you are trying to maximize. The previous metric $J(\theta)$ did not have any explicit dependence on new and old policy parameters, say θ_{k+1} and θ_k, respectively. There is an alternative formulation for the policy objective using importance sampling. You can state this without mathematical derivation, as shown here:

354

$$J(\theta,\theta_k)=E_{a\sim\pi_{\theta_k}(a|s)}\left[\frac{\pi_\theta(a|s)}{\pi_{\theta_k}(a|s)}A^{\pi_{\theta_k}}(s,a)\right]\tag{8-43}$$

Here, θ is the parameter of the revised/updated policy, and θ_k is the parameter of the old policy. You are trying to take the largest possible step to go from the old policy parameter θ_k to the revised policy with the parameter θ, such that the KL divergence between the new and old policies is not too large. Rephrasing, you need to find a new policy that increases the objective to the maximum extent without going out of the trust region around the old policy, as defined by $D_{KL}(\theta\|\theta_k)\le\delta$. In mathematical terms, you can summarize the maximization problem as follows:

$$\theta_{k+1}=\arg\max_\theta J(\theta,\theta_k)$$

$$s.t.D_{KL}(\theta\|\theta_k)\le\delta$$

$$where, J(\theta,\theta_k)=E_{a\sim\pi_{\theta_k}(a|s)}\left[\frac{\pi_\theta(a|s)}{\pi_{\theta_k}(a|s)}A^{\pi_{\theta_k}}(s,a)\right]$$

The advantage $A^{\pi_{\theta_k}}(s,a)$ is defined as before:

$$A^{\pi_{\theta_k}}(s,a)=Q^{\pi_{\theta_k}}(s,a)-V^{\pi_{\theta_k}}(s)$$

Or, when you roll out by one step and V is parameterized by another network with parameter ϕ, here is the equation:

$$A^{\pi_{\theta_k}}(s_t,a_t)=r(s_t,a_t)+V^{\pi_{\theta_k}}(s_{t+1};\phi)-V^{\pi_{\theta_k}}(s_t;\phi)$$

This is the theoretical representation of the objective maximization in TRPO. However, with the Taylor series expansion of the objective $\theta_{k+1}=\arg\max_\theta J(\theta,\theta_k)$ and the KL constraint $D_{KL}(\theta\|\theta_k)\le\delta$, coupled with Lagrangian duality using convex optimization, you can get an approximate update expression. The approximation can break the guarantee of KL divergence being bounded, for which a backtracking line search is added to the update rule. Lastly, it involves the inversion of an $n\,x\,n$ matrix, which is not easier to compute. In that case, the conjugate gradient algorithm is used. At this point, you have a practical algorithm to calculate the update using TRPO.

The mathematical derivation of the approach is fairly involved, and the final algorithm has many tweaks to make it efficient. I am not getting into the details, as this is beyond the scope of this book. Interested readers can refer to the original paper for more.

Proximal Policy Optimization Algorithm (PPO)

Proximal policy optimization (PPO) is also motivated by the same question as TRPO. "How can we take the maximum possible step size in policy parameters without going too far and getting to a worse policy than the original one before the update?"

The PPO-clip variant does not have KL divergence. It depends on clipping the gradients in the objective function such that the update has no incentive to move the policy too far from the original step. The exact details of how this clipping works are explained in the next chapter, which does a deep dive into the PPO algorithm.

PPO is simpler to implement and has empirically been shown to perform as good as TRPO. The details are in the 2017 paper titled "Proximal Policy Optimization Algorithms."[3]

The objective using the PPO-clip variant is as follows:

$$J(\theta,\theta_k) = \min\left(\frac{\pi_\theta(a|s)}{\pi_{\theta_k}(a|s)}A^{\pi_{\theta_k}}(s,a), g\left(\epsilon, A^{\pi_{\theta_k}}(s,a)\right)\right)$$

where:

$$g(\epsilon, A) = \begin{cases} (1+\epsilon)A, & A \geq 0 \\ (1-\epsilon)A, & A < 0 \end{cases}$$

Rewrite $J(\theta, \theta_k)$, when advantage A is +ve, as shown here:

$$J(\theta,\theta_k) = \min\left(\frac{\pi_\theta(a|s)}{\pi_{\theta_k}(a|s)}, (1+\epsilon)\right)A^{\pi_{\theta_k}}(s,a)$$

---

[3] https://arxiv.org/pdf/1707.06347.pdf

When the advantage is +ve, you want to update the parameters so that the new policy $\pi_\theta(a|s)$ is higher than the old policy $\pi_{\theta_k}(a|s)$. But instead of increasing it too far, you clip the gradients to ensure that the new policy increase is within $1 + \varepsilon$ times the old policy.

Similarly, when the advantage is -ve, you get the following:

$$J(\theta,\theta_k) = \min\left(\frac{\pi_\theta(a|s)}{\pi_{\theta_k}(a|s)}, (1-\epsilon)\right) A^{\pi_{\theta_k}}(s,a)$$

In other words, when the advantage is -ve, you want the parameters to be updated so that you decrease the policy probability for that (s, a) pair. However, instead of decreasing all the way, you clip the gradients so that the new policy probability does not go down below $(1 - \varepsilon)$ times the old policy probability.

In other words, you are clipping the gradients to ensure that the policy updates leave the policy probability distribution to be within $(1 - \varepsilon)$ to $(1 + \varepsilon)$ times the old probability distribution. ϵ acts as a regularizer. It is easy to implement PPO as compared to TRPO. You can follow the same pseudocode given in Figure 8-9 for A2C with just one change, to swap the objective $J(\theta)$ with the objective in the previous paragraphs.

This time, instead of coding it, you can use SB3 (`stable-Baselines3`), which was introduced previously. As discussed earlier, SB3 has implementations for many of the popular and newest algorithms.

This code will follow the same pattern as before, except that you will not be explicitly defining the policy network. You will also not write the training step of calculating loss and stepping through the gradient. You can find the complete code to train and record the performance of `CartPole` training using PPO in `8.c-ppo_sb3.ipynb`.

I will now walk through the code snippet of creating an agent, training it on `CartPole`, and evaluating the performance, as given in Listing 8-8. You will use almost everything from SB3 library. First, you import the required packages. This is followed by creating the environment. Next, you create an MLP policy that comes as a default in SB3. In a previous chapter, you learned how to customize the policy if a need arises. At this point, you just call the `learn` function on the model, which spins up all the required machinery to train and optionally log the progress in many different ways. Once the agent is trained, you evaluate the agent by again using the SB3 provided `evaluate_policy` function. The `evaluate_policy` function runs the agent over 100 episodes (`n_eval_episodes=100`) and calculates two metrics: 1) the mean value of return per episode, signifying the overall quality of the agent's performance, and 2) the standard deviation signifying the variability of the first metric, that is, the mean value of return per episode.

You want the mean value to be as high as possible and sthe tandard deviation to be as low as possible, which will signal a high quality (high mean value) and consistent (low standard deviation) performance of the agent.

In Listing 8-8, you can see that the agent has trained well. The accompanying notebook carries the usual additional code to record a video in order to share the trained model on the HuggingFace hub.

Listing 8-8. PPO Agent for CartPole Using a Stable Baselines3 Implementation

```
import gymnasium as gym
from stable_baselines3 import PPO
from stable_baselines3.ppo.policies import MlpPolicy
from stable_baselines3.common.evaluation import evaluate_policy
from stable_baselines3.common.monitor import Monitor
# other imports omitted for brevity

def make_env(env_name):
    env = gym.make(env_name, render_mode="rgb_array")
    return env

env_name = 'CartPole-v1'

# create a SB3 default MLP policy
model = PPO(MlpPolicy, env, verbose=0)

# Train the agent for 30000 steps
model.learn(total_timesteps=30000)

# Evaluate the trained agent
mean_reward, std_reward = evaluate_policy(Monitor(model), env, n_eval_
episodes=100)
print(f"mean_reward:{mean_reward:.2f} +/- {std_reward:.2f}")
```

If you want to know more about the details of PPO, continue to read. I have a full chapter devoted to PPO, as this is a state-of-the-art algorithm and has recently been used to fine-tune Large Language Models under the RLHF approach.

I want to also use this opportunity to stress once more the point of getting to know the popular RL libraries well. While walking through the implementations of various algorithms in this book, you should note that it is equally important to get comfortable with the popular RL implementations and to learn to use them for your specific needs. The code in this chapter highlights the key concepts and the notebooks accompanying this book help you do a deep dive into the concepts along with a complete end-to-end training process. However, these are by no means production grade. Libraries like SB3 and others introduced previously have highly optimized code with the ability to leverage the GPU and multiple cores to run many agents in parallel.

Curiosity-Driven Learning

Having studied most of the popular policy gradient algorithms, you can now focus on how reward quality impacts the learning in all these algorithms. You have seen that all algorithms use reward-to-go to shape the policy, whether it be REINFORCE, actor-critic, TRPO, or PPO.

Sometimes, these *extrinsic rewards* are given all the time, like the score in a video game or how close a robotic arm is to an object it's trying to reach. But often, especially in real-life situations, these outside rewards are rare or don't exist, making it hard for the agent to learn because it only gets feedback when it reaches a specific goal. This is like trying to find a goal by randomly guessing, which usually doesn't work unless the environment is very simple.

Humans also deal with rare rewards. For example, an entrepreneur's ultimate reward is making a profit, but that might take years. Yet, they keep going, driven by their own motivation or curiosity. This inner drive helps them explore and learn new things, even if the reward is far off. In the same way, in reinforcement learning, when outside rewards are rare, the agent's own/intrinsic motivation or curiosity becomes vital. This *intrinsic reward* encourages the agent to explore and learn, which might help it achieve rewards later. This the what the authors of a 2017 paper titled "Curiosity-driven Exploration by Self-Supervised Prediction"[4] showed as to how adding intrinsic rewards can help improve learning.

---

[4] https://arxiv.org/pdf/1705.05363.pdf

There are two common ways to model these intrinsic rewards 1) encourage agents to explore unseen/novel states, or 2) encourage the agent to perform actions that reduce the error/uncertainty in the agent's ability to predict the consequence of its own actions (i.e., its knowledge about the environment or transition function $p(s_{t+1}|s_t, a_t)$).

The authors proposed a very interesting approach, as shown in Figure 8-13.

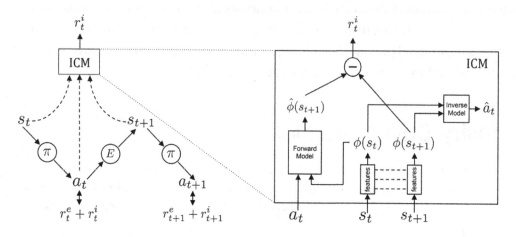

Figure 8-13. *Intrinsic Curiosity Module (ICM) formulation*
Source: https://arxiv.org/pdf/1705.05363.pdf

I first focus on the Intrinsic Curiosity Module (ICM) on the right side of Figure 8-13. ICM takes the current state s_t, the action a_t and the next state observed s_{t+1} as input. It converts the high dimensional state—the image of a game—and passes both the current and the next state through a network to map the high dimensional space to a smaller dimension feature space. The state s_t is mapped to $\phi(s_t)$ and the state s_{t+1} is mapped to $\phi(s_{t+1})$ using the same feature extractor. $\phi(s_t)$ *and* $\phi(s_{t+1})$ are fed as inputs to another network, called the *Inverse Model,* which uses these two inputs to predict the action a_t that made the system transition from s_t to s_{t+1}. This way, the feature extractor learns to represent the states into a smaller dimension space, focusing only on the parts of the state that are most relevant or influenced by the action a_t. The irrelevant parts of s_t, which would not have had any influence in making system transition from s_t to s_{t+1} due to the action a_t, are weeded out from the feature extractor submodule during training. Mathematically, this can be expressed by a set of equations as follows:

$$\hat{a}_t = g\left(s_t, s_{t+1}; \theta_I\right)$$

$$\min_{\theta_I} L_I(\hat{a}_t, a_t)$$

where θ_I are the parameters of the feature extractor and inverse model jointly, as shown in Figure 8-13. L_I is a loss that depends on the nature of action. For example, for a discrete action space, it will be the usual cross-entropy loss.

The third submodule is called the forward model and it takes the feature $\phi(s_t)$ and the action a_t to predict the next state $\hat{\phi}(s_{t+1})$ feature of the next state s_{t+1}. The error between the predicated next state feature and the actual next state feature is minimized and called intrinsic reward r^i. This error or intrinsic reward will be lower for the states already visited and higher for unvisited states. The higher the error, the higher the intrinsic reward to the agent to explore the new state s_{t+1} via action a_t. Mathematically, this can be represented as follows:

$$\hat{\phi}(s_{t+1}) = f\left(\phi(s_t), a_t; \theta_F\right)$$

where θ_F are the parameters of the forward model inside the ICM. The loss you minimize and the intrinsic reward are represented as follows:

$$L_F = \frac{1}{2} \| \hat{\phi}(s_{t+1}) - \phi(s_{t+1}) \|_2^2$$

$$r_t^i = \frac{\eta}{2} \| \hat{\phi}(s_{t+1}) - \phi(s_{t+1}) \|_2^2$$

They are the same quantities, except for a scaling factor η for intrinsic reward.

The usual policy gradient algorithm is now trained, not just on the extrinsic reward r^e from the environment, but on a combined extrinsic and intrinsic reward $r^e + r^i$. In practice, all three submodules of the ICM module and the policy parameters are trained together via a combined loss using this equation:

$$\min_{\theta_P, \theta_I, \theta_F} \left[-\lambda \cdot E_{\pi(s_t; \theta_P)}\left[\Sigma_t r_t\right] + (1-\beta) L_I + \beta \cdot L_F \right]$$

where $0 \le \beta \le 1$ is a scalar that weighs the inverse model loss against the forward model loss and $\lambda > 0$ is a scalar that weighs the importance of the policy gradient loss against the importance of learning the intrinsic reward signal.

In Figure 5 of the paper, the authors show that curiosity-driven exploration learning offers much better performance in all the settings of dense rewards, sparse rewards, and very sparse rewards. Interested readers can refer to the paper for more details. The `https://pathak22.github.io/noreward-rl/` website has a demo and source code implementation using TensorFlow and OpenAI Gym environment.

A year after this paper came out, a team from OpenAI published a paper in 2018 titled "Large-Scale Study of Curiosity-Driven Learning"[5] which further extends the results to even more extreme situations of zero extrinsic rewards. The paper, based on a large-scale study, narrowed down a few qualities that a good feature space should have:

- **Compact:** The features should be low dimensional and filter out irrelevant parts of the observation space.

- **Sufficient:** The features should contain all the important information.

- **Stable:** Non-stationary rewards make it difficult for reinforcement agents to learn. Exploration bonuses introduce non-stationarity since what is new becomes old and boring with time. In a dynamics-based curiosity formulation such as the ICM approach, there are two sources of non-stationarity: the forward dynamics model is evolving over time as it is trained, and the features are changing as they learn. The former is intrinsic to the method, and the latter should be minimized where possible.

The feature space was represented by a model that learned to follow the random initialized model. The error between the feature space value produced via a random model and the learned model acts as guidance on whether the state is a new one or something that the agent has already visited. The error will be lower for old states and higher for newer states. As before, while you reduce this error to train the feature space model, you also use the error as a form of intrinsic reward. This approach works even in environments with no extrinsic rewards.

This brings you to the end of this chapter.

---

[5] `https://arxiv.org/pdf/1808.04355.pdf`

Summary

This chapter introduced the alternate approach of directly learning a policy instead of going through the learning state/action values first and then using these values to find the optimal policy.

First, thew chapter looked at the derivation of REINFORCE, the most basic of the policy gradient methods. After the initial derivation, you looked at a couple of variance reduction techniques such as rewards-to-go and the use of baselines. You also implemented a version of REINFORCE from scratch.

This led you to look at the actor-critic family, which combined value-based methods to learn state values acting as baselines inside REINFORCE with rewards-to-go. The chapter showed two approaches for reward-to-go estimation—one used the MC estimation and the other was the approach of $TD(0)$. The policy network (actor) with the state value network (critic) gives you the ability to combine the best of value-based methods and policy gradient methods. The chapter briefly touched on the asynchronous version known as A3C. You implemented actor-critic from scratch. The chapter also showed the improved learning stability that actor-critic algorithms have compared to REINFORCE.

After discussing the vanilla policy gradient approaches, you looked at a couple of advanced policy optimization techniques, such as trust region policy optimization (TRPO) and proximal policy optimization (PPO). I talked about the key motivations and approaches using each of these two techniques and demonstrated how these can be leveraged using standard RL libraries such as SB3.

Finally, you turned your attention to the setting when extrinsic rewards may be sparse or completely absent. I discussed two papers that applied the concept of curiosity-driven learning to use intrinsic rewards in addition to the external extrinsic rewards. I discussed the formulation of this approach in detail and explained how the performance of the agent improves significantly, especially when extrinsic rewards are sparse.

CHAPTER 9

Combining Policy Gradient and Q-Learning

So far in this book, in the context of deep learning combined with reinforcement learning, Chapters 6 and 7 explained deep Q-learning with its variants. You looked at policy gradients in Chapter 8. Neural network training requires multiple iterations, and Q-learning, an off-policy approach, enables you to reuse sample transitions multiple times, giving you sample efficiency. However, Q-learning can be unstable at times. Further, it is an indirect way of learning. Instead of learning an optimal policy directly, you first learn q-values and then use these action values to learn optimal behavior. Chapter 8 looked at the approach of learning a policy directly, giving you much better improvement guarantees. However, all the policy learning algorithms in Chapter 7 were on-policy. You used a policy to interact with the environment and made updates to the policy weights to increase the probability of good trajectories/actions while reducing the probability of bad ones. After each update the policy is changed, making the previous sample rollouts useless. Therefore, you had to discard the old ones and collect new trajectory rollouts after each update, making the process an on-policy leaning.

This chapter looks at combining the benefits of both approaches—that is, off-policy learning while learning the policy directly. I first talk about the tradeoffs of Q-learning versus policy gradient methods. After this, you will look at three popular approaches of combining Q-learning with policy gradients: deep deterministic policy gradients (DDPG), twin delayed DDPG (TD3), and soft actor-critic (SAC). I largely follow the notations, approach, and sample code documented in the OpenAI Spinning Up library.[1] CleanRL[2] is another good library to explore for code implementations of these.

---

[1] https://spinningup.openai.com/
[2] https://github.com/vwxyzjn/cleanrl

N. Sanghi, *Deep Reinforcement Learning with Python*, https://doi.org/10.1007/979-8-8688-0273-7_9

Tradeoffs in Policy Gradient and Q-Learning

Chapter 6 looked at DQN, the deep learning version of Q-learning. In Q-learning, an off-policy method, you collect transitions from an exploratory behavior policy and then use these transitions in a batch stochastic gradient update to learn the q-values. As you learn the q-values, you improve the policy by taking the max over the q-values of all possible actions in a state to choose the best action. This was the equation:

$$w_{t+1} = w_t + \alpha \cdot \frac{1}{N} \sum_{i=1}^{N} \left[r_i + \gamma \cdot \max_{a_i'} \tilde{q}\left(s_i', a_i'; w_t^-\right) - \hat{q}\left(s_i, a_i; w_t\right) \right] \cdot \nabla_{w_t} \hat{q}\left(s_i, a_i; w_t\right) \quad (9\text{-}1)$$

Note the max. By taking max, that is, $\max_{a_i} \tilde{q}\left(s_i', a_i'; w_t^-\right)$, you are defining the TD(0) target value as $r_i + \gamma \cdot \max_{a_i} \tilde{q}\left(s_i', a_i'; w_t^-\right)$, which is the maximum value under best action a_i'. This forces the current state-action q-value $\hat{q}\left(s_i, a_i; w\right)$ to update the weights to reach the higher target. In other words, you are guiding network updates to follow the Bellman optimality equation. The learning is off-policy because the Bellman optimality equations hold no matter what policy was followed. It needs to be satisfied for all the (s, a, r, s') transitions, no matter which policy these were generated from. You can reuse the transitions with the replay buffer, which makes this learning very sample-efficient. However, there are a few issues that Q-learning suffers from.

The first one is about using Q-learning in its current form for continuous actions. Look at $\max_{a_i} \tilde{q}\left(s_i', a_i'; w_t^-\right)$. All the examples in Chapters 6 and 7 had discrete actions. Do you know why? How do you think you will perform *max* when the action space is continuous and multidimensional, for example, moving multiple joints of a robot together?

When the actions are discrete, it is easy to take *max*. You input state s to the model, and you get $Q(s, a)$ for all the possible actions, that is, $Q(s, a_1), \cdots Q(s, a_m)$. With a limited number of discrete actions, choosing the *max* is easy. Figure 9-1 shows a sample model.

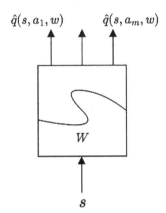

$\hat{q}(s, a_1, w)$ $\hat{q}(s, a_m, w)$

W

s

Figure 9-1. *Generic model for DQN learning with discrete actions*

Now imagine the actions being continuous! How would you take the maximum? To find $\max_{a_i} \tilde{q}\left(s_i', a_i'; w_t^-\right)$, you will have to run another optimization algorithm, which will find the q-value maximizing a_i'. And you will have to do for every value of the state s in the training samples. You can imagine how expensive this process would be—running maximizing action search for every transition in the batch as part of policy improvement.

The second issue is that of learning the wrong objective. You want an optimal policy, but you do not do this directly under DQN. You learn the action-value functions and then use *max* to find the optimal q-value/optimal action.

The third issue is that DQN is also not stable at times. There are no theoretical guarantees, and you are trying to update the weights using the semi-gradient update covered in Chapter 5. You are essentially trying to follow the process of supervised learning with changing targets. What you learn can impact the new trajectory generation, which in turn can impact your quality of learning. All the training progress curves for the average return per episode that you saw for DQN in previous chapters did not continually improve. They were choppy and required careful adjustment of hyperparameters to ensure that the algorithm settled toward a good policy.

Finally, the fourth issue is that DQN learns a deterministic policy. You use an exploratory behavior policy to generate samples and add exploration while the agent learns a deterministic policy. Experiments, especially in robotics, have shown that some amount of stochastic policy is better, as the modeling of the world and manipulations one does with joints are not always perfect. You need to have some amount of randomness to adjust for imperfect modeling or conversion of action values to actual robot joint movements. Also, deterministic policies are a limiting case of stochastic policies.

Turn your attention now to policy gradient methods. In policy gradient methods, you input the state and the output you get is the policy, that is, the probability of actions for discrete actions or the parameters of a probability distribution in the case of continuous actions. The policy gradients allowed you to learn the policies for both discrete and continuous actions. However, learning continuous actions is not feasible in DQN. Figure 9-2 shows the model used in policy gradients.

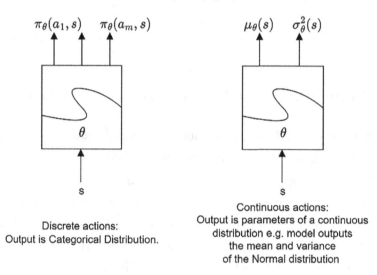

Figure 9-2. Policy networks for policy gradient methods

Chapter 8 showed discrete actions, but the process works out fine for continuous actions, as discussed in that chapter.

Further, using the policy gradient approach, you directly learn to improve the policy instead of the roundabout way of first learning the value functions and then using them to find the optimal policy. The vanilla policy gradient did suffer from a collapse to a bad region, and you saw approaches like TRPO and PPO controlling the step size to improve the guarantee of policy gradients, leading to a better policy learning.

Policy gradients, unlike DQN, learn a stochastic policy, and thereby exploration is built into the policy you are trying to learn. However, the biggest lacuna in the policy gradient method was that it is an on-policy approach. Once you use the transitions to calculate the gradient update, the model has moved to a new policy. The earlier transitions are not relevant anymore in this updated policy world. You need to discard the previous transitions after an update and generate new trajectory/transitions to train the model on. This makes the policy learning very sample inefficient.

You did use value learning as part of policy gradients using the actor-critic methods in which the policy network was the actor trying to learn the best actions, and the value network was the critic informing the policy network of how good or bad the actions are. However, even using the actor-critic methods, the learning was on-policy. You used the critic to guide the actor. The critic is based on the current policy. Therefore, as the policy changes with updates, you need to keep discarding all the current sample transitions after each update to the policy (and/or value) networks.

Is there a way you could learn the policy directly but somehow leverage Q-learning to learn off-policy? And could you this so for the continuous action space? This is covered in this chapter. You will learn how to combine Q-learning with policy gradients to come up with algorithms that are off-policy and work well for continuous actions.

General Framework to Combine Policy Gradient with Q-Learning

This section looks at continuous action policies. There are two networks—one for learning the optimal action for a given state, the actor network. Suppose the policy network is parameterized by θ, and the network learns a policy that produces the action $a = \mu_\theta(s)$, which maximizes $Q(s, a)$. In mathematical notation:

$$\max_{a'} Q^*(s', a') \approx Q^*(s', \mu_\theta(s)) \tag{9-2}$$

The second network, the critic network, will also take state s as one input and the optimal action from the first network, $\mu_\theta(s)$, as the other input to produce the q-value $Q_\phi(s, \mu_\theta(s))$. Figure 9-3 shows the interaction of the two networks conceptually.

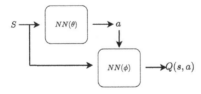

Figure 9-3. *Combining the policy and Q-learning. You learn the policy directly as well as Q using two networks in which the action output from the first network (actor) is fed into the second network (critic), which learns Q(s, a)*

To ensure exploration, take action a, which is exploratory. This is similar to the approach taken in Q-learning, where you learned a deterministic policy but generated transitions from an exploratory ε-greedy policy. Similarly, here, while you learn **a**, you add a bit of randomness $\varepsilon \sim N(0, \sigma^2)$ and use the **a** + **ε** action to explore the environment and generate trajectories.

As with Q-learning, you will use a replay buffer to store the transitions and reuse the these during learning. This is one of the big benefits common to all the approaches discussed in this chapter. It will make policy learning off-policy and hence sample efficient.

In Q-learning, you had to use a target network that was a copy of the Q-network. The reason was to provide some kind of stationery target while learning q-values. You can revisit the discussion of target networks in Chapters 5 and 6. In those approaches, the target network weights were updated with the online/agent network weights periodically. Here, you will also use a target network. However, the approach for updating the target network weights for algorithms in this chapter is that of *polyak averaging* (*exponential averaging*), as given by the following equation:

$$\phi_{target} \leftarrow \rho \cdot \phi_{target} + (1 - \rho)\phi \tag{9-3}$$

You will also be using a target network for the policy network. It is for the same reason as in q-value networks, that is, providing stable targets. With stable targets, you can carry out the supervised learning style of gradient descent and adjust the weights to the approximate $Q(s, a)$.

With this background, you are ready to look at the first algorithm—deep deterministic policy gradient (DDPG).

Deep Deterministic Policy Gradient

In 2016, in a paper titled "Continuous Control with Deep Reinforcement Learning,"[3] authors from DeepMind introduced the DDPG algorithm. The authors had the following points to make about their approach:

- While DQN solves high-dimensional state space, it can only handle discrete and low-dimensional action space. DQN cannot be applied to continuous and high-dimensional action domains, such as physical control tasks like robots.

---

[3] https://arxiv.org/pdf/1509.02971.pdf

- Discretizing action space is not an option due to the *curse of dimensionality*. Assume that you have a robot with seven joints and each joint can move in the range $(-k, k)$. Do a coarse discretization of each joint having three possible values $\{-k, 0, k\}$. Even with this coarse discretization, the total combinations of discrete actions in all seven dimensions work out to $3^7 = 2187$. Instead, if you decide to divide the discrete range for each joint in ten possible values in the range $(-k, k)$, you get $10^7 = 10$ *million* options. This is the *curse of dimensionality,* where the set of possible action combinations grows exponentially with each new dimension/joint.

- The DDPG algorithm is as follows:

  - *Model free*: You do not know the model. You learn it from the interaction of the agent with the environment.

  - *Off-policy*: DDPG, like DQN, uses an exploratory policy to generate transitions and learns a deterministic policy.

  - *Continuous and high-dimensional action space*: DDPG works only for a continuous action domain and works well with high-dimensional action spaces.

  - *Actor-critic*: This means you have an actor (policy network) and critic, the action-value (q-value) network.

  - *Replay buffer*: Like DQN, DDPG uses a replay buffer to store transitions and uses them to learn. This breaks the temporal dependence/correlation of training examples, which could otherwise mess up the learning.

  - *Target network*: Like DQN, DDPG uses a target network to provide stable targets for the q-value to learn. However, unlike DQN, it does not update the target network by copying over the weights of the online/agent/main network periodically. Rather, it uses the polyak/exponential average to keep moving the target network a little bit after every update to the main network.

Now turn your attention to the network architecture and losses you calculate. First you'll look at the Q-learning part, and then you will look at the policy learning network.

Q-Learning in DDPG (Critic)

In DQN, you calculated a loss that was minimized by gradient descent. The loss is given by Equation 6-3, reproduced here:

$$L = \frac{1}{N} \sum_{i=1}^{N} \left[r_i + \left((1 - done_i) . \gamma . \max_{a'} \hat{q}\left(s_i', a'; w_t^-\right) \right) - \hat{q}\left(s_i,\, a_i; w_t\right) \right]^2 \qquad (9\text{-}4)$$

Let's rewrite the equation. Drop subindex **i** and **t** to reduce the clutter of notation. Change the summation to expectation to emphasize that what you usually want is an expectation, but it is estimated by average over samples under Monte Carlo. Eventually in the code you have sums, but they are Monte Carlo estimations of some expectation. Also replace w_t^- , the target network weights, with ϕ_{targ}. Similarly, replace the main weights w_t with ϕ. Further, move the weights from inside the function parameter to the sub-indices on the function, that is, $Q_\phi(...) \leftarrow Q(...; \phi)$. With all these symbol changes, Equation 9-4 now looks like this:

$$L(\phi, D) = \underset{(s,a,r,s',d) \sim D}{E} \left[\left(Q_\phi(s,a) - \left(r + \gamma(1-d) \max_{a'} Q_{\phi_{targ}}(s', a') \right) \right)^2 \right] \qquad (9\text{-}5)$$

This is still the DQN formulation where you take a max of discrete actions in state s' to get $\max_{a'} Q_{\phi_{targ}}(s', a')$. In continuous space, you cannot take *max*, and hence you have another network (actor) to take input state s and produce the action, which maximizes $Q_{\phi_{targ}}(s', a')$. That is, you replace $\max_{a'} Q_{\phi_{targ}}(s', a')$ with $Q_{\phi_{target}}\left(s', \mu_{\theta_{targ}}(s')\right)$ where $a' = \mu_{\theta_{targ}}(s')$ is the target policy. The updated loss expression is as follows:

$$L(\phi, D) = \underset{(s,a,r,s',d) \sim D}{E} \left[\left(Q_\phi(s,a) - \left(r + \gamma(1-d) \max_{a'} Q_{\phi_{targ}}\left(s', \mu_{\theta_{targ}}(s')\right) \right) \right)^2 \right] \qquad (9\text{-}6)$$

This is the updated mean squared Bellman error (MSBE) that you will implement in code and then do back propagation to minimize the loss function. Note that this is only a function of ϕ and so the gradient of $L(\phi, D)$ is with respect to ϕ. As discussed earlier, in the code you will replace the expectation with the sample average being the MC estimate of the expectation. Next let's look at the policy learning part.

Policy Learning in DDPG (Actor)

In the policy learning part, you are trying to learn $a = \mu_\theta(s)$, a deterministic policy that gives the action that maximizes $Q_\phi(s, a)$. As the action space is continuous and you assume that the Q function is differentiable with respect to the action, you can just perform gradient ascent with respect to the policy parameters to solve.

$$\max_\theta J(\theta, D) = \max_\theta \underset{s \sim D}{E}\left[Q_\phi\left(s, \mu_\theta(s)\right)\right] \tag{9-7}$$

As the policy is deterministic, the expectation in Equation 9-7 does not depend on the policy, which is different than what you saw in stochastic gradients in the previous chapter. The expectation operator there depended on policy parameters because the policy was stochastic, which in turn impacted the expected q-value.

You can take the gradient of J about θ to get the following:

$$\nabla_\theta J(\theta, D) = \underset{s \sim D}{E}\left[\nabla_a Q_\phi(s, a)\big|_{a=\mu_\theta(s)} \nabla_\theta \mu_\theta(s)\right] \tag{9-8}$$

This is straight application of chain rule. Also note that you do not get any $\nabla \log(\ldots)$ terms inside the expectation, as the state s over which the expectation is being taken is coming from the replay buffer, and it has no dependence on the parameter θ with respect to which gradient is being taken.

Further, in a 2014 paper titled "Deterministic Policy Gradient Algorithms,"[4] the authors showed that Equation 9-8 is the policy gradient; that is, the gradient of the policy's performance. You are advised to read through both papers to get a deeper theoretical understanding of the math behind DDPG.

As discussed earlier, to aid in exploration, while you learn a deterministic policy, you will use a noisy exploratory version of the learned policy to explore and generate transitions. You do this by adding a mean zero Gaussian noise to the learning policy. Adding noise is a standard process in ML and it acts as a regularizer, making sure that the learned policy has good generalization capabilities.

---

[4]http://proceedings.mlr.press/v32/silver14.pdf

Pseudocode and Implementation

At this point, you are ready to see the complete pseudocode. Refer to Figure 9-4 for this.

Deep Deterministic Policy Gradient

---

1. Input initial policy parameters θ, Q-function parameters ϕ, empty replay buffer D

2. Set target parameters equal to online parameters $\theta_{\text{targ}} \leftarrow \theta$ and $\phi_{\text{targ}} \leftarrow \phi$

3. **repeat**

4. Observe state s and select action $a = clip(\mu_\theta(s) + \alpha \cdot \epsilon, a_{Low}, a_{High})$, where $\epsilon \sim N$, and α is the scale of noise equal to range of action values.

5. Execute a in environment and observe next state s', reward r, and done signal d

6. Store (s, a, r, s', d) in Replay Buffer D

7. if s' is terminal state, reset the environment

8. if it's time to update **then**:

9.   for as many updates as required:

10.   Sample a batch $B = (s, a, r, s', d)$ from replay Buffer D:

11.   Compute targets:

$$y(r, s', d) = r + \gamma(1 - d)Q_{\phi_{\text{targ}}}(s', \mu_{\theta_{\text{targ}}}(s'))$$

12.   Update Q function with one step gradient descent on ϕ:

$$\nabla_\phi \frac{1}{|B|} \sum_{(s,a,r,s',d)\in B} (Q_\phi(s, a) - y(r, s', d))^2$$

13.   Update Policy with one step gradient Ascent on θ:

$$\nabla_\theta \frac{1}{|B|} \sum_{s\in B} Q_\phi(s, \mu_\theta(s))$$

14. Update target networks using polyak averaging:

$$\phi_{targ} \leftarrow \rho\phi_{targ} + (1 - \rho)\phi$$

$$\theta_{targ} \leftarrow \rho\theta_{targ} + (1 - \rho)\theta$$

Figure 9-4. *Deep deterministic policy gradient algorithm*

Gymnasium Environments Used in Code

Turning to the implementation, you will be using two environments in this chapter to run the code. The first one is a pendulum swing environment called Pendulum-v1. Here, the state is a three-dimensional vector giving the angle of the pendulum (i.e., its *cos* and *sin* components), with the third dimension being the angular velocity (theta-dot): $\left[\cos(\theta),\ \sin(\theta),\ \dot{\theta}\right]$.

The action is a single continuous value in the range of [-2.0, 2.0]—the torque being applied to the pendulum. The idea is to balance the pendulum in an upright position for as long as possible. See Figure 9-5. You may also refer to the documentation inside the Gymnasium library.[5]

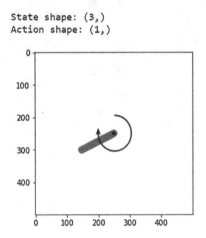

Figure 9-5. *Pendulum environment from the Gymnasium library*

You will train the algorithm on another environment—the continuous version of Lunar-Lander called LunarLanderContinuous-v2. In this environment, you try to land the lunar module on the moon between the flags. The state vector is eight-dimensional: [x_pos, y_pos, x_vel, y_vel, lander_angle, lander_angular_vel, left_leg_ground_contact_flag, right_leg_ground_contact_flag].

---

[5] https://gymnasium.farama.org/environments/classic_control/pendulum/

The action is two-dimensional floats: [main engine, left-right engines].

- Main engine: Range (-1..0) is for engine off, and range (0,1) is the engine throttle from 50% to 100% power. The engine can't work with less than 50 percent power.

- Left/Right – lateral boosters: Range (-1.0, -0.5) fires the left booster, range (+0.5, +1.0) fires the right booster, and range (-0.5, 0.5) is for both boosters off.

Figure 9-6 shows a snapshot of the environment.

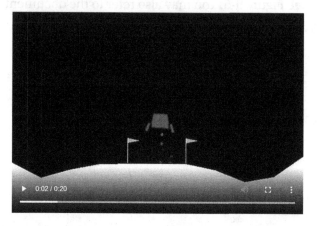

Figure 9-6. *Lunar-lander continuous from the Gymnasium library*

Code Listing

Now turn your attention to the actual code for implementing the DDPG pseudocode in Figure 9-4. The code is from the 9.a-ddpg.ipynb file. This section first explains the Q and policy networks, followed by loss calculation and then the training loop. Finally, you will learn about the code and test the performance of the trained agent.

Policy Network Actor

First let's look at the *actor/policy* network. Listing 9-1 shows the policy network code in PyTorch. You define a simple neural network with two hidden layers of size 256, each with ReLU activation. If you look at the forward function, you will notice that the final layer (self.actor) is passed through the tanh activation. Tanh is a squashing function; it re-maps the values in $(-\infty, \infty)$ to a squashed range of $(-1, 1)$. You then multiply this squashed value with the action limit (`self.act_limit`) so that the continuous output from `MLPActor` is within the valid range of action values that the environment accepts. You create a network class by extending the PyTorch `nn.Module` class, which requires you to define a `forward` function with input state *s* as input, thus producing the action value *a* as the network output.

Listing 9-1. Policy Network in PyTorch from 9.a-ddpg.ipynb

```
class MLPActor(nn.Module):
    def __init__(self, state_dim, act_dim, act_limit):
        super().__init__()
        self.act_limit = act_limit
        self.fc1 = nn.Linear(state_dim, 256)
        self.fc2 = nn.Linear(256, 256)
        self.actor = nn.Linear(256, act_dim)

    def forward(self, s):
        x = self.fc1(s)
        x = F.relu(x)
        x - self.fc2(x)
        x = F.relu(x)
        x = self.actor(x)
        x = torch.tanh(x)   # to output in range(-1,1)
        x = self.act_limit * x
        return x
```

Q-Network Critic Implementation

Next, you look at the Q-network (*critic*). This is also a simple two-layer hidden network with ReLU activation and then a final layer with output dimension equal to one, the dimensionality of action space for the environment. The final layer does not have any activation, enabling the network to produce any value as output of the network. This network is outputting the q-value, and that is why you need to have a possible range of $(-\infty, \infty)$.

Listing 9-2 shows the code for the critic network in PyTorch. This is pretty similar to the implementation of the actor/policy network, except for the minor differences discussed earlier.

Listing 9-2. Q Critic Network in PyTorch from 9.a-ddpg.ipynb

```
class MLPQFunction(nn.Module):
    def __init__(self, state_dim, act_dim):
        super().__init__()
        self.fc1 = nn.Linear(state_dim+act_dim, 256)
        self.fc2 = nn.Linear(256, 256)
        self.Q = nn.Linear(256, 1)

    def forward(self, s, a):
        x = torch.cat([s,a], dim=-1)
        x = self.fc1(x)
        x = F.relu(x)
        x = self.fc2(x)
        x = F.relu(x)
        q = self.Q(x)
        return q
```

Combined Model-Actor-Critic Implementation

Once both the networks are defined, you combine them into a single class that allows you to manage the online and target networks in a more modular way. This is just for better code organization and nothing more. The class combining the two networks is implemented as MLPActorCritic. In this class, you also define a get_action function, which takes the state and a noise scale. It passes the state through the policy network to get $\mu_\theta(s)$, and then it adds a noise (zero mean Gaussian noise) to give a noisy action for exploration. This function implements Step 4 of the algorithm:

$$a = clip\left(\mu_\theta\left(s\right) + \alpha \cdot \epsilon, a_{Low}, a_{High}\right) \text{where} \, \epsilon \sim N, \text{and} \, \alpha \text{ is the scale of noise}$$

The get_action method is the one you will be using to get the action from a policy instead of directly calling MLPActor from Listing 9-1. Listing 9-3 shows the implementation of MLPActorCritic in PyTorch.

Listing 9-3. MLPActorCritic in PyTorch from 9.a-ddpg.ipynb

```
class MLPActorCritic(nn.Module):
    def __init__(self, observation_space, action_space):
        super().__init__()
```

```
        self.state_dim = observation_space.shape[0]
        self.act_dim = action_space.shape[0]
        self.act_limit = action_space.high[0]

        #build Q and policy functions
        self.q = MLPQFunction(self.state_dim, self.act_dim)
        self.policy = MLPActor(self.state_dim, self.act_dim, self.act_limit)

    def act(self, state):
        with torch.no_grad():
            return self.policy(state).numpy()

    def get_action(self, s, noise_scale):
        a = self.act(torch.as_tensor(s, dtype=torch.float32))
        a += noise_scale * self.act_limit * np.random.randn(self.act_dim)
        return np.clip(a, -self.act_limit, self.act_limit)
```

Experience Replay

As with DQN, you will again use experience replay in DDPG to store the prior transition. The implementation of experience replay stays the same as that for DQN. I am not listing the code for it. Interested readers can review the code from the Jupyter notebook called 9.a-ddpg.ipynb.

Q-Loss Implementation

This section investigates the Q-loss calculation. You are essentially implementing the equations from Steps 11 and 12 of the pseudocode.

$$y(r,s',d) = r + \gamma(1-d)Q_{\phi_{\text{targ}}}\left(s',\mu_{\theta_{\text{targ}}}(s')\right)$$

$$Q_{\text{loss}} = \frac{1}{|B|}\sum_{(s,a,r,s',d)\in B}\left(Q_\phi(s,a) - y(r,s',d)\right)^2$$

Listing 9-4 gives the PyTorch implementation. You first convert the batch of (s, a, r, s', d) into PyTorch tensors. Next, you calculate $Q_\phi(s, a)$ using the batch of (s, a) and pass it through the *policy* network. The code that implements this is predicted_qvalues = agent.q(states, actions).

Following this, you calculate the target $y(r, s', d)$ as per the previous expression. You use `torch.no_grad()` for efficiency purposes to disable the gradient calculation while computing the target, as you do not want to adjust the target network weights using auto-diff from PyTorch. Rather, you adjust the target network weights manually using polyak averaging. Stopping the calculation of unwanted gradients can speed up the training, and it ensures that you do not have any unintended side-effects of a gradient step impacting the weights that you want to keep frozen or adjust manually. Finally, you calculate the Q_{Loss} as per the previous expression.

PyTorch carries out the back propagation to calculate the gradient. You do not need to calculate the gradient explicitly in the code.

Listing 9-4. compute_q_loss function in PyTorch from 9.a-ddpg.ipynb

```
def compute_q_loss(agent, target_network, states, actions,
                                    rewards, next_states, done_flags,
                                    gamma=0.99):

    # code for converting numpy array to torch tensors
    ...
    ...
    # get q-values for all actions in current states
    # use agent network
    predicted_qvalues = agent.q(states, actions)

    # Bellman backup for Q function
    with torch.no_grad():
        q_next_state_values = target_network.q(next_states,
                                    target_network.policy(next_states))
        target = rewards + gamma * (1 - done_flags) * q_next_state_values

    # MSE loss against Bellman backup
    loss_q = F.mse_loss(predicted_qvalues, target)

    return loss_q
```

Policy Loss Implementation

Next, you need to calculate the policy loss as per Step 13 of the pseudocode.

$$\text{Policy}_{\text{Loss}} = -\frac{1}{|B|}\sum_{s \in B} Q_\phi\left(s, \mu_\theta\left(s\right)\right)$$

This is a straightforward computation. It is just a three-line code implementation in PyTorch. Listing 9-5 contains the code in PyTorch. Note the -ve sign in the loss. The algorithm needs to do gradient ascent on the policy objective, but auto-differentiation libraries like PyTorch and TensorFlow implement gradient descent. Multiplying the policy objective by -1.0 makes it a loss, and gradient descent on a loss is the same as gradient ascent on a policy objective.

Listing 9-5. compute_policy_loss function in PyTorch from 9.a-ddpg.ipynb

```
def compute_policy_loss(agent, states):

    # convert numpy array to torch tensors
    states = torch.tensor(states, dtype=torch.float)

    predicted_qvalues = agent.q(states, agent.policy(states))

    loss_policy = - predicted_qvalues.mean()
```

One-Step Update Implementation

Next you define a function called one_step_update that gets a batch of (s, a, r, s', d) and computes the Q-loss followed by back propagation, and then a similar step of policy-loss computation followed by the gradient step. It finally carries out an update to the target network weights using polyak averaging. Essentially, this step with the previous two functions compute_q_loss and compute_policy_loss, implements Steps 11 to 14 of the pseudocode.

Listing 9-6 shows the PyTorch code for one_step_update. The first step is to calculate the Q-loss and carry out gradient descent on the critic/Q network weights. This is followed by computing the policy loss and gradient descent on the actor/policy network weights. In certain versions of DDPG, the update to policy is delayed as compared to update to Q-network, that is, one update of policy is carried out for every "k" updates of the Q network. The logic to control the "k" is by using the function input parameter called update_every. This implementation uses a value of 1 so that every update of Q is followed by an update of the policy as well. Lastly, you update the target network weights using polyak averaging.

Listing 9-6. one_step_update function in PyTorch from 9.a-ddpg.ipynb

```
def one_step_update(agent, target_network, q_optimizer, policy_optimizer,
                                  states, actions, rewards, next_
                                  states, done_flags,
                                  step, update_every=1, gamma=0.99,
                                  polyak=0.995):

    #one step gradient for q-values
    loss_q = compute_q_loss(agent, target_network, states,
                                          actions, rewards,
                                          next_states,
                                          done_flags, gamma)
    q_optimizer.zero_grad()
    loss_q.backward()
    q_optimizer.step()

    #one step gradient for policy network
    if step % update_every == 0:
        loss_policy = compute_policy_loss(agent, states)
        policy_optimizer.zero_grad()
        loss_policy.backward()
        policy_optimizer.step()

    # update target networks with polyak averaging
    for params, params_target in zip(agent.parameters(), target_network.
    parameters()):
        params_target.data.mul_(polyak)
        params_target.data.add_((1-polyak)*params.data)

    return loss_q.item(), loss_policy.item()
```

DDPG: Main Loop

The final step is the implementation of the DDPG algorithm, which uses the earlier one_step_update function. It creates the optimizers and initializes the environment. It keeps stepping through the environment using the current online policy. It also maintains a history of q_loss, policy_loss, and average return per episode so that you can plot the trends of these three key metrics during the progress of training.

Initially, for the first start_steps=10000, it takes a random action to explore the environment, and once enough transitions are collected, it uses the current policy with scaled Gaussian noise to selection actions. The transitions are added to the ReplayBuffer, dropping the earliest one from buffer if the buffer has reached full capacity. Once start_steps count is reached, it starts gradient updates inside the main training loop. This happens using the one_step_update function defined in Listing 9-6. You then run the training loop for a total of 20,000 steps, the first 10,000 to collect transitions to fill the replay buffer, followed by the next 10,000 steps of gradient updates.

The loop also has code to update the plots of training progress for Q-loss, policy-loss, and average return per episode. The update frequency of graphs is controlled with the eval_freq parameter. The training loop code is shown in Listing 9-7.

Listing 9-7. DDPG main training loop in PyTorch from 9.a-ddpg.ipynb

```
def ddpg(env_fn, seed=5,
         total_steps=20000, replay_size=20000, gamma=0.98,
         polyak=0.995, policy_lr=0.001, q_lr=0.001, batch_size=256, start_
         steps=10000,
         update_every=1, act_noise=0.1, num_test_episodes=3, eval_
         freq = 500):

    torch.manual_seed(seed)
    np.random.seed(seed)
    env, test_env = env_fn(), env_fn()
    loss_q_history, loss_policy_history, return_history, length_history =
    [], [], [], []
    state_dim = env.observation_space.shape
    act_dim = env.action_space.shape[0]
    act_limit = env.action_space.high[0]
```

```python
agent = MLPActorCritic(env.observation_space, env.action_space)
target_network = deepcopy(agent)

# Experience buffer
replay_buffer = ReplayBuffer(replay_size)

#optimizers
q_optimizer = Adam(agent.q.parameters(), lr=q_lr)
policy_optimizer = Adam(agent.policy.parameters(), lr=policy_lr)

state, _ = env.reset(seed=seed)

for t in range(total_steps):
    if t >= start_steps:
        with torch.no_grad():
            action = agent.get_action(state, act_noise)
    else:
        action = env.action_space.sample()

    next_state, reward, terminated, truncated, _ = env.step(action)
    # some environments do not terminate and therefore we get
    # truncated signal from env
    # we will use both terminated or truncated to reset the game
    done = terminated

    # Store experience to replay buffer
    replay_buffer.add(state, action, reward, next_state, done)

    # dont ever forget this step :)
    state = next_state

    # End of trajectory handling
    if terminated or truncated:
        state, _ = env.reset()

    # Update handling
    if t >= start_steps:
        states, actions, rewards, next_states, done_flags = \
        replay_buffer.sample(batch_size)
```

```
      loss_q, loss_policy = one_step_update(
                    agent, target_network, q_optimizer, policy_
                    optimizer,
                    states, actions, rewards, next_states,
                    done_flags,
                    t, update_every, gamma, polyak)

      loss_q_history.append(loss_q)
      loss_policy_history.append(loss_policy)

   # Statistic updates
   if t >= start_steps and t % eval_freq == 0:
   # Code to update the graphs
        # refer to notebook 9.a-ddpg.ipynb for details

 return agent, loss_q_history, loss_policy_history, \
            return_history, length_history
```

There are additional utility codes used to record the trained agent videos. The rest of the code trains the agent and then records the performance of the trained agent. You run the algorithm first for the Pendulum environment. These code versions are interesting, but as they are incidental to the objective of learning about DDPG, I do not dive into the details of these code implementations. However, interested readers may want to check out the relevant library documentation and step through the code. Figure 9-7 shows the graph as training progresses. If you watch the video of the trained agent in the 9.a-ddpg. ipynb notebook, you will see that the agent quickly learns to balance the pole in an upright position, within 10,000 steps of gradient updates.

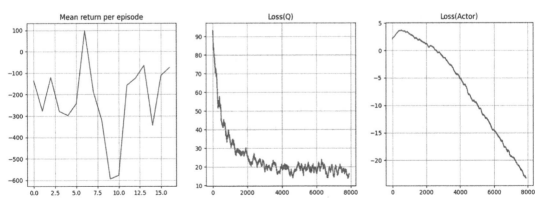

Figure 9-7. *Training progress for Pendulum using DDPG*

You can also run the same code train an agent for Lunar-Lander Continuous environment. You can refer to the notebook for observing the training graphs and videos.

Next, you look at twin delayed DDPG, also known as TD3. It has some other enhancements and tricks to address some of the stability and convergence speed issues seen in DDPG.

Twin Delayed DDPG

Twin delayed DDPG (TD3) was proposed in 2018 in a paper titled "Addressing Function Approximation Error in Actor-Critic Methods."[6] DDPG suffers from the overestimation bias that you saw in Q-learning in Chapter 4 (in the section "Maximization Bias and Double Q-learning"). You saw the approach of double DQN in Chapter 7 to address the bias by decoupling the maximizing action and maximum q value. In the previously mentioned paper, the authors show that DDPG also suffers from the same overestimation bias. They propose a variant of double Q-learning, which addresses this overestimation bias in DDPG. The approach uses the following modifications:

- *Clipped double Q-learning*: TD3 uses two independent Q-functions and takes the minimum of the two while forming targets under Bellman equations, that is, the targets in Step 11 of the DDPG pseudocode in Figure 9-4. This modification is the reason the algorithm is called *twin*.

- *Delayed policy updates*: TD3 updates the policy and target networks less frequently as compared to the Q-function updates. The paper recommends one update to the policy and target networks for every two updates of the Q function. It means carrying out updates in Steps 13 and 14 of the DDPG pseudocode in Figure 9-4 once for every two updates of the Q-function in Steps 11 and 12. This modification is the reason to call this algorithm *delayed*.

- *Target policy smoothing*: TD3 adds noise to the target action, making it harder for the policy to exploit Q-function estimation errors and control the overestimation bias.

---

[6] https://arxiv.org/pdf/1802.09477.pdf

Target-Policy Smoothing

The action used to calculate the target $y(r, s', d)$ is based on the target networks. In DDPG, you calculated $a'(s') = \mu_{\theta_{targ}}(s')$ in step 11 of Figure 9-4. However, in TD3, you carry out the target policy smoothening by adding noise to the action. To the deterministic action $\mu_{\theta_{targ}}(s')$, you add a mean zero Gaussian noise with some clip ranges. The action is then further clipped using *tanh* and multiplied with max_action_range to make sure that action values fit within the range of accepted action values.

$$a = clip\left(\mu_{\theta}(s) + \alpha \cdot \epsilon, a_{Low}, a_{High}\right) \tag{9-9}$$

where $\epsilon \sim N$, and α is the scale of noise equal to range of action values.

Q-Loss (Critic)

You'll use two independent Q-functions. The minimum of the two Q functions is used in the TD(0) target and this target is then used for the combined Q-loss, which is minimized to update the parameters of the two agent Q networks. Expressing the target mathematically looks like this:

$$y(r, s', d) = r + \gamma(1 - d)\min_{i=1,2} Q_{\phi_{targ,i}}\left(s', a'(s')\right) \tag{9-10}$$

This first uses Equation 9-9 to find the noisy target action $a'(s')$. This in turn is used to calculate the target q-values: the q-value of the first and second Q target networks $Q_{\phi_{targ,1}}\left(s', a'(s')\right)$ and $Q_{\phi_{targ,2}}\left(s', a'(s')\right)$

The common target in Equation 9-10 is used next to find the losses for both Q-networks, as shown here:

$$Q_{Loss,1} = \frac{1}{B}\sum_{(s,a,r,s',d)\in B}\left(Q_{\phi_1}(s, a) - y(r, s', d)\right)^2 \tag{9-11a}$$

And,

$$Q_{Loss,2} = \frac{1}{B}\sum_{(s,a,r,s',d)\in B}\left(Q_{\phi_2}(s, a) - y(r, s', d)\right)^2 \tag{9-11b}$$

The losses are added and then minimized independently to train the Q_{ϕ_1} and Q_{ϕ_2} networks (i.e., two online Critic networks).

$$Q_{Loss} = \sum_{i=1,2} Q_{Loss,i} \tag{9-12}$$

Policy Loss (Actor)

The policy loss calculation remains unchanged, the same as what was used in DDPG.

$$Policy_{Loss} = -\frac{1}{B}\sum_{s\in B} Q_{\phi_1}\left(s, \mu_\theta(s)\right) \tag{9-13}$$

Note that you use only Q_{ϕ_1} in the equation. As with DDPG, also note the -ve sign. You need to do gradient ascent, but PyTorch does gradient descent. You can convert the ascent to descent using a -ve sign.

Delayed Update

Update the online policy and the agent network weights in a delayed manner—one update for every two updates of the online Q networks Q_{ϕ_1} and Q_{ϕ_2}.

Pseudocode and Implementation

At this point, you are ready to see the complete pseudocode. Refer to Figure 9-8.

Twin Delayed DDPG (TD3)

1. Input initial policy parameters θ, Q-function parameters ϕ_1 and ϕ_2, empty replay buffer D

2. Set target parameters equal to online parameters $\theta_{targ} \leftarrow \theta$, $\phi_{targ,1} \leftarrow \phi_1$ and $\phi_{targ,2} \leftarrow \phi_2$

3. **repeat**

4. Observe state s and select action $a = clip(\mu_\theta(s) + \alpha \cdot \epsilon, a_{Low}, a_{High})$, where $\epsilon \sim N$, and α is the scale of noise equal to range of action values.

5. Execute a in environment and observe next state s', reward r, and done signal d

6. Store (s, a, r, s', d) in Replay Buffer D

7. if s' is terminal state, reset the environment

8. if it's time to update **then**:

9.   for j in range (as many updates as required):

10.   Sample a batch $B = (s, a, r, s', d)$ from replay Buffer D:

11.   Compute target actions:

$$a'(s') = \text{clip}\left(\mu_{\theta_{\text{targ}}}(s') + \text{clip}(\alpha \cdot \epsilon, -c, c), a_{Low}, a_{High}\right), \qquad \epsilon \sim \mathcal{N}(0, \sigma)$$

12.   Compute action targets:

$$y(r, s', d) = r + \gamma(1 - d) \min_{i=1,2} Q_{\phi_{\text{targ},i}}(s', a'(s'))$$

13.   Update Q function with one step gradient descent on ϕ:

$$\nabla_\phi \frac{1}{|B|} \sum_{(s,a,r,s',d) \in B} (Q_{\phi_i}(s, a) - y(r, s', d))^2, \qquad \text{for } i = 1, 2$$

14.   if j mod policy_update == 0:

15.   Update Policy with one step gradient Ascent on θ:

$$\nabla_\theta \frac{1}{|B|} \sum_{s \in B} Q_{\phi_1}(s, \mu_\theta(s))$$

16.   Update target networks using polyak averaging:

$$\phi_{targ,i} \leftarrow \rho\phi_{targ,i} + (1 - \rho)\phi_i, \qquad \text{for } i = 1, 2$$

$$\theta_{targ} \leftarrow \rho\theta_{targ} + (1 - \rho)\theta$$

Figure 9-8. *Twin delayed DDPG algorithm*

Code Implementation

Now you walk through the code implementation. As with DDPG, you will run the algorithm on Pendulum and Lunar-Lander. Most of the code is similar to that of DDPG except for the three modifications covered earlier. Accordingly, I will only walk through just the highlights. You can find the complete code for the PyTorch version in the 9.b-td3.ipynb file.

Combined Model-Actor-Critic Implementation

You'll first look at the agent networks. The individual Q-network (critic) MLPQFunction and policy-network (actor) MLPActor are the same as before. However, the agent where you combine the actor and critic, that is, MLPActorCritic, sees a minor change. You now have two Q-networks in line with the "twin" part of TD3. Listing 9-8 contains the code for MLPActorCritic.

Listing 9-8. MPLActorCritic in PyTorch from 9.b-td3.ipynb

```
class MLPActorCritic(nn.Module):
    def __init__(self, observation_space, action_space):
        super().__init__()
        self.state_dim = observation_space.shape[0]
        self.act_dim = action_space.shape[0]
        self.act_limit = action_space.high[0]

        #build Q and policy functions
        self.q1 = MLPQFunction(self.state_dim, self.act_dim)
        self.q2 = MLPQFunction(self.state_dim, self.act_dim)
        self.policy = MLPActor(self.state_dim, self.act_dim, self.
        act_limit)

    def act(self, state):
        with torch.no_grad():
            return self.policy(state).numpy()

    def get_action(self, s, noise_scale):
        a = self.act(torch.as_tensor(s, dtype=torch.float32))
        a += noise_scale * self.act_limit * np.random.randn(self.act_dim)
        return np.clip(a, -self.act_limit, self.act_limit)
```

Q-Loss Implementation

The replay buffer remains the same. The next change is in the way Q-loss is calculated. You implement target policy smoothing and clipped double Q-learning as per Equations 9-9 to 9-12. Listing 9-9 contains the code for `compute_q_loss` in PyTorch. Similar to DDPG, you first convert all the inputs from NumPy array to PyTorch tensors, moving these to GPU if required. `q1 = agent.q1(states, actions)` and `q2 = agent.q2(states, actions)` pass the batch of `(state,action)` tuples through the two Q-networks for the online agent and obtain the two q-values—the twin part of the TD3. Next, you target the policy to form `action_target`, which are the noisy actions with smoothing for next state in line with Equation 9-9. `action_target` is then passed through the target Q-networks to obtain `q1_target` and then to `q2_target`, followed by the minimum of the two values to get `q_target`. Once `q_target` is obtained, you form the $TD(0)$ target, as shown in Equation 9-10. Next, you form the joint MSE loss between `q1` and `target` as well as between `q2` and `target` in line with Equations 9-11 and 9-12. The calculations not involving the online q-network are carried out under `torch.no_grad()` to stop gradient accumulation, speed up the calculation, and save memory. This is because `q_loss` is used to adjust only the weights of online q-networks.

Listing 9-9. Q-Loss in PyTorch from 9.b-td3.ipynb

```
def compute_q_loss(agent, target_network, states, actions,
                               rewards, next_states, done_flags,
                               gamma, target_noise, noise_clip,
                               act_limit):

    # code to convert numpy array to torch tensors
    # refer to 9.b-td3.ipynb for actual code

    # get q-values for all actions in current states
    # use agent network
    q1 = agent.q1(states, actions)
    q2 = agent.q2(states, actions)

    # Bellman backup for Q function
    with torch.no_grad():
```

```
        action_target = target_network.policy(next_states)

        # Target policy smoothing
        epsilon = torch.randn_like(action_target) * target_noise *
        act_limit
        epsilon = torch.clamp(epsilon, -noise_clip, noise_clip)
        action_target = action_target + epsilon
        action_target = torch.clamp(action_target, -act_limit, act_limit)

        q1_target = target_network.q1(next_states, action_target)
        q2_target = target_network.q2(next_states, action_target)
        q_target = torch.min(q1_target, q2_target)
        target = rewards + gamma * (1 - done_flags) * q_target

    # MSE loss against Bellman backup
    loss_q1 = F.mse_loss(q1, target)
    loss_q2 = F.mse_loss(q2, target)
    loss_q = loss_q1+loss_q2

    return loss_q
```

Policy-Loss Implementation

Policy loss calculation remains the same except that you use only one of the Q-networks, which is the first one with weights ϕ_1.

One-Step Update Implementation

The implementation of the one_step_update function is also pretty similar, except for the fact that you need to carry out the gradient of q-loss with respect to the combined network weights of the Q1 and Q2 networks passed into the function as q_params. Further, you carry out online policy updates and updates to target Q-networks and the target policy network once every two updates of the online Q-network update. Listing 9-10 contains the implementation of one_step_update in PyTorch.

Listing 9-10. one_step_update in PyTorch from 9.b-td3.ipynb

```python
def one_step_update(agent, target_network, q_optimizer, policy_optimizer,
                    states, actions, rewards, next_states, done_flags,
                    step, update_every,
                    gamma, polyak, target_noise, noise_clip, act_limit):

    #one step gradient for q-values
    loss_q = compute_q_loss(agent, target_network, states, actions,
                                            rewards, next_states, done_flags,
                                            gamma, target_noise,
                                            noise_clip, act_limit)
    q_optimizer.zero_grad()
    loss_q.backward()
    q_optimizer.step()

    loss_policy_ret = None
    # Update policy and all target networks after `update_every` gradient
    steps of Q-networks
    if step % update_every == 0:
        #one step gradient for policy network
        loss_policy = compute_policy_loss(agent, states)
        policy_optimizer.zero_grad()
        loss_policy.backward()
        policy_optimizer.step()
        loss_policy_ret = loss_policy.item()

        # update target networks with polyak averaging
        for params, params_target in zip(agent.parameters(), target_
        network.parameters()):
            params_target.data.mul_(polyak)
            params_target.data.add_((1-polyak)*params.data)

    return loss_q.item(), loss_policy_ret
```

TD3 Main Loop

The next change is to the frequency of updates. Unlike DDPG, in TD3 you update the online policy and target weights once for every two updates of Q-networks, which is controlled inside the one_step_update function shown in Listing 9-10.

Now run TD3 first for the Pendulum environment and then for the Lunar-Lander Gymnasium environment. Figure 9-9 shows the graph as training progresses. If you watch the video of the trained agent in the 9.b-td3.ipynb notebook, you will see that the agent learns to balance the pole in an upright position fairly quickly, within 10,000 steps of gradient updates. Also, if you compare the average return per episode for DDPG, as shown in Figure 9-7, with the return under TD3 shown in Figure 9-9, you can see that under TD3, the return improves almost monotonically as training progresses, while in DDPG there is an initial dip before the return starts moving up. This points to the stability of the learning process displayed in TD3 due to the multiple enhancements implemented in TD3 over the DDPG versions. Interested readers can refer to the original paper of TD3 to check out the benchmark studies that the authors of TD3 did with respect to other algorithms.

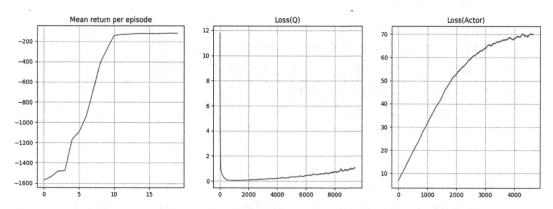

Figure 9-9. *Training progress for Pendulum using TD3*

Soon you will look at the last algorithm in this chapter, an algorithm called *soft actor-critic (SAC)*. However, before you do that, the chapter takes a short detour to explain something called the *reparameterization trick* that SAC uses.

Reparameterization Trick

The reparameterization trick is a "change in the variable" approach that is used in variational auto-encoders (VAEs). There, it is needed to propagate gradients through nodes, which are stochastic. Reparameterization is also used to bring down the variance of gradient estimates. The second reason is explored here. This deep dive follows a blog post by Goker Erdogan[7] with additional analytical derivations and explanations.

Suppose you have a random variable x, which follows a normal distribution with mean θ and unit variance. Let the distribution be parameterized by mean θ as follows:

$$x \sim p_\theta(x) = N(\theta, 1) = \frac{1}{\sqrt{2\pi}} e^{-\frac{1}{2}(x-\theta)^2} \qquad (9\text{-}14)$$

First, you draw samples from it. Next, you assume that you do not know θ and use these samples from the true distribution to find the estimated value of θ, using mean square error (MSE) $J(\theta)$ with the help of gradient descent.

$$J(\theta) = E_{x \sim p_\theta(x)}\left[x^2\right]$$

The focus is on finding two different approaches for determining the estimate of the derivate/gradient $\nabla_\theta J(\theta)$.

Score/Reinforce Way

Since the parameter \theta over which you want to find the gradient also impacts the distribution over which expectation is carried out in the expression for $J(\theta)$, you can follow the log trick, which is what you did with REINFORCE and policy gradient in Chapter 8. You saw that it has high variance, and that's what I want to demonstrate for a simple example distribution shown earlier. Take the derivative of $J(\theta)$ with respect to θ.

$$\nabla_\theta J(\theta)$$

$$= \nabla_\theta E_{x \sim p_\theta(x)}\left[x^2\right]$$

---

[7] http://gokererdogan.github.io/2016/07/01/reparameterization-trick/

$$= \nabla_\theta \int p_\theta(x) x^2 dx$$

$$= \int \nabla_\theta p_\theta(x) x^2 dx$$

$$= \int \frac{p_\theta(x)}{p_\theta(x)} \nabla_\theta p_\theta(x) x^2 dx$$

$$= \int p_\theta(x) \frac{\nabla_\theta p_\theta(x)}{p_\theta(x)} x^2 dx$$

$$= \int p_\theta(x) \nabla_\theta \log p_\theta(x) x^2 dx$$

$$= E_{x \sim p_\theta(x)} \left[\nabla_\theta \log p_\theta(x) x^2 \right]$$

Next, use Monte Carlo to form an estimate of $\nabla_\theta J(\theta)$ using samples.

$$\nabla_\theta \hat{J}(\theta) = \frac{1}{N} \sum_{i=1}^{N} \nabla_\theta \log p_\theta(x_i) x_i^2$$

Substituting the expression for $p_\theta(x)$ from Equation 9-14 and taking a log followed by gradient w.r.t. θ, you get the following:

$$\nabla_\theta \hat{J}(\theta) = \frac{1}{N} \sum_{i=1}^{N} (x_i - \theta) x_i^2 \tag{9-15}$$

Reparameterization Trick and Pathwise Derivatives

The second approach is that of the reparameterization trick. You will redefine random variable x as a composition of a constant and a normal distribution that has no parameter θ. Let x be defined as follows:

$$x = \theta + \epsilon \quad \text{where} \quad \epsilon \sim N(0,1)$$

Note that the previous reparameterization leaves the distribution for x unchanged.

$$p_\theta(x) = N(\theta,1)$$

Now calculate: $\nabla_\theta J(\theta) = \nabla_\theta E_{x \sim p_\theta(x)}\left[x^2\right]$

$$= \nabla_\theta E_{x \sim p_\theta(x)}\left[x^2\right]$$

$$= \nabla_\theta E_{\epsilon \sim N(0,1)}\left[(\theta + \epsilon)^2\right]$$

As the expectation does not depend on θ, you can move the gradient inside without running into the *log* issue (i.e., finding log inside the derivate) shown in the previous method.

$$\nabla_\theta J(\theta)$$

$$= \nabla_\theta \int p(\epsilon)(\theta + \epsilon)^2 \, d\epsilon$$

$$= \int p(\epsilon)\nabla_\theta (\theta + \epsilon)^2 \, d\epsilon$$

$$= \int p(\epsilon)2(\theta + \epsilon) d\epsilon$$

$$= E_{\epsilon \sim N(0,1)}\left[2(\theta + \epsilon)\right]$$

Next, convert the expectation to the Monte Carlo (MC) estimate to get the following:

$$\nabla_\theta \hat{J}(\theta) = \frac{1}{N}\sum_{i=1}^{N} 2(\theta + \epsilon_i) \tag{9-16}$$

Take a moment to compare the two estimates of $\nabla_\theta J(\theta)$ given in Equations 9-15 and 9-16. Remember these two are the estimates of the true gradient $\nabla_\theta J(\theta)$ and therefore may show different convergence properties toward the actual value. This convergence will be a function of the number of samples of the true distribution you have. Both will converge to the true distribution as N, the number of training samples increase. However, one of them will converge to the true value a lot faster and in a more stable way, while the other one will take a large number of training samples. Can you guess which one is better? Ponder this for a few minutes before proceeding.

Experiment

Equations 9-15 and 9-16 are used to calculate the mean and variance of the estimates using two approaches. They do so using different values of N to calculate the estimate as per Equations 9-15 and 9-16, and then repeat the experiment 10,000 times for each value of N to calculate the mean and variance of the gradient estimates under two approaches. This experiment will show that the mean is the same for both equations. In other words, they estimate the same value, but the variance of the estimate in Equation 9-15 is higher than the variance of the estimate in Equation 9-16 by almost one order of magnitude. It's higher by a factor of 21.75 in this case.

To carry out this experiment, follow the steps listed here:

- Freeze: $\theta = 2$

- Generate samples for $x \sim N(\theta, 1)$ and use these samples in Equation 9-15 to calculate the REINFORCE estimate of $\nabla_\theta J(\theta)$.

- Generate samples for $\epsilon \sim N(0, 1)$ and use these samples in Equation 9-16) to calculate the reparameterized estimate of $\nabla_\theta J(\theta)$.

You can see the details of the experiment and code in `9.c-reparam.ipynb`. When you run the code for different values of N, you'll get the convergence graphs of mean and the variance of estimates for different values of N, as shown in Figure 9-10.

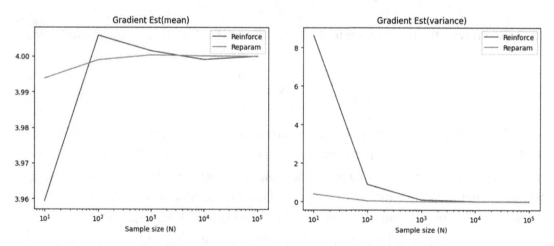

Figure 9-10. *Convergence of the reinforce(log trick) and reparameterization approaches to gradient estimates*

Do these make sense? You can derive the mean and variance of two gradient estimates that are given in Equations 9-15 and 9-16 analytically. Start with Equation 9-15, which is an estimate of $\nabla_\theta J(\theta)$ under the reinforce/log-trick approach, and find its expected value.

$$E\left[\nabla_\theta \hat{J}(\theta)\right] = E\left[\frac{1}{N}\sum_{i=1}^{N}(x_i - \theta)x_i^2\right]$$

For now, assume that $x_i \sim N(\mu, \sigma^2)$.

Taking expectation operation inside the summation, you get:

$$E\left[\nabla_\theta \hat{J}(\theta)\right] = \frac{1}{N}\sum_{i=1}^{N}E\left[(x_i - \theta)x_i^2\right]$$

$$= \frac{1}{N}\sum_{i=1}^{N}E\left[x_i^3\right] - \theta.E\left[x_i^2\right]$$

$$= \frac{1}{N}\sum_{i=1}^{N}E\left[x_i^3\right] - \theta.E\left[x_i^2\right]$$

Using the standard results for a normal distribution $x_i \sim N(\mu, \sigma^2)$, you get:

$$E\left[\nabla_\theta \hat{J}(\theta)\right] = \frac{1}{N}\sum_{i=1}^{N}\left(\mu^3 + 3 \cdot \mu \cdot \sigma^2\right) - \theta\left(\mu^2 + \sigma^2\right)$$

$$= \left(\mu^3 + 3 \cdot \mu \cdot \sigma^2\right) - \theta\left(\mu^2 + \sigma^2\right)$$

In this case, under reinforce, the samples x_i come from $N(\theta, 1)$. Accordingly, substituting, $\mu = \theta$ and $\sigma = 1$, you get:

$$E\left[\nabla_\theta \hat{J}(\theta)\right] = \left(\theta^3 + 3 \cdot \theta.1\right) - \theta\left(\theta^2 + 1^2\right)$$

In the experiment, you took $\theta = 2$. Accordingly, making a further substitution of $\theta = 2$, you get:

$$E\left[\nabla_\theta \hat{J}(\theta)\right] = \left(2^3 + 3.2.1\right) - 2.\left(2^2 + 1\right) = 14 - 10 = 4$$

As you can see from Figure 9-10, the mean of the estimate of $\nabla_\theta J(\theta)$ under reinforce converges to a value of 4 as derived analytically. But it does so only for larger Ns, the number of samples. At $N = 100$, the mean is incorrect and well above the value of 4. This indicates that the variance of the reinforce estimate must be high. Now analytically derive the variance of the reinforce estimate to show that indeed is the case.

$$Var\left(\nabla_\theta \hat{J}(\theta)\right) = var\left(\frac{1}{N}\sum_{i=1}^{N}(x_i - \theta)x_i^2\right)$$

By using the property of variance of the sum of the random variables, you get:

$$Var\left(\nabla_\theta \hat{J}(\theta)\right) = \frac{1}{N^2}\sum_{i=1}^{N} var\left(x_i^3 - \theta.x_i^2\right)$$

$$= \frac{1}{N^2}\sum_{i=1}^{N} E\left[\left(x_i^3 - \theta.x_i^2\right)^2\right] - \left(E\left[x_i^3 - \theta.x_i^2\right]\right)^2$$

$$= \frac{1}{N^2}\sum_{i=1}^{N} E\left[x_i^6 + \theta^2 x_i^4 - 2.\theta.x_i^3 x_i^2\right] - \left(E\left[x_i^3 - \theta.x_i^2\right]\right)^2$$

Using the standard results for the normal distribution, you get:

$$Var\left(\nabla_\theta \hat{J}(\theta)\right) = \frac{1}{N}\left[\begin{array}{l} \mu^6 + 15\mu^4\sigma^2 + 45\mu^2\sigma^4 + 15\sigma^6 + \theta^2\left(\mu^4 + 6\mu^2\sigma^2 + 3\sigma^4\right) \\ -2.\theta\left(\mu^5 + 10\mu^3\sigma^2 + 15\mu\sigma^4\right) - \left(\mu^3 + 3\mu\sigma^2 - \theta.\left(\mu^2 + \sigma^2\right)\right)^2 \end{array}\right]$$

Substituting, putting $\mu = \theta = 2$, and $\sigma = 1$, you get:

$$Var\left(\nabla_\theta \hat{J}(\theta)\right) = \frac{1}{N}\left[\begin{array}{l} 2^6 + 15.2^4 + 45.2^2 + 15 + 4\left(2^4 + 6.2^2 + 3\right) \\ -4\left(2^5 + 10.2^3 + 15.2\right) - \left(2^3 + 3.2 - 2\left(2^2 + 1\right)\right)^2 \end{array}\right]$$

$$= \frac{1}{N}\left[64 + 240 + 180 + 15 + 4\left(16 + 6.4 + 3\right) - 4\left(32 + 10.8 + 15.2\right) - \left(8 + 3.2 - 2\left(4 + 1\right)\right)^2\right]$$

$$= \frac{1}{N}\left[64 + 240 + 180 + 15 + 4\left(16 + 6.4 + 3\right) - 4\left(32 + 10.8 + 15.2\right) - \left(8 + 3.2 - 2\left(4 + 1\right)\right)^2\right]$$

$$Var\left(\nabla_\theta \hat{J}(\theta)\right) = \frac{87}{N}$$

Referring to right-side graph in Figure 9-10, you can see that N = 100. You get a variance of estimate around 0.9, which tallies with the analytically derived result of 87/100=0.87.

Having derived the mean and variance of the reinforce gradient estimate given in Equation 9-15 analytically, next turn your attention to a similar analytical derivation of the mean and variance for the reparametrized gradient estimate given in Equation 9-16. First calculate the mean of the estimate:

$$\mathrm{E}\left[\nabla_\theta \hat{J}(\theta)\right] = \mathrm{E}\left[\frac{1}{N}\sum_{i=1}^{N} 2(\theta + \epsilon_i)\right]$$

Note that under reparameterization, you now have $\epsilon_i \sim N(0,1)$ and θ is a constant equal to 2.

$$\mathrm{E}\left[\nabla_\theta \hat{J}(\theta)\right] = \frac{1}{N}\sum_{i=1}^{N} 2\mathrm{E}\left[(\theta + \epsilon_i)\right]$$

$$= \frac{1}{N}\sum_{i=1}^{N} 2.\theta$$

$$\mathrm{E}\left[\nabla_\theta \hat{J}(\theta)\right] = 2.\theta = 4$$

The estimate's mean is 4 and this is what you see in left side graph in Figure 9-10. You can also observe that convergence of mean around 4 is lot smoother for estimates using the reparameterization approach given in Equation 9-16 versus the convergence of estimate's mean under Equation 9-15 using the reinforce approach.

You can proceed with a similar analysis to calculate the variance of the reparametrized estimate of the gradient in Equation 9-16.

$$Var\left(\nabla_\theta \hat{J}(\theta)\right) = Var\left(\frac{1}{N}\sum_{i=1}^{N} 2(\theta + \epsilon_i)\right)$$

$$= \frac{1}{N^2}\sum_{i=1}^{N} Var\left(2(\theta + \epsilon_i)\right)$$

$$= \frac{1}{N^2}\sum_{i=1}^{N} Var\left(2\epsilon_i\right)$$

$$= \frac{1}{N^2} \sum_{i=1}^{N} 4.Var(\epsilon_i)$$

$$= \frac{1}{N^2} \sum_{i=1}^{N} 4.1$$

$$Var\left(\nabla_\theta \hat{J}(\theta)\right) = \frac{4}{N}$$

Again, refer to the right-side graph in Figure 9-10. Notice that variance of the estimate under the reparameterization approach is smaller than the variance under the reinforce/log-trick approach. This is exactly what the analytical derivations also show. A variance of 4/N for the reparameterization approach versus a variance of 87/N for the reinforce approach.

In summary, you have shown that for REINFORCE, the gradient estimate using Equation 9-15 has the following mean and variance:

$$\text{mean} = 4$$

$$\text{variance} = \frac{87}{N}$$

And, under the REPARAMETERIZED gradient using Equation 9-16, the gradient estimate has the following mean and variance:

$$\text{mean} = 4$$

$$\text{variance} = \frac{4}{N}$$

You can see that the gradient estimate has the same mean under both approaches. However, the variance in the reparametrized approach is a lot smaller, by one order of magnitude.

Readers may wonder how this really helps you and how to use the lower variance approach. Suppose you have a policy network that takes state s as input, and the network is parameterized by θ. The policy network produces the mean and variance of the policy, that is, a stochastic policy with normal distribution whose mean and variance are the outputs of the network, as shown in Figure 9-11.

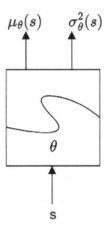

Continuous actions:
Output is parameters of a continuous
distribution e.g. model outputs
mean and additionally variance
of the Normal distribution

Figure 9-11. *Stochastic policy network*

Define the action a as follows:

$$a \sim N\left(\mu_\theta(s), \sigma_\theta^2(s)\right) \tag{9-17}$$

Reparametrize action a to decompose the deterministic and stochastic part out so that the stochastic part does not depend on the network parameters θ.

$$a \sim N\left(\mu_\theta(s), \sigma_\theta^2(s)\right) \text{ where: } \epsilon \sim N(0,1) \tag{9-18}$$

Reparameterization allows you to calculate the policy gradients that have lower variance as compared to using a non-parametrized approach log-trick. In addition, reparameterization gives you an alternative way to flow the gradient back through the network by separating out the stochastic part into a nonparameterized part. You will use this approach in the soft actor-critic algorithm.

It is a very common approach when you need to build neural networks using stochastic nodes and need to flow back the gradients. The foundational paper that formalized the concept is "Gradient Estimation Using Stochastic Computation Graphs" by Schulman et al., 2015.[8] There has been a lot more work after this paper was published.

---

[8] https://arxiv.org/abs/1506.05254

Interested readers may want to refer to this paper and then refer to the various subsequent works on the topic of stochastic computation graphs.

Entropy Explained

Just one more thing before you start digging into the details of SAC: let's revisit entropy. I talked about entropy as a regularizer as part of the REINFORCE code walk-through in the previous chapter. You will be doing something similar. Let's look at what entropy is.

Suppose you have a random variable x following some distribution $P(x)$. The entropy of x is defined as follows:

$$H(P) = \mathop{E}_{x \sim P}\left[-logP(x)\right] \tag{9-19}$$

Suppose you have a coin with $P(Head) = \rho$ and $P(Tail) = 1 - \rho$. Calculate the entropy H for different values of $\rho \; \varepsilon \; (0, 1)$.

$$H(x) = -\left[\rho \, log\rho + (1-\rho) log(1-\rho)\right]$$

You can plot the curve of $H(x)$ versus ρ, as shown in Figure 9-12.

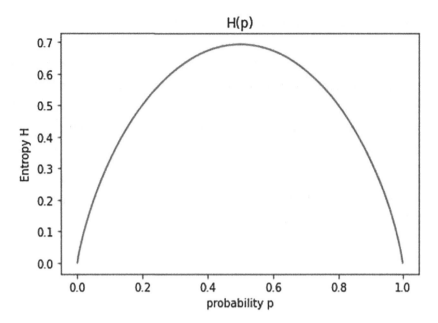

Figure 9-12. *Entropy of a Bernoulli distribution as a function of p, which is the probability of getting 1 in a trial*

You can see that entropy H is the maximum for $\rho = 0.5$, that is, when you have maximum uncertainty between getting a 1 or a 0. In other words, by maximizing the entropy, you are ensuring that the stochastic action policy has a broad distribution and does not collapse into a sharp peak too early. A sharp peak will reduce the exploration of state action space. Readers may refer to the `9.d-entropy_bernoulli.ipynb` notebook for the code that generated Figure 9-12. You can modify the code to try various other probability distributions and vary the parameters to find the points that maximize the entropy.

Soft Actor-Critic

The soft actor-critic came about around the same time as TD3. Like DDPG and TD3, SAC also uses an actor-critic construct with off-policy learning for continuous controls. However, unlike DDPG and TD3, SAC learns a stochastic policy. Hence, SAC forms a bridge between the deterministic policy algorithms like DDPG and TD3 with the stochastic policy optimization. The algorithm was introduced in a 2018 paper titled

"Soft Actor-Critic: Off-Policy Maximum Entropy Deep Reinforcement Learning with a Stochastic Actor."[9]

It uses the clipped double-Q trick like TD3, and due to its learning stochastic policy, it indirectly benefits from the target policy smoothing without an explicit need to add noise to the target policy.

A core feature of SAC is the use of entropy as part of the maximization. To quote the authors of the paper:

> "In this framework, the actor aims to simultaneously maximize expected return and entropy; that is, to succeed at the task while acting as randomly as possible."

SAC vs. TD3

Here are the similarities:

- Both approaches use mean squared Bellman error (MSBE) minimization toward a common target.

- A common target is calculated using target Q networks that are obtained using polyack averaging.

- Both approaches use clipped double Q, which consists of a minimum of two q-values to avoid overestimation.

Here are the differences:

- SAC uses entropy regularization, which is absent in TD3.

- The TD3 target policy is used to calculate the next-state actions, while in SAC, you use the current policy to get actions for the next states.

- In TD3, the target policy uses smoothing by adding random noise to actions. However, in SAC, the policy learned is a stochastic one that provides a smoothing effect without any explicit noise addition.

---

[9] https://arxiv.org/abs/1801.01290

Q-Loss with Entropy-Regularization

Entropy measures randomness in the distribution. The higher the entropy, the flatter the distribution. A sharp peaked policy has all its probability centered around that peak, and hence it will have low entropy. With entropy regularization, the policy is trained to maximize a tradeoff between the expected return and entropy with α controlling the tradeoff. The policy is trained to maximize a tradeoff between the expected return and entropy, a measure of randomness in the policy.

$$\pi^* = \arg\max_{\pi} \, \underset{\tau \sim \pi}{E} \left[\sum_{t=0}^{\infty} \gamma^t \left(R(s_t, a_t, s_{t+1}) + \alpha H\left(\pi(\cdot | s_t) \right) \right) \right] \tag{9-20}$$

In this setting, V^π is changed to include the entropy from every timestep.

$$V^\pi(s) = \underset{\tau \sim \pi}{E} \left[\sum_{t=0}^{\infty} \gamma^t \left(\left(R(s_t, a_t, s_{t+1}) + \alpha . H\left(\left(\pi(\cdot | s_t) \right) \right) \right) \right) \middle\| s_0 = s \right] \tag{9-21}$$

In addition, Q^π is changed to include the entropy bonuses from every timestep except the first.

$$Q^\pi(s,a) = \underset{\tau \sim \pi}{E} \left[\sum_{t=0}^{\infty} \gamma^t R(s_t, a_t, s_{t+1}) + \alpha \sum_{t=1}^{\infty} \gamma^t H\left(\pi(\cdot | s_t) \right) \middle| s_0 = s, a_0 = a \right] \tag{9-22}$$

With these definitions, V^π and Q^π are connected by the following:

$$V^\pi(s) = \underset{a \sim \pi}{E} \left[Q^\pi(s,a) + \alpha H\left(\pi(\cdot | s) \right) \right] \tag{9-23}$$

The Bellman equation for Q^π is as follows:

$$\begin{aligned} Q^\pi(s,a) &= \underset{s' \sim P, a' \sim \pi}{E} \left[R(s,a,s') + \gamma \left(Q^\pi(s',a') + \alpha H\left(\pi(\cdot | s') \right) \right) \right] \\ &= \underset{s' \sim P}{E} \left[R(s,a,s') + \gamma V^\pi(s') \right] \end{aligned} \tag{9-24}$$

The right side is an expectation that you convert into a sample estimate.

$$Q^\pi(s,a) \approx r + \gamma \left(Q^\pi(s', \widetilde{a}') - \alpha \log \pi(\widetilde{a}' | s') \right), \quad \widetilde{a}' \sim \pi(\cdot | s') \tag{9-25}$$

In this equation, (s, a, r, s') all come from the replay buffer, and $\tilde{a}'$ comes from sampling the online/agent policy. *In SAC, you do not use the target network policy at all.*

Like TD3, SAC uses clipped double Q and minimizes the mean squared Bellman error (MSBE). Putting it all together, the loss functions for the Q-networks in SAC are as follows:

$$L(\phi_i, D) = \mathop{E}_{(s,a,r,s',d)\sim D}\left[\left(Q_{\phi_i}(s,a) - y(r,s',d)\right)^2\right], \quad i=1,2 \tag{9-26}$$

where the target is given by the following:

$$y(r,s',d) = r + \gamma(1-d)\left(\min_{i=1,2} Q_{\phi_{targ,i}}(s',\tilde{a}') - \alpha\log\pi_\theta(\tilde{a}'|s')\right), \tilde{a}' \sim \pi_\theta(\cdot|s') \tag{9-27}$$

Convert the expectations to sample averages.

$$L(\phi_i, D) = \frac{1}{|B|}\sum_{(s,a,r,s',d)\in B}\left(Q_{\phi_i}(s,a) - y(r,s',d)\right)^2, \quad \text{for } i=1,2 \tag{9-28}$$

The final Q-loss you will minimize is as follows:

$$Q_{Loss} = L(\phi_1, D) + L(\phi_2, D) \tag{9-29}$$

Policy Loss with the Reparameterization Trick

The policy should choose actions to maximize the expected future return and future entropy, that is, $V^\pi(s)$.

$$V^\pi(s) = \mathop{E}_{a\sim\pi}\left[Q^\pi(s,a) + \alpha H(\pi(\cdot|s))\right]$$

Rewrite this as follows:

$$V^\pi(s) = \mathop{E}_{a\sim\pi}\left[Q^\pi(s,a) - \alpha\log\pi(a|s)\right] \tag{9-30}$$

The authors of the paper use reparameterization along with the squashed Gaussian policy.

$$\widetilde{a}_\theta(s,\xi) = \tanh\big(\mu_\theta(s) + \sigma_\theta(s) \odot \xi\big), \quad \xi \sim \mathcal{N}(0,I) \tag{9-31}$$

Combining the previous two Equations 9-30 and 9-31 and noting that the policy network is parameterized by θ, the policy network weights, you get the following:

$$\mathop{\mathrm{E}}_{a \sim \pi_\theta}\Big[Q^{\pi_\theta}(s,a) - \alpha \log \pi_\theta(a|s)\Big] = \mathop{\mathrm{E}}_{\xi \sim \mathcal{N}}\Big[Q^{\pi_\theta}\big(s,\widetilde{a}_\theta(s,\xi)\big) - \alpha \log \pi_\theta\big(\widetilde{a}_\theta(s,\xi)|s\big)\Big] \tag{9-32}$$

Next, substitute the function approximator for Q, taking the minimum of the two Q-functions.

$$Q^{\pi_\theta}\big(s,\widetilde{a}_\theta(s,\xi)\big) = \min_{i=1,2} Q_{\phi_i}\big(s,\widetilde{a}_\theta(s,\xi)\big) \tag{9-33}$$

The policy objective is accordingly transformed into the following:

$$\max_\theta \mathop{\mathrm{E}}_{\substack{s \sim \mathcal{D} \\ \xi \sim \mathcal{N}}}\Big[\min_{i=1,2} Q_{\phi_i}\big(s,\widetilde{a}_\theta(s,\xi)\big) - \alpha \log \pi_\theta\big(\widetilde{a}_\theta(s,\xi)|s\big)\Big] \tag{9-34}$$

Like before, you should use minimizers in PyTorch/TensorFlow. Accordingly, you'll introduce a -ve sign to convert maximization to a loss minimization.

$$Policy_{Loss} = -\mathop{\mathrm{E}}_{\substack{s \sim \mathcal{D} \\ \xi \sim \mathcal{N}}}\Big[\min_{i=1,2} Q_{\phi_i}\big(s,\widetilde{a}_\theta(s,\xi)\big) - \alpha \log \pi_\theta\big(\widetilde{a}_\theta(s,\xi)|s\big)\Big] \tag{9-35}$$

Also convert the expectation to an estimate using samples to get the following:

$$Policy_{Loss} = -\frac{1}{|B|}\sum_{s \in B}\Big(\min_{i=1,2} Q_{\phi_i}\big(s,\widetilde{a}_\theta(s)\big) - \alpha \log \pi_\theta\big(\widetilde{a}_\theta(s)|s\big)\Big) \tag{9-36}$$

Pseudocode and Implementation

At this point, you are ready to see the complete pseudocode. Refer to Figure 9-13.

Soft Actor Critic (SAC)

1. Input initial policy parameters θ, Q-function parameters ϕ_1 and ϕ_2, empty replay buffer D

2. Set target parameters equal to online parameters $\phi_{targ,1} \leftarrow \phi_1$ and $\phi_{targ,2} \leftarrow \phi_2$

3. **repeat**

4. Observe state s and select action $a \sim \pi_\theta(\cdot|s)$

5. Execute a in environment and observe next state s', reward r, and done signal d

6. Store (s, a, r, s', d) in Replay Buffer D

7. if s' is terminal state, reset the environment

8. if it's time to update **then**:

9.   for j in range (as many updates as required):

10.   Sample a batch $B = (s, a, r, s', d)$ from replay Buffer D:

11.   Compute target for Q functions:

$$y(r, s', d) = r + \gamma(1 - d) \left(\min_{i=1,2} Q_{\phi_{targ,i}}(s', \tilde{a}') - \alpha \log \pi_\theta(\tilde{a}'|s') \right), \quad \tilde{a}' \sim \pi_\theta(\cdot|s')$$

12.   Update Q function with one step gradient descent on ϕ:

$$\nabla_{\phi_i} \frac{1}{|B|} \sum_{(s,a,r,s',d)\in B} \left(Q_{\phi_i}(s, a) - y(r, s', d) \right)^2, \quad \text{for } i = 1, 2$$

13.   Update policy by one step of gradient ascent using:

$$\nabla_\theta \frac{1}{|B|} \sum_{s\in B} \left(\min_{i=1,2} Q_{\phi_i}(s, \tilde{a}_\theta(s)) - \alpha \log \pi_\theta \left(\tilde{a}_\theta(s)| s \right) \right)$$

where $\tilde{a}_\theta(s)$ is a sample from $\pi_\theta(\cdot|s)$ which is differentiable wrt θ via the reparameterization trick.

14.   Update target networks using polyak averaging:

$$\phi_{targ,i} \leftarrow \rho\phi_{targ,i} + (1 - \rho)\phi_i, \quad \text{for } i = 1, 2$$

Figure 9-13. *Soft actor-critic algorithm*

With all the mathematical derivations and pseudocode, it is time to dive into the implementation in PyTorch. The implementation using PyTorch is in the `9.e-sac.ipynb` file. Like before, you'll train the agent using SAC on the Pendulum and Lunar-Lander environments.

Policy Network-Actor Implementation

Let's first look at the actor network. The actor network implementation takes the state as input, which is the same as before. However, the output has two components:

- Either squashed (i.e., passing the action value through tanh) deterministic action a, which is $\mu_\theta(s)$, or a sample action a from the distribution $N\left(\mu_\theta(s), \sigma_\theta^2(s)\right)$. The sampling uses the reparameterization trick, and PyTorch implements this for you as `distribution.rsample()`. You can read about it at `https://pytorch.org/docs/stable/distributions.html` under the topic, "Pathwise Derivative."

- The second output is the log probability that you will need for calculating entropy inside the Q-loss, as per Equation 9-27. Since you are using squashed/tanh transformation, the log probability needs to apply a change of variables for random distribution using the following:

$$f_Y(y) = f_X\left(g^{-1}(y)\right)\left|\frac{d}{dy}\left(g^{-1}(y)\right)\right|$$

The code uses some tricks to calculate a numerically stable version. You can find more details in the original paper.

Listing 9-11 lists the code for `SquashedGaussianMLPActor`, as discussed earlier. The neural network continues to be the same as before: two hidden layers of size 256 units with ReLU activations.

Listing 9-11. SquashedGaussianMLPActor in PyTorch from 9.e-sac.ipynb

```
LOG_STD_MAX = 2
LOG_STD_MIN = -5

class SquashedGaussianMLPActor(nn.Module):
    def __init__(self, state_dim, act_dim, act_limit):
        super().__init__()
        self.act_limit = act_limit
        self.fc1 = nn.Linear(state_dim, 256)
        self.fc2 = nn.Linear(256, 256)
```

```python
        self.mu_layer = nn.Linear(256, act_dim)
        self.log_std_layer = nn.Linear(256, act_dim)

    def forward(self, s):
        x = F.relu(self.fc1(s))
        x = F.relu(self.fc2(x))
        mu = self.mu_layer(x)
        log_std = self.log_std_layer(x)
        log_std = torch.tanh(log_std)
        log_std = LOG_STD_MIN + 0.5 * (LOG_STD_MAX - LOG_STD_MIN) * (log_
        std + 1)
        return mu, log_std

    def get_action(self, x):
        mean, log_std = self(x)
        std = torch.exp(log_std)
        # Pre-squash distribution and sample
        pi_distribution = Normal(mean, std)
        x_t = pi_distribution.rsample()
        y_t = torch.tanh(x_t)
        pi_action = self.act_limit * y_t
        log_prob = pi_distribution.log_prob(x_t)
        # Enforcing Action Bound
        log_prob -= torch.log(self.act_limit * (1 - y_t.pow(2)) + 1e-6)
        log_prob = log_prob.sum(-1, keepdim=True)
        mean = torch.tanh(mean) * self.act_limit
        return pi_action, log_prob, mean
```

Q-Network, Combined Model, and Experience Replay

The Q-function network MLPQFunction, which combines the actor and critic into an agent in the MLPActorCritic class, as well as the ReplayBuffer implementations, are mostly the same. Accordingly, I do not list these parts of the code here.

Q-Loss and Policy-Loss Implementation

Next, you'll look at `compute_q_loss` and `compute_policy_loss`. This is a straight implementation of Steps 11 to 13 of the SAC pseudocode in Figure 9-13. If you compare these steps with Steps 11 to 14 of TD3 in Figure 9-8, you will see lots of similarities, except for the fact that the action is sampled from the online network and SAC has an additional term of entropy in both losses. These changes are minor, and hence I do not list the code explicitly here.

One-Step Update and SAC Main Loop

Again, the code for `one_step_update` and the overall training algorithm follows a similar pattern as before. For SAC, like TD3, the policy and target networks are updated once for every "x" update of the online Q-network. However, unlike TD3 where the update is only done once, SAC updates it "x" times in a loop.

Once you run and train the agent, you'll see similar outcomes as DDPG and TD3. Figure 9-14 shows the graphs as training progresses. You can see that performance characteristics match those of TD3 shown in Figure 9-9.

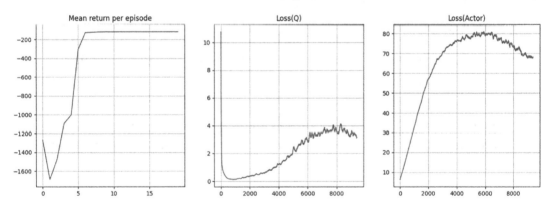

Figure 9-14. *Training progress for Pendulum using SAC*

This chapter uses simple environments, and hence all three continuous control algorithms (DDPG, TD3, and SAC) in this chapter behave well. Consult the various papers referenced in this chapter to do a deep dive into the official performance comparisons of various approaches.

This brings you to the end of this chapter of continuous control in actor-critic settings. By now, you have seen model-based policy iteration approaches, deep learning–based Q-learning (DPN) approaches, policy gradients for discrete actions, and policy gradients for continuous control. This covers most of the popular approaches of reinforcement learning. You have one more major topic to learn about before you finish your journey: that of using model learning in a model-free world and efficient model exploration for environments where you know the model, but it is too complex or vast to be explored exhaustively.

Summary

In this chapter, you learned about the actor-critic methods for continuous control where off-policy Q-learning types are combined with policy gradients to derive off-policy continuous control actor-critic methods. Q-learning has the benefits of being sample efficient, off-policy, and indirect, but it also has the drawbacks of being unstable and difficult to apply to continuous actions. Policy gradients have the benefits of being direct, stable, and suitable for continuous actions, but they also have the drawbacks of being on-policy and sample inefficient.

The common framework for combining Q-learning and policy gradient consists of two networks: an actor network that learns a deterministic policy that maximizes the Q-value, and a critic network that learns the Q-value using the Bellman equation. The framework also employs a replay buffer, a target network, and an exploratory policy.

You first looked at deep deterministic policy gradients, which was introduced in 2016. It was the first continuous control algorithm. DDPG is an actor-critic method with deterministic continuous control policies.

Next, you looked at TD3 (twin delayed DDPG), which came out in 2018 and addressed some of the stability and inefficiency challenges found in DDPG. Like DDPG, it also learned a deterministic policy in an off-policy setting with the actor-critic architecture. It used clipped double Q-learning, delayed policy updates, and target policy smoothing.

Lastly, you looked at Soft Actor-Critic (SAC), which bridged the DDPG style of learning with stochastic policy optimization using entropy. SAC is an off-policy stochastic policy optimization using the actor-critic setup. It uses entropy regularization, the reparameterization trick, and clipped double Q-learning.

## CHAPTER 10

Integrated Planning and Learning

Studying topics separately followed by learning about them together has been a recurring theme in this book. We first looked at model-based algorithms in Chapter 3. In this setup, we knew the model dynamics of the world the agent was operating in. The agent used the knowledge of model dynamics along with Bellman equations to first carry out the evaluation/prediction task to learn the state or state-action values. It then followed this up by improving the policy to get the optimal behavior, which was called *policy improvement/policy iteration*. Once you know the model, you can plan ahead to carry out the evaluation/improvement steps. This is called the *planning phase*.

Chapter 4 started exploring a model-free regime. *Model-free* means you do not know the model. You learn the model by interacting with it. This is called *learning*. In the model-free setup, you learned about the Monte Carlo (MC) and temporal difference (TD) methods. I compared the merits and demerits of using both approaches. Next, you combined MC and TD into a single unified approach using *n-step* and *eligibility traces*.

Chapters 5 through 7 extended the MC and TD methods from Chapter 4 to non-tabular, continuous space, large-scale problems using function approximation and deep learning. All the methods covered in Chapters 3 through 7 are called *value-based methods,* wherein the state or state-action values are learned, followed by policy improvement using these learned state/state-action values.

Chapter 8 focused on an alternate approach, namely, *policy-based methods*. This method involves learning the optimal policy directly without going through the intermediary step of learning the state/action values. You learned about Reinforce, Actor-Critic, as well as TRPO and PPO. Initially, this gave the impression that the value-based and policy-based methods were distinct. However, just like combining MC and TD into one under eligibility traces, Chapter 9 combined value-based Q-learning and policy learning to get the best of both worlds under actor-critic (AC) methods like DDPG, TD3, and SAC.

© Nimish Sanghi 2024
N. Sanghi, *Deep Reinforcement Learning with Python*, https://doi.org/10.1007/979-8-8688-0273-7_10

This chapter continues recombining different approaches. Specifically, the chapter combines the model-based approaches and model-free approaches to make the algorithms more powerful and sample efficient. This approach leverages the best of both of them and is the main emphasis of this chapter.

You will also study the exploration-exploitation dilemma in more detail to go beyond just blindly following e-greedy policies. You will look at simpler setups to gain a stronger understanding of the exploration-exploitation dilemma. This will be followed by a deep dive into a "guided forward looking tree search approach," called *Monte Carlo tree search* (MCTS).

Finally, you will look at the AlphaGo and AlphaZero approaches as examples of MCTS application.

Model-Based Reinforcement Learning

What is model-free RL? In model-free RL, you do not know the model. Rather, you learn the value function and/or policy by having the agent interact with the environment, that is, learning by experience. This is what you saw in Chapters 4 through 9 of this book. In comparison, the model-based RL here is the one under which you learn the model by having the agent interact with the environment, again learning by experience. The learned model is used to plan the value function and policy—something similar to what you saw in Chapter 3. Chapter 3 assumed prior knowledge of the model and assumed it to be perfect. However, in model-based RL, you learn the model and then use that knowledge. However, the knowledge may be incomplete; in other words, you may not know the exact transition probabilities or complete distribution of rewards. You learn parts of the system dynamics that you experience as the agent interacts with the environment. You base this knowledge on limited interactions, and hence without an exhaustive interaction covering all the possibilities, the knowledge of the model is incomplete.

What do I mean by *model* here? This means to have estimates of the transition probabilities $P(s_{t+1}|s_t, a_t)$ and the reward $R(s_t, a_t)$. The agent interacts with the world/environment and forms an impression of the world. Pictorially it can be represented as shown in Figure 10-1.

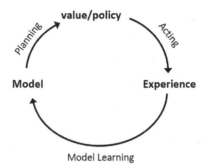

Figure 10-1. *Model-based reinforcement learning. The agent learns the model based on its interaction with the environment, and that model is used for planning*

The advantages are that learning becomes sample efficient. You can use the learned model to plan effectively, as compared to the model-free approach. Further, the model interaction is all about taking an action a_t while in state s_t and observing the outcome of the next state s_{t+1} and the reward r_{t+1}. You can use supervised learning machinery to learn from the interactions with the real world, where a given (s_t, a_t) is the sample input and s_{t+1} and/or r_{t+1} are the sample targets.

$$s_1, a_1 \rightarrow r_2, s_2$$

$$s_2, a_2 \rightarrow r_3, s_3$$

$$\vdots \quad \vdots \quad \quad \vdots$$

$$s_{T-1}, a_{T-1} \rightarrow r_T, s_T$$

As with any other regular supervised machine learning setup, you can learn from the previous samples that the agent gathers from its interaction with the environment.

- Learning $s_t, a_t \rightarrow r_{t+1}$ is a regression problem with a loss like mean squared loss.

- Learning the transition dynamics $s_t, a_t \rightarrow s_{t+1}$ is a density estimation problem. You could learn a discrete categorical distribution, a Gaussian, or a mixture of Gaussian model parameters. The loss could be a KL-divergence loss.

If you extend the learning to Bayesian learning, you can also reason about the model uncertainties—how sure or unsure you are about the learned model transitions and reward functions. The Bayesian approach to learning produces not just a point estimate but a whole distribution of the estimate that gives you the capability to reason about the strength of the estimation. A narrow probability distribution for the estimate means that a large part of the probability is centered around the peak; that is, you have strong confidence in the estimate. Conversely, a broad distribution for the estimate reflects a higher uncertainty about the estimate.

However, nothing comes free in life. This two-step approach of first learning an imperfect representation of the model followed by using this imperfect representation to plan or find the optimal policy introduces two sources of error. First, the learned representation of the model dynamics may be inaccurate. Second, learning a value function from an imperfect model may have its own inaccuracies.

The learned model can be represented as a "table lookup" (similar to what you saw in Chapters 2 and 3), linear expectation model, or linear Gaussian model. There can be even more complex model representations like the Gaussian process model or deep belief network models. It depends on the nature of the problem, ease of data collection, and so on. Deciding on the right representation requires domain expertise. Let's look at a simple example of learning a table lookup model.

First, look at the expressions that you will use to learn the reward and transition dynamics. It is a simple averaging method that you will use here. To estimate the transition probability, you take the average of the number of times the transition s_t, $a_t \rightarrow s_{t+1}$ was seen and divide it by the total number of times the agent saw itself in s_t, a_t.

$$\hat{p}(s,a) = \frac{1}{N(s,a)} \sum_{t=1}^{T} 1(S_t = s, A_t = a, S_{t+1} = s') \tag{10-1}$$

Here, $1(S_t = s, A_t = a, S_{t+1} = s')$ is an indicator function. Indicator functions take a value of 1 when their condition inside the brackets is true and a value of 0 otherwise. In summary, this indicator function is counting the number of times (s, a) leads to a transition to the next state, $S_{t+1} = s'$.

Similarly, you can define the reward learning as an average.

$$\hat{r}(s,a) = \frac{1}{N(s,a)} \sum_{t=1}^{T} 1(S_t = s, A_t = a) R_t \tag{10-2}$$

Let's look at a simple environment, shown in Figure 10-2, in which there are only two states, and you observe a set (eight) of the agent's interaction with the environment. Assume no discounting—that is, $\gamma = 1$.

Suppose you see $(A, 0, B, 0)$. This means that the agent starts in state A, observes a reward of 0, followed by state B, and a reward of 0, and finally followed by a transition to the terminal state.

Let's say the eight transitions/interactions collected by the agent are as follows:

A, 0, B, 0

B, 1

B, 1

B, 1

B, 1

B, 1

B, 1

B, 0

Apply Equations 10-1 and 10-2 to construct the model as follows.

You saw only one transition from A to B, and you got a reward of 0. So, you conclude that $P(B|A)$ is 1 and $R(state = A) = 0$. Note that, in the previous example, I am not explicitly showing actions to keep things simple. You could think of this as taking a random action in each state or you could think of this as a Markov reward process (MRP) instead of a full-blown MDP.

You saw eight transitions from B, all of them leading to the terminal state. In two instances, the reward was 0 while in the remaining six cases it was 1. This can be modeled as saying that $P(\text{Terminal} | B) = 1$. The reward $R(B) = 1$ with 0.75 (6/8) probability and $R(B) = 0$ with 0.25 probability.

Figure 10-2 shows the model you learned from these eight interactions.

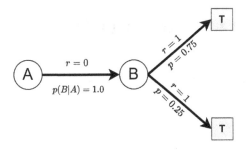

Figure 10-2. *Table lookup model learned from the environment interaction*

This is one of the many parameterized model representations where you explicitly learn the model dynamics and then discard the sample interactions with the real world. However, there is another approach that was used in DQN: the nonparameterized model. In the nonparameterized model, you store the interactions in a buffer and then sample from the buffer. An example from the previous eight interactions would be to store tuples of (state, reward, next_state) in a list (remember in this example, you do not have actions, as it is an MRP).

$$D = \left[(A,0,B), (B,0,T), (B,1,T), (B,1,T), (B,1,T), (B,1,T), (B,1,T), (B,1,T), (B,0,T) \right]$$

The first two values in buffer D come from one transition, that is, A, 0, B, 0. The rest of the seven entries in the buffer are the remaining seven transitions from the previous example.

A broad categorization of model learning is as follows:

- *Parameterized*:

  - Table lookup model

  - Linear expectation model

  - Linear Gaussian model

  - Gaussian process model

  - Deep belief network model

- *Nonparameterized*: Store all interactions (s, a, r, s') in a buffer and then later do sampling from this buffer to generate example transitions.

Planning with a Learned Model

Once you know the model, you can use it to carry out planning using the value or policy iteration in Chapter 3. In these methods, you do one-step rollouts using Bellman equations.

However, there is another way the learned model can be used. You can sample from it and use these samples under the *MC* or *TD* approach of learning. While carrying out the *MC* or *TD* style of learning in this case, the agent does not interact with the real environment. Rather, it interacts with the *model of the environment* that it has approximated from past experiences. To an extent, it is still called *planning*. You are using the model learned to plan and not planning from direct interaction with the real world.

Keep in mind that the model you learn is not exact. It is based on a partial interaction the agent has with the environment. It is usually not complete or thorough. I talked about this in the previous section. The errors in the learned model would affect the quality of learning. The algorithms optimize learning with respect to the model you have formed and not the real-world model. This can limit the quality of learning. If your confidence in the learned model is not too high, you could go back to the model-free RL methods that you saw before. Or you could use the Bayesian approach to reason about the model uncertainties. The Bayesian approach is not something I cover more in this book. Interested readers can refer to various advanced RL texts and papers to learn more about it.

Until now, you have seen that model-based RL offers the benefit of learning the model and thereby making learning more sample efficient. However, it comes at a cost of the model estimates not being accurate, which in turn limits the quality of learning. Is there a way to combine model-based and model-free learning into a single unified framework and take advantage of both approaches? This is what you will look at next.

Integrating Learning and Planning (Dyna)

There are two kinds of experiences that you can have: a real experience where the agent gets the next state and reward by interacting with the real environment, and a simulated experience where the agent uses the model it learned to create more simulated experiences. A simulated experience is easier and cheaper to create, especially in robotics. Fast robot simulators can produce samples that may not match the real world completely, but that allow you to simulate and generate agent behavior faster than the real-world interaction of robots. However, the simulations may not be precise, which is why learning from real-world experiences can help. We can change the diagram from Figure 10-1 to include this step, as shown in Figure 10-3.

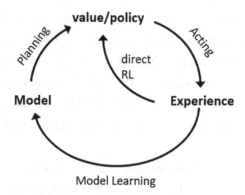

Figure 10-3. *Dyna architecture*

In Dyna, the agent interacts with the environment (acting) to produce real experiences. These real experiences are used to learn the model of the world and directly improve the value/policy like a model-free RL setup. This is the same as what you saw in Chapters 4 through 8. The model learned from real-world interactions is used to generate simulated transitions. These simulated transitions are used to improve the value/policy. This is called step planning, as you are using the model of the world to generate experiences. You can carry out the planning step by using the simulated experience multiple times for each step of "acting" and generate the new samples from the real world.

Let's look at a concrete implementation of Dyna—Tabular Dyna Q. In this, you assume that the state and actions are discrete and form a small set so that you can use a tabular representation for the $Q(s, a)$, which are the q-values. You use the q-learning approach under TD to learn and improve policies. This is similar to the approach we used in Chapter 4 of Q-learning, the off-policy TD control. You will have some additional steps to learn from the simulated experience as well. In Chapter 4, you only used the real experiences to learn q-values, but you will now learn the model and use that to generate additional simulated experiences to learn from. Figure 10-4 gives the complete algorithm.

TABULAR DYNA Q

Initialize:

State-action values $Q(s, a) = 0$ for all $s \in S$ and $a \in A$.

Initialize model (s, a) for all $s \in S$ and $a \in A$.

policy $\pi = \epsilon - \text{greedy}$ policy with some small $\epsilon \in [0, 1]$

learning rate (step size) $\alpha \in [0, 1]$

discount factor $\gamma \in [0, 1]$

Loop for each episode:

Start state S

Loop for each step till episode end:

Choose action A based on ϵ-greedy policy

Take action A and observe reward R and next state S'

If S' not terminal:

$$Q(S, A) \rightarrow Q(S, A) + \alpha \cdot [R + \gamma \cdot \max_{a'} Q(S', a') - Q(S, A)]$$

else:

$$Q(S, A) \rightarrow Q(S, A) + \alpha \cdot [R - Q(S, A)]$$

Model $(S, A) \leftarrow R, S'$ (we are assuming deterministic env)

$S \leftarrow S'$

Loop repeat n times:

$S \leftarrow$ a random previously seen state

$A \leftarrow$ random action previously taken in S

$R, S' \leftarrow Model(S, A)$

$$Q(S, A) \leftarrow Q(S, A) + \alpha \cdot [R + \gamma \cdot \max_{a'} Q(S', a' - Q(S, A)]$$

Return policy π based on final Q values.

Figure 10-4. *Tabular Dyna Q*

You can see from the pseudocode in Figure 10-4 that tabular Dyna Q is similar to the tabular Q-learning of Figure 4-16, except that in Dyna Q, for every step in the real environment, you also carry out additional n steps using a simulated experience. In this case, it samples from the previous transitions seen in the real world. As you increase n, the number of episodes required to converge and learn an optimal value will decrease.

Consider a case where the maze has zero rewards for every transition, except for the last terminal/goal state transition with a reward value of 1. An example of such a maze would be the maze on the left side of Figure 10-5. Let's also assume that the initial q-values are 0 for every (s, a) pair. In regular Q-learning, you learn the goal value of 1 when the first episode terminates. Then the value of 1 will propagate slowly—one cell/level per episode—to eventually reach start state. The q-values and optimal policy will start to converge over additional episodes after that. However, in the case of Dyna Q, you generate n additional examples per the real step in the environment. This will speed up the convergence. You will see policy convergence much before the vanilla q-learning approach. Refer to Chapter 8 of the book *Reinforcement Learning: An Introduction*[1] for a more theoretical and detailed explanation of this faster convergence.

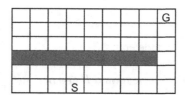

Figure 10-5. *Maze with obstruction in the middle. The reward is given only when the agent reaches the goal state, G. All other transitions have a reward of 0*

Let's apply the previous tabular Dyna Q pseudocode to the environment you saw in Chapter 4. You will modify code from `4.d-qlearning.ipynb` to incorporate model learning and then n steps of planning (*learning from a simulated experience*) for each step of direct RL-based learning from actual experience. The complete code is given in `10.a-dynaQ.ipynb`. In the walkthrough, I highlight the key differences between Dyna Q code and Q-learning code from Chapter 4.

First, rename the agent from `QLearningAgent` to `DynaQAgent`. It's just a name change, and the code remains the same except for an additional dictionary to store the real-world transitions. Add the two lines of code to the `__init__` function shown in bold in Listing 10-1.

Listing 10-1. Dyna Q Agent from 10.a-dqnaQ.ipynb

```
class DynaQAgent:
    def __init__(self, alpha, epsilon, gamma, get_possible_actions, n):
        self.get_possible_actions = get_possible_actions
        self.alpha = alpha
        self.epsilon = epsilon
        self.gamma = gamma
```

---

[1]http://incompleteideas.net/book/the-book.html

```python
        self.n = n
        self._Q = defaultdict(lambda: defaultdict(lambda: 0))
        self.buffer = {}

    def get_Q(self, state, action):
        return self._Q[state][action]

    def set_Q(self, state, action, value):
        self._Q[state][action] = value

    # Q learning update step
    # same as that of 4.d-qlearning.ipynb
    def update(self, state, action, reward, next_state, done):
    pass
            # code omitted #

    # get best A for Q(S,A) which maximizes the Q(S,a) for actions
    in state S
    # same as that of 4.d-qlearning.ipynb
    def max_action(self, state):
        # code omitted #
        return np.random.choice(np.array(best_action))

    # choose action as per ε-greedy policy for exploration
    # same as that of 4.d-qlearning.ipynb
    def get_action(self, state):
        # code omitted #
```

Second, add logic to follow the planning part of the pseudocode from Figure 10-4, shown here:

The loop repeats *n* times:

$S\leftarrow$A random previously seen state

$A\leftarrow$Random action previously taken in S

$R, S' \leftarrow Model(S, A)$

$$Q(S,A) \rightarrow Q(S,A) + \alpha \cdot \left[R + \gamma \cdot \max_{a'} 1\, Q(S', a') - Q(S,A) \right]$$

This is added to the `train_agent` function, where you train the agent. Listing 9-2 shows the revised implementation of the function. The new addition to the code is toward the end of the function. The first one is to add the transition to the agent's buffer so that the agent can use the prior experience buffer to estimate the model transition dynamics. The second change is that of an *n* step of planning using these simulated experiences that the agent collected from prior interactions with the real environment. These changes are highlighted in bold in Listing 10-2.

Listing 10-2. Function train_agent from 10.a-dynaQ.ipynb

```
#training algorithm
def train_agent(env, agent, episode_cnt = 10000,
                tmax=10000, anneal_eps=True):
    episode_rewards = []
    for i in range(episode_cnt):
        G = 0
        state,_ = env.reset()
        for t in range(tmax):
            action = agent.get_action(state)
            next_state, reward, done, truncated, _ = env.step(action)
            done = done or truncated
            agent.update(state, action, reward, next_state, done)
            G += reward
            if done:
                episode_rewards.append(G)
                # to reduce the exploration probability epsilon over the
                # training period.
                if anneal_eps:
                    agent.epsilon = agent.epsilon * 0.99
                break
            # add the experience to agent's buffer (i.e. agent's model
            estimate)
            agent.buffer[(state,action)] = (next_state, reward, done)
            state = next_state
            # plan n steps through simulated experience
            for j in range(agent.n):
```

```
        state_v, action_v = random.choice(list(agent.buffer))
        next_state_v, reward_v, done_v = agent.
        buffer[(state_v,action_v)]
        agent.update(state_v, action_v, reward_v, next_state_v, done_v)

    return np.array(episode_rewards)
```

Next, run Dyna Q against the maze as well as the taxi world environment from Gymnasium. Looking at the training curves in the Python notebook, note that convergence takes fewer episodes under Dyna Q when compared to vanilla Q-learning. In the Q-learning training curves given in `4.d-qlearning.ipynb`, you can see that the agent takes about 100-150 steps before it reaches the optimal policy. While the Dyna Q agent in `10.a-dynaQ.ipynb` takes about 20 steps or so. This establishes the claim that Dyna Q (or in general, any Dyna architecture) is sample-efficient.

Dyna Q and Changing Environments

Let's now consider the case where you first learn the optimal policy for a maze using Dyna Q. After some steps, you change the environment and make it harder, as shown in Figure 10-6. The left figure shows the original grid for which the agent learns to navigate from start S to goal G through the opening on the right side of the gray brick wall. After the agent has learned the optimal behavior, the grid changes. The opening on the right side closes, and a new opening appears on the left side. The agent, when trying to navigate to the goal through the earlier path, now sees that the original path is blocked. The opening on right side of the original maze is now closed. This is due to the change you just made.

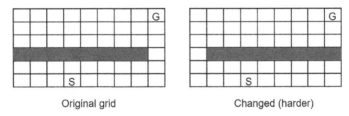

Original grid   Changed (harder)

Figure 10-6. *Dyna Q with grid made more difficult midway after the agent learns to navigate through the opening on the right side. The agent learns to navigate the grid through the new opening on the left side of the brick wall*

Sutton and Barto, in their book *Reinforcement Learning: An Introduction,* show that Dyna Q will take a while to learn the changed environment. Once the environment is changed midway, there are a number of episodes in which the agent keeps going to the right but finds the path blocked, and hence it needs lots of additional steps under the ε-greedy policy to learn the alternate route of taking the opening on the left side to reach the goal.

Let's consider the second case that they show in their book. In the second case, the environment is changed midway and made simpler, that is, a new opening is introduced to the right without closing the original opening on the left. Figure 10-7 shows the maze before and after the change.

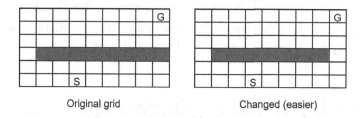

Original grid   Changed (easier)

Figure 10-7. *Dyna Q with grid made simpler after the agent has learned the optimal policy for the left grid. The Dyna Q agent fails to discover the shorter path through the new right gap*

By running a Dyna Q-learning algorithm on this setup, you can see that Dyna Q does not learn the new opening on the right side of the brick wall. The new path through the right of the brick wall offers a shorter path to the goal state. Dyna Q has already learned the optimal policy, and it has no incentive to explore for a changed environment. A random/chance exploration of the new opening depends on the exploratory policy, which in turn depends on the ε exploration.

There is a revised algorithm known as Dyna Q+ to fix this problem, as explained in the following section.

Dyna Q+

Dyna Q+ is a typical example of the exploration/exploitation dilemma that needs to be well thought out in all reinforcement learning setups. If you start to exploit the knowledge (i.e., the model or the policy) early, you risk the chance of not getting to know better paths. The agent greedily becomes happy with whatever it has managed to learn. On the other hand, if the agent explores too much, it is wasting time looking at suboptimal paths/options even when it already has the most optimal solution. Unfortunately, the agent has no direct way of knowing that it has reached the optimal policy, and hence it needs to use other heuristics/ways to balance the exploration-exploitation dilemma.

In Dyna Q+, you encourage the exploration of unseen states by adding a reward to the states over and above what you observed. This additional reward term encourages exploration for the states that have not been visited in the real world in a while.

In the simulated planning part, you add $\kappa\sqrt{\tau}$ to the reward; that is, the reward r becomes $r + \kappa\sqrt{\tau}$ where κ is a small constant and τ is the time since the transition under question was seen in the real-world exploration. It encourages the agent to try these transitions and discover the changes in the environment effectively, with some lag as controlled by κ. In Figure 10-7, the agent initially learns to go through the left opening. It carries out the exploration as per the revised term, but finds that the only way to the goal is through the left opening. However, after the environment is changed by making a new opening on the right side, the agent on the subsequent exploration of this part of maze will discover the new opening and eventually the fact that it is a shorter path to the goal. The agent will revise its optimal behavior to follow the opening on the right side. As the time of the last visit to a part of grid increases, the $\kappa\sqrt{\tau}$ term in the reward grows. At some stage, it may grow so much that it overshadows the reward from the current optimal behavior, forcing the agent to explore the unvisited part once more.

Expected vs. Sample Updates

You have seen various ways to combine learning and planning. You can see that learning and planning are both about the way the value functions are updated. The first dimension is about what is being updated, whether a state value (v) or an action value (q). The other dimension is the width of the update. In other words, does the update happen based on a single sample seen (sample update), or is the update based on all possible transitions (expected update) using the transition probabilities for the next state given the current state and action, that is, $p(s_{t+1}|s_t, a_t)$? The third and last dimension is whether the update is carried out for any arbitrary policy v_π, q_π or for the optimal policy v^*, q^*. Let's look at various combinations and map them back to what you have seen so far in the book. You will look at seven different types of updates you have seen so far.

The first one is $v_\pi(s)$ being updated using the expected update. The update is for a value function. It is an expected update covering all the possible transitions from a given state using the transition probability distribution $p(s', r|s, a)$. The action is carried out using the current policy agent. This is policy evaluation using dynamic programming, which you saw in Equation 3-6.

$$v_{k+1}(s) \leftarrow \sum_a \pi(a|s) \sum_{s',r} p(s',r|s,a)\big[r + \gamma v_k(s')\big]$$

The second one is the update of $v_\pi(s)$ using the sample update. The update is for the value function and is based on the sample the agent sees following a policy π. This is policy evaluation under $TD(0)$ that you saw in Equation 4-5. The value of state (s) is updated based on the action taken by the agent, as per policy π, the subsequent reward (r), and the next state (s') seen by the agent. I expressed the Equation 4-5 in an iterative form to highlight the update from step k to step $k + 1$.

$$v_{k+1}(s) \leftarrow v_k(s) + \alpha\big[r + \gamma v_k(s') - v_k(s)\big]$$

The third one is $v^*(s)$ being updated using max over all the possible actions. The value of state (s) is updated based on the expectation of all the possible next states and rewards. The update is done by taking the max over all the possible actions, that is, the optimal action at a given point in time. It is not based on the current policy the agent is following. This is an approach of value iteration using dynamic programming as per Equation 3-8.

$$v_{k+1}(s) \leftarrow \max_a \sum_{s',r} p(s',r|s,a)\big[r + \gamma v_k(s')\big]$$

The following ones are about changes to Q values. The fourth approach is that of $q_\pi(s, a)$ being updated using expected update. The value being updated is q-value and is updated in the expectation based on all the next states and rewards possible from a given state-action pair. The update is based on the current policy the agent is following. This is q-policy evaluation using dynamic programming as shown in Equation 3-2. Here Equation 3-2 is expressed in iterative form.

$$q_{k+1}(s,a) \leftarrow \sum_{s',r} p(s',r|s,a)\Big[r + \gamma \sum_{a'} \pi(a'|s') q_k(s',a')\Big]$$

The fifth one is that of $q_\pi(s, a)$ being updated using the samples. The update for the q-value is based on the samples and on the current policy the agent is following—in other words, an on-policy update. This is q-value iteration using dynamic programming. Here is the iterative version of Equation 3-4.

$$q_{k+1}(s,a) \leftarrow \sum_{s',r} p(s',r|s,a)\left[r + \gamma \max_{a'} q_k(s',a')\right]$$

The sixth one is $q^*(s, a)$ being updated using all the possible states and max over all the possible actions in the next state. This is an update for the q-value and based on the expectation of all possible following state and action pairs. It is carried out by taking a max over actions and hence updating for an optimal policy and not the policy the agent is currently following. This is SARSA using a model-free setup, as shown in Equation 4-7. The following equation is a rewrite of the same equation to match the notations in this section.

$$q_{k+1}(s,a) \leftarrow q_k(s,a) + \alpha\left[r + \gamma q_k(s',a') - q_k(s,a)\right]$$

The seventh approach is about $q^*(s, a)$ being updated using the sample action and then taking the max over all the possible actions from the next state. The update is of the q-value using a sample. The update is based on a *max* over all possible actions, that is, an off-policy update. It is not the update for the current policy the agent is following. This is the Q-learning you studied in Chapter 4 and then extended to deep networks under DQN in Chapter 6. Here is Equation 4-10 rewritten to match the notations in this chapter.

$$q_{k+1}(s,a) \leftarrow q_k(s,a) + \alpha\left[r + \gamma \max_{a'} q_k(s',a') - q_k(s,a)\right]$$

All these updates show how DP and model-free worlds are linked. In DP, since you know the model dynamics, you do a wide sweep over all the possible transitions $p(s_{t+1}|s_t, a_t)$. In the model-free world, you do not know the model, and hence you do sample-based MC or TD updates. The difference between DP and model-free is all about the update being done over the expectation or done over samples.

It also goes to show that once you start learning model dynamics, you can mix and match any of the previous approaches under the Dyna setup. Just like earlier, you can combine the model-free and model-based approaches in a unified single setup. This is similar to what you did by combining value-based DQN and policy gradient into a unified approach under actor-critic as well as the way you combined a one-step sample (TD) and a multistep sample (MC) into a single framework using eligibility traces.

Another way to organize the RL algorithms is to look at a two-dimensional world where the vertical axis is the length/depth of update, with TD(0) being at one end and MC at other end. The second/horizontal dimension can be visualized as the width of the update with the dynamic programming based on the expected updates at one end and the sample-based MC/TD updates at the other end. Figure 10-8 shows this distinction.

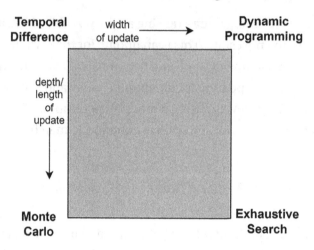

Figure 10-8. *Reinforcement learning approaches in a nutshell*

You can further refine this unification by adding more dimensions. A third possible dimension, orthogonal to the width and depth of the update, would be off-policy versus on-policy updates. You could also add a fourth dimension, which is the purely value-based approach or purely policy-learning approach, or a mix of both, that is, actor-critic.

Up to now, you have looked at combining learning and planning in what is called the *backward view*. You used planning (e.g., Dyna Q) to update the value functions, but there is no planning involved while choosing an action in the real world. It is kind of a passive planning approach, where planning was used to generate additional steps/synthetic examples to train and improve the model/policy/value functions. In a following section, you will look at *forward planning*, which you carry out planning at decision time. In other words, you use the model knowledge (learned or given) to look ahead and then take action based on what you think is the best action. You will look at this approach in the context of the Monte Carlo tree search algorithm (MCTS). However, before you do that, the next section revisits the exploration versus exploitation issue.

Exploration vs. Exploitation

In reinforcement learning, you always need to balance between using current knowledge and exploring more to gain new knowledge. Initially you have no or very little knowledge about the world (environment). You explore a lot and start improving your belief about the world. As your belief improves and strengthens, the agent starts exploiting that belief along with gradually reducing its exploration.

Up until now you have looked at exploration in different forms. The first encounter with exploration was in the form of ε-greedy strategies, where the agent took the best action based on its current belief with probability $(1 - \varepsilon)$, and it explored randomly with probability ε. You gradually reduced ε as learning you progressed. You saw this in all the DQN approaches. If you look at the Chapter 6 and 7 notebooks, you will notice a epsilon_schedule function, as shown in Listing 10-3. At the start of learning, you start with ε = 1 and over a period reduce it to 0.05 to have a small amount of exploration toward the end of training. This is called an *epsilon reduction schedule*. The reduction can be linear or follow any other curve, similar to what you do for α, the learning rate. In supervised learning, as part of stochastic gradient descent, you vary α as per a schedule called the *learning rate schedule*. There are various approaches to varying α. You can use most of those approaches for varying ε if you have ε-greedy strategies.

Listing 10-3. Epsilon Schedule for Exploration

```
def epsilon_schedule(start_eps, end_eps, step, final_step):
    return start_eps + (end_eps-start_eps)*min(step, final_step)/final_step
```

The second approach toward exploration was in the form of learning stochastic policies in policy gradient methods where you always learned a policy distribution $\pi_\theta(a|s)$. Unlike with DQN, you did not take any "max" over actions to learn a single optimal action. This ensures that you are not learning the deterministic action. Accordingly, different non-zero probabilities for all the actions ensured exploration.

Another approach you saw was the use of entropy regularization in policy gradient and actor-critic methods that forced enough exploration to ensure that the agent did not prematurely commit to exploitation without doing enough exploration of unknown parts of the policy/action space. Refer to Chapter 8 for details.

In the next section, you look at a simple setup to study the tradeoffs between exploration versus exploitation more formally and the various strategies for efficient exploration.

Multi-Arm Bandit

This section considers an environment that is known as a *multi-arm bandit*. It's based on the concept of multiple casino slot machines stacked together. The agent does not know the individual reward (success rate) for each slot machine. The agent's job is to choose a machine and pull the slot machine lever to receive a reward. This cycle repeats for as long as you want. The agent needs to explore to try all the machines and form a belief about the individual reward distribution for each of the machines. As its belief gets stronger, it must start exploiting its knowledge/belief more to pull the best machine's lever and reduce exploration. Figure 10-9 shows the setup.

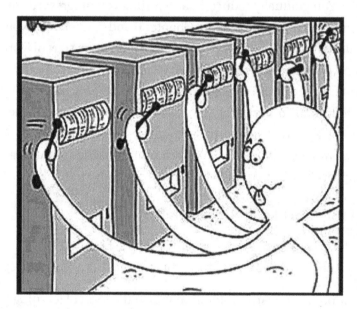

Figure 10-9. *Noncontextual multi-arm bandit*

The reason we study multi-arm bandit is that it is a simple setup. Further, it can be extended to contextual multi-arm bandit, in which the distribution of rewards for each slot machine is not fixed. Rather, it depends on the context at a given point in time, for example, the "current state" of the agent. The contextual multi-arm bandit has many practical use cases. You will look at one such example here.

Assume you are an entrepreneur and have decided to build a competing business to Google AdWords. As part of the design, suppose you have ten advertisements, and you want to show one advertisement to the user in the browser window based on the context (e.g., content) of the page. The current page is the context and it shows one of the ten advertisements is the action. When the user clicks the advertisement, the user gets a reward of 1; otherwise, the reward is 0. The agent has the history of how many times a user clicked a shown advertisement within the same context, that is, clicking the shown advertisement on a web page with similar content. The agent wants to take a series of actions with the goal to increase/maximize the chances of users clicking the shown advertisement. This is an example of a contextual multi-arm bandit.

Another example is an online store showing other product recommendations based on what the user is currently viewing. The online store hopes that the user will find some of the recommendations interesting and may click them. Yet another example is choosing a drug out of K possible options, administering the drug to a patient to choose the best one, and reaching that conclusion over a maximum of T tries.

You can further extend the bandit framework to consider full-scale MDPs. But each extension makes the formal analysis even more laborious and cumbersome. Accordingly, this chapter studies the simplest setup of non-contextual bandit to help you appreciate the basics. Interested readers can study this further with many good resources available on the web.

Regret: Measure the Quality of Exploration

Here, you define the quality of exploration. Consider that the system is in state S. In non-contextual bandit, it is a frozen initial state that remains the same every time. For contextual bandit, the state S may change over time. Further, suppose you follow a policy $\pi_t(s)$ at time t to choose an action. Also consider that you know the optimal policy $\pi^*(s)$.

The regret at time t, that is, the regret of choosing an action from the nonoptimal policy, is defined in Equation 10-3. It is the expected reward from following an optimal policy minus the reward from following a specific policy.

$$regret = \eta = \operatorname*{E}_{a \sim \pi^*(s)} \left[r(s,a) \right] - \operatorname*{E}_{a \sim \pi_t(s)} \left[r(s,a) \right] \tag{10-3}$$

You can sum it over all timesteps to get the total regret over T steps to obtain the expression given in Equation 10-4.

$$\eta = \sum_{t=1}^{T} \left(\operatorname*{E}_{a \sim \pi^*(s)} \left[r(s,a) \right] - \operatorname*{E}_{a \sim \pi_t(s)} \left[r(s,a) \right] \right) \tag{10-4}$$

This is non-contextual bandit, where the state (s) remains the same at every time step—it remains fixed as the initial state. Therefore, you can simplify Equation 10-4 as shown in Equation 10-5.

$$\eta = T. \underset{a \sim \pi^*(s)}{E} \left[r(s,a) \right] - \sum_{t=1}^{T} \left(\underset{a \sim \pi_t(s)}{E} \left[r(s,a) \right] \right) \qquad (10\text{-}5)$$

This example uses Equation 10-5 to compare various sampling strategies. You will see three commonly used exploration strategies in the following sections, with the help of code examples. You can find the complete code in 10.b-explore-vs-exploit.ipynb.

Let's first describe the setup you will use. Consider that the multi-arm bandit has K actions. Each action has a fixed probability θ_k, $k \in [1..K]$ of producing a reward of 1 and a $(1 - \theta_k)$ probability of producing a reward of 0. The optimal action is the value of k (lowercase k), which has a maximum probability of success.

$$\theta^* = \theta_k$$

Listing 10-4 shows the code for the bandit. You store K as self.n_actions and all the θ_k, as self._probs. The pull function takes as input the k action, returning a reward of 1 or 0 based on the probability distribution of the action, that is, θ_k. The optimal_reward function returns the max of self._probs.

Listing 10-4. Bandit Environment from 10.b-explore-exploit.ipynb

```
class Bandit:
    def __init__(self, n_actions=5):
        self._probs = np.random.random(n_actions)
        self.n_actions = n_actions

    def pull(self, action):
        if np.random.random() > self._probs[action]:
            return 0.0
        return 1.0

    def optimal_reward(self):
        return np.max(self._probs)
```

Next, we will look at three exploration strategies starting with the ε-greedy exploration strategy.

Epsilon Greedy Exploration

This is similar to the ϵ-greedy strategy you saw in previous chapters. The agent tries different actions and forms an estimate $\hat{\theta}_k$ of the unknown actual success probabilities θ_k for the bandit environment's different actions. It takes the action k with $\hat{\theta}^* = \max_k \hat{\theta}_k$ with the $(1 - \epsilon)$ probability. In other words, it exploits the knowledge gained so far to take the action that it considers best based on $\hat{\theta}_k$ estimates. Further, it takes a random action with the probability ϵ to explore.

Let's keep $\epsilon = 0.01$ as the exploration probability and keep it constant throughout the experiment. Listing 10-5 shows the code implementing this strategy. First, you define RandomAgent as the base class. There are two arrays to store the cumulative count of success and failure outcomes of each action. You store these counts in arrays: self. success_cnt and self.failure_cnt.

The update(action, reward) function updates these counts based on which action was taken and the outcome. The reset() function resets these counts. It does so to run the same experiment multiple times and plot the average values to smoothen out the variance. The get_action() function is where the exploration strategy is implemented. It will use different approaches to balance the explore versus exploit based on the type of strategy. For RandomAgent, the actions are always selected at random from the set of possible actions.

Next, you implement the ϵ-exploration agent in the EGreedyAgent class. This is implemented by extending RandomAgent and overriding the get_action() function. As explained, the action with the highest probability is chosen with the probability $(1 - \epsilon)$, and a random action is chosen with the probability ϵ.

Listing 10-5. Epsilon Greedy Exploration Agent from 10.b-explore-vs-exploit.ipynb

```
class RandomAgent:
    def __init__(self, n_actions=5):
        self.success_cnt = np.zeros(n_actions)
        self.failure_cnt = np.zeros(n_actions)
        self.total_pulls = 0
        self.n_actions = n_actions
```

```python
    def reset(self):
        self.success_cnt = np.zeros(n_actions)
        self.failure_cnt = np.zeros(n_actions)
        self.total_pulls = 0

    def get_action(self):
        return np.random.randint(0, self.n_actions)

    def update(self, action, reward):
        self.total_pulls += 1
        if reward == 1:
            self.success_cnt[action] += 1
        else:
            self.failure_cnt[action] += 1

class EGreedyAgent(RandomAgent):
    def __init__(self, n_actions=5, epsilon=0.01):
        super().__init__(n_actions)
        self.epsilon = epsilon

    def get_action(self):
        estimates = self.success_cnt / (self.success_cnt+self.failure_
        cnt+1e-12)
        if np.random.random() < self.epsilon:
            return np.random.randint(0, self.n_actions)
        else:
            return np.argmax(estimates)
```

In this setup, you are not changing the ϵ value, and hence the agent never learns to execute a perfect optimal policy. It continues to explore with the ϵ probability forever. Therefore, the regret will never go down to 0, and you expect the cumulative regret as given by Equation 10-5 to grow linearly with the slope of growth defined by the value of ϵ. The bigger the ϵ value, the steeper the slope of growth.

Upper Confidence Bound Exploration

This section looks at the strategy called *upper confidence bound* (UCB). In this approach, the agent tries different actions and records the number of successes (α_k) or failures (β_k). It calculates the estimates of success probabilities $\hat{\theta}_k$ that will be used to choose the action with the highest success estimate $\theta^* = \hat{\theta}_k$. This is the "exploit" part for action k.

$$\text{exploit}_k = \hat{\theta}_k = \frac{\alpha_k}{\alpha_k + \beta_k}$$

It also calculates the exploration needed that favors the least visited actions, as shown here:

$$\text{explore}_k = \sqrt{\frac{2.\log t}{\alpha_k + \beta_k}}$$

where t is the total number of steps taken so far. The explore and exploit are added to define the score for each action k, as shown in Equation 10-6.

$$\text{score}_k = \text{explore}_k + \text{exploit}_k = \frac{\alpha_k}{\alpha_k + \beta_k} + \lambda \sqrt{\frac{2.\log t}{\alpha_k + \beta_k}} \tag{10-6}$$

The agent then picks the action k with the maximum value of *score$_k$*. In Equation 10-6, λ controls the relative importance of explore versus exploit. You will use $\lambda = 1$ in the code.

The approach in UCB is to compute an upper confidence bound (upper confidence interval) to ensure that you have a 0.95 (or any other confidence level) probability that the actual estimate is within the UCB value. As the number of trials increases, and as you take a particular action again and again, the uncertainty in the estimate goes down and the score in Equation 10-6 will approach the estimate $\hat{\theta}_k$. The actions that have not been visited at all will have *score* = ∞ because $\alpha_k + \beta_k = 0$ in Equation 10-6, making score infinite. The UCB is shown to have a close to optimal growth bound. I do not go into the mathematical proof of this. You can refer to advanced texts on bandit problems.

Now, you'll walk through the implementation of a UCB agent. Listing 10-6 gives the code. As explained, you again extend the `RandomAgent` class shown in Listing 10-5 and override the `get_action` function to choose the action as per the UCB Equation 10-6.

Note that to avoid dividing by zero, a small constant 1e-12 is added to the denominators where self.success_cnt+self.failure_cnt $(\alpha_k + \beta_k)$ appears.

Listing 10-6. UCB Exploration Agent from 10.b-explore-vs-exploit.ipynb

```
class UCBAgent(RandomAgent):
    def get_action(self):
        exploit = self.success_cnt / (self.success_cnt+self.failure_
        cnt+1e-12)
        explore = np.sqrt(2*np.log(np.maximum(self.total_pulls,1))/ \
                  (self.success_cnt+self.failure_cnt+1e-12))
        estimates =  exploit + explore
        return np.argmax(estimates)
```

Thompson Sampling Exploration

This section looks at the final strategy, that of Thompson sampling. This is based on the idea of choosing an action based on the probability of it being the one that maximizes the expected reward. While UCB is based on the frequentist notion of probability, Thompson sampling is based on the Bayesian idea.

Initially, you have no prior knowledge of which is a better action, that is, which action has the highest success probability θ_k. Accordingly, you form an initial belief that all the θ_k values are uniformly distributed in range $(0, 1)$. This is called *prior* in Bayesian terminology. You express this with a beta distribution[2] of $\hat{\theta}_k \sim \text{Beta}(\alpha = 1, \beta = 1)$. In the first iteration, the agent samples a value for each action k using the Beta distribution $\hat{\theta}_k$ and chooses the action k that has with the maximum sampled value $\hat{\theta}_k$.

Next, it plays out that action (i.e., calls the pull function to perform the chosen action) and observes the outcome (success or failure). Based on the outcome, it updates the posterior distribution of $\hat{\theta}_k$ for the action k that was chosen. As the success and failure counts are updated for the different action k in each step, the agent continues to update its belief of the specific $\hat{\theta}_k$ in each step k. This is the posterior distribution of the unknown value θ_k for all the actions $k \in K$ under the Bayesian framework.

$$\hat{\theta}_k \sim Beta(\alpha = \alpha_k + 1, \beta = \beta_k + 1) \ \ \forall k \in [1..K] \tag{10-7}$$

---

[2]https://en.wikipedia.org/wiki/Beta_distribution

The cycle goes on. As you go through multiple steps, α_k and β_k, the count of success and failure for action k, increase. The *Beta* distribution becomes narrower with a peak around the mean value of $\alpha_k/(\alpha_k + \beta_k)$.

Now look at the implementation of Thompson sampling in code, shown in Listing 10-7. As before, you extend the `RandomAgent` class and override the `get_action` method to implement the *Beta* distribution Equation 10-7.

Listing 10-7. Thompson Sampling Exploration from 10.b-explore-vs-exploit.ipynb

```
class ThompsonAgent(RandomAgent):
    def get_action(self):
        estimates = np.random.beta(self.success_cnt+1, self.failure_cnt+1)
        return np.argmax(estimates)
```

Comparing Different Exploration Strategies

Having implemented the various exploration strategies, you will now evaluate the regret of each of these strategies and plot these regrets as t – the number of steps increase. You implement a `get_regret` function to take `n_steps=10000` steps using the bandit environment and a chosen exploration strategy. You then carry out the experiment `n_trials=50` times to reduce the variance and store the average cumulative regret as a function of the step count. Listing 10-8 shows the code that implements `get_regret`, which acts on a single agent/strategy.

Listing 10-8. get_regret Function from 10.b-explore-vs-exploit.ipynb

```
def get_regret(env, agent, n_steps=10000, n_trials=10):
    score = np.zeros(n_steps)
    optimal_r = env.optimal_reward()

    for trial in range(n_trials):
        agent.reset()
        for t in range(n_steps):
            action = agent.get_action()
            reward = env.pull(action)
            agent.update(action, reward)
            score[t] += optimal_r - reward
```

```
    score = score / n_trials
    score = np.cumsum(score)
    return score
```

Having defined all the machinery, you run the experiment for the three exploration-exploitation agents—ε-greedy, UCB, and Thompson exploration. The regret values for these three agents as a function of step count are plotted. The plotting code and the code to run the experiment is shown in Listing 10-9.

Listing 10-9. Code to Run Experiment and Plot the Regret Graph from 10.b-explore-vs-exploit.ipynb

```
def plot_regret(scores):
    for k,v in scores.items():
        plt.plot(v)
    plt.legend([k for k,v in scores.items()])
    plt.ylabel("regret")
    plt.xlabel("steps")
    plt.show()

n_actions = 5
n_steps=10000
n_trials=50
epsilon = 0.01

agents = {
    #"Random" : RandomAgent(n_actions),
    "EpsilonGreedy(eps=0.01)" : EGreedyAgent(n_actions, epsilon),
    "UCB": UCBAgent(n_actions),
    "Thompson": ThompsonAgent(n_actions) }

env = Bandit(n_actions)
scores = {}

for name, agent in agents.items():
    score = get_regret(env, agent, n_steps, n_trials)
    scores[name] = score

plot_regret(scores)
```

Running the experiment in Listing 10-9 produces the regret graph shown in Figure 10-10.

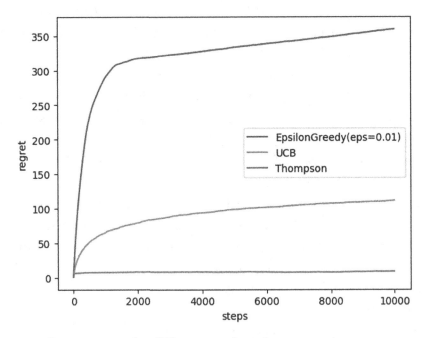

Figure 10-10. *Regret curve for different exploration strategies*

You can see that UCB and Thompson have sublinear growth, while the ε-greedy regret continues to grow at a faster pace (at a linear growth with a rate of growth determined by the value of ε).

Now you'll move forward to the discussion of Monte Carlo tree search, where you will use UCB to carry out explore versus exploit decisions. The choice of UCB in MCTS is based on the current trends shown in popular MCTS implementations.

Planning at Decision Time and Monte Carlo Tree Search

The way Dyna uses planning is called *background planning*. The model is used to generate simulated samples that are then fed to the algorithm, just like with real samples. These additional simulated samples help improve the value estimates further over and above the estimates done using real-world interactions. When an action has to be selected in, say, state S_t, there is no planning at this decision time.

The other way to plan is to begin and end the plan while in state S_t to look ahead, kind of like playing out the various scenarios in your head using the model learned so far and using this plan to decide on action A_t to be taken at time t. Once the action A_t is taken, the agent finds itself in a new state S_{t+1}. The agent again looks ahead, plans, and takes a step. This cycle goes on. This is called *decision-time planning*. Depending on the time available, the further/deeper the agent can plan from the current state, the more powerful and helpful the plan will be in making a decision about what action to take at time t. Hence, decision-time planning is most helpful when fast responses are not required. In board games like Chess and Go, the decision-time plan can look, for instance, a dozen moves ahead before deciding on which action to take. However, if a fast response is required, then background planning is most suitable.

These forward search algorithms select the best possible action by looking ahead. They build a search tree of possible options starting from the current state S_t that the agent is in. The possible options in a given state-action pair may depend on the model of the MDP the agent has learned from prior interactions, or it may be based on the model of MDP given to the agent in the beginning. For example, in the case of board games like Chess and Go, there are clearly defined the rules of the game, and therefore you know the environment in complete, exact detail. Figure 10-11 shows an example of a search tree.

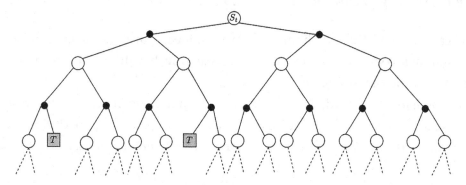

Figure 10-11. *Search tree starting from current state S_t*

However, if the branching factor is high, you cannot build an exhaustive search tree. In the game of Go, the average branching factor is around 250. Even if you build a three-level deep exhaustive tree, the third level will have $250^3 = 15,625,000$ nodes (15 million nodes). And to take possible advantage of looking far ahead, you'll want to go a lot deeper than two or three levels. That is where a *simulation-based search tree* is formed instead of an exhaustive one. You use a *rollout policy* to simulate multiple trajectories and form an MC estimate of action values. These estimates are used to choose the best action at root state S_t. This helps improve the policy. After the best action is taken, the agent finds itself in a new state, and it starts all over again to roll out the simulation tree from that new state. This cycle goes on. The number of rollout trajectories tried depends on the time constraint within which the agent needs to decide on an action.

MCTS is an example of decision time planning. MCTS is exactly as described earlier with some additional enhancements to accumulate value estimates and direct the search to the most rewarding part of the tree. The basic version of MCTS consists of four steps.

1. *Selection/tree traversal*: Starting from the root node, use a tree policy to traverse the tree and reach a leaf node, that is, a node that's not fully expanded.

2. *Expansion*: Add child nodes (actions) to the leaf node and choose one of them.

3. *Simulation/rollout*: Play the game until termination, starting from the previously selected child node. Use a fixed rollout policy to play this out, and it could be a random rollout policy as well.

4. *Back propagation*: Use the results of playout to update all the nodes from termination to all the way up to the root node by following (backtracking) the path that was traversed.

The cycle is repeated multiple times, always starting from the current root node until you run out of time allotted for search or you run it for a fixed number of iterations. This is called the MCTS "anytime algorithm" as it can be stopped anytime, and it will always return a valid search answer. The quality of the search returned by MCTS will improve more if the time allotted is increased. At this point, an action with the highest value is selected based on the accumulated statistics. Figure 10-12 shows a schematic of MCTS.

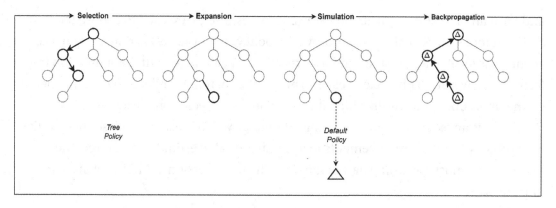

Figure 10-12. *One iteration of MCTS consisting of four steps*

Next, the agent performs the chosen action in a real environment. This moves the agent to a new state, and the previous MCTS cycle starts once more, this time with the new state being the root state from which the simulated based tree is built.

Let's look at a simple, made-up example. You will use the UCB to measure the following form:

$$UCB1(S_i) = V_i + C\sqrt{\frac{\ln N}{n_i}}, \text{ where } C = 2$$

$$V_i = \frac{Score(S_{n_i})}{n_i}$$

*UCB*1 will be used to decide which node to choose in the expansion phase. You start with the root node S_0, and for each node you keep two statistics—the total score (S) and the number of visits (n) to that node. The initial setup is as follows:

$$\boxed{S_0} \quad \begin{array}{l} S = 0 \\ n = 0 \end{array}$$

You expand the tree at the root node by adding all the available actions from S_0, let's say a_1 and a_2. These actions lead to two new states, S_1 and S_2, respectively. You initialize the score and count statistics for these two nodes. The tree at this point looks like this:

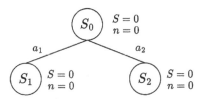

This is the initial tree. You now start executing the steps of MCTS. The first step is tree traversal using $UCB1$ values. You are at S_0 and you need to choose a child node with the highest $UCB1$ value. Let's calculate the $UCB1$ for S_1. As $n = 0$, the exploration term in $UCB1$, that is, $\sqrt{\dfrac{\ln N}{n}}$, is infinite. This is the case with S_2. As part of the tree policy, let's choose the node with a smaller subindex in the case of ties, which leads to a selection of S_1. As S_1 has not been visited before, that is, as $n = 0$ for S_1, you do not expand and instead move on to Step 3 of rollout. You use a rollout policy to simulate random actions from S_1 until termination. Let's assume that you observe a value of 20 at termination, as shown here:

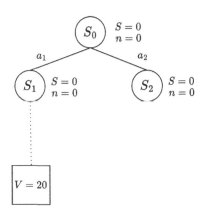

The next step is the fourth one, that is, to back propagate the value from the terminal state all the way up to the root node. You update the visit count and the total score for all the nodes falling in the path from the terminal state to the root node. And then you erase the rollout part from S_1. At this point, the tree looks like the one shown. Note the updated statistics for nodes S_1 and S_0.

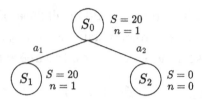

This brings you to the end of the first iteration of MCTS steps. You will now run one more iteration of MCTS steps, again starting from S_0. The first step is a tree traversal starting from S_0. You need to select one of the two nodes, that is, S_1 or S_2 based on which has a higher $UCB1$ value. For S_1, the $UCB1$ value is given by the following:

$$UCB(S_1) = 20 + 2.\sqrt{\frac{\ln 1}{1}} = 20$$

The 1 in $ln\ 1$ comes from the value of n at the root node, which tells you the total number of trials so far. And the 1 in the denominator inside the square root comes from the value of n for node S_1. The $UCB(S_2)$ continues to be ∞ as the n for S_2 is still 0. So, you choose node S_2. That ends the tree traversal step since S_2 is a leaf node. You carry out a rollout from S_2. Let's assume that the rollout from S_2 produces a terminal value of 10. That completes the rollout phase. The tree at this point looks like this:

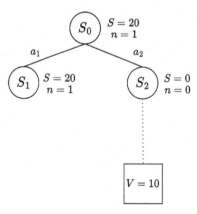

The next step is to backpropagate the terminal value of 10 all the way back to the root state and in the process update the statistics of all the nodes in the path—S_2 and S_0 in this case. You then erase the rollout part. At the end of backpropagation step, the updated tree looks like this:

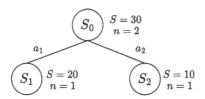

Let's carry out one more iteration of MCTS, and this will be the final iteration. Again start from S_0 and recalculate the $UCB1$ values of child nodes. As a reminder, you calculate $V_i = (\text{score of node } i)\backslash n_i$

$$UCB(S_1) = \frac{20}{1} + 2.\sqrt{\frac{\ln 2}{1}} = 21.67$$

$$UCB(S_2) = \frac{10}{1} + 2.\sqrt{\frac{\ln 2}{1}} = 11.67$$

The UCB value of S_1 is higher, you choose that. You have reached a leaf node, and that ends the tree traversal part step of MCTS. You are at S_1, and it has been visited before, so you now expand S_1 by simulating all possible actions from S_1 and the resulting states due to these actions. The tree at this point looks like this:

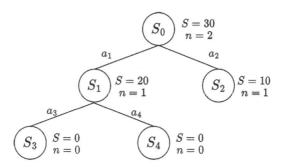

You need to choose one leaf node between S_3 and S_4 followed by rollout from that chosen node. As before, the $UCB1$ for both S_3 and S_4 is ∞ since these nodes have not been visited yet, resulting in $n = 0$ for each of these two nodes. As per the tree policy, you choose the node with a lower subindex in the case of a tie. This means you choose S_3 and do a rollout from S_3. Assume that the terminal state at the end of the rollout produces a terminal value of 0. This completes the rollout step of MCTS. The tree looks like this:

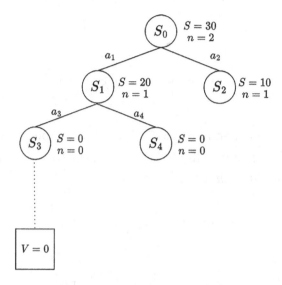

The final step is to backpropagate the value of $V = 0$ all the way back to the root node. You update the statistics of the nodes in the path—nodes S_3, S_1, and S_0. The tree looks like this:

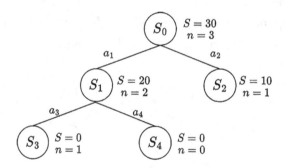

At this point, let's say you have run out of time. The algorithm now needs to choose the best action at S_0 and take that step in the real world. At this point, you will only compare the state values (score/n) values, and for S_1 and S_2, both the average values are 10. It is a tie. You can have additional rules to break the tie. However, let's say you always choose a random action to break a tie while making the final choice of the best action. And let's say you randomly selected S_1. The agent takes action a_1 and lands at S_1, assuming a deterministic setup. You now make S_1 the new root node and carry out multiple iterations of MCTS from S_1, eventually choosing the best action to take at S_1. This way, you step through the real world.

This is the essence of MCTS. There are many variations of MCTS that I do not go into. The main objective was to familiarize you with a simple and basic MCTS setup as well as walk you through the computations to solidify your understanding.

Example Uses of MCTS

Before I conclude this chapter, this section briefly talks about a recent and famous use case of MCTS—AlphaGo and AlphaZero, which helped design an algorithm whereby a computer beat human experts in the game of Go. The section also looks at AlphaFold with MCTS, which helps predict the structure of large protein complexes. It also briefly touches upon the various domains in which MCTS has recently been used.

AlphaGo

AlphaGo was architected by Deep Mind in 2016[3] to beat human experts in the game of Go. A standard search tree would test everything possible from a given position. However, games with high branching factors cannot use this approach. If a game has a branching factor of b (i.e., the number of legal moves at any position is b) and the total game length (i.e., depth or number of sequential moves until the end) is d, then there are b^d possible sequences of moves. Evaluating all the options is not feasible for any moderate board game. Considering the game of chess with $(b \approx 35, d \approx 50)$ or the game of Go with $(b \approx 250, d \approx 150)$, an exhaustive search is impossible. But the effective search space can be reduced with two general approaches.

- **You can reduce the depth with position evaluation.** After expanding the search tree for some depth, replace the state at a leaf node with an approximate evaluation function $v(s) \approx v^*(s)$. This approach worked for board games such as chess, checkers, and Othello, but did not work for Go due to Go's complexity.

- **You can reduce the breadth of the tree from full expansion at a node with sampling actions from a policy $p(s)$.** Monte Carlo rollouts do not branch at all; they play out an episode from a leaf node and then use the average return as an estimate of $v(s)$. You saw this

---

[3] https://deepmind.google/technologies/alphago/

approach in MCTS in which the rollout values were used to estimate the $v(s)$ of the leaf node and the policy was further improved by using the revised estimate of $v(s)$.

In this academic paper on AlphaGo, the authors combined MCTS with deep neural networks to achieve super-human performance. They passed the board image as a 19×19-pixel image to a convolutional neural network (CNN) and used this to construct the representation of the board position, that is, the *state*.

They used expert human moves to carry out supervised learning and learn two policies, that is, $p(s)$.

- **SL Policy Network** (p_σ): The authors used supervised learning to learn the policy network p_σ. This was trained using the CNN style network with 19x19 board positions as input state (s) with board positions being based on expert human moves.

- **Rollout Policy** (p_π): The same data was used to train another network, called fast policy p_π, which acted as the rollout policy in the "rollout" phase of MCTS. The original MCTS used a random rollout policy. However, as noted earlier, you can use a policy model if it is very fast.

Next, the authors trained a reinforcement learning (RL) network p_ρ, initialized from the SL policy network, which tries to improve the SL policy network to have more winnings, that is, using policy optimization to improve with the objective of more wins. It is similar to what you learned in Chapter 8 about policy optimization.

Lastly, another network v_θ is trained to predict the value of a state $v(s)$ for self-plays by the RL policy network p_ρ. Figure 10-13 shows a high-level diagram of the neural network training.

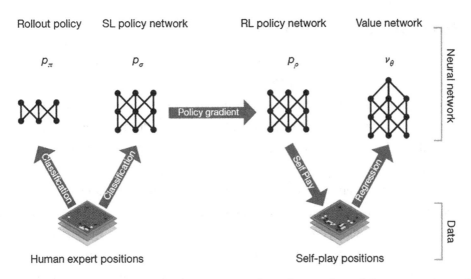

Figure 10-13. *Neural network training pipeline (reproduced from Figure 1 of AlphaGo paper)[4]*

Finally, the policy network and value network are combined to implement[4] MCTS and to carry out forward search. The authors used a UCB-like approach to traverse the expanded part of the tree. Once a leaf node (S_L) was reached, they used one sample from the SL policy network p_σ to expand the leaf node (step 2 of MCTS). The leaf node was evaluated in two ways: one by using value network $v_\theta(S_L)$ and another by a fast rollout until termination using the rollout network p_π to get outcome z_L. These two values were combined using a mixing parameter λ.

$$V(S_L) = (1 - \lambda)v_\theta(s_L) + \lambda z_L$$

The fourth step of MCTS is carried out to update the score and the visit counts at each node, as you saw in the example in the previous section.

Once the search is complete, the algorithm chooses the best action from the root node, and then the cycle of MCTS goes on from the new state.

---

[4] https://research.google/pubs/mastering-the-game-of-go-with-deep-neural-networks-and-tree-search/

AlphaGo Zero and AlphaZero

A year later, in 2017 the authors came out with a further refinement in a paper titled "Mastering the Game of Go Without Human Knowledge."[5]

There were a few enhancements to introduce an algorithm based solely on reinforcement learning, without human data, guidance, or domain knowledge beyond game rules. AlphaGo became its own teacher: a neural network was trained to predict AlphaGo's own move selections and the winner of AlphaGo's games. This neural network improved the strength of tree search, resulting in higher quality move selection and stronger self-play in the next iteration. Figure 10-14 shows a high-level schematic of the setup under AlphaGo Zero. The program plays a game $s_1, ..., s_T$ against itself. In each position s_t, a Monte-Carlo tree search (MCTS) α_θ is executed using the latest neural network f_θ. Moves are selected according to the search probabilities computed by the MCTS, $a_t \sim \pi_t$. The terminal position s_T is scored according to the rules of the game to compute the game winner z.

---

[5] https://www.nature.com/articles/nature24270

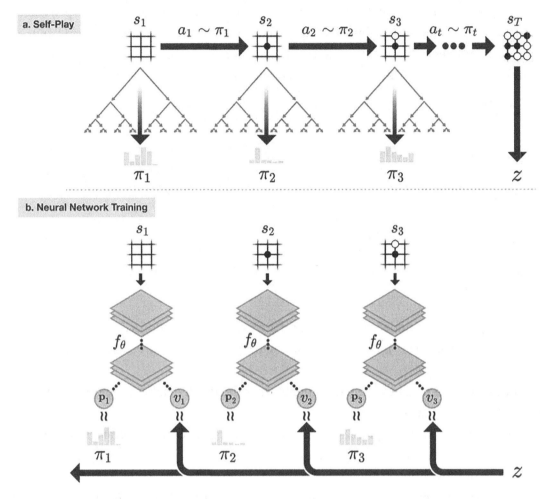

Figure 10-14. *Self-play RL in AlphaGo Zero (Figure 1 of the reference paper)*

Next, you look at neural network training in AlphaGo Zero. The neural network takes the raw board position s_t as its input, passes it through many convolutional layers with parameters θ, and outputs both a vector p_t, representing a probability distribution over moves, and a scalar value v_t, representing the probability of the current player winning in position s_t. The neural network parameters θ are updated to maximize the similarity of the policy vector p_t to the search probabilities π_t, and to minimize the error between the predicted winner v_t and the game winner z. The new parameters are used in the next iteration of self-play.

Now turn your attention to the MCTS part, as shown in Figure 10-15. In Step a, each simulation traverses the tree by selecting the edge with maximum action-value Q, plus an upper confidence bound U that depends on a stored prior probability P and visit count N for that edge (which is incremented once traversed).

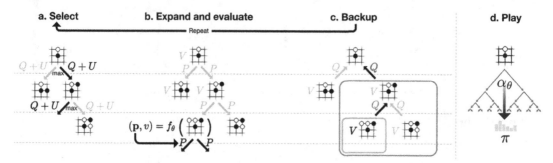

Figure 10-15. *Monte-Carlo tree search in AlphaGo Zero (Figure 2 of the reference paper)*

In Step b, the leaf node is expanded and the associated position s is evaluated by the neural network $(P(s, \cdot), V(s)) = f_\theta(s)$; the vector of P values is stored in the outgoing edges from s.

In Step c, action-values Q are updated to track the mean of all evaluations V in the subtree below that action.

Once the search is complete, search probabilities π are returned, proportional to $N^{1/\tau}$, where N is the visit count of each move from the root state and τ is a parameter controlling temperature.

Soon after this paper, the authors generalized the approach in a paper titled "Mastering Chess and Shogi by Self-Play with a General Reinforcement Learning Algorithm."[6] Starting from random play, and given no domain knowledge except the game rules, within 24 hours AlphaZero achieved a superhuman level of play in the games of chess and shogi (Japanese chess) as well as Go, and convincingly defeated a world-champion program in each case.

About two years later in 2019, the DeepMind team came up with another variation to the same approach in a paper titled "Mastering Atari, Go, Chess and Shogi by Planning with a Learned Model,"[7] which extended the capabilities of AlphaZero.

---

[6] https://arxiv.org/abs/1712.01815

[7] https://arxiv.org/abs/1911.08265

AlphaFold with MCTS

AlphaFold[8] is an AI system developed by DeepMind that predicts a protein's 3D structure from its amino acid sequence. It regularly achieves accuracy competitive with experiment. AlphaFold can predict the structure of single- and multiple-chain proteins with very high accuracy. However, the accuracy decreases with the number of chains, and the available GPU memory limits the size of protein complexes that can be predicted.

In a paper titled "Predicting the Structure of Large Protein Complexes Using AlphaFold and Monte Carlo Tree Search,"[9] the authors search for an optimal path using MCTS, as shown in Figure 10-15. Starting from a randomly selected chain (node), chains are added randomly to expand the path, thereby creating new nodes. From these expansions, complete assemblies are simulated. Simulations are stopped when no additional subunits can be added. The simulated assemblies are scored by their cumulative mpDockQ (multiple-interface predicted DockQ; average interface plDDT times the logarithm of the number of interface contacts) score, and the scores are backpropagated to yield support for the previous selections. The path with the most support is selected, creating a complex that is the most likely to be correct. Refer to Figure 2 of the paper for the schematics of the MCTS in this context.

Use of MCTS in Other Domains

Some other uses of MCTS are as follows:

- Chemical synthesis, as you saw in previous subsection on AlphaFold.

- Planning: For optimal planning in logistics and robotics, automated planning is one of the major domains of application of the MCTS algorithm outside games. The planning problem is typically formulated as MDP. MCTS has also been used for planning in robotics. A decentralized variant of MCTS has been suggested for multi-robot scenarios. The trees in compressed forms are periodically sent to other robots, which results in an update of joint distribution over the policy space.

---

[8] https://alphafold.ebi.ac.uk/

[9] https://pubmed.ncbi.nlm.nih.gov/36224222/

- Security: Identifying the best patrolling schedules for security forces in attacker-defender scenarios. A recent application of the Monte Carlo Tree Search (MCTS) is in Security Games (SG). In essence, SG uses game theory methods to find the most efficient patrolling schedules for security forces, such as the police, secret service, and security guards. They do this while protecting a set of potential targets from attackers, like criminals, terrorists, and smugglers. The importance and use of SGs has increased recently due to a rise in terrorist threats worldwide.

- Scheduling: The combination of MCTS (Monte Carlo Tree Search) with additional heuristics is a common approach to large computational problems. MCTS has been proposed for generating initial solutions for a multimode resource-constrained multi-project scheduling problem. These solutions are then modified through a local search, guided by carefully crafted hyper-heuristics.

- Vehicle routing: Transport problems such as Capacitated VRP (CVRP) are a special class of problems that often involve planning and scheduling.

This brings you to the end of this chapter. The AlphaGo explanation was at a conceptual level. Interested readers may want to refer to the original paper.

Summary

This chapter carried out the last unification, that of combining model-based planning with model-free learning. Under integrated planning and learning, you learned how to combine model-based and model-free approaches to reinforcement learning, and how to balance exploration and exploitation. The chapter also talked about how different value-based and policy-based methods can be unified by varying the width and depth of the updates.

You studied the concept of learning a model of the environment dynamics and rewards from interaction and using the model for planning and policy improvement covering different types of model representations, such as table lookup, linear, and Gaussian models, and the sources of error in model learning.

Next, you looked at a specific algorithm, Dyna Q, which combines Q-learning with model learning and planning, and its extension, Dyna Q+, which encourages exploration of states that have not been visited for a long time.

Throughout the book you have seen the balance between exploitation and exploitation that is needed, and it becomes even more important for integrated planning when applied to problems with very large state spaces. Under the topic of a multi-arm bandit, you saw a simple setup of multiple slot machines with unknown reward probabilities to study the tradeoffs between exploration and exploitation, and compared different strategies such as epsilon greedy, UCB, and Thompson sampling. You also learned about the notion of regret as a measure of quality of exploration, and saw how different strategies perform on the cumulative regret.

Having understood the various strategies for balancing exploration versus exploitation, you studied Monte Carlo Tree Search (MCTS). MCTS is a simulation-based search algorithm that uses a tree policy to guide the exploration and a rollout policy to estimate the value of leaf nodes. You looked at a simple synthetic example to understand the details of the steps involved in MCTS.

Finally, you looked at the application of MCTS and showed how it can be applied to games like Go and Chess. The chapter also touched upon AlphaGo and AlphaZero approaches, which combine MCTS with deep neural networks to achieve superhuman performance in various games.

CHAPTER 11

Proximal Policy Optimization (PPO) and RLHF

This is an exciting chapter. It starts from the foundations and builds toward one of the most exciting uses of RL. Have you have used ChatGPT or another Large Language Model (LLM) and found it amazing how these models seem to follow your prompts and complete a task that you describe in English? Apart from the machinery of generative AI and transformers-driven architecture, RL also plays a very important role. Proximal Policy Optimization (PPO) using human annotated (or machine annotated) preferences over pairs of sentences is used as a reward model to fine-tune LLMs to follow the human preferences. This makes LLMs safer and better. This chapter starts with the basics, building toward a deeper understanding of PPO, which even after so many years is still the state-of-the-art policy-based optimization technique in RL. This is followed by a quick overview of LLMs—the architecture, the training process, and the overall LLM ecosystem. The chapter walks through a complete demo of RLHF tuning on a LLM using the state-of-the-art approaches.

Along the way, the chapter also touches upon concepts like prompt engineering, fine-tuning, parameter efficient fine-tuning, and various approaches of using LLMs for increasing complex and human-like tasks. These tasks have resulted in the Large Language Models being called foundational models or even new-age CPUs.

Chapter 8 briefly touched upon the issue of policy, deviating a lot during the gradient update thereby showing instability. It briefly talked about the intuition that controlling the size of update in the policy space gives you better control versus controlling the update in network parameter space. Chapter 8 introduced two approaches following this concept of control in policy/probability space. First, it talked about Trust Region

461

© Nimish Sanghi 2024
N. Sanghi, *Deep Reinforcement Learning with Python*, https://doi.org/10.1007/979-8-8688-0273-7_11

Policy Optimization (TRPO) and then there was a brief discussion of Proximal Policy Optimization (PPO). These both control the update size by making sure that each update does not lead to a revised policy that is significantly different from the current policy.

I first discuss why this is important. Refer to Figure 11-1 and imagine that the agent is trying to learn a policy to climb the hill with a very uneven surface. There are steep valleys on the right side of the path to the top. The agent needs to be cautious and take careful steps while exploring the neighborhood. A step in the wrong direction could cause the agent to fall off the cliff and it would be hard to recover and climb back to the top or to the previous point in the path. The red circle in Figure 11-1 shows the *trust region* around the current policy, point which is safe and does not lead to any disastrous consequences. If the change in policy is within this circle, the agent will continue to see improvement in the policy while being safe from the cliff.

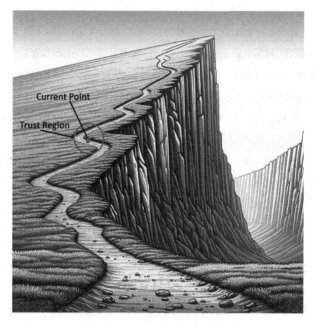

Figure 11-1. *Trust region around a point in policy space*

This is the concept on which TRPO and PPO are based. In TRPO, you put this as a hard constraint as expressed in Equation 11-1. You first form a policy objective that has dependence on the current and the new policy parameters, θ_k and θ, respectively. To do this, you must resort to using importance sampling, which was covered in a previous chapter.

$$J(\theta,\theta_k) = E_{a \sim \pi_{\theta_k}(a|s)} \left[\frac{\pi_\theta(a|s)}{\pi_{\theta_k}(a|s)} A^{\pi_{\theta_k}}(s,a) \right] \tag{11-1}$$

In order to make sure that new policy π_θ stays with the trust region centered on the current policy π_{θ_k} shown in Figure 11-1, you use a hard constraint of Kullback-Liebler distance between two policies being within some small value δ.

$$\theta_{k+1} = \arg\max_\theta J(\theta,\theta_k) \ s.t. D_{KL}(\theta \| \theta_k) \le \delta$$

and $J(\theta, \theta_k)$ is defined in 11-1 \hfill (11-2)

The theoretical update in TRPO as per Equations 11-1 and 11-2 is not easy to carry out. Calculating KL involves expectations set across all possible states and actions. With the help of the second order Taylor approximation and by replacing the KL divergence with a sample-based estimation, the final TRPO updates require estimating the policy gradient as shown in Equation 11-3:

$$\hat{g}_k = \frac{1}{D_k} \sum_{r \in D_k} \sum_{t=0}^{T} \nabla_\theta \log \pi_\theta(a_t|s_t)\big|_{\theta_k} \cdot \hat{A}_t \tag{11-3}$$

Next, you calculate $\hat{x}_k = \hat{H}_k^{-1}\hat{g}_k$ where $\hat{H}_k$ is the Hessian of the sample average KL divergence. The key point to note is that Hessian is the second-order derivative of the policy maximization objective. If the policy network has n parameters, $\hat{H}_k$ will be a $n \times n$ matrix. You need to calculate the inverse of $\hat{H}_k$ to get $\hat{x}_k$ and this inverse is a very slow process, of the order of n^3 computations. The last part of TRPO is the update of parameter, as per Equation 11-4.

$$\theta_{k+1} = \theta_k + \alpha^j \sqrt{\frac{2\delta}{\hat{x}_k^T \hat{H}_k \hat{x}_k}} \cdot \hat{x}_k \tag{11-4}$$

Where $j \in \{0, 1, 2, ...K\}$ is the smallest value that improves the sample loss and satisfies the KL divergence constraint.

---

[1]https://arxiv.org/pdf/1502.05477.pdf

The derivation of the TRPO equation is fairly involved. If you are interested, you can refer to the TRPO paper[1] for more details. However, there are two key take aways. First, constraining a policy update within a trust region using KL divergence ensures that each update to the policy avoids a disastrous drop into bad regions and using KL divergence as a hard constraint like TRPO requires calculating and inverting a very large $n \times n$ matrix $\hat{H}_k$.

Soon after the TRPO paper, the lead author of that paper came up with PPO approach. The paper's introduction reads:

> "We propose a new family of policy gradient methods for reinforcement learning, which alternate between sampling data through interaction with the environment and optimizing a "surrogate" objective function using stochastic gradient ascent. Whereas standard policy gradient methods perform one gradient update per data sample, we propose a novel objective function that enables multiple epochs of minibatch updates. The new methods, which we call proximal policy optimization (PPO), have some of the benefits of trust region policy optimization (TRPO), but they are much simpler to implement, more general, and have better sample complexity (empirically). Our experiments test PPO on a collection of benchmark tasks, including simulated robotic locomotion and Atari game playing, and we show that PPO outperforms other online policy gradient methods, and overall strikes a favorable balance between sample complexity, simplicity, and wall-time."

As you can see, PPO follows policy updates in kind of a trust region on the policy space, just like TRPO. Yet PPO is much simpler and has better sample efficiency. Therefore, it seems to have the best of both the worlds, making it the state-of-the-art direct policy optimization approach since its introduction in 2017.

Having introduced PPO, the next section gets into a step-by-step derivation of PPO and its theoretical basis. If you want to skip ahead, you can skip the sections on PPO and move to the topic of RLHF and fine-tuning LLMs.

Theoretical Foundations of PPO**

The next set of action sections will walk the you through the core foundations of Score function, Hessian Matrix, Fisher Information Matrix, and natural gradients, leading to the PPO in its current form. A section is devoted to discussing the practical implementation details of PPO. This deep dive is optional. You can skip ahead to the sections on LLM- and RLHF-based instruction fine-tuning without any loss of continuity. However, if you go through the PPO deep dive, it is time well spent, as PPO is a popular and very efficient algorithm in the domain of direct policy optimization.

I have to introduce a few key concepts from statistics and optimization theory to set the foundations for natural policy optimization family of algorithms (to which TRPO and PPO belong).

Score Function and MLE Estimator

Most of the machine learning models revolve around finding the parameters of a probability distribution that maximize the joint probability of all the data seen—the training data. This section derives how this comes about and, in the process, introduces the Score function.

Assume that you have a probability model of the unknown parameter θ signified as $x \sim p_X(x; \theta)$ where x, a random variable, follows this distribution. Now say you know the type of distribution and the only thing unknown is θ, which is the parameter of the probability distribution. You also have some samples drawn using this distribution as $x_1, x_2, x_3, ...x_n$. Calculate the joint probability of this data conditioned on the unknown parameter θ, as shown in Equation 11-5.

$$P\left(x_1, x_2, ..., x_n | \theta\right) = \prod_{i=1}^{n} p_X\left(x_i | \theta\right) \tag{11-5}$$

Take the log on both sides of the Equation 11-5 to get:

$$\log P\left(x_1, x_2 | ... | x_n | \theta\right) = \sum_{i=1}^{n} \log p_X\left(x_i | \theta\right) \tag{11-6}$$

The log is the probability of observing a specific value of $x = x_i$. In machine learning, you usually form something called Negative Log Loss (NLL), which is nothing but -ve of the expression in Equation 11-6 with an average taken over a number of samples. This is shown in Equation 11-7.

$$NLL = -\frac{1}{n} \cdot \log P\left(x_1, x_2 | ... | x_n | \theta\right) = -\frac{1}{n} \cdot \sum_{i=1}^{n} \log p_X\left(x_i | \theta\right) \tag{11-7}$$

In most of the machine learning models, including even Large Language Models and other generative AI models, you should try to maximize the evidence, that is, the θ that will maximize the probability of observing the data seen so far, the training data. You convert the maximization to minimization using NLL. In other words, you minimize NLL with respect to θ using stochastic gradient descent. This requires you to take the gradient of NLL with respect to θ, as shown in Equation 11-8.

$$\nabla_\theta NLL = -\frac{1}{n} \cdot \sum_{i=1}^{n} \nabla_\theta \log p_X\left(x_i|\theta\right) \qquad (11\text{-}8)$$

In Stochastic gradient, you adjust θ with $\alpha \cdot \nabla_\theta$NLL, where α is known as the learning rate. The update expression is shown in Equation 11-9.

$$\theta = \theta + \alpha \cdot \nabla_\theta NLL \qquad (11\text{-}9)$$

This is how most machine learning models are trained. Now turn your attention to something called the score function. Look again at the expression in Equation 11-8—the term inside the summation, $\nabla_\theta \log p_X(x_i|\theta)$, is known as the *score function*. The score function is the derivative of the log likelihood function. In Equation 11-8 you have a specific sample in the form of x_i. However, while defining the score function, you use the unsampled random variable x. The definition of the score function is shown in Equation 11-10.

$$\text{score}\left(\theta\right) = \nabla_\theta \log p\left(x|\theta\right) \qquad (11\text{-}10)$$

In traditional statistics, you also use the score function to form the Most Likelihood Estimator (MLE). Under MLE, you take the form in Equation 11-8 and find the θ (analytically), which will make derivative ∇_θNLL = 0. The θ, so found, will minimize the NLL and therefore maximize the probability of the training data.

The score function holds significant importance across various domains, including statistics, probability, and machine learning, primarily due to its fundamental role in estimating and optimizing parameters. In statistics, the score function—often represented as the derivative of the log-likelihood function with respect to a parameter—is crucial for assessing the sensitivity of the likelihood function to the parameter. This sensitivity analysis is vital in maximum likelihood estimation, where the goal is to find parameter values that maximize the likelihood function, thereby providing the best fit for the data. In probability theory, the score function aids in understanding the behavior of probability distributions, especially in the context of small changes in parameters, thus enhancing the understanding of stochastic models.

In the realm of machine learning, the score function is particularly useful in gradient-based optimization algorithms, as shown in Equations 11-6 to 11-10. It serves as a guide to navigate the parameter space in order to find optimal parameters that minimize or maximize a chosen objective function, such as a loss function in supervised learning. This is crucial for training models where the objective is to learn from data by adjusting parameters and improve predictive accuracy. Furthermore, in reinforcement learning, the score function plays a key role in policy gradient methods, where it helps estimate the gradient of expected rewards with respect to policy parameters, enabling the development of strategies that maximize cumulative rewards. Overall, the score function is a versatile tool that enhances your ability to make statistical inferences, understand probabilistic models, and develop efficient machine learning algorithms.

Look at one of the important properties of the score function. What is its expected value $E[score(\theta)]$? Go through the derivation:

$$E\left[score(\theta)\right] = E\left[\nabla_\theta \log p(x|\theta)\right]$$

$$= \int \nabla_\theta \log p(x|\theta) p(x|\theta) dx$$

$$= \int \frac{\nabla_\theta p(x|\theta)}{p(x|\theta)} p(x|\theta) dx$$

$$= \int \nabla_\theta p(x|\theta) dx$$

Using the linearity property, you get:

$$= \nabla_\theta \int p(x|\theta) dx$$

$$= \nabla_\theta (1) = 0$$

$$\text{i.e.,} E\left[\nabla_\theta \log p(x|\theta)\right] = 0 \qquad (11\text{-}11)$$

The expected value of the score function is zero. From MLE point of view, you can view this result as saying that MLE estimator in limit will give you an unbiased estimate of the unknown parameter θ of the underlying distribution that's generating the data.

The actual estimate of θ will depend on the evidence, that is, the data you see, and you use MLE estimator to get an estimate. In statistics, whenever you use data to form some estimate, you are always interested in understanding the variance of this estimate. How much will it change as you see new data or repeat the estimation exercise with another batch of data? Therefore, the next section looks at the covariance of the score function $\nabla_\theta \log p(x|\theta)$.

Fisher Information Matrix (FIM) and Hessian

This section is interested in the covariance of the score function. Whenever you use the MLE estimator, you are also interested in the sensitivity of this estimate based on the data samples used. This can be inferred from the covariance of the score function. Remember that score function is a vector—it is a derivate of log likelihood which is scalar with respect to the unknown parameters of the distribution, a vector θ. The dimensions of the score function and θ will match. The covariance, accordingly, will be a matrix. Derive the expression for this covariance, that is, the Fisher Information Matrix, F:

$$\text{covariance} = F = \mathrm{E}_{p(x|\theta)}\left[\left(s(\theta)-0\right)\left(s(\theta)-0\right)^T\right]$$

$$F = \mathrm{E}_{p(x|\theta)}\left[\nabla_\theta \log p(x|\theta)\nabla_\theta \log p(x|\theta)^T\right] \tag{11-12}$$

In machine learning where you replace expectations with sample estimates, you can get the estimate $\hat{F}$ as shown:

$$\hat{F} = \frac{1}{N}\sum_{i=1}^{N}\nabla_\theta \log p(x_i|\theta)\nabla_\theta \log p(x_i|\theta)^T \tag{11-13}$$

The Fisher Matrix has an interesting property. The negative expected Hessian of log likelihood is equal to the Fisher Information Matrix, F. If you have not heard of Hessian H, it is a square matrix and is the second-order partial derivative of a scalar function—in this case, the likelihood function. The relationship between F and H is as shown here:

$$\mathrm{E}_{p(x|\theta)}\left[H_{\log p(x|\theta)}\right] = -F$$

Or written the other way:

$$F = -\mathrm{E}_{p(x|\theta)}\left[H_{\log p(x|\theta)}\right] \tag{11-14}$$

The derivation of this property involves a bit of math and interested readers can refer to any standard text on statistics for a step-by-step derivation. There are two insights that you can draw from the expression in Equation 11-13 or 11-14. The first one is that the Fisher Information Matrix can be used to make an estimate about the second derivative of the MLE estimator or the Negative Loss function (NLL). The second derivate, the Hessian H, tells you about the sensitivity of θ found using the NLL minimization over samples and you can calculate this sensitivity using F as a replacement in place of hessian. The second insight is about doing gradient descent in machine learning using second-order approximation instead of the first-order approximation as shown in Equation 11-9. The use of the property in Equation 11-14 will help you develop practical algorithms that do gradient descent using second order and using the Fisher Matrix instead of Hessian. The Fisher Information Matrix is also linked to KL divergence. I explore these in the next section, when I talk about the Natural Gradient method, which forms the basis for TRPO and PPO.

Natural Gradient Method

To recap the motivation of TRPO and PPO, you saw earlier in the chapter that to make sure that updates to the policy stay within a defined difference, you need to put constraints on the delta change in the probability distribution space (i.e., policy) instead of the parameters of the probability distribution. TRPO does that by way of a hard constraint on KL divergence between the old and new policy, as described in Equations 11-1 to 11-4. PPO does this without using a hard constraint and you are working your way toward that. To further motivate this point, look at two Gaussian distributions with mean -2 and +2 and variance of 0.49 (standard deviation of 0.7). You plot the two distributions as shown in the left side of Figure 11-2. You can see that the two distributions are far apart with negligible overlap. Keeping the means fixed at -2 and +2, change the variance from 0.49 to 4.0. You can see the second version of the graph on the right side of Figure 11-2 and confirm that the two distributions have a significant overlap, even when the mean distance between the two distributions has remained constant.

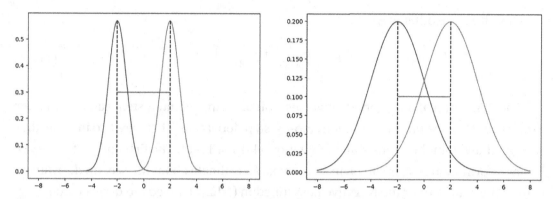

Figure 11-2. *KL divergence changes even when the mean distance remains constant*

The KL divergence due to significant overlap will be lower in the right-side distributions in Figure 11-2. Therefore, operating in parameter space will not give you the ability to control the similarity or dissimilarity of two distributions. Putting a bound on the KL divergence gives you a better way to control the change in distribution as you update its parameters.

Now you'll work your way through establishing the relationship between KL divergence and the Fisher Information Matrix. Look at the expression of KL divergence between two policies, the current one with parameter θ and the revised one with parameter θ′.

$$KL\left[p\left(x|\theta\right)\|p\left(x|\theta'\right)\right]=E_{p(x|\theta)}\left[\log p\left(x|\theta\right)\right]-E_{p(x|\theta)}\left[\log p\left(x|\theta'\right)\right]$$

Taking the first derivate of this expression with respect to θ′, the parameter of the updated distribution, you get:

$$\nabla_{\theta'}KL\left[p\left(x|\theta\right)\|p\left(x|\theta'\right)\right]=\nabla_{\theta'}E_{p(x|\theta)}\left[\log p\left(x|\theta\right)\right]-\nabla_{\theta'}E_{p(x|\theta)}\left[\log p\left(x|\theta'\right)\right]$$

The first term on the right side of the expression will be 0, as the derivative is with respect to θ′ but the expression $E_{p(x|\theta)}[\log p(x|\theta)]$ has no dependence on θ′.

Now expand the second term on right side from the expectation operation to an explicit integral:

$$\nabla_{\theta'}KL\left[p\left(x|\theta\right)\|p\left(x|\theta'\right)\right]=-\int p\left(x|\theta\right)\nabla_{\theta'}\log p\left(x|\theta'\right)dx$$

Take the second derivate of the expression to get:

$$\nabla_{\theta'}^2 KL\big[p(x|\theta)\|p(x|\theta')\big] = -\int p(x|\theta)\nabla_{\theta'}^2 \log p(x|\theta')\,dx$$

The left expression—the second derivate of the scalar value against the parameter vector—is nothing but a Hessian, the Hessian of KL divergence between old and updated distribution with respect to the parameters of the updated distribution. Now evaluate the Hessian at $\theta' = \theta$ to get:

$$H_{KL[(p(x|\theta)\|p(x|\theta'))]} = -\int p(x|\theta)\nabla_{\theta'}^2 \log p(x|\theta')\big|_{\theta'=\theta}\,dx$$

$$= -\int p(x|\theta)H_{\log p(x|\theta)}\,dx$$

$$= -E_{p(x|\theta)}\Big[H_{\log(x|\theta)}\Big]$$

Substituting Equation 11-14, you get:

$$H_{KL[(p(x|\theta)\|p(x|\theta'))]} = F \tag{11-15}$$

Equation 11-15 is an important result. It tells you that the Fisher Information Matrix F is the Hessian of KL divergence between two distributions $p(x|\theta)$ and $p(x|\theta')$, with respect to θ', evaluated at $\theta' = \theta$. How will you use this? Can you think of it?

Look at $KL[p(x|\theta)\|\,[p(x|\theta')]$ and switch the variables, as $\theta' = \theta + d$ to explicitly define the change in parameters from old value θ to new value θ' with a change vector d. Now derive a second order, Taylor series expansion of KL divergence around the original parameter vector θ. I shorten the probability function notation to keep the focus on the core manipulations:

$$KL\big[p_\theta \| p_{\theta'}\big] \approx KL\big[p_\theta \| p_\theta\big] + \big(\nabla_{\theta'}KL\big[p_\theta \| p_{\theta'}\big]\big|_{\theta'=\theta}\big)^T d + \frac{1}{2}d^T\nabla_{\theta'}^2 KL\big[p_\theta \| p_{\theta'}\big]\big|_{\theta'=\theta} d$$

The first term, $KL[p_\theta\|p_\theta]$, will be 0, as the KL distance between the same distributions is 0. The second term has $\nabla_{\theta'}KL[p_\theta\|p_{\theta'}]|_{\theta'=\theta}$, which is equal to $E_{p(x|\theta)}[\nabla_\theta \log p(x|\theta)]$, which according to Equation 11-11, is also 0. Expression $\nabla_{\theta'}^2 KL[p_\theta \| p_{\theta'}]\big|_{\theta'=\theta}$, by Equation 11-15 is the Fisher Information Matrix F. Making these substitutions, you get:

$$KL\big[p_\theta \| p_{\theta'}\big] \approx \frac{1}{2}d^T F d \tag{11-16}$$

With this result, you now know that if you need to constrain the KL divergence during a parameter update, you can do so with the help of Equation 11-16. Say your loss function is L(θ) and you want to find the optimal θ that minimizes this loss function subject to KL divergence being within a constant c. You can express this mathematically as follows:

$$d^* = \arg \min_{d \ \text{s.t.} KL[p_\theta \| p_{\theta+d}]=c} L(\theta+d)$$

You can express this minimization problem with constraint in Lagrangian form as follows:

$$d^* = \arg \min_d L(\theta+d) + \lambda \left(KL[p_\theta \| p_{\theta+d}] - c \right)$$

Using Equation 11-15 to replace $KL[p_\theta \| p_{\theta+d}]$, you get:

$$d^* \approx \arg \min_d L(\theta+d) + \frac{1}{2}\lambda d^T F d - \lambda c$$

Next, approximate L(θ + d) by its first-order Taylor expansion L(θ + d) = L(θ) + ∇$_\theta$L(θ)^{T}d to get:

$$d^* \approx \arg \min_d \left[L(\theta) + \nabla_\theta L(\theta)^T d + \frac{1}{2}\lambda d^T F d - \lambda c \right]$$

To find the d^*, you set the derivate of right side with respect to d equal to 0, to get:

$$\frac{\partial}{\partial d}\left[L(\theta) + \nabla_\theta L(\theta)^T d + \frac{1}{2}\lambda d^T F d - \lambda c \right] = 0$$

$$\Rightarrow \nabla_\theta L(\theta) + \lambda d F = 0$$

Solving for d, you get:

$$d = -\frac{1}{\lambda}F^{-1}\nabla_\theta L(\theta) \tag{11-17}$$

Where the Natural Gradient $\tilde{\nabla}_\theta$ is defined as:

$$\tilde{\nabla}_\theta L(\theta) = F^{-1}\nabla_\theta L(\theta) \tag{11-18}$$

Equation 11-18 is an important result that forms the basis for TRPO and PPO as well as many other policy algorithms based on natural gradients. Natural gradients convert the gradient in the parameter space to a gradient in distribution space by using F^{-1}, which captures the curvature of the probability graph around the parameter θ.

Using the result from Equation 11-18, you can define the natural gradient descent by using the following steps:

1. Do a forward pass and calculate the loss $L(\theta)$.

2. Compute the Fisher Information Matrix F estimate using Equation 11-13.

3. Compute the gradient $\nabla_\theta L(\theta)$.

4. Calculate the natural gradient $\tilde{\nabla}_\theta L(\theta) = F^{-1} \nabla_\theta L(\theta)$ using Equation 11-18.

5. Update the parameter $\theta = \theta - \alpha \tilde{\nabla}_\theta L(\theta)$.

With natural gradients, the size of the update is controlled in the distribution space instead of in the parameter space and it is achieved by using natural gradients.

This has been a long section, full of math and derivations. You do not need to understand everything in the first reading. You can also skip the whole section—just remember the concept of a natural gradient and how it is related to a regular gradient, as per Equation 11-18.

The next section visits TRPO briefly in the context of this understanding and then goes into the details of how PPO implements that KL constraint, what are the implementation level details, and so on.

Trust Region Policy Optimization (TRPO)

Having derived the natural gradient, you can now better appreciate the TRPO Equations 11-1 to 11-3. They implement the KL divergence constraint and ensure improvement of policy on each update using something called a *line search*, as explained in Equation 11-4. TRPO also calculates the Hessian $\hat{H}_k$ of the KL divergence by calculating the Hessian on the sample average KL divergence. Next, it calculates $\hat{x}_k = \hat{H}_k^{-1} \hat{g}_k$ using the conjugate gradient approach and then $\hat{x}_k$ is used to update the policy parameters, as shown in Equation 11-4. Note that TRPO calculates the Hessian of the sample average KL divergence instead of using the Fisher Information Matrix as a drop-in replacement.

Even with all the optimization in implementation, TRPO is a complex and exacting algorithm. It may not have been very popular, but it laid the foundation for authors to come up with PPO, which is covered next.

PPO Deep Dive**

PPO is inspired by the same challenge as TRPO: how can developers update a policy as much as possible with the data they have now, without going too far and risking a drop in performance? While TRPO tries to solve this problem with a complicated method based on second-order derivatives, PPO is a group of methods that use first-order derivatives and some other techniques to keep new policies similar to the old ones. PPO methods are much easier to implement, and they seem to work as well as or better than TRPO in practice. There are two primary variants of PPO—PPO-Penalty and PPO-Clip.

Like TRPO, PPO-Penalty updates the policy network with an approximate solution to a KL-constrained problem, but instead of enforcing the constraint strictly, it adds a penalty term for the KL divergence in the objective function, and it automatically adapts the penalty coefficient during training so that it has the right magnitude.

Instead of a KL divergence term in the objective or a constraint, PPO-Clip uses a special clipping technique in the objective function to prevent the new policy from deviating too much from the old policy. This chapter focuses only on PPO-Clip. Readers may want to quickly review the section on PPO from Chapter 8 before proceeding further.

PPO CLIP Objective

This section first talks about the clipped objective as defined in Equation 8 of the PPO paper.[2] This objective is reproduced in Equation 11-19.

$$L^{CLIP}(\theta) = \hat{E}_t \left[\min\left(r_t(\theta)\hat{A}_t, clip(r_t(\theta), 1-\epsilon, 1+\epsilon)\hat{A}_t \right) \right] \qquad (11\text{-}19)$$

---

[2] https://arxiv.org/pdf/1707.06347.pdf

In Equation 11-19, $r_t(\theta)$ is the importance sampling ratio and $\hat{A}_t$ is the usual Advantage function, which you studied about in Chapter 8. Advantage $\hat{A}_t$ helps reduce the variance when used as a drop-in replacement for rewards-to-go value in the policy optimization algorithms. $r_t(\theta)$ is the ratio of the probability of a given action in a given state under the new policy and under the old policy, as shown in Equation 11-20. Epsilon (ϵ) is the hyperparameter and is usually kept at a value of 0.2.

$$r_t(\theta) = \frac{\pi_\theta(a_t|s_t)}{\pi_{\theta_{old}}(a_t|s_t)} \tag{11-20}$$

Note that $r_t(\theta)$ is 1 when $\theta = \theta_{old}$, that is, when old and new policies are the same. Let's unpack the expression $clip\left(r_t(\theta), 1-\epsilon, 1+\epsilon\right)\hat{A}_t$. Note the range of values. $r_t(\theta)$, being the ratio of two probabilities, would lie between 0 and 1. $CLIP$, as the name suggests, clips the ratio $r_t(\theta)$ in the range $[1-\epsilon, 1+\epsilon]$, thereby removing any incentive for r_t to move outside of this range. This limits policy updates to be within a range controlled by the parameter ϵ.

In Equation 11-19, you use the minimum of clipped and unclipped as the objective. Depending on whether the value of $\hat{A}_t$ is positive or negative, the behavior of the min function is different.

Now consider the case of positive $\hat{A}_t$. Consider three different values of r_t; below $1-\epsilon$, in the range $[1-\epsilon, 1+\epsilon]$, and above $1+\epsilon$. To keep the notation uncluttered, you can drop the sub-index t. When r is below $1-\epsilon$, the clipped value will be $(1-\epsilon)\hat{A}$ and the unclipped value will be $r\hat{A}$, which is lower than the clipped value. Therefore, the surrogate objective function L^{CLIP} will be $r\hat{A}$. When r lies in $[1-\epsilon, 1+\epsilon]$, clipped and unclipped portions will be the same, thus making $L^{CLIP} = r\hat{A}$. For $r > 1+\epsilon$, clipped value will be $(1+\epsilon)\hat{A}$ and it will be less than $r\hat{A}$, making L^{CLIP} equal to $(1+\epsilon)\hat{A}$. You can see these in the following table as well as in Figure 11-3.

You can carry out similar analysis for negative $\hat{A}_t$. Once more consider three different values of r_t; below $1-\epsilon$, in the range $[1-\epsilon, 1+\epsilon]$, and above $1+\epsilon$. When r is below $1-\epsilon$, the clipped value will be $(1-\epsilon)\hat{A}$ and the unclipped value will be $r\hat{A}$. As $\hat{A}_t$ is negative, $(1-\epsilon)\hat{A}$ will be smaller and will be the L^{CLIP}. When r lies in $[1-\epsilon, 1+\epsilon]$, clipped and unclipped portions will be the same, making $L^{CLIP} = r\hat{A}$. For $r > 1+\epsilon$, the clipped value will be $(1+\epsilon)\hat{A}$ and it will be more than $r\hat{A}$, since $\hat{A}$ is negative. Therefore, in this case L^{CLIP} will be equal to $r\hat{A}$. Again, refer to the following table, as well as Figure 11-3 for a summary and plot of how L^{CLIP} changes for different values of ration r.

	$r < 1 - \epsilon$	$r \in [1 - \epsilon, 1 + \epsilon]$	$r > 1 + \epsilon$
$\hat{A} \geq 0$	$r\hat{A}$	$r\hat{A}$	$(1+\epsilon)\hat{A}$
$\hat{A} < 0$	$(1-\epsilon)\hat{A}$	$r\hat{A}$	$r\hat{A}$

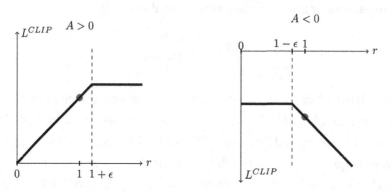

Figure 11-3. L^{CLIP} plot for different values of ratio r and Advantage $\hat{A}$

You can see that, in Figure 11-3, whenever the change in probability ratio will make L^{CLIP} improve, you ignore the change beyond a threshold, $1 + \epsilon$ for positive $\hat{A}$ and $1 - \epsilon$ for negative $\hat{A}$. You include the ratio without any clipping whenever the change is making the objective worse.

Advantage Calculation

Like the other policy gradient approaches in Chapter 8, calculating Advantage remains the same. There are various ways to calculate the Advantage. The original PPO paper suggested using two types of Advantage functions, as shown in Equations 11-21 and 11-22.

$$\hat{A}_t = -V(s_t) + r_t + \gamma r_{t+1} + \cdots + \gamma^{T-t+1} r_{T-1} + \gamma^{T-t} V(s_T) \tag{11-21}$$

$$\hat{A}_t = \delta_t + (\gamma\lambda)\delta_{t+1} + \cdots + \cdots + (\gamma\lambda)^{T-t+1}\delta_{T-1}$$

$$\text{Where } \delta_t = r_t + \gamma V(s_{t+1}) - V(s_t) \tag{11-22}$$

Equation 11-21 is the n-step return with $n = T$ in this case. Equation 11-22 is the Generalized Advantage Estimate, which reduces to Equation 11-21 for $\lambda = 1$ and to TD(0) for $\lambda = 0$.

Value and Entropy Loss Objectives

As with many other Actor-Critic algorithms that involve Advantage calculations, you need a network that can take state to produce state-value. PPO also has a value network that is fitted as part of the loss objective:

$$L^{VF} = \hat{E}_t\left[\left(V_\phi\left(s_t\right) - V_t^{target}\right)^2\right]$$

(11-23)

As you have seen a few times before, you add entropy term $S[\pi_\theta](s_t)$ to encourage exploration. With all these, the objective function that you want to maximize is given in Equation 11-24. Remember that the objective in Equation 11-24 is for maximization, which can be converted into a minimization by multiplying by -1.0.

$$L^{CLIP+VF+S} = L^{CLIP} - L^{VF} + S$$

(11-24)

Implementation Details of PPO

Having explained the theoretical foundations of the PPO algorithm, this section will now go through its implementation, as seen in CleanRL. CleanRL library documentation[3] and actual code file[4] will be the primary source.

CleanRL library authors documented all the implementation details in a blog, which you can read on the ICLR website.[5] There are 37 details that the blog talks about, 13 of which are core details, with the remaining 24 relevant to specific variants such as the Atari environment, robotics tasks, LSTM version, and so on. This section goes over the 13 core details.

---

[3] https://docs.cleanrl.dev/rl-algorithms/ppo/
[4] https://github.com/vwxyzjn/cleanrl/blob/master/cleanrl/ppo.py
[5] https://iclr-blog-track.github.io/2022/03/25/ppo-implementation-details/

1. Vectorized Environment

PPO leverages an efficient paradigm known as the *vectorized architecture,* which features a single learner that collects samples and learns from multiple environments. You can see this in Lines 162 to 164 of the implementation file `ppo.py`,[6] as shown in Listing 11-1.

Based on the command-line argument `args.num_envs`, many environments are created in parallel and wrapped with `gym.vector.SyncVectorEnv`. The `SyncVectorEnv` in Gymnasium is a subclass of `gymnasium.vector.VectorEnv`. `VectorEnv` enables running of multiple independent copies of the same environment in parallel. Vector environments can provide a linear speed-up in the steps taken per second through sampling multiple subenvironments at the same time. To prevent terminated environments waiting until all subenvironments have terminated or truncated, the vector environments auto-reset subenvironments after they are terminated or truncated. The vector environments batch observations, rewards, terminations, truncations, and info for each parallel environment. In addition, the `step()` function expects to receive a batch of actions for each parallel environment.

Listing 11-1. Vectorized Environment: Lines 162-164 - ppo.py

```
envs = gym.vector.SyncVectorEnv(
    [make_env(args.env_id, i, args.capture_video, run_name) for i in
    range(args.num_envs)],
)
```

The command-line arguments `num_envs` and `num_steps` determine the number of steps to run in each environment per policy rollout, and together they control the batch size. The batch size is defined as `args.batch_size = int(args.num_envs * args.num_steps)`. The `total_timesteps` argument controls the total number of environment steps that will be taken during the training. This is broken down into several iterations, with each iteration having `batch_size steps`. Each batch is used multiple times to train the model with the number of repetitions controlled by a command-line argument `args.update_epochs`. A high-level breakdown of this training loop is shown in Listing 11-2.

---

[6] https://github.com/vwxyzjn/cleanrl/blob/master/cleanrl/ppo.py

Listing 11-2. PPO Training Loops from ppo.py

```
args.batch_size = int(args.num_envs * args.num_steps)
args.minibatch_size = int(args.batch_size // args.num_minibatches)
args.num_iterations = args.total_timesteps // args.batch_size

## initialize vectorized environment
## initialize Agent (actor and critic network)
## initialize optimizer, loggers etc.

for iteration in range(1, args.num_iterations + 1):
    for step in range(0, args.num_steps):
        ## step through each of the environments (args.num_steps)
        ## add the step, action, reward and done flag to batch_size
        ## flatten (args.num_envs x args.num_steps) samples to args.
        batch_size

    for epoch in range(args.update_epochs):
        for start in range(0, args.batch_size, args.minibatch_size):
            ## use data from batch to train agents -
            ## actor and critic with exploration and entropy
```

2. Parameter Initialization

In general, the weights of hidden layers use orthogonal initialization of weights with scaling $np.sqrt(2)$, and the biases are set to 0, as shown in Listing 11-3.

Listing 11-3. Parameter Initialization: Lines 94-97 – ppo.py

```
def layer_init(layer, std=np.sqrt(2), bias_const=0.0):
    torch.nn.init.orthogonal_(layer.weight, std)
    torch.nn.init.constant_(layer.bias, bias_const)
    return layer
```

3. Adam Optimizer's Epsilon Parameter

PPO sets the epsilon parameter to 1e-5, which is different from the default epsilon of 1e-8 in PyTorch and 1e-7 in TensorFlow. Note that PPO uses the same optimizer for both policy and value network, as shown in Listing 11-4.

Listing 11-4. Optimizer: Lines 168 – ppo.py

```
optimizer = optim.Adam(agent.parameters(), lr=args.learning_rate, eps=1e-5)
```

4. Adam Learning Rate Annealing

The Adam optimizer's learning rate can be constant or set to decay. By default, the hyperparameters for training agents playing Atari games set the learning rate to linearly decay from 2.5e-4 to 0 as the number of timesteps increases. In MuJoCo, the learning rate linearly decays from 3e-4 to 0. Listing 11-5 shows the code details.

Listing 11-5. Optimizer: Lines 187-190 – ppo.py

```
if args.anneal_lr:
    frac = 1.0 - (iteration - 1.0) / args.num_iterations
    lrnow = frac * args.learning_rate
    optimizer.param_groups[0]["lr"] = lrnow
```

5. Generalized Advantage Estimation

Although the PPO paper uses the abstraction of Advantage estimate in the PPO's objective, as shown in Listing 11-6, the PPO implementation uses a Generalized Advantage Estimation, as shown in Equation 11-22. The gamma (γ) is set to 0.99 and lambda (λ) is set to 0.95. Both of these can be controlled as command-line parameters.

Listing 11-6. GAE: lines 229-230 – ppo.py

```
delta = rewards[t] + args.gamma * nextvalues * nextnonterminal - values[t]
advantages[t] = lastgaelam = delta + args.gamma * args.gae_lambda *
nextnonterminal * lastgaelam
```

6. Mini-Batch Updates

While learning, the PPO implementation randomly arranges the training data of size N*M and splits it into smaller batches to calculate the gradient and improve the policy. This is presented in Listing 11-3.

7. Normalization of Advantages

Using GAE to compute the advantages, PPO then normalizes them by taking away their mean and dividing them by their standard deviation. This normalization is done at the minibatch level, not the whole batch level, as shown in Listing 11-7.

Listing 11-7. Normalization: Lines 261-262 – ppo.py

```
if args.norm_adv:
    mb_advantages = (mb_advantages - mb_advantages.mean()) / \
      (mb_advantages.std() + 1e-8)
```

8. Clipped Surrogate Objective

PPO clips the objective, which is controlled via the command-line argument `clip_coef`, and it is set by default to 0.2. Listing 11-8 shows the code.

Listing 11-8. Clipping Objectives and Gradients: Lines 264-290 – ppo.py

```
# Policy loss
pg_loss1 = -mb_advantages * ratio
pg_loss2 = -mb_advantages * torch.clamp(ratio, 1 - args.clip_coef, 1 +
args.clip_coef)
pg_loss = torch.max(pg_loss1, pg_loss2).mean()

# Value loss
newvalue = newvalue.view(-1)
if args.clip_vloss:
    v_loss_unclipped = (newvalue - b_returns[mb_inds]) ** 2
    v_clipped = b_values[mb_inds] + torch.clamp(
        newvalue - b_values[mb_inds],
        -args.clip_coef,
        args.clip_coef,
    )
    v_loss_clipped = (v_clipped - b_returns[mb_inds]) ** 2
    v_loss_max = torch.max(v_loss_unclipped, v_loss_clipped)
    v_loss = 0.5 * v_loss_max.mean()
```

```
else:
        v_loss = 0.5 * ((newvalue - b_returns[mb_inds]) ** 2).mean()
entropy_loss = entropy.mean()
loss = pg_loss - args.ent_coef * entropy_loss + v_loss * args.vf_coef

optimizer.zero_grad()
loss.backward()
nn.utils.clip_grad_norm_(agent.parameters(), args.max_grad_norm)
optimizer.step()
```

9. Value Function Loss Clipping

PPO clips the value function like the PPO's clipped surrogate objective, and it is controlled via the same command-line argument, `clip_coef`. Listing 11-8 shows how this clipping is done.

10. Overall Loss and Entropy Bonus

As shown in Listing 11-8, the overall loss is calculated as `loss = policy_loss - entropy * entropy_coefficient + value_loss * value_coefficient`, which maximizes an entropy bonus term. Note that the policy parameters and value parameters share the same optimizer.

11. Global Gradient Clipping

For each update iteration in an epoch, PPO rescales the gradients of the policy and value network so that the "global l2 norm" (i.e., the norm of the concatenated gradients of all parameters) does not exceed 0.5, as contained in Listing 11-8. The value of 0.5 can be controlled via the command-line argument `max_grad_norm`.

12. Debug Variables

The PPO implementation comes with several debug variables, which are:

- `policy_loss`: The mean policy loss across all data points
- `value_loss`: The mean value loss across all data points
- `entropy_loss`: The mean entropy value across all data points

- `clipfrac`: The fraction of the training data that triggered the clipped objective

- `approxkl`: The approximate Kullback–Leibler divergence, measured by `(-logratio).mean()`

In the CleanRL implementation, these are logged to the weights and biases if that option is enabled from the command line. The logging code is from lines 300 to 309 in the `ppo.py` file.

13. Shared and Separate MLP Networks for Policy and Value Functions

You can either use a shared network with separate heads or separate networks. CleanRL implementations separate the network approach in the `ppo.py` file, as shown in Listing 11-9.

Listing 11-9. Actor-Critic Agent Networks Used in PPO: Lines 100-126 – ppo.py

```
class Agent(nn.Module):
    def __init__(self, envs):
        super().__init__()
        self.critic = nn.Sequential(
            layer_init(nn.Linear(np.array(envs.single_observation_space.
            shape).prod(), 64)),
            nn.Tanh(),
            layer_init(nn.Linear(64, 64)),
            nn.Tanh(),
            layer_init(nn.Linear(64, 1), std=1.0),
        )
        self.actor = nn.Sequential(
            layer_init(nn.Linear(np.array(envs.single_observation_space.
            shape).prod(), 64)),
            nn.Tanh(),
            layer_init(nn.Linear(64, 64)),
            nn.Tanh(),
            layer_init(nn.Linear(64, envs.single_action_space.n),
            std=0.01),
        )
```

```
def get_value(self, x):
    return self.critic(x)

def get_action_and_value(self, x, action=None):
    logits = self.actor(x)
    probs = Categorical(logits=logits)
    if action is None:
        action = probs.sample()
    return action, probs.log_prob(action), probs.entropy(), self.
    critic(x)
```

Running CleanRL PPO

You saw PPO running in Chapter 8. Refer to the `8.c-ppo_sb3.ipynb` notebook for running PPO on `CartPole`, to record videos, log training details to weights and biases, as well as ways to share trained agents using HuggingFace.

Asynchronous PPO

A group of authors of a paper titled "Sample Factory: Egocentric 3D Control from Pixels at 100000 FPS with Asynchronous Reinforcement Learning"[7] introduced a high-throughput training system with a focus on very efficient synchronous and asynchronous implementations of policy gradients (PPO). Figure 11-4 shows the schematic and the main components of the library.

---

[7] https://arxiv.org/pdf/2006.11751.pdf

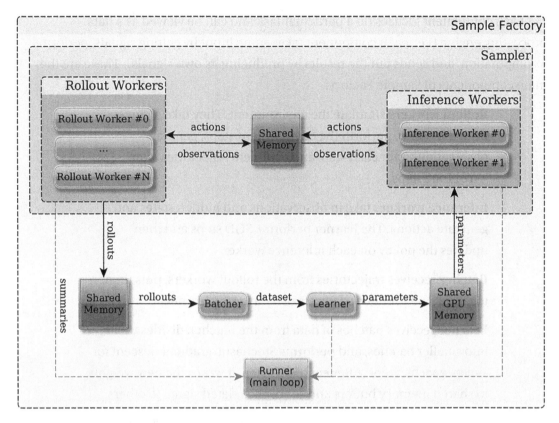

Figure 11-4. *Simple Factory library[8]*

Sample Factory is based on the concept that RL training can be divided into several mostly separate components, each one dedicated to a specific task. This allows for a modular design, where these components can be sped up/parallelized separately, enabling you to reach the highest performance on any RL task. Components interact asynchronously by sending and receiving messages with components typically living on different event loops in different processes. Instead of explicitly sending the data between components (i.e., by serializing observations and sending them across processes), the library chooses to send the data through shared memory buffers. Each time a component needs to send data to another component, it writes the data to a shared memory buffer and sends a signal containing the buffer ID (essentially a pointer to data). This massively reduces the overhead of message passing.

---

[8] https://www.samplefactory.dev/06-architecture/overview/

Each component focuses on a particular task and can be viewed as a data processing engine (i.e., each component takes some input by getting signals, performs a computation, and sends out the results by producing its own signals). These are the main components of Sample Factory:

> **Rollout workers** simulate the environment. They take actions from the policy, perform the environment steps, generate observations after each step, and complete trajectories after `--rollout` steps.

> **Inference workers** take in observations and hidden states and generate actions. The learner performs SGD steps and then updates the policy on each inference worker.

> **Batcher** receives trajectories from the rollout workers, puts them together, and produces datasets of data for the learner.

> **Learner** receives batches of data from the batcher, divides them into smaller batches, and performs Stochastic gradient descent for `--num_epochs` times. After each SGD step, it writes the new weights to shared memory buffers and sends the related signal to others.

> **Runner** bootstraps the whole system, receives all sorts of statistics from other components, and takes care of logging and summary writing.

> **Sampler**, despite being a separate component that can communicate with other signals, usually acts as a simple layer around rollout/inference workers and connects them to the rest of the system.

Interested readers can refer to the Sample Factory Documentation[9] details for using this library. Listing 11-10 shows a sample run.

Listing 11-10. Asynchronous PPO with Sample Factory Library

```
python -m sf_examples.mujoco.train_mujoco --env=mujoco_ant \
--experiment=Ant --train_dir=./train_dir
```

---

[9] https://www.samplefactory.dev/

This completes the deep dive of PPO. In upcoming sections, you will look at the application of PPO in Large Language Models.

Large Language Models(**)

Having look ed PPO in detail, now you'll see its latest use in making Large Language Models by following human prompts to carry out specific tasks. OpenAI demonstrated the use of RL with human feedback using PPO in a seminal paper in March 2022, titled "Training Language Models to Follow Instructions With Human Feedback".[10] The current models deployed by OpenAI, ChatGPT-3.5, and ChatGPT-4 use this approach. Even all other models that follow a prompt—an instruction from a user to complete some tasks like summarizing a text, generate new content, identify themes, or extract information from text—use some form of fine-tuning to make these models follow these instructions. RLHF was the first technique that was successfully demonstrated by OpenAI and still continues to be a very popular approach for fine-tuning models to make these models follow instructions.

Before diving into the details of this paper and the approach of RLHF, I quickly recap Large Language Models. I also discuss various emerging trends in the world of LLMs. Most of the information in this section is orthogonal to understanding RLHF. Therefore, this section is optional and readers with prior knowledge of LLMs or wanting to jump right into RLHF may skip this section on LLMs.

A language model is a system that can learn the statistical patterns of natural language and generate text that resembles the input data. Language models have many applications, such as speech recognition, machine translation, text summarization, question answering, and conversational agents. However, most traditional language models are limited by the size of their vocabulary and the length of their input and output sequences. They also rely on supervised learning, which requires large amounts of labeled data for each specific task.

Large Language Models are a new paradigm in natural language processing, where a single model is trained on a massive amount of unlabeled text data from diverse domains and sources, such as books, websites, news articles, social media posts, and

---

[10] https://arxiv.org/abs/2203.02155

more. These models can achieve impressive performance on various natural language tasks, even without any fine-tuning on task-specific data, by simply using the text itself as supervision. Large Language Models can also generate coherent and fluent text across different genres and styles, demonstrating a high level of generalization and creativity.

One of the key innovations that enabled the development of Large Language Models is the Transformer architecture, which was introduced in a paper "Attention Is All You Need" by Vaswani et al. (2017).[11] The Transformer architecture introduced a novel approach to handling sequential data, moving away from the limitations of previous models like recurrent neural networks (RNNs) and long short-term memory (LSTM) networks. It achieved this with self-attention mechanisms, which allow the model to weigh the importance of different parts of the input data differently. This innovation enabled the Transformer to process all parts of the input data in parallel, significantly improving efficiency and effectiveness in handling long-range dependencies in text.

The Transformer is based on the concept of attention, which allows the model to focus on the most relevant parts of the input and output sequences, regardless of their length or position. The Transformer also uses a self-attention mechanism, which enables the model to learn the dependencies and relations among the words in a sequence. The Transformer consists of two main components—an encoder, which encodes the input sequence into a latent representation, and a decoder, which generates the output sequence from the latent representation. Figure 11-5 shows the schematic of the Transformer model proposed in the paper.

---

[11] https://arxiv.org/abs/1706.03762

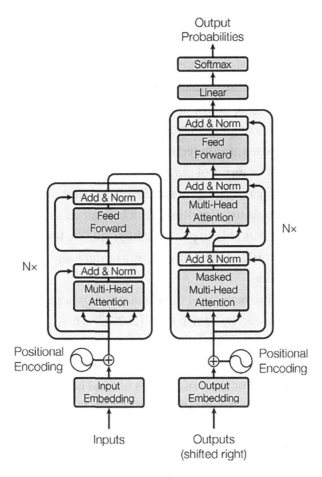

Figure 11-5. *Transformer model[12]*

The Transformer architecture was further extended in a paper titled "Improving Language Understanding by Generative Pre-Training" by Radford et al. (2018),[13] who proposed the Generative Pretrained Transformer (GPT) model. GPT is a variant of the Transformer that only uses the decoder component and is trained as an autoregressive language model, which means that it predicts the next word in a sequence given the previous words. GPT is pretrained on a large corpus of text using the objective of masked language modeling, where some of the words in the input sequence are randomly

---

replaced with a special token and the model has to recover the original words. This way, the model can learn the general patterns and structures of natural language from the data.

However, GPT still faced some limitations, such as the fixed size of its vocabulary and the difficulty of scaling up to larger models and datasets. These challenges were addressed in a paper titled "Language Models Are Unsupervised Multitask Learners" by Radford et al. (2019),[14] who introduced the GPT-2 model, which was an order of magnitude larger than GPT in terms of parameters and data. GPT-2 used a sub-word tokenization method, which allowed it to handle a larger and more diverse vocabulary, as well as a more efficient training algorithm, which reduced the computational cost and memory usage. GPT-2 was able to generate high-quality text on various topics and domains, without any task-specific fine-tuning or data, showing remarkable generalization and versatility.

The GPT-2 model was further scaled up as discussed in a paper titled "Language Models Are Few-Shot Learners" by Brown et al. (2020),[15] who presented the GPT-3 model, which was the largest language model at the time of its publication, with 175 billion parameters and 45 terabytes of text data. GPT-3 achieved state-of-the-art results on several natural language benchmarks, such as natural language understanding, natural language generation, and commonsense reasoning, surpassing the performance of many specialized models that were fine-tuned on task-specific data. GPT-3 also demonstrated the ability to perform few-shot learning, where the model could adapt to a new task by simply providing a few examples of the desired input and output in the query, without any explicit fine-tuning or data. This is known as *prompting* or *prompt engineering*.

This scale-up to very large sizes eventually led to these models being referred to as *Large Language Models* (LLMs) or also as *Foundation Models*. The term reflects not just their physical size in terms of parameters but also their extensive understanding of language and the capability to perform a wide range of NLP tasks with little to no task-specific training.

---

[14] https://openai.com/research/better-language-models
[15] https://arxiv.org/abs/2005.14165

The GPT series of models belong to a broader class of AI models, known as generative models, which aim to learn the underlying distribution of the data and generate new samples from it, mimicking the training data. Generative models have many applications beyond natural language processing, such as computer vision, audio synthesis, and even code. Generative models can also be used for data augmentation, anomaly detection, domain adaptation, and more. Some of the common types of generative models are variational autoencoders, generative adversarial networks, and autoregressive models, such as GPT. A schematic of autoregressive training for Large Language Models is shown in Figure 11-6. In autoregressive training, the input text itself acts as the model output and there is no manual step of human-generated labels. The input sequence of words shifted by one place is used as the label/output sequence. This way, the LLM is trained to predict the next word in the sequence given the sequence of words so far in the sentence. Using this approach, a very large model is trained on a very large corpus of text, almost close to 2-6 trillion word tokens. In a nutshell, this is what LLMs are all about.

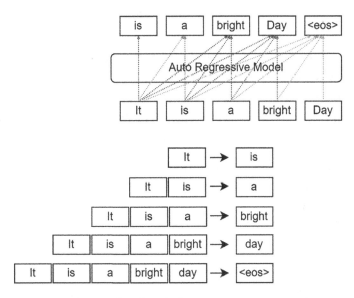

Figure 11-6. *Autoregressive style of training*

However, in its raw form when trained in an autoregressive way only, LLMs can complete a sentence or text but they cannot efficiently follow additional instructions given to the model. RLHF is the process by which a raw trained (pretrained) LLM is further trained to produce output preferred by humans that can follow human instructions much more effectively.

Currently, the pace of release of newer fine-tuned LLMs for different tasks and domains is exponentially increasing. Hardly a week goes by when there is a new announcement of a new model or technique that makes LLMs even more powerful.

This ability of models to follow human instructions (also called "prompts") has made these models powerful and given rise to a new area of expertise, called *prompt engineering*. Researchers are spending countless hours refining the way prompts are generated. You will learn about a few popular approaches but I will not go into the detail, as the focus in this chapter is at an introductory level.

Prompt Engineering

Due to the effectiveness of LLMs's ability to follow instructions/prompts and carry out various NLP tasks, prompt engineering has become a new discipline when it comes to LLMs. Some experts in the AI field have gone to the extent to say that the English language is the new programming language. The same LLM can produce outputs from very poor quality to extremely high and relevant quality based purely on prompts. Prompts are natural language instructions or questions that specify the task, the input, the output format, and sometimes additional information or hints for the LLM. Prompt engineering aims to leverage the general knowledge and capabilities of LLMs without requiring any or minimal fine-tuning on task-specific data.

For example, if you want to use an LLM to perform sentiment analysis on a given sentence, you can use different types of prompts, such as the following:

- Classify the sentiment of the sentence "I love this movie" as positive, negative, or neutral.

- How does the speaker feel about the movie? Happy, sad, or indifferent?

- Sentence: I love this movie

- Sentiment: [MASK]

The quality and effectiveness of the prompts may vary depending on the LLM, the task, the input, and the output. Prompt engineering involves experimenting with different prompt styles, formats, and structures to find the optimal one that maximizes the performance and accuracy of the LLM. A prompt contains any of the following elements:

- Instruction: The description of the task that the model is being asked to do.

- Context: The external information or additional context to help the model produce more accurate/relevant responses.

- Input data: The actual question or task that you are asking the model to complete.

- Output format: The type or format of the output.

Prompting Techniques

Using this structure, there are various ways to guide the LLMs to provide the desired type of output for a given task. It is a rapidly changing area with newer techniques emerging on a daily basis. This section visits some of the more popular approaches being practiced at the time of writing this book.

Zero-shot prompting is the simplest type, where you produce instructions and input data only. There is no additional context or guidance. An example of zero-shot prompting would be:

> """Instruction: Classify the sentiment of the sentence as positive or negative:
>
> Sentence: I love this movie
>
> Sentiment:
>
> """

The model will get these three lines of input and it will complete it with either "positive" or "negative" based on the sentence. In this specific case, you expect the LLM response to be "positive" as "I love this movie" is a sentence that signifies a positive sentiment.

Few-shot prompting is also called *in-context learning*. With this approach, additional context in the form of examples is provided to the model. In the case of the sentiment classifier, you could provide the following examples that show the mapping of sample sentences to the sentiment these signify. An example prompt using this approach would look like this:

> """Instruction: Classify the sentiment of the sentence as positive or negative:
>
> Context: Move was worth the money // positive
>
> What a waste of the time // negative
>
> Though slow at times, it is still worth a watch // positive
>
> Sentence: I love this movie
>
> Sentiment:
>
> """

We have expanded the prompt by providing three examples of "sentence -> sentiment".

The next one is *chain of thought prompting*. It came from the Google team early in 2022, in a paper titled "Chain-of-Thought Prompting Elicits Reasoning in Large Language Models."[16] It is a very simple idea but one with a lot of power packed in it. The idea adds intermediate steps of reasoning, leading to a final solution for each of the few-shot examples and then asks the model to solve another user-defined problem. The LLM uses the patterns of examples to complete the prompt by giving the intermediate reasoning steps and a final answer for the unsolved problem. The paper shows that such an approach increases the correctness of the answer significantly over the naïve few-shot approach. An example is as follows:

> """
>
> The odd numbers in this group add up to an even number: 4, 8, 9, 15, 12, 2, 1.
>
> A: Adding all the odd numbers (9, 15, 1) gives 25. The answer is False.

---

[16] https://arxiv.org/pdf/2201.11903.pdf

The odd numbers in this group add up to an even number: 15, 32, 5, 13, 82, 7, 1.

A:

""""

Another example from the paper is shown in Figure 11-7.

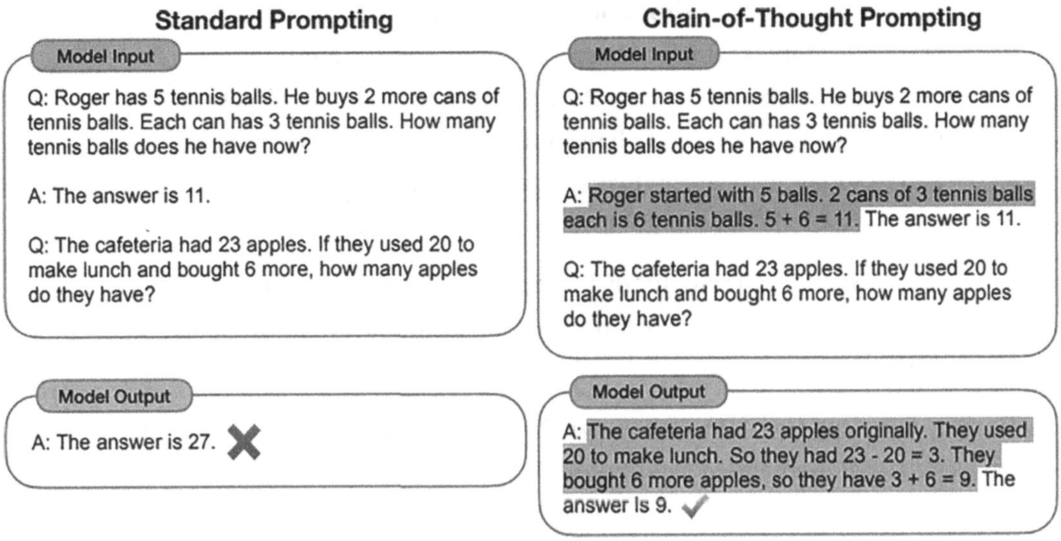

Figure 11-7. *Chain of thought (CoT) promptingSource: CoT paper[17]*

After this paper came out, there was an explosive growth in the number of papers extending the original CoT idea with different variants/enhancements. Some examples are as follows:

- **Self-consistency improvement to CoT**: Use of multiple CoT passes with majority voting.

- **Zero-shot chain-of-thought**: Involves adding "Let's think step by step" to the original prompt. An example prompt is:

---

[17] https://arxiv.org/pdf/2201.11903.pdf

""

Q: A juggler can juggle 16 balls. Half of the balls are golf balls, and half of the golf balls are blue. How many blue golf balls are there?

A: Let's think step by step.

""

- **Automatic chain-of-thought**: The original CoT requires hand-crafted examples which the zero-shot CoT tried to eliminate by replacing the examples with the instructions "Let's us think step by step". However, it can still produce errors. Auto-CoT[18] uses an automated way to select diverse examples that help increase the guidance to the LLM in the form of additional context.

- **Many more**: There are many more approaches. At the time of writing this book, I counted an additional 20+ variations and improvements to the CoT approach.

Having discussed CoT, the next section looks at yet another very popular approach, known as *Retrieval Augmented Generation* (RAG).

RAG and Chat Bots

Retrieval Augmented Generation (RAG) was introduced by Meta in 2020[19] before the emergence of LLMs. This approach can be applied to LLMs for tasks involving specific external knowledge such as Wikipedia or other large corpora, to enhance the generation of text by LLMs. For example, to generate a biography of a person, you can use a RAG prompt like this:

- Retrieve relevant information about the person from Wikipedia or other sources.

- Use the retrieved information to write a coherent and informative biography that covers the person's life, achievements, and legacy.

---

[18] https://arxiv.org/pdf/2210.03493.pdf
[19] https://arxiv.org/abs/2005.11401

The RAG approach consists of two main components—a retriever and a generator. The retriever is responsible for finding the most relevant documents or passages from the external knowledge sources given a query or a partial generation. The generator is a LLM that takes the retrieved information as context and produces the final text output.

The RAG approach can improve the quality, diversity, and specificity of the generated text, especially for tasks that require factual or domain-specific knowledge. However, it also introduces some challenges, such as how to select the best sources of knowledge, how to deal with noisy or incomplete information, and how to ensure the consistency and coherence of the generated text.

An example RAG schematic from LangChain—a library that helps manage complex multi-step complex and dynamic prompt driven LLM interactions—is shown in Figure 11-8.

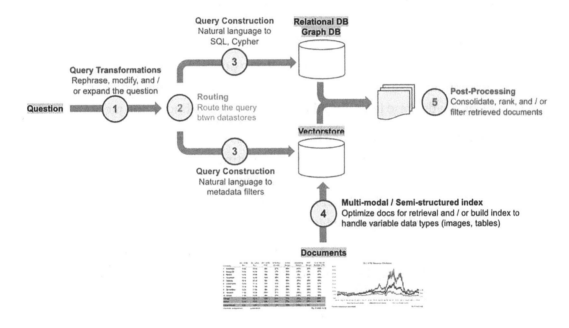

Figure 11-8. *RAG pipelineSource: LangChain library blog*[20]

---

[20] https://blog.langchain.dev/deconstructing-rag/

RAG and LLMs are increasingly being used to power chatbots that provide customer service or answer specific queries. If you use Bing or Google search today, apart from the links, you also see a well-crafted responses at the top of the page. This is an example of RAG-driven search and response. It can also be used to provide customer service, answer internal employee queries about HR policies, and so on. Let's walk through an example of customer query asking about some product details. The complete interaction between the customer and chatbot can be broken into the following steps:

- Receive the customer query and analyze its intent and domain. For example, if the customer asks about the features of a product, the intent is information seeking and the domain is the product category.

- Retrieve relevant documents or passages from a large-scale knowledge source, such as Wikipedia or a product database, using RAG. RAG uses a neural retriever to find the most similar documents to the query, and a neural generator to reran them based on their usefulness for answering the query. A very popular approach right now is to use cosine similarity between the LLM-based embeddings of knowledgebase items and the LLM base embedding of the query.

- Generate a natural language response to the customer query using the retrieved documents as additional context for the LLM. The LLM can use the documents to provide factual or domain-specific information, as well as to personalize the response based on the customer profile or preferences.

- Evaluate the quality and relevance of the generated response using metrics such as ROUGE, BLEU, or human feedback. If the response is satisfactory, send it to the customer. If not, repeat the previous steps with a different set of documents or a different LLM model.

The chatbots also have a way to handle multi-turn chats between the system and user, wherein the previous query from the user and the response from the system are maintained as a list of past interactions and added to the context at each turn of interaction with the system.

LLMs as Operating Systems

All these approaches have taken the LLM use to a level across such diverse sets of tasks that LLMs are now being called *new age CPUs*. Check out the very informative talk by Andrej Karpathy, one of the top voices and experts when it comes to AI.[21] Figure 11-9 shows a slide from Andrej's talk that propositions that LLMs are going to be the new operating systems.

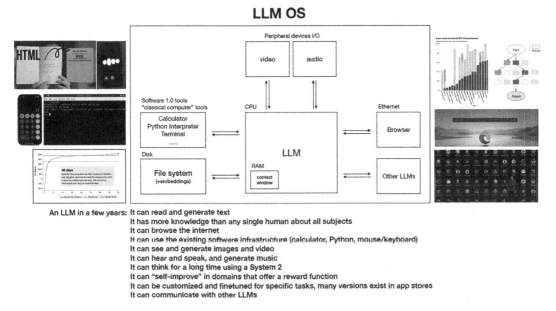

Figure 11-9. *LLMs as operating systems*
Source: Andrej Karpathy talk

Fine-Tuning

LLM fine-tuning involves modifying the parameters of a pretrained LLM to make it more suitable for a specific downstream task. Unlike prompt engineering, which relies on using natural language or special tokens as inputs and outputs, fine-tuning modifies the internal weights of the LLM to optimize its performance on the task.

Fine-tuning is required when the task-specific data is very different from the general domain data that the LLM was trained on, or when the task requires a high level of accuracy and robustness that cannot be achieved by prompt engineering alone.

---

[21] https://www.youtube.com/watch?v=zjkBMFhNj_g

For example, if the task is to classify legal documents or generate biomedical summaries, the LLM may not have enough knowledge or linguistic skills to handle these domains and tasks without fine-tuning. Furthermore, some tasks may have complex output formats or evaluation metrics that are difficult to implement with prompt engineering, such as machine translation or question answering. The process of fine-tuning LLMs can be broken into the following high-level steps:

- Select a pretrained LLM that is suitable for the task and the data. For example, if the task is related to natural language understanding, a masked language model like BERT may be a good choice. If the task is related to natural language generation, an autoregressive model like GPT may be a better option.

- Define a task-specific architecture that includes an input layer, an output layer, and optionally some intermediate layers that connect the LLM with the input and output. The input layer should match the format and vocabulary of the LLM, and the output layer should match the format and vocabulary of the task. The intermediate layers can be used to transform or enhance the representations from the LLM, such as adding attention, pooling, or classification layers.

- Prepare a task-specific dataset that consists of input-output pairs that represent the examples and labels of the task. The dataset should be split into training, validation, and test sets, and should be balanced and representative of the task domain and difficulty.

- Fine-tune the LLM and the task-specific layers on the training set using a suitable optimization algorithm, such as stochastic gradient descent or Adam. The objective function should be defined according to the task, such as cross-entropy loss for classification or generation or mean squared error for regression. The learning rate, batch size, number of epochs, and other hyperparameters should be tuned to optimize the performance on the validation set and avoid overfitting or underfitting.

- Evaluate the fine-tuned model on the test set using appropriate metrics, such as accuracy, precision, recall, F1-score, BLEU, ROUGE, or METEOR. The results should be compared with the baseline models or the state-of-the-art methods to assess the effectiveness of the fine-tuning technique.

Fine-tuning LLMs has its own challenges. The first one is *data scarcity*. There may not be enough task-specific data to fine-tune the LLM effectively, especially for low-resource languages or domains. This may lead to poor generalization or catastrophic forgetting of the LLM's general knowledge. Some techniques to address this challenge are data augmentation, transfer learning, multi-task learning, or meta-learning, which can increase the diversity and quantity of the data, leverage the knowledge from related tasks or domains, or adapt the model to new tasks quickly and efficiently.

The second challenge is that of *data noise*. The task-specific data may contain errors, inconsistencies, or ambiguities that may affect the quality and reliability of the fine-tuning process. This may lead to incorrect or misleading outputs or reduced robustness of the model. Some techniques to address this challenge are data cleaning, data filtering, data validation, or data quality assessment, which can remove or correct the noisy data, select or prioritize the high-quality data, or monitor and evaluate the data quality throughout the fine-tuning process.

The third challenge is that of *model complexity*. The LLMs have a large number of parameters that are difficult to fine-tune on limited computational resources or within a reasonable time. This may limit the scalability or feasibility of the fine-tuning technique. Some techniques to address this challenge are model compression, model pruning, model distillation, and model quantization, which can reduce the size or complexity of the model, remove or simplify the redundant or irrelevant parameters, or approximate the model with a smaller or faster one. You will investigate one of the approaches in the next section, called Parameter Efficient Fine Tuning (PEFT).

The capability of fine-tuning models is increasingly being offered by all the cloud platforms that provide LLM services, such as Azure, AWS, Databricks, Google Vertex AI, and so on. It is a fast-growing list of vendors that provide these capabilities of low-code-no-code approach to RAG, fine-tuning, RLHF, and so on.

Parameter Efficient Fine-Tuning (PEFT)

You studied the challenges of fine-tuning LLMs in the previous section. Full fine-tuning is not feasible to train on common hardware when models get bigger and bigger. Moreover, it is very costly to store and deploy fine-tuned models separately for each downstream task, because fine-tuned models have the same size as the original pretrained model. Parameter-Efficient Fine-tuning (PEFT) approaches aim to solve both issues.

PEFT approaches only adjust a small number of (additional) model parameters while keeping most parameters of the pretrained LLMs fixed, thereby greatly reducing the computational and storage costs. This also avoids the problems of catastrophic forgetting, a behavior seen during the full finetuning of LLMs. PEFT approaches have also proven to be better than fine-tuning in the low-data regimes and generalize better to out-of-domain scenarios.

Another benefit of PEFT methods is that they make the models more portable, as users can modify the models with small checkpoints that only take up a few MBs of storage, instead of the large checkpoints of full fine-tuning. The small number of trained weights from PEFT methods are added on top of the pretrained LLM, so the same LLM can be used for multiple tasks by adding small weights, without having to change the whole model.

There are two common approaches of PEFT. The first one is prompt-based fine-tuning. A prompt can either explain a task or give an example of a task that you want the model to learn. Soft-prompting methods do not require manually creating these prompts, but instead add learnable parameters to the input embeddings that can be adjusted for a specific task while keeping the pretrained model's parameters fixed. This makes it quicker and simpler to fine-tune Large Language Models (LLMs) for new downstream tasks. A good place to refer to the various techniques of this approach is the documentation of the PEFT library from HuggingFace.[22]

The other very popular approach is LoRA (Low Rank Adaptation), which was introduced in 2021 in a paper titled "LoRA: Low-Rank Adaptation of Large Language Models."[23] This approach has an even more efficient version called QLoRA (Quantized LoRA). QLoRA was introduced in 2023 in a paper titled "QLoRA: Efficient Finetuning of Quantized LLMs."[24]. Let's briefly look into both of these approaches.

---

[22] https://huggingface.co/docs/peft/main/en/task_guides/prompt_based_methods
[23] https://arxiv.org/pdf/2106.09685.pdf
[24] https://arxiv.org/pdf/2305.14314.pdf

Low Rank Adaptation (LoRA) freezes the pretrained model weights and injects trainable rank decomposition matrices into each layer of the Transformer architecture, greatly reducing the number of trainable parameters for downstream tasks. Figure 11-10 shows the schematic of the LoRA approach from the paper.

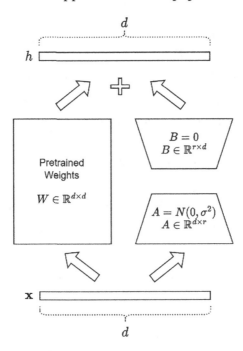

Figure 11-10. *LoRA approach*

Suppose one of the linear layers somewhere in the Transformer has a weight matrix W of size $d \times d$. As part of fine-tuning, you freeze this original pretrained weight matrix W and build another path from the input vector $x \in R^d$ to output vector $h \in R^d$. You do this by passing x additionally through two weight matrices $A \in R^{d \times r}$ and then through $B \in R^{r \times d}$. Here d is the dimension of the input and output vectors and r is a number lot lower than the value of d. The weight matrix W and the product A. B have the same dimensions, that of $d \times d$.

Originally, vector x can be represented as $h = W. x$. With the addition of a new path as per Figure 11-10, the new expression for h is h = (W + A. B). x = W. x + (A. B). x, that is, the original flow from x to h and an additional flow. During fine-tuning, W is kept constant and only weights of A and B are allowed to change during backpropagation. Weights of A are initialized with a zero mean Gaussian, and B is initialized with zeros. With $B = 0$ at the start of fine-tuning, you get $h = W. x$. As fine-tuning progresses, A

and B matrix weights change, allowing for LLM to learn new tasks in an efficient way. As an example, consider the case that $d = 512$ and $r = 10$. The number of parameters in W will be $d \times d = 512 \times 512 = 2,62,144$. The combined number of parameters in A and B is $2 \times d \times r = 2 \times 512 \times 10 = 10,240$, which is about 3.9 percent of the number of parameters in W.

Note that the memory footprint of a model's parameters is very heavy during training compared to the memory footprint of the model during inference. The reason for a heavy memory footprint during training is because during training you need to maintain additional information to allow for backpropagation and optimizer gradient history tracking, information such as hidden layer activations, backprop gradients, and the running average of first moment and second moment gradient values. Though LoRA marginally increases the memory footprint of model parameters during inference, it offers significant memory advantage and data volume requirements during fine-tuning. This is because, under LoRA, the number of parameters being tuned are in the range of 3-5 percent of the actual model parameters. In a nutshell, LoRA offers these key advantages:

- A pretrained model can be shared and used to build many small LoRA modules for different tasks. You can freeze the shared model and efficiently switch tasks by replacing the matrices A and B in Figure 11-10, thus reducing the storage requirement and task-switching overhead significantly.

- LoRA makes training more efficient and lowers the hardware barrier to entry by up to three times when using adaptive optimizers, since you do not need to calculate the gradients or maintain the optimizer states for most parameters. Instead, you only optimize the injected, much smaller low-rank matrices.

- Simple linear design allows for merging the trainable matrices with the frozen weights when deployed, introducing no inference latency compared to a fully fine-tuned model.

- LoRA is orthogonal to many prior methods and can be combined with many of them.

You can refer to the paper and other detailed technical blogs for more details.

QLoRA is a further refinement of LoRA in which model parameters are quantized with different precision and numerical representation to offer even more memory efficiency. To quote from the paper:

> QLoRA reduces the average memory requirements of finetuning a 65B parameter model from >780GB of GPU memory to <48GB without degrading the runtime or predictive performance compared to a 16- bit fully fine-tuned baseline. This marks a significant shift in accessibility of LLM finetuning: now the largest publicly available models to date fine-tunable on a single GPU.

Source: QLoRA paper

Another approach, named Weight-Decomposed LowRank Adaptation (DoRA), was introduced in a 2024 paper titled "DoRA: Weight-Decomposed Low-Rank Adaptation."[25] It decomposes the pretrained weights into magnitude and direction components for fine-tuning, especially with LoRA to efficiently update the direction component. Interested readers may refer to the paper for further details.

PEFT is seeing many additional approaches. I hope the discussion here has given the you an exposure to explore further.

Chaining LLMs Together

Another way to leverage the power of LLMs for various downstream tasks is to chain them together in a multistep pipeline. This means that instead of using a single LLM for a given task, you can use multiple LLMs, each performing a different subtask, and pass the output of one LLM as the input of another LLM. For example, suppose you want to generate a summary of a long document. You can first use an LLM to extract the most relevant sentences from the document, then use another LLM to paraphrase those sentences into a concise form, and finally use another LLM to merge those sentences into a coherent summary. This way, you can use the different strengths and specialties of different LLMs, and reduce the complexity of each subtask.

---

[25] https://arxiv.org/pdf/2402.09353.pdf

One of the key benefits of chaining LLMs together is that it allows you to customize the pipeline according to the specific needs and characteristics of the task. For instance, you can choose the best LLM for each subtask based on its performance, domain, language, or architecture. You can also fine-tune or adapt each LLM with task-specific data or parameters, such as adapter layers or prompt embeddings, to improve its accuracy and generalization. Moreover, you can control the length and quality of the intermediate outputs by setting appropriate hyperparameters, such as beam size, temperature, or length penalty, for each LLM.

However, chaining LLMs together also poses some challenges and limitations. One of the main challenges is how to ensure the compatibility and coherence of the intermediate outputs and inputs across different LLMs. For example, if the output of one LLM contains tokens or symbols that are not recognized by the input of another LLM, it may cause errors or degradation in the final output. Similarly, if the output of one LLM does not match the expected format or style of the input of another LLM, it may lead to inconsistency or confusion in the final output. Therefore, it is important to carefully select and preprocess the LLMs that are part of the pipeline, and to monitor and evaluate the intermediate outputs and inputs for any potential issues.

Another challenge is how to optimize the overall performance and efficiency of the pipeline, considering the tradeoffs between speed, memory, and quality. For example, using more LLMs in the pipeline may increase the quality of the final output, but it may also increase the latency and the computational cost of the task. Similarly, using larger or more complex LLMs may improve the accuracy of each subtask, but it may also require more resources and time to process the data. Therefore, it is important to balance the number and size of the LLMs in the pipeline, and to leverage techniques such as caching, pruning, or quantization to speed up the inference and reduce the memory footprint of the LLMs.

A third challenge is how to explain and monitor the behavior and outcomes of the pipeline, considering the complexity and opacity of the LLMs involved. For example, it may be difficult to trace back the source of errors or biases in the final output, as they may arise from any of the LLMs or their interactions. Similarly, it may be hard to understand the rationale or logic behind the decisions or predictions made by the LLMs, as they may rely on hidden or latent features that are not interpretable by humans. Therefore, it is important to apply methods and tools for explainability and MLOps, such as attribution, visualization, debugging, or logging, to the pipeline, and to evaluate the performance and robustness of the LLMs on various metrics and benchmarks.

LangChain[26] and LlamaIndex[27] are currently two popular libraries that help you manage the complex multi-step solutions built using LLMs, including production monitoring, logging, deployments, and so on.

Auto Agents

An Auto Agent or an LLM Agent is a term that refers to LLM applications that can perform complex tasks by using an architecture that integrates LLMs with essential modules like planning and memory. In LLM agents, an LLM acts as the main controller or "brain" that directs a series of operations needed to finish a task or user request. The LLM agent needs key modules such as planning, memory, and tool usage.

As an example, take a look at a 2023 paper titled "ChemCrow: Augmenting Large-Language Models with Chemistry Tools."[28] To quote from the paper:

> *ChemCrow, an LLM chemistry agent designed to accomplish tasks across organic synthesis, drug discovery, and materials design. By integrating 18 expert-designed tools, ChemCrow augments the LLM performance in chemistry, and new capabilities emerge. Our agent autonomously planned and executed the syntheses of an insect repellent, three organo-catalysts, and guided the discovery of a novel chromophore....Using a variety of chemistry related packages and software, a set of tools is created. These tools and a user input are then given to an LLM. The LLM then proceeds through an automatic, iterative chain-of-thought process, deciding on its path, choice of tools, and inputs before coming to a final answer.*

Source: ChemCrow paper

A schematic of the process is shown in Figure 11-11. The purpose of the figure is to illustrate the complex ways LLMs are being integrated into various processes. In scenarios like these, LLMs almost act like the CPU or operating system, as shown in Figure 11-9.

---

[26] https://www.langchain.com/

[27] https://www.llamaindex.ai/

[28] https://arxiv.org/pdf/2304.05376.pdf

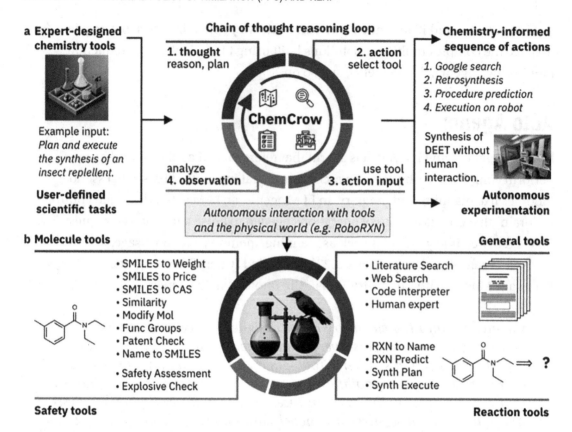

Figure 11-11. ChemCrow overview

Note that the key difference between chaining LLMs vs LLM agents together is that in simple chaining, the flow of information is in one direction only, from the first LLM to the last. However, in LLM agents, there is an iterative flow of information back and forth between various LLMs and the decision of this flow is also orchestrated by LLMs.

Another interesting paper in this category is the one from Microsoft titled "AutoGen: Enabling Next-Gen LLM Applications via Multi-Agent Conversation."[29] It is accompanied by a great open-source library to try things out. Interested readers can refer to the autogen library on GitHub.[30]

---

[29] https://arxiv.org/pdf/2308.08155.pdf
[30] https://github.com/microsoft/autogen

Multimodal Generative AI

Multimodal Generative AI is a branch of artificial intelligence that aims to create systems that can generate diverse and complex outputs across multiple modalities, such as text, images, video, and audio. Such systems can leverage the power of Large Language Models (LLMs) to understand natural language instructions and queries as well as combine them with other generative models that can produce high-quality content in different domains and formats.

One of the main components of multimodal generative AI is the ability to fuse and transform information from different sources and modalities, and to maintain coherence and consistency across them. For example, a multimodal generative system can take a textual description of a scene and generate a corresponding image, or vice versa. Alternatively, it can take an image and a caption and generate a relevant audio narration or take a piece of music and generate lyrics that match its mood and style.

Another component of multimodal generative AI is the ability to reason and plan overly complex scenarios and tasks that involve multiple steps and subgoals. For example, a multimodal generative system can take a high-level instruction, such as "create a short video tutorial on how to make a paper airplane," and decompose it into a sequence of actions, such as "find a suitable paper sheet," "fold it in half," "make a crease," and so on. Then, it could execute each action by generating the appropriate visual and verbal outputs, such as showing the paper sheet on the screen, narrating the folding steps, or highlighting the creases.

Some of the interesting applications and examples of multimodal generative AI include:

- **Content creation and editing:** Multimodal generative systems can help users create and edit various types of content, such as images, videos, podcasts, blogs, or books, by providing suggestions, feedback, or corrections based on natural language commands or queries. For example, a user could ask a system to "add a sunset filter to the image," "change the voice of the narrator to a female one," or "rewrite the sentence to make it more concise".

- **Education and entertainment:** Multimodal generative systems can provide engaging and personalized learning and entertainment experiences, such as interactive games, simulations, stories, or quizzes, by generating content that adapts to the user's preferences, goals, and feedback. For example, a system could create a customized adventure game for a user, where the user can explore a virtual world, interact with characters, and influence the plot by using natural language commands or queries.

- **Communication and collaboration:** Multimodal generative systems can facilitate communication and collaboration among users or between users and machines, by providing translation, summarization, transcription, or annotation services across different languages and modalities. For example, a system could translate a speech from one language to another, summarize a long video into a short text, transcribe an audio conversation into a text chat, or annotate an image with relevant labels and captions.

This completes the optional section on LLM and emerging trends in LLM world. The chapter now returns to application of RL for fine-tuning LLM to follow human instructions. As discussed, the fine-tuning of LLM for RLHF could be a full parameter fine-tuning or a LoRA based approach.

RL with Human Feedback

One of the challenges of training Large Language Models (LLMs) is to ensure that they can follow natural language instructions and generate outputs that are aligned with human preferences and expectations. However, obtaining high-quality labeled data for every possible instruction and output combination is impractical and costly. Moreover, LLMs may encounter novel or ambiguous instructions at inference time, which require generalization and adaptation skills that are not easily learned from supervised data alone.

A promising solution to this challenge is to leverage human feedback as a form of a reinforcement learning (RL) signal, which can guide LLMs to improve their performance on a given instruction by rewarding or penalizing their outputs based on human judgments. This technique, known as reinforcement learning from human feedback (RLHF), can enable LLMs to learn from their own interactions with humans, rather than relying on predefined labels or rules.

The OpenAI team, in a paper titled "Training Language Models to Follow Instructions with Human Feedback" by Ouyanf et al.(2022),[31] introduced a successful application of this concept and the resulting family of LLMs, which was called InstructGPT. The approach used was to fine-tune GPT-3 to follow a broad class of written instructions. This technique uses human preferences as a reward signal to fine-tune LLM models. The overall approach can be broken into three steps as shown in Figure 11-12.

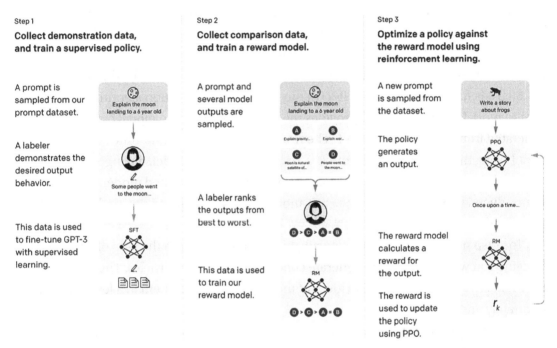

Figure 11-12. *Three steps of RLHF*[32]

Step 1 is *supervised fine-tuning (SFT)*, which involves starting from a pretrained model such as GPT-3 that has been trained on a very large corpus of data from the Internet. The pretraining data is usually 1-2 trillion tokens of text data scraped from the Internet and other open-source datasets. The pretrained model is fine-tuned on a

[31] https://arxiv.org/pdf/2203.02155.pdf
[32] Figure 2 from the RLHF paper - https://arxiv.org/pdf/2203.02155.pdf

dataset of human-written summaries. These examples serve as direct demonstrations of the task you want the model to perform. The model learns from this dataset, by adjusting its parameters to minimize the difference between its outputs and the correct summaries in the dataset using supervised learning approach.

Step 2 involves *reward modelling (RM)*. Reward modeling involves creating a model that can predict the quality of the model's outputs based on human preferences. This is achieved by collecting a new dataset where humans compare pairs of model-generated outputs and indicate which one they prefer. The reward model is then trained to predict these human preferences, effectively learning what constitutes a "better" summary as a binary classification task. The most successful approach to reward modelling is to start with the SFT model from Step 1 and incorporate an additional linear layer on the final transformer layer to generate a scalar prediction representing the reward value.

Step 3 is *RL fine-tuning using PPO* or any other RL optimization algorithm. The reward model learned in Step 2 provides reward feedback for the model output generated from the model in Step 1. The reward is then used to fine-tune the model from Step 1 using the PPO algorithm from RL. To ensure that the model, during RL fine-tuning, does not deviate too much from the original SFT fine-tuned model, a KL penalty is introduced between the reference SFT model from step one and the model being fine-tuned in this step.

To grasp the concept better, let's walk through the steps with the help of an example. Imagine you want your LLM to generate concise summaries of articles. The goal is for the model to understand the core ideas of the text and present them in a few sentences accurately and coherently.

In Step 1, you collect a dataset of articles paired with their concise summaries written by human experts. A pretrained LLM is then fine-tuned on this dataset, learning how to condense articles into a few informative sentences. For instance, if one article is about a new scientific discovery in space exploration, its summary might highlight the discovery, its significance, and potential implications, all in a few sentences.

After the SFT step, you present the model with new articles and generate summaries. Human reviewers are then asked to compare pairs of summaries for the same article, selecting which summary they find more concise and informative. These preferences are used to train a reward model that learns to score summaries based on their conciseness, accuracy, and coherence.

With the reward model in place, the SFT-tuned LLM generates summaries for a new set of articles. These summaries are then scored by the reward model. The LLM uses these scores to adjust its summarization strategy, focusing on generating summaries that are more likely to be concise, accurate, and informative according to the reward model's criteria. And it does so with the PPO policy optimization approach. Over time, this iterative process results in a model that generates high-quality summaries that align closely with human preferences.

Latest Advances in LLM Alignment

Since its inception in 2022, RLHF has significantly impacted the way artificial intelligence models are trained and improved. As discussed in previous section, RLHF initially involved steps like supervised fine-tuning (SFT), reward modeling, and policy optimization. Over time, this process has seen substantial advancements, particularly in automating and enhancing efficiency, leading to the advent of Direct Preference Optimization (DPO).

Originally, RLHF relied heavily on SFT, where a model was fine-tuned on a dataset of human-labeled examples to learn a desired behavior. However, as the field evolved, researchers sought ways to automate and refine this process, reducing the reliance on extensive human-labeled datasets, which are resource-intensive and potentially biased.

One significant development was the replacement of SFT with outputs from other, more advanced models. This approach leverages the knowledge encapsulated in these pretrained models, transferring it to the model being trained. For instance, using the outputs of a larger or more specialized language model as a target for fine-tuning can provide a richer, more nuanced understanding of complex tasks without direct human annotation.

The reward model step in RLHF, where a model is trained to predict the reward (or quality) of an output based on human judgments, has also seen remarkable automation advancements. Instead of manually labeling data to train these reward models, researchers have started using other Large Language Models (LLMs) to automate the generation of reward signals. These LLMs can assess the quality of outputs based on criteria learned from vast datasets, effectively acting as proxies for human judgment. This automation not only streamlines the training process but also scales it, enabling more frequent updates and refinements to the reward models.

Recently, in a paper titled "Direct Preference Optimization: Your Language Model is Secretly a Reward Model" by Rafailov et al. (2023),[33] a team of Stanford Researchers came up with a pivotal advancement. Direct Preference Optimization (DPO) directly optimizes the model's parameters to align with human preferences, bypassing the need for a separate reward model. This approach integrates preference modeling directly into the training process, making the optimization more efficient and potentially more aligned with human values.

DPO fundamentally relies on comparing pairs of model-generated outputs and adjusting the model parameters to increase the probability of generating outputs that align with human preferences. A key technical derivation in DPO is the use of preference-based gradients, where the gradient of the preference score with respect to the model parameters is computed. This allows for direct optimization of the model in a direction that is more likely to produce preferred outcomes.

The benefits of DPO over traditional RLHF and its components like SFT and separate reward models are manifold. First, DPO streamlines the training process by integrating preference learning directly into model optimization, eliminating the need for separate reward model training. This integration can lead to more efficient use of computational resources and faster iteration cycles.

Secondly, by optimizing directly for human preferences, DPO can more closely align model outputs with what is considered beneficial or desirable by humans, potentially reducing the risk of misaligned incentives. Moreover, DPO allows for continuous improvement and adaptation of models as new preference data becomes available, maintaining the model's relevance and effectiveness over time.

Fine-tuning LLMs is a very active area of research. Readers are well advised to search on the Internet and blogs on this topic to stay in touch with the latest advances.

Libraries and Frameworks for RLHF

Some popular frameworks, libraries, and cloud services providing RLHF fine-tuning are listed in this section. This is also a rapidly evolving ecosystem with almost something new every week. Therefore, use the following list as a starting reference only.

---

[33] https://arxiv.org/abs/2305.18290

VertexAI from Google

VertexAI simplifies and streamlines the process of fine-tuning Large Language Models (LLMs) using Reinforcement Learning from Human Feedback (RLHF). With a user-friendly interface and pre-built containers, VertexAI eliminates the need for complex local setups, allowing you to focus on your data and task. It leverages Google Cloud's powerful compute resources to train LLM on large datasets efficiently. It manages the entire RLHF fine-tuning process. You can upload your data, define your fine-tuning objective, and track progress through a central dashboard. VertexAI, with the help of training pipelines, automates training, allowing you to focus on data quality and human feedback loops.

You pay only for the resources you use, and you can share your project and collaborate with team members in a secure environment, facilitating smoother development and iteration.

SageMaker from AWS Using Trlx

AWS provides RLHF capabilities in SageMaker using an open-source library named `trlX`. The `trlX` library stands out as a tool for fine-tuning Large Language Models (LLMs) with Reinforcement Learning from Human Feedback (RLHF). Unlike generic frameworks, `trlX` focuses specifically on the unique needs of RLHF, simplifying and automating critical tasks like reward model training, policy optimization, and distributed training. It also easily integrates with popular LLM libraries like HuggingFace Transformers, eliminating the need for extensive coding from scratch. Further leveraging the power of distributed training across multiple GPUs or TPUs allows for efficient handling of large datasets and models. `trlX` has an API capable of production-ready RLHF with PPO and Implicit Language Q-Learning ILQL at the scales required for LLM deployment (e.g., 33 billion parameters).

TRL Library from HuggingFace

TRL from HuggingFace is a full stack library that provides a set of tools to train Transformer language models with reinforcement learning, from the Supervised Fine-tuning step (SFT), the Reward Modeling step (RM), to the Proximal Policy Optimization (PPO) step. The library is integrated with HuggingFace Transformers library. `trlX` is a fork of the TRL library.

Walkthrough of RLHF Tuning

This section explains a code example that tunes an LLM using RLHF. As LLM fine-tuning requires a GPU with large configuration and may take hours to train, you can just walk through the code without executing it. If you have access to a GPU either using Google Colab or any other cloud GPU provider, you can try to run the code.

An example notebook that comes with the TRL library from HuggingFace is used here. You can find more about it in the `examples` section of the TRL library documentation.[34] This section uses the example contained in the `gpt2-sentiment.ipynb` notebook.[35] I also copied the notebook over to the code accompanying this book as `11.b-gpt2-sentiment.ipynb`.

In this notebook, you fine-tune GPT-2 (small) to generate positive movie reviews based on the IMDB dataset. The model gets the start of a real review and is tasked to produce positive continuations. To reward positive continuations, you use a BERT classifier to analyze the sentiment of the produced sentences and use the classifier's outputs as reward signals for PPO training.

As you may have noticed from the description of the notebook, you do not train your own reward model. Rather, you use a pretrained classifier that classifies a review and outputs the logits for the negative and positive class. You will use the logits for a positive class as a reward signal for the language model.

In the notebook, you install the required libraries, import the modules you will be using, and log in to Weights and Biases to record the run statistics of the experiment. This is shown in Listing 11-11.

Listing 11-11. Install and Import Step from gpt2-sentiment.ipynb

```
%pip install transformers trl wandb
import torch
from tqdm import tqdm
import pandas as pd
tqdm.pandas()
from transformers import pipeline, AutoTokenizer
```

---

[34] https://huggingface.co/docs/trl/index
[35] https://github.com/huggingface/trl/blob/main/examples/notebooks/gpt2-sentiment.ipynb

```
from datasets import load_dataset
from trl import PPOTrainer, PPOConfig, AutoModelForCausalLMWithValueHead
from trl.core import LengthSampler

import wandb
wandb.init()
```

You can also create a configuration for the PPO trainer, as shown in Listing 11-12.

Listing 11-12. PPO Configuration from gpt2-sentiment.ipynb

```
config = PPOConfig(
    model_name="lvwerra/gpt2-imdb",
    learning_rate=1.41e-5,
    log_with="wandb",
)

sent_kwargs = {"return_all_scores": True, "function_to_apply": "none",
"batch_size": 16}
```

In this notebook, you do not carry out the SFT step of RLHF. Using the HuggingFace (hf) Transformers library and the HF model-hub, you load a GPT-2 model that was additionally trained on IMDB dataset for one epoch. The IMDB dataset contains 50,000 movie reviews annotated with "positive/negative" feedback indicating the sentiment. The IMBD dataset is a dataset containing movie reviews. As can be seen in Listing 11-12, PPOConfig initialization takes in model_name="lvwerra/gpt2-imdb" hosted on the HF model hub.[36] The details of fine-tuning the dataset and script can be found at https://huggingface.co/lvwerra/gpt2-imdb.

To do RLHF fine-tuning, you again use the IMBD dataset. During RLHF training, data is passed through a pretrained BERT-based reward model. These rewards are used to fine-tune gpt2-imdb using PPO. Listing 11-13 shows the code for preparing the IMBD dataset for training. You first load the tokenizer for GPT-2 to tokenize the text before passing it through the GPT-2 LLM. You download the train split of the IMBD dataset, rename the text column to review, and filter out the reviews that are smaller than 200 characters. Next, for each review in the dataset, you sample a random number of tokens

---

[36] https://huggingface.co/lvwerra/gpt2-imdb

ranging from 2 to 8 from the beginning of the review, and then store the tokenized version in a column named input_ids and the decoded text version in a column named query. You also retain all the previous columns in the dataset, namely review and label. You can explore the HuggingFace dataset at https://huggingface.co/datasets/imdb. Define a collator function to collate the data in a format appropriate for batch training in PPO. If some of these concepts are not clear, refer to the HuggingFace documentation.

Listing 11-13. Dataset Preparation for RLHF from gpt2-sentiment.ipynb

```
def build_dataset(config, dataset_name="imdb",
    input_min_text_length=2, input_max_text_length=8):

    tokenizer = AutoTokenizer.from_pretrained(config.model_name)
    tokenizer.pad_token = tokenizer.eos_token
    # load imdb with datasets
    ds = load_dataset(dataset_name, split="train")
    ds = ds.rename_columns({"text": "review"})
    ds = ds.filter(lambda x: len(x["review"]) > 200, batched=False)

    input_size = LengthSampler(input_min_text_length, input_max_text_length)

    def tokenize(sample):
        sample["input_ids"] = tokenizer.encode(sample["review"])[: input_size()]
        sample["query"] = tokenizer.decode(sample["input_ids"])
        return sample

    ds = ds.map(tokenize, batched=False)
    ds.set_format(type="torch")
    return ds

dataset = build_dataset(config)
def collator(data):
    return dict((key, [d[key] for d in data]) for key in data[0])
```

Now you are ready to carry out the third step in RLHF, that of PPO training. You load the gpt2-imbd model and make two copies—a reference model that will not change during PPO training and another one that will be fine-tuned under PPO. Next, you initialize the PPO trainer using the PPOConfiguration, the two GPT2-imdb models, and the dataset with the collator function, as shown in Listing 11-14. You also load the pretrained reward model, `distilbert-imdb` classifier from HF model hub, and prepare the sentiment pipeline. As discussed earlier, the `distilbert-imdb` model outputs the logits for the negative and positive class and you use the logits for a positive class as a reward signal for the language model.

Listing 11-14. Model Loading Including Reward Model and PPOtrainer Initialization from gpt2-sentiment.ipynb

```
# Load pretrained GPT2 language models
model = AutoModelForCausalLMWithValueHead.from_pretrained(config.
model_name)
ref_model = AutoModelForCausalLMWithValueHead.from_pretrained(config.
model_name)
tokenizer = AutoTokenizer.from_pretrained(config.model_name)
tokenizer.pad_token = tokenizer.eos_token

# Initialize PPO trainer
ppo_trainer = PPOTrainer(config, model, ref_model,
                         tokenizer, dataset=dataset, data_collator=collator)

#Load Bert Classifier
device = ppo_trainer.accelerator.device
if ppo_trainer.accelerator.num_processes == 1:
    device = 0 if torch.cuda.is_available() else "cpu"
    sentiment_pipe = pipeline("sentiment-analysis", model="lvwerra/
    distilbert-imdb", device=device)
```

You are now ready to train the model. Training consists of a loop with the following main steps:

1. Get the query responses from the policy network (GPT-2).

2. Get sentiments for query/responses from BERT.

3. Optimize policy with PPO using the (`query, response, reward`) triplet.

Listing 11-15 shows the complete code of training loop.

Listing 11-15. Training Loop from gpt2-sentiment.ipynb

```
output_min_length = 4
output_max_length = 16
output_length_sampler = LengthSampler(output_min_length, output_max_length)

generation_kwargs = {
    "min_length": -1,
    "top_k": 0.0,
    "top_p": 1.0,
    "do_sample": True,
    "pad_token_id": tokenizer.eos_token_id,
}

for epoch, batch in tqdm(enumerate(ppo_trainer.dataloader)):
    query_tensors = batch["input_ids"]

    #### Get response from gpt2
    response_tensors = []
    for query in query_tensors:
        gen_len = output_length_sampler()
        generation_kwargs["max_new_tokens"] = gen_len
        response = ppo_trainer.generate(query, **generation_kwargs)
        response_tensors.append(response.squeeze()[-gen_len:])
    batch["response"] = [tokenizer.decode(r.squeeze()) for r in response_
    tensors]

    #### Compute sentiment score
    texts = [q + r for q, r in zip(batch["query"], batch["response"])]
    pipe_outputs = sentiment_pipe(texts, **sent_kwargs)
    rewards = [torch.tensor(output[1]["score"]) for output in pipe_outputs]

    #### Run PPO step
    stats = ppo_trainer.step(query_tensors, response_tensors, rewards)
    ppo_trainer.log_stats(stats, batch, rewards)
```

Since we have initialized Weights and Biases (wandb). We will be able to track the training progress and various other metrics on wandb.

Once the training is over, you can inspect the PPO trained model output and compare it to the output of the `reference gpt2-imdb` model. The notebook contains the code to carry out the inspection. You can see that response (after) is generally better and tilts toward a positive tone, as compared to the reference model. A sample comparison is shown in Figure 11-13. For a detailed walkthrough, refer to the `gpt2-sentiment.ipynb` notebook.

	query	response (before)	response (after)	rewards (before)	rewards (after)
0	Oh dear,	what are I saying?! I fast-forwarded through	I must say that I are hanging my head on this	-0.858954	-1.007609
1	I've seen	it, as well.<br	three million dialogue throughout, and	1.996807	2.240883
2	Hi: This movie is a turkey though when it comes to	/>I also like that movie. It's so funny	-0.438191	2.415630
3	I'm a writer	and I'm not going to be asked to	, not a screenwriter. I've written	-0.655991	-0.724324
4	If you	absolutely love sensitive romance, the plot a...	are looking at the cinematography, the acting,	2.221309	0.148751
5	OMG this	casting cast. Obi cult breezy, this is	movie was totally wonderful, I it was the ide...	-1.533139	2.590190
6	It's	unrealistic; the guy who was supposed to be E...	a very good film. It reminds us about over	-2.097017	2.835831
7	There is a really	awful laptop game! I used to	interesting story that set us the journey. Th...	-2.341743	2.282939
8	This is	my favorite part about	a well thought well	2.554794	2.734139
9	Wasn't	Wasn't it clichéd?<|endoftext|>	anyone else interested in this movie? It's a ...	-1.790802	2.631960
10	This film is another of director Tim	Burton's masterpieces	Curry's best bombs	2.622917	2.544106
11	I thought this movie	was excellent. I actually laughed 6 times and...	was perfect, and I believe it's almost overlo...	2.548022	2.601913
12	This early John Wayne	films looked like an abandoned police beating	film is a realistic portrayal of what	-1.742279	2.609762
13	I was	given an experience-a big one, almost 25	very happy with all the reflections and this ...	2.250709	2.558540
14	Embarrassingly, I	am more at a strict conformity after getting ...	had never seen a movie before. There was one ...	-2.021666	-1.803383
15	I am a fan	of living on simple islands, and we have visi...	of many things and learned how to appreciate ...	1.791297	2.324461

Figure 11-13. *Three steps of RLHF[37]*

You can see that rewards on the PPO trained models are higher than the rewards for the same sentences using the reference `gpt2-imdb` model.

This was a toy example of how RLHF works and even on this toy dataset, it can take close to two to three hours. It should make you appreciate that even a fine-tuning of LLM is an expensive process when done on state-of-the-art LLM with large datasets. Almost all models today go through something similar to make the LLMs follow instructions, even for code analysis and code completion like the GitHub AutoPilot.

This completes this chapter, which was a PPO deep dive and its application in RLHF fine-tuning for LLMs.

---

[37] Figure 2 from the RLHF paper - `https://arxiv.org/pdf/2203.02155.pdf`

Summary

This chapter provided a comprehensive overview of the Proximal Policy Optimization (PPO) algorithm, a state-of-the-art policy gradient method for reinforcement learning. The chapter started by reviewing the foundations of policy gradient methods, such as the policy objective function, the importance sampling ratio, the trust region, and the natural gradient.

The chapter then derived the PPO algorithm step by step, explaining how it uses a clipped surrogate objective function to control the policy update size and avoid large deviations from the current policy. The chapter also discussed the practical implementation details of PPO, such as the vectorized environment, the parameter initialization, the Adam optimizer, the generalized advantage estimation, the mini-batch updates, the normalization of advantages, the value function loss clipping, the global gradient clipping, and the debug variables.

Next, the concept of Reinforcement Learning from Human Feedback (RLHF) was introduced. RLHF is a technique to fine-tune Large Language Models (LLMs) using human preferences as a reward signal. There was a brief overview of the Transformer architecture and the GPT series of LLMs, which are generative models that can learn from massive amounts of unlabeled text data and generate high-quality text across different domains and tasks. The chapter also discussed topics like prompt engineering, fine-tuning and PEFT, chaining LLMs together, or even making multiple LLMs work together in an autonomous way, almost like an operating system.

That was followed by a discussion of RLHF and how it can be applied to train LLMs to follow natural language instructions and generate outputs that align with human expectations. The chapter discussed the three main steps of RLHF: supervised fine-tuning, reward modeling, and policy optimization. The chapter also described the latest advances in LLM alignment, such as Direct Preference Optimization (DPO), which bypasses the need for a separate reward model and directly optimizes the model parameters to match human preferences.

The chapter concluded with a walkthrough of RLHF tuning using the TRL library from HuggingFace, which provides a set of tools to train transformer language models with RLHF. The chapter showed you how to use the TRL library to fine-tune a GPT-2 model to generate positive movie reviews based on the IMDB dataset. The discussion also covers the use of a pretrained BERT classifier as a reward model to score the model's outputs based on their sentiment, and how to use PPO to optimize the model's parameters using the reward signals. It concluded with an inspection of the model's outputs and a comparison of the reference model outputs.

## CHAPTER 12

Multi-Agent RL (MARL)

The book so far has covered most of the popular RL approaches, including the state-of-the-art PPO with its application in Large Language Models for RLHF fine-tuning. You may have noticed that the focus has always been on only one agent in the environment that learns to act optimally using RL training algorithms. However, there is a whole range of settings with more than one agent. These agents in the environment—either individually or in a collaborative manner—try to achieve some goal. A setup involving multiple agents in the same environment is the focus of this chapter. It is called Multi-Agent RL (MARL). MARL is a very fascinating and vast topic. To do proper justice to the topic would require a complete book of its own. This chapter introduces the key topics with simple examples and finally concludes the chapter with an example of applying learning in MARL setup for a simple environment. I will introduce various terms and concepts and make some assertions without getting into depth. The key purpose is to introduce MARL with the expectation that readers interested in MARL will refer to other detailed MARL-related books.

In environments characterized by the presence of multiple decision-makers, agents are required to not only understand and adapt to the dynamics of the environment but also to predict and react to the actions of other agents. This capability is critical for any system where cooperation, competition, or a mixture of both is necessary for achieving optimal outcomes. A schematic of MARL is shown in Figure 12-1.

© Nimish Sanghi 2024
N. Sanghi, *Deep Reinforcement Learning with Python*, https://doi.org/10.1007/979-8-8688-0273-7_12

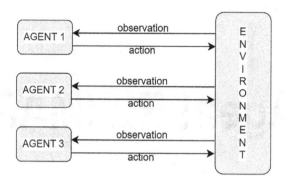

Figure 12-1. *Schematic of Multi-Agent RL (MARL)*

Let's look at some examples of MARL in different domains:

1. *Autonomous driving*: MARL is used to simulate and train autonomous vehicles to navigate complex traffic scenarios, where each vehicle (agent) must consider the actions of others to avoid collisions and optimize traffic flow.

2. *Robotics*: In collaborative robotics, MARL enables multiple robots to work together on tasks such as assembly lines or search-and-rescue missions, coordinating their actions for efficiency and effectiveness.

3. *Smart grid management*: MARL can optimize energy distribution and consumption in smart grids, with each agent representing a consumer, producer, or storage system, working together to balance supply and demand.

4. *E-commerce*: Online platforms use MARL to model interactions between buyers, sellers, and recommendation systems, optimizing for user satisfaction and revenue.

5. *Multiplayer online games*: Game developers use MARL to create more intelligent and adaptable non-player characters (NPCs) that can interact with human players or other NPCs in complex game environments.

6. *Financial markets*: MARL models the behavior of traders in financial markets, helping to understand market dynamics and develop strategies for trading and investment.

The next question that comes to mind is how do these agents learn in a setup like this? It is actually very similar to the RL you have been seeing in the book so far. Like the single-agent case, the agents learn their behaviors by trying different actions and receiving rewards, or returns, for their outcomes. Figure 12-2 illustrates the basic MARL training loop. Each of the agents picks an individual action, and the combination of these actions is called *joint action*. The joint action affects the state of the environment according to the environment dynamics, and the agents get individual rewards from this change, as well as individual observations about the new environment state. This loop repeats until a stopping criterion is met such as one agent losing a game or all agents collaboratively reaching an end goal, such as picking all the articles from the floor, or agents unloading all the items from a delivery truck and placing them properly on the warehouse shelves. A full run of this loop from the initial state to the final state is called an *episode*. The data generated from multiple separate episodes—that is, the observed observations, actions, and rewards in each episode—are used to constantly improve the agents' policy/reward. After the agent has been trained, in actual runs, the agents may still act independently, as shown in Figure 12-1, or jointly, as shown in Figure 12-2. In a subsequent section you will take a look at the various combinations of how to train and how to execute after training.

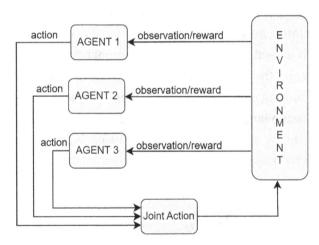

Figure 12-2. *Training loop in Multi-Agent RL (MARL)*

Look again at Figure 12-2. Could you still solve it using what you have learned so far for single agent setup? Theoretically, there is a way. Think of the three agents as a combined single agent. And think of combining individual rewards into a collective single reward. The same view can be extended to observations and actions as well.

With this approach you can recast a multi-agent setup to a single agent setup. Though theoretically possible, it has a few challenges when viewed as a single combined-agent RL. I briefly talk about these challenges.

Suppose each agent has a single dimension action with five discrete values and further suppose that you are combining three such agents together. What is the action space for the combined agent? It will be 5x5x5=125 actions. The same applies to state/observation as well. Therefore, using a single agent RL approach leads to exponential growth of the possible action space and state/observation space.

Another challenge is when agents act sequentially, like in a game of Chess or Go. Each agent takes a turn to act and the action depends on the action of other player. In a setup like this, the agents' actions are not independent. There is no easy way to look at it as a single agent RL setup.

The third challenge is with the combination of state/observations. Consider the case of a bunch of self-driving cars. If they are treated as a single agent RL, you need to combine the local observation of each car together into a big observation vector. For two cars which are, say, a few kilometers apart, the decision to steer each car will mostly depend on the local state of that car. The state/observation of the car far away will have no bearing (unless it involves a police chase!). However, by combining the states of both the car together, you are providing a whole lot of irrelevant information, thus making learning tougher or almost impossible.

Therefore, there is a need to have a separate approach and a separate set of algorithms that recognize the presence of multiple agents and their unique challenges and then carry out planning effectively. Let's talk about these key challenges. The point is that treating all agents as a large combined single agent RL entity has a set of challenges. However, depending on the abstraction used, treating each agent as an individual entity under MARL has a different set of challenges.

Key Challenges in MARL

Moving and changing targets: I mentioned this problem when I introduced RL. In RL, unlike supervised learning where the training data is given beforehand, new data/observations appear as agents learn a behavior and explore the environment. The training data is not fixed, so learning a target is like aiming at a moving target. You saw this in the TD(0) method of bootstrapping approach and how you treated the targets as constants. This problem is even bigger in MARL. Since there are multiple agents, each agent tries to adjust to the policies of the other agents and vice versa, in a cycle. This creates an unstable cyclic learning dynamic that requires MARL algorithms to deal with this non-stationary aspect in a more reliable way than single agent RL algorithms.

The best policy and the balance: In a single agent RL, the best policy means getting the highest expected returns in each state. But when there are multiple agents together, the return for each agent depends on the return for the others. So, what is the best result? Is it for one agent, all agents together or relative to what? How do you combine the returns of different agents? There may be more than one policy with the same total return but different individual returns. If the environment is cooperative or competitive, how does that affect the idea of the best? The whole idea of the best depends on the context and that needs to be abstracted in MARL.

Reward assignment: In RL, reward assignment, also known as temporal credit assignment in RL, is the challenge of determining which of the previous actions contributed to the reward agent receives at a given instance. In MARL, this challenge is even more complicated because it also requires identifying who performed the action that resulted in the reward. With only this state/action/reward information, it can be very hard to distinguish the influence of each agent on the received reward, especially if an agent did not contribute to the reward since its action was irrelevant. Although theoretical ideas based on counterfactual reasoning can address this problem, it is still an open problem how to do multi-agent reward assignment in an effective and scalable way.

Scaling for number of agents: In a MARL environment, as the number of agents grows, the number of combinations of agent actions also increase and could do so in an exponential way. The current state-of-the-art MARL approaches does not handle a large number of agents. It is an active area of research.

Apart from these aspects, there are a few other issues that arise in MARL. When you have more than one agent in the system, they may also communicate with each other. This communication usually helps in stabilizing the learning process; however, if the messages/information being communicated between agents contains irrelevant information or noise, it could impede the learning process. Another aspect is the robustness and stability of the learning process. With a higher number of agents interacting and affecting each other in multiple ways, it becomes tricker to ensure stability. MARL algorithms have to take special care of this aspect.

MARL Taxonomy

Continuing with introduction of MARL and its components, you'll look next at the hierarchy of MARL. Figure 12-3 shows one such possible classification. You have been mostly learning about Markov Decision Processes (MDP) introduced in Chapter 2 which involves single agent in the environment. It is based on the action and response of the environment to the action; the agent can move from one state to another. In Chapter 10, I introduced multi arm bandit, which is a special case of MDP with one agent and one state. After the agent takes an action, the agent gets a reward based on the action and then the agent state is reset to the beginning.

Just like multi arm bandit was the foundational building block of MDP, repeated normal form game is the foundational block of MARL. By the way, in this chapter, I use "environment" and "game" interchangeably. The word "game" comes from the discipline of "game theory" which is the foundational concept on which MARL models are built. Normal form game defines a single interaction between two or more agents. The way it works is that each agent i from the set of n agents, select an action a_i as per its policy π_i. The resulting actions from all the agents form a joint action, $a = (a_i, a_2, ..., a_n)$. Based on the joint action, each agent receives a reward r_i. You can consider this as one loop of the interaction, as shown in Figure 12-3.

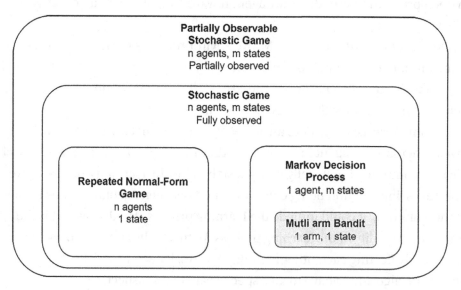

Figure 12-3. *MARL taxonomy*

Normal form games can be further sub-classified based on the type impacting the reward structure.

- *Zero-sum game:* The sum of agents' rewards is zero. In the case of a two-agent setup, the gain of one agent is a loss of the same amount for the other one.

- *Common-reward game:* All agents receive the same reward. Most cooperative behaviors would fall under this category.

- *General-sum game:* All other types that do not fit into the previous two categories.

Two-player normal form games are also known as *Matrix games* because the actions of one player can be represented as rows of a matrix and the actions of the second agent can be represented as columns of the matrix. The reward in this case is shown in the cells of the matrix as a tuple, with the first element of the tuple showing the reward of the first agent and the second element of the tuple showing the reward of the second agent. Do you remember about the game of rock-paper-scissors (RPS) discussed in Chapter 8? Figure 12-4 shows the reward and actions matrix form. It also shows a possible reward structure for a sample cooperative game as well as the famous prisoner's dilemma. Prisoner's Dilemma is a famous problem setup in game theory and can merit a complete book of its own. In cooperative/coordination, games have the same reward to all the agents, and you can have a single entry per cell, the common reward.

	R	P	S
R	(0,0)	(-1,1)	(1,-1)
P	(1,-1)	(0,0)	(-1,1)
S	(-1,1)	(1,-1)	(0,0)

Rock-Paper-Scissors

	X	Y
X	5	0
Y	0	5

Coordination Game

	silent	testify
silent	(-1,-1)	(-3,0)
testify	(0,-3)	(-2,-2)

Prisonner's Dilemma

Figure 12-4. *Example matrix games*

This is a *normal-form game. Repeated normal-form game* is nothing but repeating this in a loop similar to taking multiple repeated actions, such as with the multi arm bandit. When I talk about repeated actions, the time horizon also becomes important.

Games with finite time may demonstrate different behavior as the game progresses and gets close to the end. The agents may choose different actions closer to the end of the game versus the actions toward the start. With infinite time internal actions, after some stabilization over certain time periods, the agent actions will converge to the same distribution as kind of a steady state. As discussed earlier, with infinite interval games, you use the concept of γ, the discounting factor, to add the rewards in order to keep the sum of the rewards bounded.

Next in the hierarchy is the concept of *Stochastic games*. This setup extends the repeated-normal-form with a single state to multiple states, very similar to the extension of multi-arm-bandit to MDP. This is a full multi-agent system, defined as MARL. Just like a single agent MDP, you now have each agent acting based on current state. These actions are combined to get *joint action*. The joint action is passed over to the environment, which responds with reward and the next state for each individual agent. The agents then use the new state to reinitiate the action-reward-next state loop. Similar to MDP, stochastic games have a Markov property—that is, the probability of the next state and action for an agent being conditionally independent of the past states and joint action given the current states and joint action. It is the same Markov Property you saw for RL and a single agent. Now, it is extended from one agent to a multi-agent setup. As shown in Figure 12-3, repeated normal-form games are a special case of stochastic games with only one state. Further, the classification to zero-sum game, common-reward game or general-sum game carries over to Stochastic games as well.

Extending the generalization, you can move from fully observable stochastic games to partially observable stochastic games. While in stochastic games, the agents are able to directly observe the state of the environment, in *Partially Observable Stochastic Games* (POSG), the agent gets an observation, which is an incomplete information of the actual state. Usually for each agent i, an individual observation function O_i is defined which specifies the probabilities over the agent's possible observations o_i^t given the actual environment state s^t and joint action in previous time step a^{t-1}. This is represented as $O_i\left(o_i^t|a^{t-1},s^t\right)$. Accordingly, each agent's action/policy depends on the history of the observations now. This distinction of zero-sum games, common-reward games and general-sum reward games also extends to POSG. The common reward games are also known as decentralized POMDP. The observation function O_i can be used to present diverse situations. Consider the environment of robot soccer in which each robot may observe the full playing field but may not communicate its action to other robots especially players from rival team. Therefore, each robot has to guess the action of other robots. One example of this would be the ball-passes between robots of the same team based on the location and presence of the rival team robots. In some other cases the agents may only observe a subset of the state and joint actions. An example of this is self-driving cars. Observation functions can also be used to model uncertainly in observation by adding some random noise. This would be the case with most of the physical sensors that robots and other autonomous vehicles use to sense the environment.

Communication Between Agents

When you have more than one agent in the environment, the way they communicate also needs to be modeled. One way to model this is to consider messages as a type of action that do not impact the environment state. For an agent i, you can think of its action a_i as a composition of two types of actions—one is the environmental action x_i which impacts the state of the environment and message action m_i. m_i is usually a discrete set specifying the structure of information passed by agent i to others. With this modification, the complete action a_i of agent i is now represented as a tuple (x_i, m_i).

Like noise in observations, you could also introduce uncertainly in the communication and message passing between agents. For communication actions, the agents as part of learning need to develop an understanding of the message received by them. It could lead to an evolution of some abstract form of shared language between agents.

Mapping with Game Theory

The field of MARL borrows a lot of foundational setups from game theory as well as traditional single agent RL. The "environment" in RL is known as a *game*. I used the word "game" while defining the hierarchy in MARL.

In RL, the agent is an entity or a construct that makes decisions on how to act. In game theory, such an entity is called a *player*. What is called reward on RL, is called *payoffs* or *utility* in game theory. Policy defines the way agent acts in RL based on the information available. This is known as *strategy* in game theory, which defines the way agent/player acts based on the information available.

Solutions in MARL

What is a solution in MARL? A previous section introduced the basic environment setup for multi-agent scenarios—how they act, what ways they could pass messages, and so on. However, this on its own does not help in developing a learning process. You need one more thing, which is the objective that you are trying to optimize. In single RL, you used total discounted reward, also known as return, as a metric to be maximized. Similarly, you need to define an objective that you want to maximize in a MARL setup.

As there are multiple agents involved in MARL, the objective that you need to optimize has its own peculiarities. For common-reward games where all agents receive the same reward, a definition of a solution could be to maximize the expected return received by all agents. The concept of a solution becomes much more complex when agents receive separate rewards. Let's look at a few common ones.

Let's look at a setup of two agent zero-sum game of rock-paper-scissors that you saw in Figure 12-4. In a zero-sum game, the reward of one agent is the negative of the other agent's return. Many two-player board games such as chess and go also fit into this category of zero-sum games with sequential moves. The solution applicable in this case is known as *minimax*. Each agent learns a policy, which is maximized against the worst-case opponent—that is, an opponent which highest level of expertise. It is also called the *best response policy,* where an agent responds to the opponent's move with a best response possible. In rock-paper-scissors, the best response policy is a uniform random action of choosing one of the three possible actions with equal probability. Such an approach gives the expected reward of zero for each of the two players/agents. Let's consider two agents i and j with policies π_i and π_j. The joint policy π is denoted as a tuple of individual agent policies: $\pi = (\pi_i, \pi_j)$. For a minimax joint policy, the following conditions must hold:

$$U_i(\pi) = \max_{\pi'_i} \min_{\pi'_j} U_i(\pi'_i, \pi'_j) \tag{12-1}$$

$$= \min_{\pi_j} \max_{\pi_i} U_i(\pi'_i, \pi'_j) \tag{12-2}$$

$$U_i(\pi) = -U_j(\pi) \tag{12-3}$$

There could be more than one minimax solution to a zero-sum game; however the utility or reward, also known as minimax value of the game, remains the same under these different solutions. For non-repeated zero-sum normal-from games, the minimax solution can be obtained via linear programming. Interested readers can refer to standard texts on MARL to understand this approach.

The next type of objective is the *Nash equilibrium*, which is applicable with a general-sum game with two or more players. It is a well-studied concept in game theory. In Nash equilibrium, an agent i cannot improve its expected return by changing its policy. If the policy of other agents changes, the agent i can find another fixed point which is the new Nash equilibrium and is the best response to the current policies of all the other agents. In a general-sum game with n agents with joint policy $\pi = (\pi_1, ..., \pi_n)$ is considered to be in Nash equilibrium if:

$$\forall i, \pi_i' : U_i\left(\pi_i', \pi_{-i}\right) \leq U_i\left(\pi\right) \tag{12-4}$$

Where π_i' is a policy for agent i and π_{-i} is the policy of all other agents except the agent i. Equation 12-4 says that if a policy π is a Nash equilibrium with agent i policy as π_i then π_i is better than any other policy π_i' for the agent i considering that other agents' policies π_{-i} remaining constant.

The concept of Nash equilibrium can also be applied to zero-sum games, as they are a special case of general-sum games. In rock-paper-scissors, the only Nash equilibrium is choosing actions with uniform random probability. In some games, the equilibrium could be unique or there could be multiple Nash equilibrium. In the prisoner's dilemma shown in Figure 12-5, the Nash equilibrium is a unique and deterministic one where for each agent there is a single best action, such as "testify". Such a deterministic policy by each agent satisfies Equation 12-4. If the column prisoner chooses to stay silent, the row prisoner is better off by choosing to testify. They will get a reward of 0 under action "testify" vs a reward of -1 if they stay silent. Similarly, if the column prisoner chooses to testify, the column prisoner is again better off by testifying since that will give the row prisoner a reward of -2 vs a reward of -3 if the row prisoner chooses to remain silent. The same logic can be applied from the perspective of the column prisoner, wherein the policy to testify is the best for them as well. Therefore, there is a single deterministic joint policy of both prisoners choosing the testify action.

	silent	testify
silent	(-1,-1)	(-3,0)
testify	(0,-3)	(-2,-2)

Prisoner's Dilemma

Figure 12-5. *Prisoner's dilemma*

However, in the case of rock-paper-scissors, there is no unique Nash equilibrium. Rather it is a probabilistic one where each agent chooses to act in a uniform random way, choosing any of the three actions (rock, paper, scissors) with equal probability.

A further generalization of the Nash equilibrium policy is by allowing for correlated policy of agents where each agent's policy is no longer independent of the others. In cases like these, there may be many policies that achieve the same expected joint rewards. In situations like these, you extend the concept of equilibrium with additional criteria such as Pareto optimality and social welfare and fairness to differentiate between different equilibrium solutions.

A joint policy is *Pareto-optimal* if there is no alternative joint policy that increases the expected return for one or more agents, without decreasing the expected return for at least one agent. This means that none of the agents can gain without causing losses to other agents. In games where the reward is shared by all players, any joint policy that is Pareto-optimal has the same expected value, which is also the highest expected value that any joint policy can attain in the game.

Pareto-optimal solutions, however, make no statement about the relative distribution of joint reward among the various agents. Welfare indicates the overall number of benefits all the agents get, while fairness indicates how the benefits are divided among the agents. In common-reward games or zero-sum games there is no notion of social welfare of fairness. As an example, in common-reward games, where all agents receive the same reward, welfare and fairness are maximized when the individual agents' expected returns are maximized. In a zero-sum game, one agent's reward is a negative of the reward of the other agent. The total expected joint return is zero and therefore for zero-sum games, the minimax solution achieves the total social welfare of zero.

MARL and Core Algorithms

Having introduced MARL concepts, this section explores the core learning algorithms for MARL. At an abstract level, they mostly follow the approach you have seen so far for one-agent RL, but with some modifications. This section briefly covers algorithms, mostly in finite space of actions and states.

Value Iteration

Value iteration for one agent was introduced in Chapter 3. Figure 12-6 reproduces the pseudocode for the value iteration algorithm from Chapter 3. There are a few minor changes that you need to make to extend this algorithm to MARL scenario. The step where you calculate $v'(s)$ needs to change—that is, $v'(s) \leftarrow \max_a \sum_{s',r} p(s',r|s,a)\left[r + \gamma v(s')\right]$ is the one that you need to change. There are two changes you need to make to this step. First you calculate the right-side expression for each agent i and store it in an intermediate action-value vector $M_{i,s}$ as shown in Equation 12-5.

$$M_{i,s}(a) \leftarrow \sum_{s',r} p(s',r|s,a)\left[r_i + \gamma v_i(s')\right], \text{ for state } s \in S, \text{ joint action } a \in A \qquad (12\text{-}5)$$

The next change is to replace "max" in the original equation with a value operator v_i for each agent i, which is obtained by computing minimax solution for agent i over all the other agents in the environment. It can be represented as $v_i(s) = Value_i(M_{1,s}, ..., M_{n,s})$. The rest of the pseudocode remains same. Note that value iteration for MARL requires access to the underlying model dynamics, similar to the requirements for single-agent RL.

VALUE ITERATION

Initialize state values $v(s)$ (Terminal states are always initialized to 0)

e.g. $v(s) = 0, \forall s \in S$

define convergence threshold θ

Make a copy: $v'(s) \leftarrow v(s)$ for all s

Loop:
$\quad \Delta = 0$
$\quad$ Loop for each $s \in S$
$$v'(s) \leftarrow \max_a \sum_{s',r} p(s',r|s,a)[r + \gamma v(s')]$$
$$\Delta = \max(\Delta, |v(s) - v'(s)|)$$
$\quad\quad v(s) \leftarrow v'(s)$ for all $s \in S$, i.e. make a copy of $v(s)$
Until $\Delta < \theta$

Output a deterministic policy, breaking ties deterministically.

Initialize $\pi(s)$, an array of length $|S|$

Loop for each $s \in S$
$$\pi(s) \leftarrow \arg\max_a \sum_{s',r} p(s',r|s,a)[r + \gamma v(s')]$$

Figure 12-6. *Value Iteration for single-agent RL*

You will see this pattern repeated, where the base construct used for MARL will follow the approaches similar to the ones for single agent RL. In MARL, state value and action value calculation are followed by additional post-processing steps like minimax and Nash equilibrium to calculate the updates.

TD Approach with Joint Action Learning

Chapter 4 explained the model-free approach for single-agent RL. This section explores how similar model-free approaches can be applied to MARL. One naïve approach is to have each agent run TD(0) at each step. In other words, if an environment has n agents, you run n separate TD(0) updates at each step, essentially treating each of the n agents as an independent agent without worrying about the actions of other agents. This approach is also known as *independent Q-learning*. This approach introduces the issues of changing targets due to the actions of other agents, reward assignments, and the notion of what is a best policy, which I talked about in the section on "Challenges in MARL..

Joint action learning (JAL) is a TD-learning approach in a TD model-free world, where the algorithm estimates the expected return of joint actions and uses that to train the agents. Like the one agent Q-learning TD approach these are off-policy algorithms. One key difference is the involvement of multiple agents, which brings the notion of what is optimal equilibrium solution. As compared to single agent Q values, just learning the Q values of each agent in a given state and action—that is, $Q_i(s, a_1, ..., a_n)$, is not sufficient. Unlike single agent Q-value, you cannot take a simple max over Q values of agent i—that is, $max_{a_i} Q_i(s, a_1, ..., a_n)$, in a given state s and joint action $a = (a_1, ..., a_n)$ to get the best Q value for that given state. The max for an agent i depends on the actions of the other agents in the environment. You need to bring in the concept of equilibrium based on how agents are interacting—are they competing or cooperating, what's the relative reward structure, are they all getting same rewards, or do they vary, and so on. You read about a few options in the section on solutions in MARL. I talked about *minimax* in a zero-sum game, as well as the *Nash equilibrium* and *Correlated-equilibrium* solution objectives. Depending on the solution strategy from game theory, JAL learning has variations on how this step toward optimal equilibrium solution is taken. These family of JALs with game theory solution objectives are commonly referred to as JAL-GT family.

The high-level approach in solving these kinds of problems is to first compute Q values for each agent at a given state and all possible joint actions. Consider that you have two agents each with three actions. The Q values will be:

$$Q_i\left(s, a_{i,1}, a_{j,1}\right), Q_i\left(s, a_{i,1}, a_{j,2}\right), ..., Q_i\left(s, a_{i,3}, a_{j,3}\right),\ \ i \in \{1, 2\}$$

With each of the two agents have three actions each, the total possible joint actions will be 3x3 = 9. Therefore, there will be 9x2 = 18 Q values with nine Q-values for the first agent and nine Q-values for the second agent.

The second step is to use the minimax or Nash equilibrium solution approach to update the policy for both the agents.

The third step is to generate a new transition using the updated policy with the usual ε-greedy policy, execute the join action in the environment, and finally observe the reward and transition to a new state. This loop is repeated: Q-value calculation, update of policy, and taking an ε-greedy joint step.

As you can see, the process is very similar to tabular Q-learning in Chapter 4. The only difference is the way max is replaced with either a minimax or Nash equilibrium-based policy update. The core structure of Q-learning remains the same as the single-agent RL approach, with replacement of a simple max with a game theory driven policy update.

I will briefly describe the three Q-learning approaches using minimax, Nash equilibrium, and correlated equilibrium, the types of game settings each of these can be used for, and some key theoretical findings.

Minimax Q-Learning

Minimax Q-learning solves Q-learning using a minimax approach and it is carried out via linear programming. It can be applied to two agent zero-sum games. Like the usual assumption of single-agent Q-learning, minimax Q-learning is guaranteed to find an optimal solution under the assumption that all state and joint action combinations are tried a sufficient number of times. One common challenge with the minimax approach is that it assumes the opponent to be an expert optimal agent who is acting to defeat the other player. If the opponent is not optimal and has a weakness, the minimax solution does not allow the other agent to exploit this weakness in the opponent's strategy.

Nash Q-Learning

Nash Q-learning solves the multi-agent Q-learning by computing a Nash equilibrium and using that to update the policy. This algorithm can be applied to a more general class of settings with a finite number of agents. Though, the more agents there are, the higher the complexity in calculating the Nash equilibrium at each step. Nash Q-learning is guaranteed to learn a Nash equilibrium of the game provided all states have a global optimum across all the agents. This condition is very restrictive and not usually present in most practical games.

Correlated Q-Learning

Just like the other two, correlated Q-learning solves the policy update step by finding a correlated equilibrium. This approach has two key benefits—it can be applied to a wider space of games than what is feasible with the Nash equilibrium and it can be computed using a linear programming approach, while the Nash equilibrium requires quadratic programming.

Assumptions on Agents

In JAL-GT agents, we make assumptions on the agents depending on the solution approach of minimax, Nash, or correlated equilibrium. But what if an agent deviates from these assumptions? As an example, in minimax, we make the assumption that each opponent is an optimal worst-case agent. Accordingly, the policy that's learned is the best action against such an optimal agent. If the opponent is not optimal, a player will not learn to exploit the weakness of its opponent. There is another way to go about this. Instead of making assumptions about the behavior of the opponent, you learn the behavior based on observations. This is known as *agent modeling* and a common approach is policy reconstruction.

Casted as a supervised problem, you collect the pair of opponent's state and actions to model the policy opponent is following. You start with an assumption of uniform policy for the opponent and keep adjusting the assumed policy of opponent based on the observations of opponent's behavior each time it acts. This family of JAL algorithms is known as JAL-AM (JAL with Agent Modeling).

Policy-Based Learning

Just like policy-based algorithms for single-agent RL that you read about in Chapters 8 and 9, MARL also has a family of algorithms that are policy based. You learn a policy directly instead of learning the V or Q values and then use these values to find a policy. These are all based on gradient ascent to nudge the actions of all agents in the environment toward a higher expected return individually. Let's talk through an example. First, consider a scenario of two agents each with two actions. Look at a non-repeated normal form game, which means that there is only one state, and the episode ends after both the agents take an action. Let's write the reward matrix for the two agents as given in Equation 12-6. If agent i chooses action 2 and agent j choses action 1, the reward for agent i will be $r_{2,1}$ and for agent j it will be $u_{2,1}$.

$$R_i = \begin{bmatrix} r_{1,1} & r_{1,2} \\ r_{2,1} & r_{2,2} \end{bmatrix}, R_j = \begin{bmatrix} m_{1,1} & m_{1,2} \\ m_{2,1} & m_{2,2} \end{bmatrix} \tag{12-6}$$

As you are in a non-repeated normal-form game, you can define the policies of the two agents as follows:

$$\pi_i = (\alpha, 1-\alpha) \qquad \pi_j = (\beta, 1-\beta) \tag{12-7}$$

Where α represents the probability with which agent i chooses action 1 and $1 - \alpha$ as the probability that agent i chooses action 2. Similarly, β represents the probability with which agent j chooses action 1 and $1 - \beta$ as the probability that agent j chooses action 2. In this notation, the join action probability can be represented as $\pi = (\pi_i, \pi_j)$. Let's now write the expected reward for each agent as follows:

$$U_i(\alpha,\beta) = \alpha\beta r_{1,1} + \alpha(1-\beta)r_{1,2} + (1-\alpha)\beta r_{2,1} + (1-\alpha)(1-\beta)r_{2,2}$$

$$= \alpha\beta u + \alpha(r_{1,2} - r_{2,2}) + \beta(r_{2,1} - r_{2,2}) + r_{2,2}$$

$$\text{Where} \quad u = r_{1,1} - (r_{1,2} + r_{2,1}) + r_{2,2} \tag{12-8}$$

And,

$$U_j(\alpha,\beta) = \alpha\beta m_{1,1} + \alpha(1-\beta)m_{1,2} + (1-\alpha)\beta m_{2,1} + (1-\alpha)(1-\beta)m_{2,2}$$

$$= \alpha\beta u' + \alpha(m_{1,2} - m_{2,2}) + \beta(m_{2,1} - m_{2,2}) + m_{2,2}$$

$$\text{Where } u' = m_{1,1} - (m_{1,2} + m_{2,1}) + m_{2,2} \tag{12-9}$$

The gradient ascent approach will update both agents' policies to maximize the expected rewards of both the agents. Assume that you are at iteration k and the current joint policy is (α^k, β^k). Agent i and j will take a gradient ascent step to adjust their respective policies, which can be expressed as follows:

$$\alpha^{k+1} = \alpha^k + \eta \frac{\partial U_i(\alpha^k, \beta^k)}{\partial \alpha^k} \tag{12-10}$$

$$\beta^{k+1} = \beta^k + \eta \frac{\partial U_j(\alpha^k, \beta^k)}{\partial \beta^k} \tag{12-11}$$

Where,

$$\frac{\partial U_i(\alpha^k, \beta^k)}{\partial \alpha^k} = \beta u + (r_{1,2} - r_{2,2}) \tag{12-12}$$

And,

$$\frac{\partial U_j\left(\alpha^k, \beta^k\right)}{\partial \beta^k} = \alpha \ u' + \left(m_{2,1} - m_{2,2}\right) \tag{12-13}$$

You can combine Equations 12-12 and 12-13 into a vector form, as follows:

$$\begin{bmatrix} \dfrac{\partial U_i}{\partial \alpha} \\ \dfrac{\partial U_j}{\partial \beta} \end{bmatrix} = \begin{bmatrix} 0 & u \\ u' & 0 \end{bmatrix} \begin{bmatrix} \alpha \\ \beta \end{bmatrix} + \begin{bmatrix} \left(r_{1,2} - r_{2,2}\right) \\ \left(m_{2,1} - m_{2,2}\right) \end{bmatrix} \tag{12-14}$$

This policy learning algorithm with infinitesimal step size is referred to as infinitesimal gradient ascent (IGA). Making the left side equal zero in Equation 12-14 and solving for α, β will give you the fixed/center point $\alpha = \alpha^*$, $\beta = \beta^*$.

$$\left(\alpha^*, \beta^*\right) = \left(\frac{m_{2,2} - m_{2,1}}{u'}, \frac{r_{2,2} - r_{1,2}}{u}\right) \tag{12-15}$$

The trajectory and speed of convergence toward this fixed point depends on the relative reward matrices values shown in Equation 12-6. It can further be controlled with the help of learning rate η, as shown in Equations 12-10 and 12-11. Interested readers can refer to books on game theory and MARL for finer aspects. This completes the example of policy ascent for two agent two action non-repeated normal-form games.

The IGA can be generalized to more than two agents and two actions and is called Generalized IGA or *GIGA*. GIGA does not require knowledge of the other agents' policies and assumes that it can observe the past actions of the other agents.

No-Regret Learning

Up to this point, you have studied two approaches of learning that are MARL extensions of the single agent—that is, value-based methods and policy learning methods. There is another class of methods known as *no-regret learners*. I will briefly explain these methods now.

Let's first define the notion of "regret," which is not having chosen the best action in past episodes. You saw this while looking at the explore-exploit dilemma and various ways to balance the regret so that you explore enough but just enough to find the optimal action. Let's look at a concrete example. Figure 12-7 reproduces the prisoner's dilemma.

	silent	testify
silent	(-1,-1)	(-3,0)
testify	(0,-3)	(-2,-2)

Prisoner's Dilemma

Figure 12-7. *The prisoner's dilemma*

Let's look at a simulation of two prisoners choosing silent (S) or testify (T) based on some policy distribution for action selection that you do not know. Let's look at a series of ten episodes of the play and its outcome of reward for prisoner 1, as shown in Figure 12-8. Based on the current ten-episode run, Prisoner 1's reward over ten episodes equals -17. If prisoner 1 chose to always stay silent, the total reward over ten episodes based on the observed action of prisoner 2 would have come to -20. Similarly, if prisoner 1 chooses to always testify against the prisoner 2, the total reward for prisoner 1 over ten episodes is -10. Therefore, the "regret" for player 1 is -10+17=7. Averaging over ten episodes, the regret per episode equals 0.7.

Prisoner's Dilemma payoffs from 10 episodes

Episode →	1	2	3	4	5	6	7	8	9	10
Player 1 action	S	S	T	S	T	T	S	T	S	T
Player 2 action	S	T	T	T	S	T	T	S	S	S
Player 1 reward	-1	-3	-2	-3	0	-2	-3	-2	-1	0
Player 1 reward for always S	-1	-3	-3	-3	-1	-3	-3	-1	-1	-1
Player 1 reward for always T	0	-2	-2	-2	0	-2	-2	0	0	0

S = stay Silent   T = Testify against other prisoner

Figure 12-8. *Prisoner's dilemma payoffs after ten episodes*

Learning algorithms that aim to minimize this notion of regret are called no-regret learners. The learning approach revolves around regret of not having chosen a specific action in the past and updating their policies to assigning higher probability to actions with higher regrets in past. They are broadly classified into two categories: a) unconditional regret matching and b) conditional regret matching. As their names suggest, unconditional regret matching is based on calculating unconditional regrets and conditional regret matching is based on calculating conditional regret matching.

Deep MARL

Just like you extended finite state and action tabular RL algorithms to a richer set of environments with real valued state and action spaces using deep neural network based models, MARL can also be extended from the finite world to continuous valued high dimensional state and action space. However, the current deep MARL algorithms still work only for a finite small/moderate number of agents in an environment.

One naïve approach is to treat each agent independently and train it with the single-agent deep RL algorithms. However, these will encounter challenges such as non-stationary environments due to action of other agents, reward assignment, and the concept of what is an equilibrium or optimal action. Despite these challenges, it is a very common approach to extend the current RL approaches to MARL, like Independent DQN, Independent REINFORCE, and other independent policy gradient methods. Even A2C and PPO can be extended. Treating other agents as part of the environment favors on-policy learning approaches, such as REINFORCE, as it learns from most recent experiences, which reflect the latest behavior of other agents.

The design space of how to use deep networks as models of the state or environment and/or policy has many more knobs. The policy or state value network could share part of the network across all the agents with individual smaller top layers specific to each agent. There is also a possibility to share experience. One form of that in zero-sum games would be something called self-play wherein an agent plays with a copy of itself to collect experience and learn.

At a high level, a common way to categorize the Deep MARL (also known as Multi-Agent Deep RL) algorithms is based on the information available during training and inference/execution phases. There are four broad categories.

The first one is *Centralized Training and Execution (CTCE)* in which the learning process and agents' policy share some kind of centralized information. It essentially involves converting multi-agent RL to one large, combined single agent RL. The sharing could be in the form of observations across all agents, join policy action, and so on. When such a sharing is feasible or applicable, this approach can be used. However, it has certain issues such as conversion of individual agent reward to a combined scalar reward, exponential growth in joint-action when individual agent's actions are combined, or the infeasibility of such a communication, such as when sharing information across a fleet of self-driving cars. Even if information could be shared, the relevance of local observation of one agent may have very little information content and add more to the noise. The higher the signal-to-noise ratio, the tougher and longer the training could become.

The second class is *Decentralized Training and Execution (DTDE)*. One flavor of this that I discussed earlier is where each agent learns independently, treating all the other agents as part of the environment. I also discussed how such an approach, although scalable, could face significant challenges due to non-stationarity of environment because of other agents' action being part of the environment. Consider the extreme case of an agent taking no action at every time step, yet the environment/state it sees keeps changing at each time tick due to the actions of other agents. However, due to its scalability it is often used as a first step toward developing more complex and complete approaches to MARL.

The third class is *Centralized training and Decentralized Execution (CTDE)*. One way to look at this is that during training, shared information is used to update each agent's policies; however, the policy of each agent depends only on that agent's observation alone. There are a few other variations. CTDE is a very popular approach for Deep MARL. Remember the actor critic policy learning algorithm that you saw in Chapter 8? You could use a CTDE version of actor critic to train agents in MARL.

Consider the case of single agent A2C algorithm. The actor is the policy network. You use Critic to help train the agent by telling which action is better than the average and which one is worse than average. After the agent is trained and ready to be executed in product, you do not need the Critic any further. Therefore, one way to extend A2C to MARL is to use a centralized Critic and continue to have Actor, the policy, decentralized at each individual agent level. The Critic for each agent could have two components, one that is the history of observations of that agent in question along with some centralized information which is available to all agents during training time. It could be some abstracted centralized info or in extreme cases it could even be history of states/observations of all the agents in the environment.

There are many other ways to extend the current Deep RL algorithms to the field of MARL. While making such extensions, you must pay careful consideration to the concept of equilibrium. Deep MARL involves neural networks with large parameter sizes. Therefore, as the number of agents in the environment increases, you also have to think about the way the agent-specific network models could share some parts of the network to keep the scale of the problem manageable. The explore versus exploit dilemma also needs careful handling, especially if the usual single agent settings do not yield result.

This completes a high-level introduction of traditional and Dep MARL, the setup, the unique challenges, and how one-agent RL algorithms could be extended to MARL. Next, you will look at some common libraries that provide MARL environments and the training algorithms along with a sample application of agent training.

Petting Zoo Library

One of the challenges of MARL is finding suitable environments and algorithms to test and train the agents. Petting Zoo is a library that aims to provide a collection of high-quality MARL environments and easy-to-use interfaces for various RL frameworks. The library is developed by Farama, a research group that focuses on multi-agent reinforcement learning and its applications. Petting Zoo offers a diverse set of environments, ranging from classic games like chess and pong, to cooperative scenarios like predator-prey and pursuit-evasion, to social dilemmas like prisoner's dilemma. Petting Zoo also supports several popular RL libraries, such as Stable Baselines3, CleanRL, RLLib, and so on, allowing users to leverage existing algorithms and tools for MARL.

Most of the existing MARL environments are based on the Markov game model, which assumes that the agents act simultaneously and synchronously at each time step. However, this assumption does not capture the reality of many real-world scenarios, where the agents may act sequentially and asynchronously, such as in turn-based games, auctions, and negotiations. To address this limitation, Petting Zoo introduced a novel modeling framework called Agent Environment Cycle (AEC) for MARL environments.

This allows Petting Zoo to represent any type of game multi-agent RL can consider. In this model, agents sequentially see their observation, agents take actions, rewards are emitted from the other agents, and the next agent to act is chosen. This is effectively a sequentially stepping form of the POSG mode. A schematic is shown in Figure 12-9.

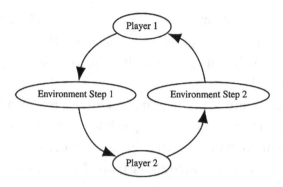

Figure 12-9. *AEC cycle*

The cycle repeats until the environment reaches a terminal state, or a predefined maximum number of steps is reached. An AEC environment can be seen as a generalization of the Markov game model, where the scheduling function can implement different types of agent interactions, such as simultaneous, sequential, or mixed. An AEC environment can also handle variable agent numbers, such as in games where agents can join or leave at any time. Furthermore, an AEC environment can model partial observability, stochasticity, and non-stationarity, which are common challenges in MARL.

Petting Zoo provides a standard API for interacting with AEC environments, which is similar to the OpenAI Gym API for single-agent RL environments. You can read more about it in paper "PettingZoo: A Standard API for Multi-Agent Reinforcement Learning."[1] Specifically refer to Appendix C3 for a detailed mathematical formulation of AEC.

You will use one of its environments named *Knights Archers Zombies* (KAZ), which I describe briefly. Knights Archers Zombies is part of the Butterfly environment. Butterfly environments are challenging scenarios created by Farama, using PyGame with visual Atari spaces. All environments require a high degree of coordination and require learning emergent behaviors to achieve an optimal policy. As such, these environments are currently very challenging to learn. Environments are highly configurable via arguments specified in their respective documentation.

In Knights Archers Zombies, Zombies walk from the top border of the screen down to the bottom border in unpredictable paths. The agents you control are knights and archers (by default, there are two knights and two archers) that are initially positioned at the bottom border of the screen. Each agent can rotate clockwise or counterclockwise

---

[1] https://arxiv.org/pdf/2009.14471.pdf

and move forward or backward. Each agent can also attack to kill zombies. When a knight attacks, it swings a mace in an arc in front of its current heading direction. When an archer attacks, it fires an arrow in a straight line in the direction of the archer's heading. The game ends when all agents die (collide with a zombie) or a zombie reaches the bottom screen border. A knight is rewarded with 1 point when its mace hits and kills a zombie. An archer is rewarded 1 point when one of their arrows hits and kills a zombie. There are two possible observation types for this environment, vectorized or image-based. This example uses the vectorized space.

Agents	agents= ['archer_0', 'archer_1', 'knight_0', 'knight_1']
Agents	4
Action Shape	(1,)
Action Values	[0,5]

You can pass the vector_state=True argument to the environment to get a vectorized environment. The observation is an (N+1)x5 array for each agent, where N = num_archers + num_knights + num_swords + max_arrows + max_zombies. And the 1 refers to the current agent. In total, there will be N+1 rows. Rows with no entities will be all 0, but the ordering of the entities will not change.

Vector breakdown: All distances are normalized to [0, 1]. Note that for positions, [0, 0] is the top-left corner of the image. Down is positive y and Left is positive x. The meaning of values in the vector for current agent are as follows:

- The first value means nothing and will always be 0.

- The next four values are the position and angle of the current agent.

- The first two values are position values, normalized to the width and height of the image, respectively.

- The final two values are heading of the agent represented as a unit vector.

For everything else other than the current agent, the respective vectors are each five wide and each row has a following breakdown:

- The first value is the absolute distance between an entity and the current agent.

- The next four values are the relative position and absolute angles of each entity relative to the current agent.

  - The first two values are position values relative to the current agent.

  - The final two values are the angle of the entity represented as a directional unit vector relative to the world.

Figure 12-10. *Knights Archers Zombies (KAZ)*

Sample Training

This section looks at a sample of running PPO training algorithm on *Knights Archers Zombies (KAZ).* You will leverage the PPO algorithm from SB3 library. The complete code to train the multiple agents in the KAZ environment is in the `12.a-ppo_ pettingzoo_kaz.ipynb` notebook. Listing 12-1 contains the training code. You first create the environment using `env_fn`, which is passed to the `train` function. Peculiar to the KAZ environment, where the agents die, you use the `ss.black_death_v3(env)` wrapper so that the number of agents stays constant. If you are using the image version of the KAZ observation space, you use a set of wrappers similar to the ones for the Atari environment in the previous chapters to reduce the image sizes, stacking of frames, and so on. For carrying out these wrapper processing, you use the library called SuperSuit as compared to Gymnasium and/or SB3 wrappers, which was used with the Atari games. If the observation space is a vectorized version, you do not need these preprocessing steps. Next, you'll create the PPO learning agent and run the training. The trained agent weights are then saved to file for future retrieval.

Listing 12-1. The train(...) function from 12.a-ppo_pettingzoo_kaz.ipynb

```
def train(env_fn, steps: int = 10_000, seed: int | None = 0, **env_kwargs):
    # Train a single model to play as each agent in an AEC environment
    env = env_fn.parallel_env(**env_kwargs)

    # Add black death wrapper so the number of agents stays constant
    # MarkovVectorEnv does not support environments with varying
    # numbers of active agents unless black_death is set to True
    env = ss.black_death_v3(env)

    # Pre-process using SuperSuit
    visual_observation = not env.unwrapped.vector_state
    if visual_observation:
        # If the observation space is visual, reduce the color channels,
        # resize from 512px to 84px, and apply frame stacking
        env = ss.color_reduction_v0(env, mode="B")
        env = ss.resize_v1(env, x_size=84, y_size=84)
        env = ss.frame_stack_v1(env, 3)

    env.reset(seed=seed)

    print(f"Starting training on {str(env.metadata['name'])}.")

    env = ss.pettingzoo_env_to_vec_env_v1(env)
    env = ss.concat_vec_envs_v1(env, 8, num_cpus=1, base_class="stable_
    baselines3")

    # Use a CNN policy if the observation space is visual
    model = PPO(
        CnnPolicy if visual_observation else MlpPolicy,
        env,
        verbose=3,
        batch_size=256,
    )
```

```
model.learn(total_timesteps=steps)
model.save(f"{env.unwrapped.metadata.get('name')}_{time.
strftime('%Y%m%d-%H%M%S')}")
print("Model has been saved.")
print(f"Finished training on {str(env.unwrapped.metadata['name'])}.")
env.close()
```

You also have a similar code to evaluate the performance of the agent. It is the usual observe/step/observe cycle using the training PPO agent and KAZ environment. The evaluation function gathers the rewards and prints a summary at the end of the evaluation.

The code to train and evaluate the code is shown in Listing 12-2. It is a simple three-step process—first train the agent, evaluate the metrics, and then show the trained agent in action.

Listing 12-2. The train and evaluate Calls from 12.a-ppo_pettingzoo_kaz.ipynb

```
env_fn = knights_archers_zombies_v10

# Set vector_state to false in order to use visual observations
# (significantly longer training time)
env_kwargs = dict(max_cycles=100, max_zombies=4, vector_state=True)

# Train a model
train(env_fn, steps=81_920, seed=0, **env_kwargs)

# Evaluate 10 games
eval(env_fn, num_games=10, render_mode=None, **env_kwargs)

# Watch 2 games
eval(env_fn, num_games=2, render_mode="human", **env_kwargs)
```

This concludes this chapter on MARL. It was a fast-paced introduction to MARL without getting into the details, as the idea was to make readers familiar with the additional complexities and challenges when multiple agents are involved in an environment. Interested readers may explore the MARL-specific literature and the Petting Zoo library documentation, which contains many other example environments and sample tutorials.

Summary

This chapter introduced the concept of MARL, where multiple agents interact within the same environment. MARL presents unique challenges, such as the need for agents to predict and react to the actions of other agents, and the exponential growth of the action and state space. Several examples of MARL in different domains were presented, including autonomous driving, robotics, and smart grid management.

The chapter also discussed key challenges in MARL, such as moving and changing targets, the definition of the best policy, reward assignment, and scaling for the number of agents. A taxonomy of MARL was presented, including normal-form games, stochastic games, and partially observable stochastic games.

The chapter also covered various solution approaches in MARL, including minimax, Nash equilibrium, and correlated equilibrium. Core algorithms for MARL, including value iteration, TD approach with joint action learning, and policy-based learning, were discussed. The chapter also introduced the concept of no-regret learning and deep MARL.

Finally, the chapter provided an example of training agents in a MARL environment using the Petting Zoo library and the PPO algorithm.

CHAPTER 13

Additional Topics and Recent Advances

This is the last chapter of the book. Throughout the book, I have discussed in-depth many foundational aspects of reinforcement learning (RL). You learned about MDP and planning in MDP using dynamic planning. You learned about model-free value methods. You learned about scaling up solution techniques using function approximation, specifically with deep learning–based approaches such as DQN. The book also covered policy-based methods such as REINFORCE, TRPO, PPO, and so on, as well as unification of value and policy optimization methods in the actor-critic (AC) approaches. You explored the possibility of combining the model-based and model-free methods. A full chapter was devoted to dissecting PPO and its application in LLM fine-tuning with the RLHF approach. The previous chapter included an introduction to Multi-Agent RL (MARL).

Most of the discussion so far has been foundational and focused on mastering reinforcement learning. However, reinforcement learning is a rapidly expanding area with lots of specialized use cases in autonomous vehicles, robotics, and similar other fields. These go beyond the problem setup used throughout this book to explain the concepts.

This last chapter covers some of the topics that I think you should be aware of at a high level. I keep the discussion at a conceptual level with links to the relevant research/academic papers, where applicable. You may use these references to extend your knowledge horizon based on your individual interest area in the field of RL. Unlike previous chapters, you will not always find the detailed pseudocode or actual code implementations. This has been done on purpose to provide a survey of some emerging areas and new advances. You should use this chapter to get a 30,000-foot understanding of different topics. Based on your specific interests, you are advised to dive deep into specific topics using the references given here as a starting point.

© Nimish Sanghi 2024
N. Sanghi, *Deep Reinforcement Learning with Python*, https://doi.org/10.1007/979-8-8688-0273-7_13

Other Interesting RL Environments

So far, the book has used some popular RL environments to explain the concepts and algorithms. The most frequently discussed one was Gymnasium, which is a fork of the OpenAI Gym. I briefly introduced a few other environments in Chapter 6 on DQN. In this section, I cover a few more and revisit one or two previous ones.

MineRL

MineRL was discussed in Chapter 6. MineRL[1] is a Python 3 library that provides a OpenAI Gym interface for interacting with the video game Minecraft, accompanied with datasets of human gameplay. Started as a research project at Carnegie-Mellon University, MineRL aims to assist in the development of various aspects of artificial intelligence within Minecraft. MineRL mainly houses *BASALT Competition Environments* and *MineRL Obtain Diamonds Environments*.

In the "Benchmark for Agents that Solve Almost-Lifelike Task" (BASALT) competition, participants solved tasks based on human judgement, instead of predefined reward functions. The goal was to produce agents that are judged by real humans to be effective at solving a given task. This calls for training on human feedback, whether it is training from demonstrations, training on human preferences, or using humans to correct agents' actions. BASALT provides a set of Gym environments paired with human demonstrations, since methods based on imitation are an important building block for solving hard-to-specify tasks.

In `MineRLBasaltFindCave-v0`, one of the BASALT environments, the agent needs to explore and find a cave. When inside a cave, the agent ends the episode. The agent is not allowed to dig down from the surface to find a cave.

In the second environment, `MineRLBasaltCreateVillageAnimalPen-v0`, after spawning in a village, the agent builds an animal pen next to one of the houses in a village. The agent uses fence posts to build one animal pen that contains at least two of the same animals. There should be at least one gate that allows players to enter and exit easily. The animal pen should not contain more than one type of animal. While building the pen, the agent should not harm villagers or existing village structures in the process.

---

[1] `https://minerl.readthedocs.io/en/latest/index.html`

In the third environment, `MineRLBasaltMakeWaterfall-v0`, after spawning in an extreme hills biome, the agent uses a water bucket to make a beautiful waterfall. Then the agent takes an aesthetic "picture" of it by moving to a good location, positioning the player's camera to have a nice view of the waterfall.

In the final and fourth BASALT environment, `MineRLBasaltBuildVillageHouse-v0`, the agent is supposed to build a house in the style of the village without damaging the village. It should be in an appropriate location, such as next to the path through the village. The agent then gives a brief tour of the house, spinning around slowly so that all the walls and roof are visible.

In the `MineRLObtainDiamondShovel-v0` environment, the agent is required to obtain a diamond shovel. The agent begins in a random starting location on a random survival map without any items, matching the normal starting conditions for human players in Minecraft. During an episode, the agent is rewarded according to the requisite item hierarchy needed to obtain a diamond shovel.

All five environments in MineRL have RGB images as an observation. This is the image from the agent's first-person perspective.

The documentation contains a set of nice tutorials to set up MineRL on Google Colab, as well as getting-started steps for doing Video-Pre-Training (VPT) using OpenAI VPT library. VPT is the process of the "Learning to Act by Watching Unlabeled Online Videos under Behavior Cloning" (BC) fine-tuning approach. You can access these Colab notebooks under the Tutorial section of the documentation. You may also refer to the `https://minerl.io/` website for more information.

Donkey Car RL

Donkey Car[2] is an open-source DIY self-driving platform for small-scale RC cars. Donkey Car is made up of several components: It is a high-level self-driving library written in Python. It was developed with a focus on enabling fast experimentation and easy contribution. It is an open-source hardware design that makes it easy for you to build your own car. It also has a simulator that enables you to use Donkey Car without hardware. The whole initiative is driven by a community of enthusiasts, developers, and data scientists who enjoy racing, coding, and discussing the future of ML, cars, and who will win the next race.

---

[2] `https://www.donkeycar.com/`

There are multiple RL environments that have been built on top of the simulator to enable RL training. I will talk about a specific one that uses a Gymnasium/OpenAI-Gym kind of API interface, making it easy to port all the algorithms that you have learned in this book over to Donkey Car. The library is called OpenAI Gym Environments for Donkey Car.[3] This library can be installed with pip using the `pip install gym-donkeycar` command. Simulator binaries for your platform also need to be installed, the details of which can be found in the documentation. The simulator has about eleven racing environments with different types of track settings. The state/observation of the car in this environment is an image of size (120,160,3) from the viewpoint of the car. Reward is a combination of game-over flag, how far the car is from the center, and its speed. Action is composed of two floating point values, `steer` and `throttle`. `Steer` refers to the steering direction of the car, with negative values indicating steering to left and positive values indicating steering to right. `Throttle` refers to acceleration.

The library also contains code to train the Donkey Car using StableBaselines3 and the DQN and PPO algorithms. Listing 13-1 shows the part of the code responsible for training the car using PPO.

Listing 13-1. PPO Training Code for DonkeyCar from `https://github.com/tawnkramer/gym-donkeycar/tree/master/examples/reinforcement_learning`

```
# make gym env
env = gym.make(args.env_name, conf=conf)

# create cnn policy
model = PPO("CnnPolicy", env, verbose=1)

# set up model in learning mode with goal number of timesteps to complete
model.learn(total_timesteps=10000)

obs = env.reset()

for i in range(1000):
        action, _states = model.predict(obs, deterministic=True)
        obs, reward, done, info = env.step(action)
        try:
                env.render()
```

---

[3] `https://gym-donkeycar.readthedocs.io/en/latest/`

```
    except Exception as e:
        print(e)
        print("failure in render, continuing...")
    if done:
        obs = env.reset()
    if i % 100 == 0:
        print("saving...")
        model.save("ppo_donkey")

# Save the agent
model.save("ppo_donkey")
print("done training")
```

Readers can refer to the library and run the training code to experiment with other algorithms. Readers may also at adding Weights and Biases (`wandb`) logging as well as hosting trained models on the Hugging Face (HF) hub.

FinRL

Reinforcement learning can also be applied to finance, especially when trading and managing portfolios. This section briefly discusses the concept and one of the libraries from this field. It is at an introductory level. If you are new to finance and you want to explore the field of RL in finance further, you may want to take a course or two on portfolio theory, trading in markets, derivates and futures, and so on. The discussion does not provide investment advice, nor does it prepare readers to trade using the RL in Finance approaches discussed here.

To motivate the discussion, consider a trader who wants to develop a trading strategy for one single stock or for a set of stocks. The pattern is that they trade once a day. Accordingly, the "action" frequency is one per day. They get the daily price of the stocks, which is commonly four prices in a day—the opening price(O), the closing price (C), the high (H) price, and low price (L). This is usually referred to as the OHLC data.

Note that sometimes traders trade multiple times in a day, such as every five minutes, every minute, every second, or even every fraction of a second in a Medium/High Frequency Trading (MFT or HFT) practice. The faster the frequency of trade, the less the benefit of using a complex machine learning algorithm to develop trading strategy. This is due to two reasons: a) most of the trading decisions require processing multiple time-series data and a complex ML model may have a latency larger than the trading time interval, and b) the higher the frequency, the less the predictive power of the algorithm due to reduction in the signal-to-noise ratio.

Moving back to the example of trading in a basket of stocks once per day, you get the daily OHLC price of the stocks. You also get the daily volume data, which is the number of stocks bought/sold in a day. The data over a period (say over many months or many years) forms a time-series. A sample of such data is shown in Figure 13-1, which plots the last five year's OHLC price along with volume date for NVIDIA stock.

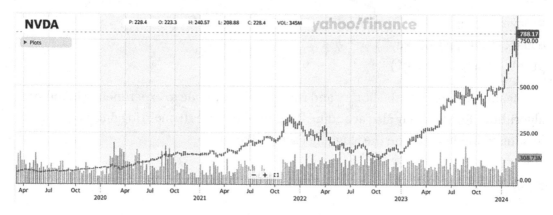

Figure 13-1. *NVIDIA price (OHLC chart) chart for five years (Mar 19 to Feb 24) Source: Yahoo Finance*

If you zoom into the data, you will notice that each day's price is denoted as a bar with two horizontal lines, as shown in Figure 13-2. Each day is represented as a vertical bar with two horizontal ticks. The vertical bar represents the high and low prices of that day. The tick on the left of the vertical bar signifies the opening price and the tick on the right of the vertical bar signifies the closing price of the day.

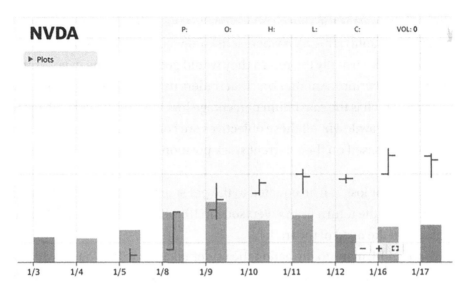

Figure 13-2. *NVIDIA chart for 10 days*
Source: Yahoo Finance

On top of this time-series of price data, traders feature analysis happens, which in the field of trading is known as *technical indicators analysis*. There are many indicators, each with a different purpose. Some simple ones are Simple Moving Average (SMA) of the closing price with the average being taken over 7 days, 10 days, 20 days, and so on, Exponential Moving Average (EMA), Moving Average Convergence/Divergence (MACD), Aroon Indicator (Aroon means dawn's early light in Sanskrit), Relative Strength Index (RSI), Average Directional Index (ADX), and Commodity Channel Index (CCI). You can think of the prices and the various indicators as part of the "state" of the environment in the RL world. Another part of the "state" is the current portfolio composition—what is the current quantity of each stock in the portfolio as well as the amount of cash the trader has in their trading account.

The traders use these time-series of past data to design an algorithm that they hope will help them achieve the financial objective of maximum return. There are a few other objectives people choose while investing. I'll briefly touch on these. The first and most common objective is the return on investment over the period of investment. It is easy to define as (final price – initial price)/initial price: $(S_T - S_0)/S_0$. Sometimes, you are not looking for just the *maximum return* but the *maximizing risk adjusted return*, which is the return you are expecting per unit of volatility. The *volatility* of a stock is measured as the standard deviation of the daily return of the stock. Volatility signifies the risks. The higher the volatility, the higher the average daily price fluctuation. Therefore, another good optimization target is the *sharpe ratio,* which is the ratio of (return from stock – risk free return) / volatility: $(r - r_f)/\sigma$ where a risk-free return is the return an investor expects with zero risk—usually the return they would get by investing in government issued bonds and securities. Another one that traders usually want to optimize is the *max drawdown,* which is the maximum percentage loss in portfolio value. They want to minimize the max drawdown. All these objectives are constructed from the daily profit or loss of the trader based on their current stock position and the price changes from one day to the next.

The daily profit or loss can be equated to the per step reward in RL. These objectives can be thought of as the return from an episode in RL which an RL agent will want to optimize based on the need of the trader.

The "action" in RL maps to the action the trader takes on a daily basis, which are buy/hold/sell. They can either increase their position in the stock or do nothing (hold or sell a portion or all of the quantity of the stock they have).

To summarize the mapping of trading to an RL setup as shown in Figure 13-3, you have:

- State: Price and other indicators of the various tocks, cash in bank and current quantity of holding in individual stocks, current portfolio composition

- Action: Buy/hold/sell decisions by the algorithm/agent/trader

- Rewards: The objective—the return over the lifetime, daily average return, volatility of the portfolio, Sharpe ratio, or max drawdown

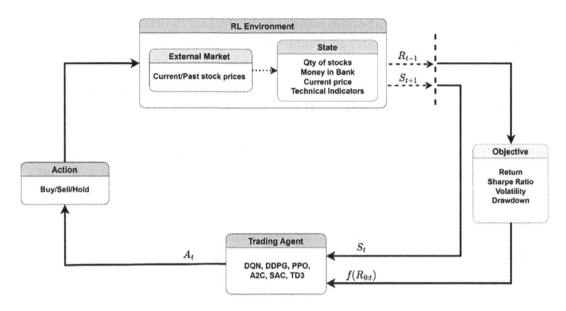

Figure 13-3. *Stock trading casted as an RL problem*

Having cast the trading in stocks as an RL problem, consider the data involved. Unlike games and other RL environments, you have a lot of historical data for financial markets. However, unlike robotics, you have no way to build a simulator that can closely resemble the actual market movements. Therefore, a common approach to applying RL or any other machine learning algorithm is to use past market data to train, validate, and test the agent. Assume you have five years' worth of data. You could use the first four years of data as the training data to do episode rollouts in the RL environment for a given trading strategy and train the agent. The next six months of data could be used as a validation set to fine-tune all the hyperparameters, using any of the hyperparameter optimization libraries and techniques discussed in earlier chapters. The last six months of data can be used to test the performance of the final strategy. Running a strategy for hyperparameter optimization and/or final testing on the past data is called *back-testing*, as it tests the effectiveness of an algorithm on past data.

Once you are ready to deploy, there are two stages. The first stage is to do something called *paper trading* wherein the strategy is deployed to generate trades and use these trades to update the state; no actual trading in the market is carried out. In other words, you deploy the trained algorithm on paper. The second and final stage of deployment is to connect the agent to the trade execution systems so that the action to the portfolio is carried out using real money and a real stock exchange. The final stage involves money, so if your strategy does not perform well, you will lose actual hard money.

A very tricky aspect of time-series analysis is data leakage, whereby the data from the future leaks into the current due to some wrong time index setting, or some other bug in the way different indicators are stacked together. At times it can be very hard to figure that out. As an example of this this problem, consider that when the agent takes an action for a given day *t*, somehow due to some bug, it has access to the price for the next day—time *t* + 1. In this case, the agent can easily find the most optimal trade to do today, as it already knows the future. Running paper trading on live market data post training alleviates this issue. If you find that paper trading gives you very different ,sub-optimal results from train and back-testing, it is a sure sign of leakage of future into past.

This completes the introduction of how RL can be applied in trading. You will now explore at a high level how all the steps can be achieved using FinRL.[4] FinRL provides a framework that supports various markets, state-of-the-art DRL algorithms, benchmarks of many quant finance tasks, live trading, and so on. The FinRL library consists of three layers, as shown in Figure 13-4.

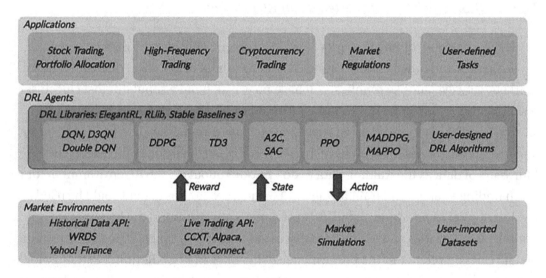

Figure 13-4. *FinRL layered architecture*
Source: FinRL documentation

The base layer is called *market environments* and is housed in a separate library called *FinRL Meta*. FinRL Meta starts by processing the market data, and then creates stock market environments. The environment monitors the variation of stock price and multiple features, the agent performs an action and gets the reward from the

---

[4]https://finrl.readthedocs.io/en/latest/index.html

environment, and then the agent adapts its strategy accordingly. By interacting with the environment, the intelligent agent will develop a trading strategy to maximize the long-term total rewards (also called Q-value). FinRL environments are based on OpenAI Gym, and they simulate the markets with real market data, using time-driven simulation. FinRL provides trading environments across many stock exchanges. FinRL Meta establishes a standard pipeline for financial data engineering in RL, ensuring that data of different formats from different sources can be incorporated in a unified framework. Further, it automates the pipeline with a data processor, which can access data, clean data, and extract features from various data sources.

You read about data leakage. FinRL employs a training-testing-trading pipeline. The DRL agent first learns from the training environment and is then validated in the validation environment for further adjustment. The validated agent is tested in historical datasets. Finally, the tested agent will be deployed in paper trading or live trading markets. This pipeline solves the information leakage problem with much more ease.

The second layer of the FinRL is that of Deep RL agents. There are three standard libraries that are used in conjunction with FinRL: a) StableBaselines3 b) RLlib, and c) Elegant RL. Elegant RL has been developed by the group responsible for FinRL.

The final layer is the actual application built on top of this to carry out trading in different markets and for any other business purposes.

FinRL has a set of nicely written tutorials with Python notebooks. Interested readers may want to go through the sequence of tutorials as suggested in the FinRL documentation.

Star Craft II: PySc2

PySC2[5] is DeepMind's Python component of the StarCraft II Learning Environment (SC2LE). It exposes Blizzard Entertainment's StarCraft II Machine Learning API[6] as a Python RL Environment. This is a collaboration between DeepMind and Blizzard to develop StarCraft II into an environment for RL research. PySC2 provides an interface for RL agents to interact with StarCraft 2, getting observations and sending actions.

The goal in StarCraft is to build a base, manage an economy, build an army, and destroy your enemies. You control your base and army from a third-person perspective, then multi-task and micro-manage your units for maximum effect. StarCraft has three

---

[5] https://github.com/google-deepmind/pysc2
[6] https://github.com/Blizzard/s2client-proto

distinct races—Terran, Protoss, and Zerg—which have different units and strategies. Starcraft II is mostly deterministic, but it does have some randomness mainly for cosmetic reasons. The two main random elements are weapon speed and update order.

Starcraft II has a very rich action and observation space. The game outputs both spatial/visual and structured elements. The structured elements are given because there is a lot of text and numbers that agents aren't expected to learn to read, especially at low resolution. It's also because it's hard to reverse replays back to exactly the same visuals that the human saw.

The SC2 action space is very large. It has hundreds of potential actions, many of which require a point on the screen or the mini map, and many of which need an extra modifier. If you were to turn the action space into a single line, it would have millions or even billions of options, most of which are invalid, and many of which are very similar. So, a simple discrete action space is not very suitable. Instead, the library made "function actions" that can combine well, without the difficulty of a random hierarchy. This is based on the idea of a C-style function call that can take some arguments of specific types. Interested readers can read more about it in the documentation on PySc2 and Blizzard ML API.

Godot RL Agents

Godot Engine[7] is a feature-packed, cross-platform game engine to create 2D and 3D games from a unified interface. It provides a set of common tools, so that users can focus on making games without having to reinvent the wheel. Games can be exported to a number of platforms, including desktop platforms (Linux, macOS, Windows), mobile platforms (Android, iOS), and web-based platforms and consoles. Godot is completely free and open-source. It is supported by the Godot Foundation.

Godot Reinforcement Learning (RL) Agents[8] is an open-source interface for developing environments and agents in the Godot Game Engine. The Godot RL Agents interface allows the design, creation, and learning of agent behaviors in challenging 2D and 3D environments with various on-policy and off-policy Deep RL algorithms. It provides a standard Gym interface, with wrappers for learning in the Ray RLlib, StableBaselines3, CleanRL and Sample Factory RL frameworks. The framework is a versatile one with the ability to create environments with discrete, continuous, and mixed action spaces.

---

[7] https://github.com/godotengine/godot
[8] https://github.com/edbeeching/godot_rl_agents

Interested readers can find out more about Godot RL agents in the 2022 paper titled "Godot Reinforcement Learning Agents."[9] Readers may also refer to the documentation to Godot RL library or Godot Engine documentation for more details on how to create your own games and how to carry out RL training. The Getting Started sections of both the libraries are well documented with respect to installation and initial explorations.

This completes the discussion on RL environments. There are many more and you could always write a customer environment of your own based on the problem you are trying to solve using RL.

Model-Based RL: Additional Approaches

In a previous chapter, you learned about model-based RL, wherein you learned the model by having the agent interact with the environment. The learned model was then used to generate additional transitions, that is, to augment the data gathered by the agent through actual interaction with the real world. This was the approach that the *Dyna algorithm* took.

While Dyna helps speed up the learning process and addresses some of the sample inefficiency issues seen in model-free RL, it is mostly used for problems with simple function approximators. It has not been successful in deep learning function approximators that require lots of samples to get trained. Too many training samples from simulators could degrade the quality of learning due to imperfections of the simulators' ability to model the world accurately. In the next section, you look at some recent approaches of combining the learned model with model-free methods within the context of deep learning. You also look at some of the most recent theoretical advances in this field.

World Models

In a 2018 paper titled "World Models,"[10] the authors proposed an approach to building a generative neural network model, which is a compressed representation of the spatial and temporal aspects of the environment, and using it to train the agent.

---

[9] https://arxiv.org/abs/2112.03636
[10] https://worldmodels.github.io/

Humans develop a mental model of the world they see around them. They do not store all the smallest possible details of the environment. Rather, they store an abstract, higher-level representation of the world that compresses spatial aspects as well as temporal aspects of the world and the relationships between the different entities of the world. They store only part of the world that they have been exposed to or is relevant for them.

Humans look at the current state or the context of the task at hand, think of the action they want to take, and predict the state that the proposed action will take them into. Based on this predictive thinking, they choose the best action. Think of a baseball batter. They have milliseconds to act, swinging the bat at the right time and in the right direction to connect with the ball. How do you think they do it? With years of practice, the batter has developed a strong internal model in which they can predict the future trajectory of the ball and can start swinging the bat at current time instance to reach that exact point in the trajectory milliseconds later. The difference between a good player and a bad player largely boils down to this predictive capability of the player based on the internal model. Years of practice make the whole thing intuitive without a lot of conscious planning in the mind. The decisions and actions humans make are based on this internal model. Jay Wright Forrester, the father of system dynamics, described a mental model as follows:

> "The image of the world around us, which we carry in our head, is just a model. Nobody in his head imagines all the world, government or country. He has only selected concepts, and relationships between them, and uses those to represent the real system."[11]

In the paper, the authors show the way they achieved the predictive internal model using a large/powerful recurrent neural network (RNN) model that they called the *world model* and using a small controller model. The reason for the small controller model is to keep the credit assignment problem within bounds, have policies iterate faster during training, and be able to use faster training algorithms like Covariance-Matrix Adaptation Evolution Strategy (CMA-ES). I talk about CMA-Es later in this chapter under the concept of derivate-free methods. The large "world model" allows it to retain all the spatial and temporal expressiveness required to have a good model of the environment and have a small controller model to keep the policy search focused on fast iterations during training.

---

[11] https://en.wikipedia.org/wiki/Mental_model

The authors explore the ability to train agents on generated world models, completely replacing the interaction with the real world. Once the agent is well trained, fusing the internal world model representation, the learned policy is transferred to the real world. They also show the benefit of adding a small random noise to the internal world model to ensure that the agent does not over-exploit the imperfections of the learned model.

Let's next look at the breakdown of the model that was used. Figure 13-5 gives a high-level overview of the pipeline. At each instant, the agent receives an observation from the environment.

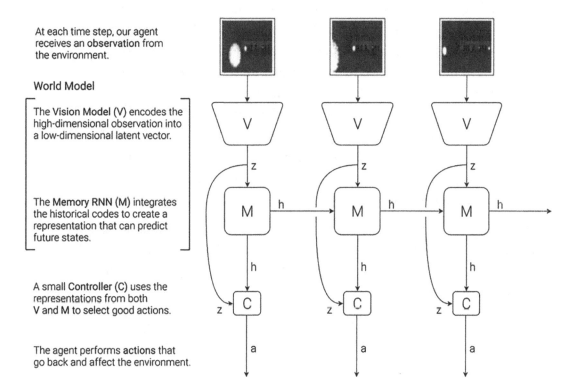

Figure 13-5. *Overview of the agent model used in the "world view"*
Source: "Recurrent World Models Facilitate Policy Evolution," 2018[12]

---

[12] https://worldmodels.github.io/

The world model consists of two modules.

- The vision model (V) encodes the high-dimensional image (observation vector) into a low-dimensional latent vector using a variational auto encoder (VAE) from deep learning. The latent vector z, the output of module V, compresses the spatial information of the observation into a smaller vector.

- The latent vector z is fed into a memory RNN (M) module to capture the temporal aspects of the environment. The M model, being an RNN network, compresses what happens over time and serves as a predictive model. Because many complex environments are stochastic and also because of the imperfections in learning, the M model is trained to predict a distribution of the next state $P(z_{t+1})$ instead of predicting a deterministic value z_{t+1}. It does so by predicting the distribution of the next latent vector z_{t+1} based on the current and past (the RNN part) as a mixture of Gaussian distribution $p(z_{t+1} | a_t, z_t, h_t)$, where h_t, the hidden state of RNN, captures the history. That is why it is called a *mixture density model based on a recurrent neural network* (MDM-RNN).

The controller model is responsible for taking the spatial information (z_t) and temporal information (h_t), concatenating them, and feeding them into a single-layer linear model. The overall flow of the information is shown in Figure 13-6 and can be broken down as follows:

1. The agent gets an observation at time t. The observation is fed into the V model, which codifies it into a smaller latent vector z_t.

2. Next, the latent vector is fed into an M model along with action a_t, which updates its internal state to produce the hidden state h_{t+1}.

3. The output of the V (z_t) and the output of the M model (h_t) are concatenated and fed into the C controller, which produces the action, a_t.

4. The action is fed into the real world and used by the M model to update its hidden state.

5. The action in the real world produces the next state/observation, and the next cycle begins.

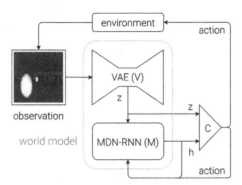

Figure 13-6. *Overview of the agent model used in the world view*
Source: "Recurrent World Models Facilitate Policy Evolution," 2018

For other details like the pseudocode, the implementation details of the networks used, and the losses used in training, refer to the previously referenced paper. In the paper, the authors also talk about how they used the predictive power to come up with hypothetical scenarios by feeding back the prediction z_{t+1} as the next real-world observation. They further show how they were able to train an agent in the "dream world" and then transfer the learning to the real world.

Imagination-Augmented Agents (I2A)

As discussed earlier, Dyna proposed a way to combine the *model-free* and *model-based* approaches. The model-free methods have higher scalability in terms of complex environments and has been seen to work well with deep learning. However, these are not sample efficient, because deep learning requires a large number of training samples to be effective. Even a simple Atari game policy training could take millions of examples to get trained. The model-based approach, on the other hand, is sample efficient. Dyna provided a way to combine the two advantages. Apart from using the real-world transitions to train the agent, the real-world transitions are also used to learn a model that is used to generate/simulate additional training examples. The problem, however, is that *model learning may not be perfect*, and unless this fact is accounted for, direct use of Dyna in a complex deep learning combined reinforcement learning does not give good results. The poor model knowledge can lead to over-optimism and poor agent performance.

As with the previous approach of using world models, the *imagination augmented agents* (I2A) method combines model-based and model-free approaches in a way that makes the combined approach work well for complex environments. I2A forms an approximate environment model and leverages it by "learning to interpret" the learned model imperfections. It provides an end-to-end learning approach to extract useful information gathered from model simulations without relying exclusively on the simulated returns. The internal model is used by agents, also referred to as *imagining,* to seek positive outcomes while avoiding adverse outcomes. The authors from DeepMind, in their 2018 paper titled "Imagination-Augmented Agents for Deep Reinforcement Learning,"[13] show that this approach can learn better with less data and imperfect models. Figure 13-7 illustrates the I2A architecture as explained in the referenced paper. I2A has the following components:

- The environment model in Figure 13-7(a), given the present information, makes predictions about the future. The imagination core (IC) has a policy net $\hat{\pi}$, which takes the current observation o_t (real observation) or $\hat{o}_t$ (imagined observation) as input and produces a rollout action $\hat{a}_t$. The observation o_t or $\hat{o}_t$ and rollout action $\hat{a}_t$ are fed into the environment model, an RNN-based network, to predict the next observation $\hat{o}_{t+1}$ and reward $\hat{r}_{t+1}$.

- Many such ICs are strung together, feeding the output of the previous IC to the next IC and producing an *imaged rollout trajectory* of length τ as shown in Figure 13-7(b). n such trajectories $\hat{T}_1, \hat{T}_2, \ldots, \hat{T}_n$ are produced. Each imagined trajectory $\hat{T}$ is a sequence of features $\left(\hat{f}_{t+1}, \ldots, \hat{f}_{t+\tau} \right)$ where t is the current time, τ the length of the rollout, and $\hat{f}_{t+i}$ the output of the environment model (i.e., the predicted observation and/or reward). As the learned model cannot be assumed to be perfect, depending solely on the predicted rewards may not be a good idea. Also, the trajectories may contain information beyond the reward sequence. Therefore, each rollout is encoded by sequentially processing the output to get an embedding on each trajectory, $e_i = \varepsilon\left(\hat{T}_i \right)$, as shown on the right side of Figure 13-7(b). The encoder is an LSTM with convolutional encoder,

---

[13] _

which sequentially processes a trajectory T. The features $\hat{f}_t$ are fed to the LSTM in reverse order, from $\hat{f}_{t+\tau}$ to $\hat{f}_{t+1}$, to mimic Bellman type backup operations. You can see this approach being represented by the direction of arrows in the encoder block of Figure 13-7(b).

- Finally, the aggregator in Figure 13-7(c) combines these individual n rollouts and feeds them as an additional context to the policy network along with the model-free path of feeding the observation directly. For the model-free path of the I2A, the authors chose a standard network of convolutional layers plus one fully connected one. The I2A therefore learns to combine information from its model-free and imagination-augmented paths; note that without the model-based path, I2As reduce to a standard model-free network. I2As can thus be thought of as augmenting model-free agents by providing additional information from model-based planning, and as having strictly more expressive power than the underlying model-free agent.

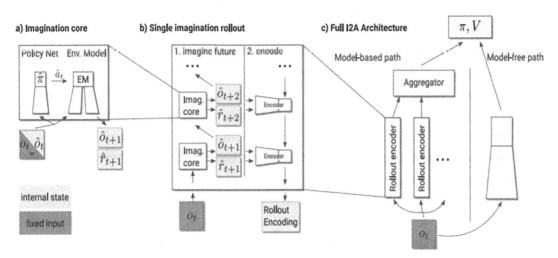

Figure 13-7. *I2A architecture. It depicts the IC in (a), the single imagination rollout in (b), and the full I2A architecture in (c)*
Source: Figure 1 from the paper "Imagination-Augmented Agents for Deep Reinforcement Learning," 2018[14]

[14] https://arxiv.org/pdf/1707.06203.pdf

The environment model shown in Figure 13-8 defines a distribution that is optimized by using a negative log likelihood loss l_{model}. The input action is broadcast and concatenated to the observation. A convolutional network transforms this into a pixel-wise probability distribution for the output image, and a distribution for the reward.

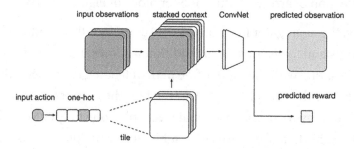

Figure 13-8. *I2A enviroment model*
Source: Figure 2 from the paper "Imagination-Augmented Agents for Deep Reinforcement Learning," 2018[15]

Training Process: The training was carried out using a fixed pretrained environment model. The training involved the remaining I2A parameters with asynchronous advantage actor-critic (A3C). An entropy regularizer was added to the policy π to encourage exploration and the auxiliary loss to distill π into the rollout policy $\hat{\pi}$. A separate hyperparameter search was carried out for each agent architecture in order to ensure optimal performance.

The authors used a puzzle environment, Sokoban, to demonstrate the performance of I2A over baselines. Sokoban is a classic planning problem, where the agent has to push a number of boxes onto given target locations. Because boxes can only be pushed (as opposed to pulled), many moves are irreversible, and mistakes can render the puzzle unsolvable. A human player is thus forced to plan moves ahead of time. The purpose of this was to confirm the expectation that artificial agents will similarly benefit from internal simulation. Readers may refer to Figure 3 of the referenced paper for a sample rendering of Sokoban environments.

Learning curves from various experiments shown in the paper are reproduced in Figure 13-9.

---

[15] https://arxiv.org/pdf/1707.06203.pdf

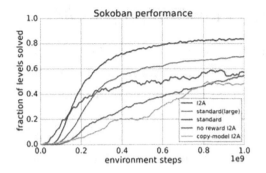 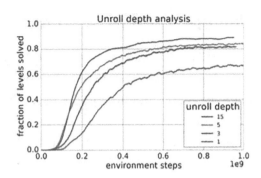

Figure 13-9. *Sokoban learning curves. Left: training curves of I2A and baselines. Right: I2A training curves for various values of imagination depth*
Source: Figure 4 from the paper "Imagination-Augmented Agents for Deep Reinforcement Learning," 2018[16]

Some interesting insights that the authors drew from this are as follows:

Figure 13-9 (right) shows that using longer rollouts, while not increasing the number of parameters, increases performance. Three unrolling steps improve the speed of learning and top performance significantly over one unrolling step; five outperforms three; and as a test for significantly longer rollouts, 15 outperforms five, reaching above 90 percent of levels solved. However, in general, the graph demonstrates diminishing returns with using I2A with longer rollouts. The authors also highlight that five steps are relatively small compared to the number of steps taken to solve a level, for which the best agents needed about 50 steps on average. This implies that even short rollouts can be highly informative. For example, they allow the agent to learn about moves it cannot recover from (such as pushing boxes against walls, in certain contexts).

In terms of data efficiency, as shown in Figure 13-9 (left), it should be noted that the environment model in I2A was pretrained with roughly about 1e8 frames. Thus, even taking pretraining into account, I2A outperforms the baselines after seeing about 3e8 frames in total.

The authors also demonstrate how a single model, which provides the I2A with a general understanding of the dynamics governing an environment, can be used to solve a collection of different tasks. Such an approach makes data efficiency even better due to reuse of the environment model to solve multiple tasks in the same environment.

The authors also show that learning the rollout encoder plays a significant role in being able to handle the imperfect model learning well.

---

[16] https://arxiv.org/pdf/1707.06203.pdf

Model-Based RL with Model-Free Fine-Tuning (MBMF)

In a 2017 paper titled "Neural Network Dynamics for Model-Based Deep Reinforcement Learning with Model-Free Fine-Tuning,"[17] the authors demonstrate yet another way to combine model-free and model-based RL. They looked at the domain of locomotion tasks. The model-free methods of training robots for locomotion suffer from high sample complexity, something seen in all deep learning–based models. The authors combined the model-free and model-based approaches to come up with sample efficient models to learn the dynamics of locomotion with varying task goals. They demonstrated that medium-sized neural network models could be combined with model predictive control (MPC) to achieve excellent sample complexity in a model-based reinforcement learning algorithm, producing stable and plausible gaits to accomplish various complex locomotion tasks. They also proposed using deep neural network dynamics models to initialize a model-free learner, in order to combine the sample efficiency of model-based approaches with the high, task-specific performance of model-free methods. With the help of experiments and learning trajectory on MuJoCo environments, the authors also showed that pure model-based approach trained on random action data can follow arbitrary trajectories with excellent sample efficiency, and that the hybrid algorithm can accelerate model-free learning further.

I first cover the Model Predictive Control (MPC). In model-based RL, a model of robot/environment dynamics is used to make predictions of the next state based on the current state and the current action. Let $\hat{f}_\theta(s_t, a_t)$ be the learned dynamics, which is parameterized by θ. It takes current state s_t and current action a_t to output an estimate of state at the next time-step $t + \Delta t$. Assume that the reward function is known and is represented as $r(s_t, a_t)$. The RL optimization problem to choose optimal actions will involve solving the following equation:

$$(a_t, \ldots, a_{t+H-1}) = \arg \max_{a_t, \ldots, a_{t+H-1}} \sum_{t'=t}^{t+H-1} \gamma^{t'-t} \cdot r(s_{t'}, a_{t'}) \tag{13-1}$$

However, due to errors in the learned model, rather than have the policy execute this complete action sequence in open-loop, the better option is to execute only the first action a_t from the sequence, and then replan at the next time-step with the updated state information s_{t+1}. This is followed by resolving the optimization problem. This approach of resolving the optimization problem at each step is known as *Model Predictive Control* (MPC).

---

[17] https://arxiv.org/pdf/1708.02596.pdf

Now consider the reward function $r_t = r(s_t, a_t)$, which is considered known since it can be calculated based on the state of the robot. The paper gives examples of the reward functions used for *moving forward* and *trajectory following*. Trajectory is shown by sparse *waypoints* that define the path the robot needs to follow. Waypoints are points on a given trajectory that, when connected with lines, approximate the given trajectory, as shown in Figure 13-10. For more details on how waypoints are used for trajectory planning, refer to any text on autonomous vehicles and robots.

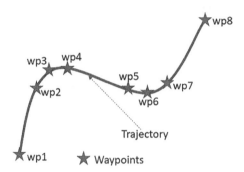

Figure 13-10. *Trajectory and waypoints*

At any point, the robot's projected sequence of actions is considered. For each action, the line segment joining the current state $\hat{s}_t$ and the estimated next state $\hat{s}_{t+1} = \hat{f}_\theta(\hat{s}_t, a_t)$ is projected on the closest line segment joining two consecutive waypoints of the trajectory. The reward is positive if the robot's movement is along the waypoint line segment and negative for a movement perpendicular to the waypoint line segment. Similarly, the paper has another set of reward functions for the *moving forward* goal.

The neural network is learning the dynamics $\hat{s}_{t+1} = \hat{f}_\theta(\hat{s}_t, a_t)$. However, this function can be difficult to learn when states s_t and s_{t+1} are very similar and therefore the action a_t seemingly has little effect on the output. This problem becomes even more pronounced as the time between states Δt becomes smaller. Therefore, instead of predicting the next state, the model predicts the difference between the two states $s_{t+1} - s_t$. With this, the next state becomes $\hat{s}_{t+1} = \hat{s}_t + \hat{f}_\theta(\hat{s}_t, a_t)$. Predicting the difference instead of the whole new state amplifies the changes and allows for small changes to be captured.

Now turn your attention to the model-based RL algorithm. Figure 13-11 illustrates this algorithm.

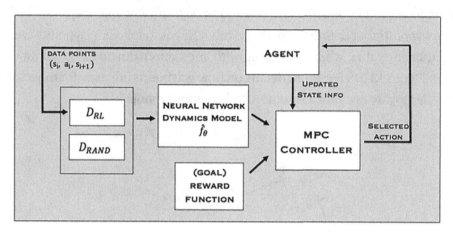

Figure 13-11. *Model-based RL*
Source: MBMF paper[18]

The algorithm works in the following manner. The reward function and dynamics model are fed into a *model predictive controller* (MPC). MPC takes in the reward and the next state to plan ahead for H steps using predicted $\hat{s}_{t+1}$ as the new input to predict $\hat{s}_{t+2}$ and so on. K randomly generated action sequences, each of length H steps, are evaluated to find the best action at initial time-step t based on the highest cumulative reward among the K sequences. The robot then takes the action a_t. At this point, the sample sequences are dropped, and complete replanning of the next K sequences, each of length H, is done at time $t + 1$. The use of this model's predictive controller (instead of an open loop approach) ensures that errors are not propagated forward. There is a replanning at each step. Figure 13-12 shows the complete algorithm.

---

[18] https://arxiv.org/pdf/1708.02596.pdf

MODEL BASED REINFORCEMENT LEARNING

1. gather dataset $\mathcal{D}_{\text{RAND}}$ of random trajectories
2. initialize empty dataset $\mathcal{D}_{\text{RL}}$, and randomly initialize $\hat{f}_\theta$
3. **for** iter=1 **to** max_iter **do**
4.   train $\hat{f}_\theta(s, a)$ by performing gradient descent:

$$\mathcal{E}(\theta) = \frac{1}{|D|} \sum_{s_t,a_t,s_{t+1} \in \mathcal{D}} \frac{1}{2}\|(s_{t+1} - s_t)-\hat{f}_\theta(s_t, a_t))\|^2,$$

$$\text{where } \mathcal{D} = \mathcal{D}_{\text{RAND}} \cup \mathcal{D}_{\text{RL}}$$

5.   **for** $t = 1$ **to** T **do**
6.   get agent's current state s_t
7.   use $\hat{f}_\theta$ to estimate optimal MPC based action sequence
8.   execute first action a_t from selected MPC action sequence $A_t^{(H)}$
9.   add (s_t, a_t, s_{t+1}) to $\mathcal{D}_{\text{RL}}$
10.   **end for**
11. **end for**

Figure 13-12. *MBMF algorithm*
Source: MBMF paper[19]

Once the model is trained, to further improve the model, the authors fine-tuned it by initializing a model-free agent with the model-based learner trained, as discussed earlier. The model-free training can be done using any of the previous approaches you learned. This is the reason for naming it *model-based with model-free fine-tuning* (MBMF).

Model-Based Value Expansion (MBVE)

In a 2018 paper titled "Model-Based Value Expansion for Efficient Model-Free Reinforcement Learning,"[20] the authors take the approach of combining the model-based and model-free approaches with a known reward function to get a more disciplined approach to value estimation.

The MBMF approach in the previous section combined the model-based (MB) approach with the model-free (MF) approach by using MB as a way to train the agent and then using the policy learned to initialize the MF agent. Model-free approaches are known to have a high capacity to learn complex behaviors but are very sample-inefficient, which can become a problem when real-world interactions are required.

---

[19] https://arxiv.org/pdf/1708.02596.pdf
[20] https://arxiv.org/pdf/1803.00101.pdf

On the other hand, model-based approaches can arrive at a near-optimal solution with learned models but under fairly restricted dynamics of the environment, such as a simple robot with limited capabilities. In this paper, authors combine the two approaches to "...*reduce sample complexity while supporting complex nonlinear dynamics by combining MB and MF learning techniques through disciplined model use for value estimation.*"

Model-based value expansion (MBE) is a hybrid algorithm that uses a dynamics model to simulate the short-term horizon and Q-learning to estimate the long-term value beyond the simulation horizon. This improves Q-learning by providing higher-quality target values for training. Splitting value estimates into a near-future MB component and a distant future MF component offers a model-based value estimate that (1) creates a decoupled interface between value estimation and model use and (2) does not require differentiable dynamics.

MVE improves value estimates for a policy π by assuming that it has access to a dynamics model $\hat{f} : S \times A \rightarrow S$ as well as the true reward function r. The improved estimate of value could be used as a critic (C) in any actor-critic algorithm. The authors in the paper modified DDPG with the improved critic from MVE to show the benefits MVE approach brings. As far as the dynamics model is concerned, the algorithm assumes that it is accurate only to certain depth H, which can be varied based on the confidence on the learned dynamics. In other words, for first H steps, you can use the dynamics model and a given policy to estimate the next state and rewards: $\hat{r}_t = r\left(\hat{s}_t, \pi\left(\hat{s}_t\right)\right)$, $\hat{s}_t = \hat{f}^{\pi}\left(\hat{s}_{t-1}\right)$. Accordingly, the H-step model value expansion (MVE) estimate for the value of a given state under policy π, $V^{\pi}(s_0)$ is as follows:

$$\hat{V}_H\left(s_0\right) = \sum_{t=0}^{H-1} \gamma^t \hat{r}_t + \gamma^H \hat{V}\left(\hat{s}_H\right) \tag{13-2}$$

The H-step MVE estimation is a composition of two components—the component estimated by learned dynamics $\sum_{t=0}^{H-1} \gamma^t \hat{r}_t$ and the tail component estimate $\gamma^H \hat{V}\left(\hat{s}_H\right)$. Since $\hat{r}_t$ is derived from actions $\hat{a}_t = \pi\left(\hat{s}_t\right)$, this is an on-policy. Therefore, in both deterministic and Stochastic cases, MVE does not require importance weights. This is in contrast to the use of traces generated by off-policy trajectories.

Figure 13-13 shows the pseudocode for the MVE algorithm. If you look carefully, you will notice that it is very similar to the regular Actor-Critic algorithms except for Step 7, which is to learn the environment dynamics, and Step 3 in how the critic is evaluated using Equation 13-2.

MODEL BASED VALUE EXPANSION (MVE)

Use model-based value expansion to enhance critic target values in a generic actor-critic method abstracted by $\ell_{\text{actor}}, \ell_{\text{critic}}$. Parameterize π, Q with θ, φ, respectively. We assume $\hat{f}$ is selected from some class of dynamics models and the space $\mathbb{S}$ is equipped with a norm.

1. **procedure** $MVE - AC$(initial θ, φ)
2.   Initialize targets $\theta' = \theta, \varphi' = \varphi$
3.   Initialize the replay buffer $\beta \leftarrow \emptyset$
4.   **while** not tired **do**
5.   Collect transitions from any exploratory policy
6.   Add observed transitions to β
7.   Fit the dynamics

$$\hat{f} \leftarrow \arg\min_{f} \mathbb{E}_{\beta}\left[\|f(S, A) - S'\|^2\right]$$

8.   **for** a fixed number of iterations **do**
9.   sample $\tau_0 \sim \beta$
10.   update θ with $\nabla_\theta \ell_{\text{actor}}(\pi_\theta, Q_\varphi, \tau_0)$
11.   imagine future transitions for $t \in [H - 1]$

$$\tau_t = \hat{f}^{\pi_{\theta'}}(\tau_{t-1})$$

12. $\forall k$ define $\hat{Q}_k$ as the k-step MVE of $Q_{\varphi'}$
13.   update φ with $\nabla_\varphi \sum_t \ell_{\text{critic}}^{\pi_{\theta'}, \hat{Q}_{H-t}}(\varphi, \tau_t)/H$
14.   update targets θ', φ' with some decay
15.   **end for**
16.   **end while**
17.   **return** θ, φ
18. **end procedure**

Figure 13-13. *Pseudocode for the MVE algorithm*
Source: MVE paper[21]

The authors show the performance of combining MBVE with DDPG versus vanilla DDPG and demonstrate significant improvement by combining MBVE with DDPG. Readers may refer to Figure 3 of the referenced paper for more information.

---

[21] https://arxiv.org/pdf/1803.00101.pdf

IRIS: Transformers as World Models

Transformers as a deep learning architecture came to scene in 2017 and today are the driving force behind the exploding AI growth, especially the transformer-driven Large Language Models in the form of auto-regressive prediction. This architecture is known as GPT, and Chapter 11 covers it. Transformers have increasingly replaced Recurrent Neural Network (RNN) architecture over the last four to five years for processing temporal or sequence-driven data. In the case of RL, the temporal sequence is the sequence of states and actions over multiple steps.

In order to learn the system dynamics and combine the benefits of the model-based approach with the model-free approach, you need a way to predict a sequence processor that can take past and present states, as well as actions to predict the next state and optionally the reward. All the papers that you have seen so far take different approaches to bringing sample efficiency while controlling the quality of solution due to imperfections in the dynamics learned inside the model-based approach versus real-world dynamics.

You also saw a need to bring sample efficiency by doing look-aheads like Monte Carlo Tree Search (MCTS) even when the dynamics were known but the search space was just too large to explore exhaustively. This is another way to think about bringing efficiency within a reasonable number of sample/computation processing.

The authors of a paper in 2023 titled "Transformers Are Sample-Efficient World Models,"[22] taking inspiration from "world model" paper's approach of learning in imagination, replaced the RNN used in that paper with a transformer-based approach and made a few other changes. The authors call their approach IRIS (Imagination with auto-Regression over an Inner Speech). They applied this technique to the Atari games benchmark. Atari games was introduced as an RL environment in Chapter 8 when covering DQN.

As per the authors, IRIS sets a new state-of-the-art in the Atari100k benchmark for methods without look-ahead search. They claim that the IRIS-based world model acquires a deep understanding of game mechanics, resulting in pixel perfect predictions in some games. The paper also illustrates the generative capabilities of the world model, providing a rich gameplay experience when training in imagination. With minimal tuning compared to existing state-of-the-art agents, IRIS opens a new path toward efficiently solving complex environments.

---

[22] https://arxiv.org/pdf/2209.00588.pdf

The authors also posit that in the future, IRIS could be scaled up to computationally demanding and challenging tasks that would benefit from the speed of its world model. Besides, its policy currently learns from reconstructed frames, but it could probably leverage the internal representations of the world model. The authors also suggest an area of research that involves combining learning in imagination with MCTS.

One more thing to highlight is the motivation for building sample efficiency. You may wonder that, when LLMs are being trained for months on datasets consisting of trillions of word tokens, why this obsession with sample efficiency in RL? One answer is that LLMs are a lot bigger model and when you take the ratio of number of training tokens and the number of model parameters, the ratio does not look that large, while the training data as a ratio of model parameter is higher in RL world. However, this is not the core point. There is an even more fundamental point to be made. LLMs deal with textual data. During training these models can be making errors and learning from those errors with some kind of gradient descent approach. The errors during training do not cause any harm or pose any threat to anyone. In contrast, RL as an approach is used a lot with physical machines such as self-driving cars, robots, and other scenarios involving human behavior. RL by design needs to collect new data as the agent learns and explores new states of the environment. Now, if these imperfect agents need to interact with the real world as part of collecting newer training data, these agents may act in a way that could be harmful to itself, to people around it, or even to property and assets that it interacts with. Even if such an interaction is possible, interacting with a real-world environment and collecting samples is a lot costlier and slower than using a software-based simulation model.

Further, you have seen that sample efficiency is higher in approaches like DQN vs policy-based learning approach. However, policy-based algorithms are a lot easier to carry out. The Actor-Critic family of approaches is one way to combine the benefits of both approaches. The model-based approach with learned dynamics is yet another efficiency step being added.

Now briefly consider the components of the IRIS approach proposed in this paper. Figure 13-14 shows the approach. The problem is formulated as a Partially Observable Markov Decision Process (POMDP) with image observations $x_t \in R^{h \times w \times 3}$, discrete actions $a_t \in \{1, ..., A\}$, scalar rewards $r_t \in R$, episode termination $d_t \in \{0, 1\}$, discount factor $\gamma \in (0, 1)$, initial observation distribution ρ_0, and environment dynamics $x_{t+1}, r_t, d_t \sim p(x_{t+1}, r_t, d_t | x_{\leq t}, a_{\leq t})$. The reinforcement learning objective is to train a policy π that learns to output actions that maximize the expected sum of rewards $E_\pi[\sum_{t \geq 0} \gamma^t r_t]$. The approach relies on the three components to learn in imagination: experience collection, world model learning, and behavior learning. The agent learns to act exclusively within its world model and uses real experience to learn the environment dynamics. The three steps are:

- `collect_experience`: Collects experience in the real environment with the agent's current policy.

- `update_world_model`: Trains the model to improve predictions for rewards, episode ends, and next observations.

- `update_behavior`: In imagination, improves the policy and value functions.

Consider the world model. How is it constructed? To process images through a transformer, you need to convert the image to a sequence of tokens just like LLMs convert a sentence into a sequence of tokens. A standard approach is something called Variational Auto Encoder (VAE). At a high level, VAE takes the image and passes it through a neural network called encoder $E: R^{h \times w \times 3} \rightarrow \{1, ..., N\}^K$. The encoder converts an input image x_t into K tokens from a vocabulary of size N, which converts the image to a smaller compressed representation $z_t = \left(z_t^1, ..., z_t^K \right) \in \{1, ..., N\}^K$ where $z_t^k = \arg\min_i \| y_t^k - e_i \|_2$, that is, the index i of the closest embedding vector in $\mathcal{E} = \{e_i\}_{i=1}^N \in R^{N \times d}$. The decoder part of the VAE, $D: \{1, ..., N\}^K \rightarrow R^{h \times w \times 3}$, turns K tokens back into a reconstructed image $\hat{x}_t$. The flow of image to a latent representation and back to a reconstructed image is shown by the green arrows in Figure 13-14.

The authors claim that the discrete autoencoder (E, D) learns a symbolic language of its own to represent high-dimensional images as a small number of tokens. This discrete autoencoder is trained on previously collected frames, with an equally weighted combination of a L_1 reconstruction loss, a commitment loss, and a perceptual loss.

The reconstructed image $\hat{x}_t$ is used in the policy network to output the action a_t as shown in the purple arrow in the Figure 13-14.

The latent representation z_t passes through the transformer along with the action a_t to predict the next time-step values like latent representation $\hat{z}_{t+1}$ of the next state, the reward $\hat{r}_t$, and the done flag $\hat{r}_t$. This next time-step prediction is done in an auto-regressive way, currently the most popular GPT way of predicting the future from the past and present.

---

[23] https://arxiv.org/pdf/2209.00588.pdf

The next state latent $\hat{z}_{t+1}$ is the imagined next state latent based on the past latent and action. This is then passed through the decoder of the VAE to get the imagined next state of the game $\hat{x}_{t+1}$, which in turn is fed into the policy network to produce the next state action a_{t+1}. As you can see from Figure 13-14, the real-world images are not used except for the start image depicted as x_0. The rest of the action is taken based on the decoded imagined images $\hat{x}$. This flow is shown as purple arrows in Figure 13-14.

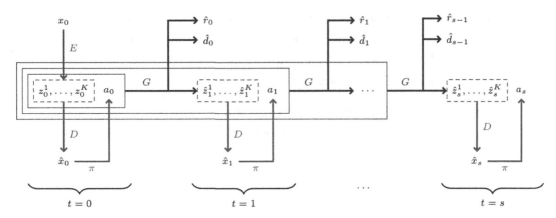

Figure 13-14. *IRIS: Transformer-based world model approach*
Source[23]

The transformer G is trained in a self-supervised manner on segments of L time-steps, sampled from past experiences. Cross-entropy loss is used for the transition and termination predictors, and a mean-squared error loss or a cross-entropy loss for the reward predictor, depending on the reward function.

Many Actor-Critic methods could be employed to train π and V in imagination. Based on the experiments carried out on various environments, the authors document the performance gains.

Causal World Models

A 2024 paper from DeepMind titled "Robust Agents Learn Causal World Models"[24] starts with a premise that causal reasoning is foundational to human intelligence and has been conjectured to be necessary for achieving human-level AI. They further state that in

---

[24] https://arxiv.org/pdf/2402.10877.pdf

recent years, this conjecture has been challenged by the development of artificial agents that achieve strong generalization without explicitly learning or reasoning on causal models. And while the necessity of causal models for solving causal inference tasks has been established, their role in decision tasks such as classification or reinforcement learning has been less clear.

In the paper, the authors establish in a model-independent way that any agent capable of robustly solving a decision task must have learned a causal model of the data generating process, regardless of how the agent was trained or the details of its architecture. It hints at an even deeper link between causality and general intelligence, as the causal model the agent learns can in turn support a much wider range of tasks beyond the original training objective. The authors conclude with the statement that:

> *"...By establishing this formal connection between causality and generalisation, our results show that causal reasoning is a fundamental ingredient for developing robust and general AI".*

Interested readers can refer to the paper for details.

Offline RL

What do the terms *online* and *offline* mean? You have seen what an on-policy and off-policy algorithm is. On-policy algorithms deal with the scenario when an agent interacts with the environment, collects the experiences, and uses experiences these to improve the agent. Immediately after that, the agent discards all the prior experiences and goes back to the cycle of interacting, collecting, and training. The agent uses the data samples from the latest policy alone to train. An off-policy approach deals with the scenario where the policy used to collect the experiences like the epsilon greedy policy is different from the policy that was learned. In DQN, you also keep storing all prior experiences in a buffer use these. A point to note is that in off-policy, while you are using the old experience data from buffer, you still periodically collect new data by interacting with the environment.

There is another approach where the agent gets the data from past interactions and then does not interact with the environment during training; it is expected to work well when deployed in the real world. Essentially it is a variation of an off-policy algorithm with no further interaction with the environment periodically during the training. The data collected before the beginning of the training is kept static and not modified during the whole process of the training.

The main motivation for such an approach is that in certain scenarios, such as robotics and autonomous vehicles, it is usually very difficult to simultaneously train and collect new experiences while the agent is getting trained. At the same time, there is an abundance of past data available for such environments that has been collected over many years. The offline approach (known as *offline RL*) can make training from past data lot more efficient. This makes offline RL a very practical and compelling proposition in the field of robotics and autonomous vehicles. However, offline RL is hard because of the difference between online interactions and a fixed dataset of recorded interactions, that is, when the learned agent takes a different action than the data collection agent, you can't determine the reward that should be given. This issue is also known as *distribution shift*. In other words, when the data seen by the agent as it gets progressively more trained is way different from the data distribution from the initial fixed set of interaction that is being used to train the agent.

The three approaches—on-policy, off-policy, and offline RL—can be summarized in the table shown in Figure 13-15. Let's spend a minute to get the terminology right. What you have studied so far in this book can be classified as online algorithms, both on-policy as well as off-policy. These are all online algorithms, as the agent keeps collecting the data during training. The difference has been about the reuse of data. In an online on-policy, after you collect the data and use it once to train the agent, it becomes stale. You throw away that data and the agent collects a completely new set of interaction data/experience from the environment for the next iteration of training. In online off-policy, while you are collecting the new interaction data, you also keep using the previously collected data—the use of the replay buffer in DQN. The premise of offline RL is never to collect new data during the training; only to use the prior collected data for training the agent and then to deploy these agents in real life, either as fully operational in production or after a short fine-tuning with any online approach.

	Only use latest policy data from current agent for training the current agent	Use a mix of past (including other agent data) and latest policy data from current agent
Current agent collects data during training	Online, on-policy RL	Online, off-policy RL
(fixed dataset) Current agent does not interact with environment during training		Offline (fully off-policy) RL

Figure 13-15. *Offline RL vs online RL*

You can find a comprehensive tutorial of offline RL in the paper "Offline Reinforcement Learning: Tutorial, Review, and Perspectives on Open Problems."[25] A visualization of the comparison between three approaches is shown in Figure 13-16. The table in Figure 13-15 and the diagram in Figure 13-16 represent the same concept—the comparison of three high-level approaches in RL. By the way, offline RL is also known as data-driven RL or batch RL.

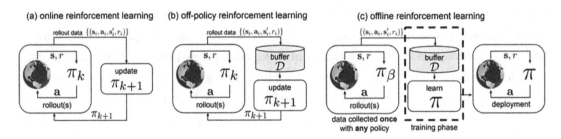

Figure 13-16. *Offline RL vs online RL*
Source: Figure 1 of the survey paper[26]

Let's now briefly look at the scenarios in which offline RL can be beneficial.

- **Robotics:** While it may be feasible to collect online data during robot training, we would like to train the robots on various skills. Each skill teaching in robotics usually requires a very large amount of interaction and doing so for a wider range of skills multiplies

---

[25] https://arxiv.org/abs/2005.01643
[26] https://arxiv.org/abs/2005.01643

the problem many folds. Instead of that, if you could use all the previously collected data to train the agent in a multi-task offline RL learning way, you could make the training lot more sample efficient. Such a pre-trained robot could easily be deployed to do a new task with a very small amount of skill specific online agent training.

- **Healthcare:** An example of RL in healthcare could be modelling the whole interaction of a doctor and a patient as a Markov MDP. Diagnostic tests and drugs and treatments prescribed by the doctors could be modelled as actions and the past medical history of the patient could be treated as states/observations. The reward could be the longevity or the health score of the individual. Training such an agent with online data is usually not feasible due to the fear of wrong diagnosis leading/treatment and other safety concerns. However, hospitals have years of historical data that could be used in an offline RL approach to train the agent without any online interaction during the training.

- **Autonomous vehicles:** It possesses the same challenge as healthcare example. Training the vehicles online and learning from the mistakes and rewards would not be feasible due to safety and accident damage concerns. However, a human driver can drive the car around and collect the data to train the agent in an offline RL way.

Offline RL makes RL start looking like supervised learning. Many such sharp distinctions are getting blurred, as researchers are finding unique ways to leverage the benefits of different approaches to overcome the weaknesses of any single approach.

Let us now talk about the existing approaches for offline RL. In theory, the usual off-policy algorithms can be used as offline RL by just switching off the real-world data collection loop during training and solely relying on the initial data collection. However, this would lead to data distribution mismatch issues, which need careful handling to reduce the mismatch issue. However, there is a fundamental mismatch issue that cannot be handled, where the initial data and observations during training are starkly different, and it is not just a marginal drift. Assume that you are training the agent on a game that has multiple levels, with each level being very different and contain new adversarial characters with unseen behavior. Further assume that you have found the ideal optimal offline RL to train the agent but train the agent only on the data seen in the first level of the game. Do you think such an agent will perform well as the agent moves progressively through levels encountering newer agents with unseen complex behaviors? The answer is a big no. These cannot be solved based entirely using offline RL. In a scenario like this, the agent trained on the offline different data could be used as a good starting point and then progressively trained on newer data in an online fashion.

Turning the focus back to extending an *off-policy* approach to *offline RL*, consider one example in which DQN has been extended. A paper titled "An Optimistic Perspective on Offline Reinforcement Learning"[27] has one solution. The paper is accompanied by a blog and code repository as well. The authors use a *DQN replay dataset,* which consists of a collection of tuples of (observation, action, reward, next observation) (approximately 50 million) encountered during training of a DQN agent, on all 60 Atari 2600 games with sticky actions enabled for 200 million frames, with the process repeated five times for each game.

A lot of approaches of extending DQN to offline RL have to do with the tampering down of optimistic Q-estimates. First, the authors tried DQN and QR-DQN in a full offline setup on the dataset. They compared the results of these approaches against the best performing online DQN agent obtained after training, that is, a fully-trained DQN. The experiments lead to the conclusion that offline DQN underperforms fully-trained online DQN on most of the games. They also found that offline QR-DQN outperforms offline DQN and fully-trained DQN on most of the games. Based on the findings, they suggested two variants toward fully offline DQN (offline RL).

The first one is called *Ensemble-DQN* and is a simple extension of DQN that trains multiple Q-value estimates and averages them for evaluation.

The second one is called *Random Ensemble Mixture (REM),* which is an easy-to-implement extension of DQN inspired by Dropout. The main idea behind REM is that you can use a weighted mix of Q-value estimates as another Q-value estimate, if you have multiple Q-value estimates available. Therefore, in each training step, REM randomly mixes different Q-value estimates and uses this random mix for stable training.

A pictorial comparison of Q-value estimates in the four approaches—DQN, QR-DQN, offline Ensemble DQN, and offline REM—is shown in Figure 13-17.

---

[27] https://offline-rl.github.io/

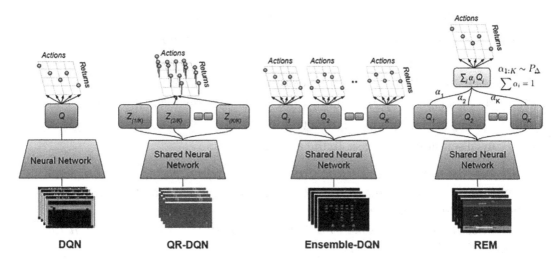

Figure 13-17. *Architecture comparison: offline DQN vs online DQN*
Source[28]

The experiments they conducted found that offline REM gave the best results; it outperforms offline DQN as well as offline QR-DQN. They also found out that the gains from offline REM are more than the gains from a fully trained online C51, a strong distributional agent.

The tutorial paper referenced earlier contains many other approaches being tried out. It is a lengthy but a very rewarding starting point for those who are interested in offline RL.

Decision Transformers

In a 2021 paper titled "Decision Transformer: Reinforcement Learning via Sequence Modeling,"[29] the authors took the approach of language modeling using transformers in LLMs. The authors describe the Decision Transformer model as an architecture that casts the problem of RL as conditional sequence modeling. Unlike prior approaches to RL that fit value functions or compute policy gradients, Decision Transformer simply outputs the optimal actions by leveraging a causally masked transformer. By conditioning an autoregressive model on the desired return (reward), past states, and actions, the Decision Transformer model can generate future actions that achieve the

---

[28] https://blog.research.google/2020/04/an-optimistic-perspective-on-offline.html
[29] https://arxiv.org/pdf/2106.01345.pdf

desired return. The authors, via experiments, show that Decision Transformer matches or exceeds the performance of state-of-the-art model-free offline RL baselines on Atari, OpenAI Gym, and Key-to-Door tasks. A schematic of the Decision Transformer model is shown in Figure 13-18. In Decision Transformer, states, actions, and returns are fed into modality-specific linear embeddings and a positional episodic time-step encoding is added. Tokens are fed into a GPT architecture, which predicts actions autoregressively using a causal self-attention mask. In the authors' words:

> *"In contrast to prior work using transformers as an architectural choice for components within traditional RL algorithms, we seek to study if generative trajectory modeling—i.e., modeling the joint distribution of the sequence of states, actions, and reward—can serve as a replacement for conventional RL algorithms."*

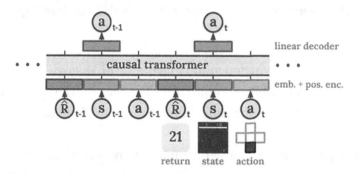

Figure 13-18. *The Decision Transformer method*
Source: Figure 1 of the Decision Transformer paper[30]

Consider the high-level architecture as well as the pseudocode of the algorithm. Figure 13-19 shows the pseudocode of the algorithm. The key idea of the architecture is that the choice of trajectory representation should enable transformers to learn meaningful patterns and should be able to conditionally generate actions at test time.

---

[30] https://arxiv.org/pdf/2106.01345.pdf

As the idea is to generate actions based on the future desired rewards instead of depending on past rewards, the authors feed the model with the returns-to-go $\hat{R}_t = \sum_{t'=t}^{T} r_{t'}$. As a result of this, the trajectory representation being fed to the transformer is given in Equation 13-3. Since returns-to-go capture the future, they are suited for autoregressive training, guiding the autoregressive model toward the desired outcome.

$$\tau = \left(\hat{R}_1, s_1, a_1, \hat{R}_2, s_2, a_2, ..., \hat{R}_T, s_T, a_T \right) \tag{13-3}$$

At test time, the desired performance (e.g., 1 for success or 0 for failure) can be specified along with the environment starting state as the conditioning information to initiate generation. After executing the generated action for the current state, the target return is decremented by the achieved reward and this process is repeated until episode termination.

With respect to the training, the last K time-steps are fed into Decision Transformer, for a total of $3K$ tokens—one for each modality: return-to-go, state, or action. The embeddings are obtained with a linear layer for each modality, which projects the raw inputs into the embedding dimension of the Decision Transformer. This is followed by a usual practice of layer normalization. If the observations are visual inputs, the observation is processed through a convolutional encoder instead of a linear layer. You saw this approach in the case of "IRIS: Transformers as World Models". Time-step encoding is learned and added to each token. However, note that this is different than the standard positional embedding used by transformers, as here one time-step corresponds to three tokens. The tokens are then processed by a GPT model, which predicts future action tokens via autoregressive modeling.

From a given dataset of offline trajectories, minibatches of sequence length K are sampled. The last layer—the prediction head corresponding to the input token s_t—is trained to predict a_t. The loss used for training is cross-entropy loss for discrete actions and mean-squared error for continuous actions.

The authors also noted that predicting the states or returns-to-go did not improve performance, although the Decision Transformer architecture allows for easy integration of the same. They have left further exploration of this for future study.

Figure 13-19 shows the pseudocode of the Decision Transformer algorithm.

DECISION TRANSFORMER ALGORITHM

```
# R, s, a, t: returns-to-go, states, actions, or timesteps
# transformer: transformer with causal masking (GPT)
# embed_s, embed_a, embed_R: linear embedding layers
# embed_t: learned episode positional embedding
# pred_a: linear action prediction layer

# main model
def DecisionTransformer(R, s, a, t):
    # compute embeddings for tokens
    pos_embedding = embed_t(t)  # per-timestep (note: not per-token)
    s_embedding = embed_s(s) + pos_embedding
    a_embedding = embed_a(a) + pos_embedding
    R_embedding = embed_R(R) + pos_embedding

    # interleave tokens as (R_1, s_1, a_1, ..., R_K, s_K)
    input_embeds = stack(R_embedding, s_embedding, a_embedding)

    # use transformer to get hidden states
    hidden_states = transformer(input_embeds=input_embeds)

    # select hidden states for action prediction tokens
    a_hidden = unstack(hidden_states).actions

    # predict action
    return pred_a(a_hidden)

# training loop
for (R, s, a, t) in dataloader:  # dims: (batch_size, K, dim)
    a_preds = DecisionTransformer(R, s, a, t)
    loss = mean((a_preds - a)**2)  # L2 loss for continuous actions
    optimizer.zero_grad(); loss.backward(); optimizer.step()

# evaluation loop
target_return = 1  # for instance, expert-level return
R, s, a, t, done = [target_return], [env.reset()], [], [1], False
while not done:  # autoregressive generation/sampling
    # sample next action
    action = DecisionTransformer(R, s, a, t)[-1]  # for cts actions
    new_s, r, done, _ = env.step(action)

    # append new tokens to sequence
    R = R + [R[-1] - r]  # decrement returns-to-go with reward
    s, a, t = s + [new_s], a + [action], t + [len(R)]
    R, s, a, t = R[-K:], ...  # only keep context length of K
```

Figure 13-19. *Decision Transformer pseudocode*
Source: Algorithm 1 of the paper[31]

---

[31] https://arxiv.org/pdf/2106.01345.pdf

592

In the paper, the authors conduct experiments to answer the following questions:

Does the Decision Transformer model perform behavior cloning on a subset of the data? The experiments suggest this to be true with the finding that Decision Transformers can be more effective than simply performing imitation learning on a subset of the dataset.

How well does Decision Transformer model the distribution of returns? The experiments show that the desired target returns and the true observed returns are highly correlated. They further show that one can prompt the Decision Transformer with higher returns than the maximum episode return available in the dataset, demonstrating that Decision Transformer is sometimes capable of extrapolation.

What is the benefit of using a longer context length? It is generally considered that the previous state (i.e., K = 1) is enough for reinforcement learning algorithms when frame stacking is used. However, the results show that that performance of Decision Transformer is significantly worse when K = 1, indicating that past information is useful for Atari games. One hypothesis is that when the current approach is representing a distribution of policies—as with sequence modeling—the context allows the transformer to identify which policy generated the actions, enabling better learning and/or improving the training dynamics.

Does Decision Transformer perform effective long-term credit assignment? The authors show that with spare reward environments like Key-to-Door, Decision Transformers are able to learn effective policies, producing near-optimal paths, despite only training on random walks. TD learning cannot effectively propagate Q-values over the long horizons involved and gets poor performance. Delayed returns minimally affect Decision Transformer due to the nature of the training process.

Can transformers be accurate critics in sparse reward settings? As Decision Transformers can produce effective policies (actors), the authors also evaluate whether Decision Transformer models can also be effective critics. Their results show that to be true.

The authors also state that they believe Decision Transformer can meaningfully improve online RL methods by serving as a strong model for behavior generation. As an example, Decision Transformer can serve as a powerful "memorization engine" and, along with powerful exploration algorithms, has the potential to simultaneously model and generate a diverse set of behaviors.

Automatic Curriculum Learning

Curriculum learning is a training strategy that originates from the way humans learn, gradually moving from simple concepts to more complex ones. This approach has been adapted in the field of artificial intelligence (AI) and, more specifically, in reinforcement learning (RL) to improve the efficiency and effectiveness of training machine learning models. Curriculum learning in RL involves structuring the learning process in stages or lessons that progressively become more challenging or complex, guiding the agent to learn more efficiently.

The core idea behind curriculum learning is to start with easier tasks or environments that capture the fundamental aspects of a problem and gradually introduce more complexity. This method stands in contrast to the traditional approach of training an agent on the full complexity of a problem from the beginning. By breaking down the learning process into manageable steps, curriculum learning aims to achieve the following:

- **Accelerate learning**: By focusing on simpler tasks initially, an agent can quickly learn basic skills that are foundational to more complex tasks, leading to faster overall learning.

- **Improve learning stability**: Starting with simpler tasks reduces the initial difficulty, helping to stabilize the learning process by providing clearer learning signals.

- **Enhance final performance**: Agents trained with a curriculum usually achieve better final performance on complex tasks because the gradual increase in difficulty allows for more effective skill acquisition.

Implementing curriculum learning involves several key components:

Task Selection: Deciding on a sequence of tasks or environments that start simple and gradually increase in complexity. This can be based on various factors such as the number of obstacles, the size of the environment, or the complexity of the interactions required.

Progression Criteria: Establishing metrics or criteria to determine when the agent is ready to progress to the next task. This could be based on performance thresholds, such as achieving a certain average reward, or learning stability measures.

Curriculum Design: Designing the curriculum itself, which involves not just selecting the tasks and progression criteria but also deciding how to modify the environment or the task parameters to create the progression of difficulty.

While curriculum learning has shown promise in various reinforcement learning applications, it also presents several challenges. The first challenge is that of *curriculum design*. Designing an effective curriculum is not trivial. It requires insight into the task at hand and an understanding of what constitutes an "easier" version of the problem. The second issue is about *adaptability*. The curriculum must be adaptable to the learning progress of the agent. A fixed curriculum might not suit all agents or learning scenarios. The next one is about *transferability*. Skills learned in earlier stages of the curriculum should transfer to later stages and the final task. Ensuring this transferability is crucial for the effectiveness of curriculum learning.

Curriculum learning has been applied successfully in various domains within reinforcement learning, such as robotics, where agents learn basic movements before tackling complex manipulation tasks, or in game playing, where AI learns to play simpler levels before advancing to more difficult ones. These applications demonstrate the versatility and potential of curriculum learning to enhance learning processes in reinforcement learning.

There are two very good resources to get started in this area. The first one is a survey paper titled "Automatic Curriculum Learning For Deep RL: A Short Survey"[32] and the second one is a blog titled "Curriculum for Reinforcement Learning."[33]

In conclusion, curriculum learning offers a promising approach to improve the training of reinforcement learning agents by structuring the learning process into a progression of increasing difficulty. By leveraging the principles of how humans learn, it seeks to make the learning process more efficient, stable, and effective. However, the challenges in designing and implementing effective curricula highlight the need for ongoing research and development in this area. As our understanding of both human learning and artificial intelligence deepens, curriculum learning in reinforcement learning is likely to become an even more powerful tool in the development of intelligent agents.

---

[32] https://arxiv.org/pdf/2003.04664.pdf
[33] https://lilianweng.github.io/posts/2020-01-29-curriculum-rl/

Imitation Learning and Inverse Reinforcement Learning

There is another branch of learning called *imitation learning*. You record the interactions of an expert and then use a supervised setting to learn a behavior that can mimic the expert. I briefly touched on this when talking about various world model approaches in this chapter.

As shown in Figure 13-20, you have an expert who looks at states s_t and produces the actions a_t. You use this data in a supervised setting with the states s_t as input to the model and the actions a_t as targets to learn a policy $\pi_\theta(a_t|s_t)$, as shown in the middle of Figure 13-20. This is the easiest way to learn a behavior, called *behavior cloning*. It's even simpler than the whole discipline of reinforcement learning. The system/learner does not analyze or reason about anything; it just blindly learns to mimic the behavior of the expert.

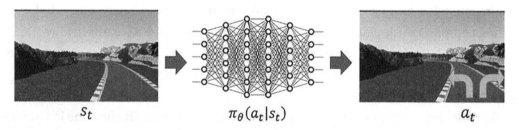

s_t $\pi_\theta(a_t|s_t)$ a_t

Figure 13-20. *Expert demonstration*

However, the learning is not perfect. Suppose you learn a near-perfect policy with some minor errors, and suppose you start in a state s_1. You follow the learned policy to take action a_1, which, say, is a minor deviation from what an expert would have taken. You keep taking these actions in a sequence, some matching the expert and some deviating a little bit from what an expert would have done. These deviations over multiple actions can add up and bring your vehicle (see Figure 13-20) to the edge of the road. Most likely, the expert would not have driven so badly to get to the edge of the road. The expert training data never saw what the expert would have done in this situation. The policy has not been trained with these kinds of situations. The learned policy will most likely take an action that could be random and definitely not designed to correct the error to bring the vehicle back toward the center of the road.

This is an *open loop problem*. Errors from every action are compounded to make the actual trajectory drift from the expert trajectory, as shown in Figure 13-21.

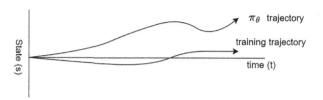

Figure 13-21. *Drift in trajectory over time*

This is also known as *distributional shift*. In other words, the distribution of states in the policy training is different from the distribution of states the agent sees when it executes the policy in an open loop with no corrective feedback.

In cases like this, with additive errors, there is an alternative algorithm known as *DAgger* (for "dataset aggregation"), in which the agent is iteratively trained by first training the agent on the data from the expert demonstration. The trained policy is used to generate additional states. The expert then provides the correct action for these generated states. The augmented data, along with the original data, are again used to fine-tune the policy. The cycle goes on, and the agent gets closer and closer to the expert in terms of following the behavior. This can be classified as *direct policy learning*. Figure 13-22 shows the pseudocode of DAgger.

DAGGER

Do till tired:
    Train $\pi_\theta(a_t|s_t)$ from human data $D = \{s_1, a_1, \ldots, s_N, a_N\}$
    Run $\pi_\theta(a_t|s_t)$ to get dataset $D_\pi = \{s_1, s_2, \ldots, s_N\}$
    Ask human/expert to label D_π with actions $\{a_1, a_2, \ldots, a_N\}$
    Aggregate: $D \leftarrow D \cup D_\pi$

Figure 13-22. *DAgger for behavior cloning*

DAgger has a human expert label the unseen states as the policy is played out, which trains the agent to recover from mistakes/drifts. DAgger is simple and efficient. However, it is just behavior cloning and not reinforcement learning. It does not reason about anything other than trying to learn a behavior to follow the expert actions. If the expert covers a large part of the state space that the agent is likely to see, this algorithm can help the agent learn a good behavior. However, anything requiring long-term planning is not suited for DAgger.

If it is not reinforcement learning, why am I talking about it here? It turns out that in many cases, having an expert give a demonstration is a good way to understand the objective that the agent is trying to achieve. When coupled with other enhancements, imitation learning is a useful approach.

Up to now, in this book you have studied various algorithms to train the agent. With some algorithms, you knew the dynamics and transitions like the model-based setups. In others, you learned in a model-free setup without explicitly learning the model, and finally in some other cases, you learned the model from interaction with the world to augment model-free learning. However, you always considered the reward to be something simple, intuitive, and well known. In some other setups, you could handcraft a simple reward function, such as learning to follow a trajectory using MBMF. But all real-world cases are not so simple. The reward at times is ill defined and/or sparse. Many of the previous algorithms will not work in the absence of a well-defined reward.

Consider a case where you are trying to train a robot to pick up a jug of water and pour the water into a glass. What will the reward be for each action in the whole sequence? Will it be 1 when the robot is able to pour the water into a glass without spilling any on the table or breaking/dropping the jug/glass? Or will you define a range of rewards based on how much water was spilled? How will you induce a behavior to have the robot learn smooth actions like a human? Can you think of a reward that can provide correct feedback to the robot on what is a good move and what is a bad move?

Now look at the alternate scenario. The robot gets to watch a human carry out the task of pouring water. Instead of learning a behavior, the robot first learns a reward function, noting all actions matching the human behavior as good and others as bad, with goodness depending on the extent of the deviation from what it saw the humans do. It can then use the learned reward function, as a next step, to learn a policy/sequence of actions to perform similar actions. This is the domain of *inverse reinforcement learning (inverse RL) coupled with imitation learning.*

Table 13-1 compares behavior cloning, direct policy learning, and inverse RL.

Table 13-1. *Types of Imitation Learning*

	Direct Policy Learning	Reward Learning	Access to Environment	Interactive Demonstrator/ Expert	Pre-Collected Demonstrations
Behavioral cloning	Yes	No	No	No	Yes
Direct policy learning	Yes	No	Yes	Yes	Optional
Inverse reinforcement learning	No	Yes	Yes	No	Yes

Inverse Reinforcement learning is the MDP setup where you know the model dynamics, but you have no knowledge of the reward function. Mathematically, it can be expressed as follows:

Given: $D = \{\tau_1, \tau_2, \ldots, \tau_m\} = \{(s_0^i, a_0^i, \ldots s_T^I, a_T^i)\} \sim \pi^*$
Goal: Learn a reward function r^*
So that $\pi^* = argmax_\pi E_\pi[r^*(s, a)]$

Figure 13-23 shows the high-level pseudocode for inverse RL. You collect sample trajectories from the expert, and you use them to learn the reward function. Next, using the learned reward function, you learn a policy that maximizes the reward. The learned policy is compared with the expert, and the difference is used to tweak the learned reward. This cycle goes on.

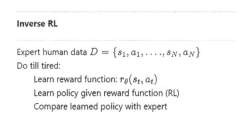

Inverse RL

Expert human data $D = \{s_1, a_1, \ldots, s_N, a_N\}$
Do till tired:
    Learn reward function: $r_\theta(s_t, a_t)$
    Learn policy given reward function (RL)
    Compare learned policy with expert

Figure 13-23. *Inverse Reinforcement learning*

Note that the inner *do loop* has a step (Step 2, "Learn policy given reward function") of learning the policy iteratively. It is a loop abstracted to a single line of pseudocode. To scale this approach when the state space is continuous and highly dimensional and also when the system dynamics are unknown, you need tweaks to make the previous approach work. In a 2016 paper titled "Guided Cost Learning: Deep Inverse Optimal Control via Policy Optimization,"[34] the authors used a sample-based approximation of Max Entropy Inverse RL.[35] You can look at the referenced papers for further details. Figure 13-24 shows the high-level diagram of the approach.

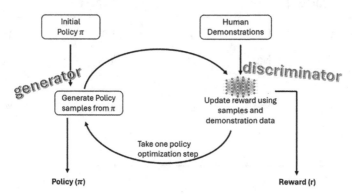

Figure 13-24.   *Guided cost learning: deep inverse optimal control via policy optimization*

The architecture can be compared to *generative adversarial networks* (GANs) from deep learning. In GAN, a *generator network* tries to generate synthetic examples, and a *discriminator network* tries to give a high score to actual samples while giving a low score to synthetic examples. The generator tries to get better in generating examples that iteratively get harder to distinguish from "real-world examples," and the discriminator gets better and better at making a distinction between synthetic versus real-world examples.

In the same way, the guided cost learning as given in Figure 13-24 can be thought of as a GAN setup, where the discriminator is getting better at giving a high reward to human observations while giving a low one to the actions/trajectories generated by the policy network. The policy network gets better and better at producing actions similar to a human expert. The uses of inverse learning coupled with imitation learning are many:

- *Making characters in animation movies*: You can move the face and lips in sync with the words the characters are speaking. The face/lip movements of a human expert are first recorded (expect demonstrations), and a policy is trained to make a character's face/lips move like a human.

---

[34] https://arxiv.org/pdf/1603.00448.pdf

[35] https://www.aaai.org/Papers/AAAI/2008/AAAI08-227.pdf

- *Parts of speech tagging*: This is based on some expert/human labels.

- *Smooth imitation learning*: To have an autonomous camera follow a game like basketball, similar to what a human operation would do, to follow the ball across the court, zoom, and pan based on certain events.

- *Coordinated multi-agent imitation learning*: To look at recordings of, say, a soccer game (human expert demonstrations) and then learn a policy to predict the next place for the players to be, based on the sequence.

Imitation learning is an expanding area. I have barely touched the surface to just introduce the topic. There are tons of good places to start exploring these topics. A good place to start this is the ICML2018 tutorial on imitation learning.[36] It is a two-hour video tutorial with slides given by two experts from Caltech.

Derivative-Free Methods

Let's move back to the regular model-free RL that you saw in major parts of the book. This section briefly covers the methods that allow you to improve policies without taking derivatives of policy $\pi_\theta(a|s)$ with respect to policy parameters θ.

Consider *evolutionary methods*. Why are they evolutionary? Because they work like natural evolution. Things that are better/fitter survive, and things that are weaker fade away.

The first method you look at is called the *cross-entropy method*. It is embarrassingly simple.

1. Pick a random/stochastic policy.

2. Roll out a few sessions.

3. Pick some percentage of sessions with higher rewards.

4. Improve the policy to increase the probability of selecting those actions.

---

[36] https://sites.google.com/view/icml2018-imitation-learning/

Figure 13-25 gives the pseudocode of the cross-entropy method for training a continuous action policy, which is assumed to be a normal distribution with d-dimensional action space. Any other distribution could also be used, but for many domains, it has been shown that normal distribution is the best in terms of balancing the expressibility of the distribution and the number of parameters of the distribution.

CROSS ENTROPY METHOD

1. Initialize $\mu \in R^d, \sigma \in R^d_{>0}$ where μ and σ are mean and standard deviation of the normal distribution: $N(\mu \in R^d, \sigma \in R^d_{>0})$
2. For `iteration` = 1, 2, ...
3.   Sample n parameters $\theta_i \sim N\left(\mu, diag\left(\sigma^2\right)\right)$
4.   For each θ_i, perform one rollout to get return $R\left(\tau_i\right)$
5.   Select the top k of θ, and fit a new diagonal Gaussian to those samples using maximum-likelihood
6.   Update μ, σ

***Figure 13-25.** Cross-entropy method*

A similar approach, called the *covariance matrix adaptation evolutionary strategy* (CMA-ES), is popular in the graphics world for an optimal gait of characters. In the cross-entropy method, you fit a diagonal Gaussian to the top $k\%$ of rollouts. However, in CMA-ES, you optimize the covariance matrix, and it amounts to the second-order model learning as compared to the first-order in the usual derivative methods.

A major drawback of the cross-entropy methods is that they work well for a relatively low-dimensional action space such as the `CartPole,` `Lunar-Lander`, and so on. Can the evolutionary strategy (ES) be made to work for deep net policies with high-dimensional action spaces? In a 2017 paper titled "Evolution Strategies as a Scalable Alternative to Reinforcement Learning,"[37] the authors show that ES can reliably train neural network policies, in a fashion well suited to be scaled up to modern distributed computer systems for controlling robots in the MuJoCo physics simulator.

This section conceptually walks through the approach that was taken in this paper. Consider a probability distribution of policy parameters: $\theta \sim P_\mu(\theta)$. Here, θ means the parameters of the policy, and these parameters follow some probability distribution $P_\mu(\theta)$ parameterized by μ.

The goal is to find the policy parameter θ such that it generates trajectories that maximize the cumulative return. This is similar to the objective of the policy gradient approach.

---

[37] https://arxiv.org/pdf/1703.03864.pdf

Goal: Maximize $E_{\theta \sim P_\mu(\theta), \tau \sim \pi_\theta} \left[R(\tau) \right]$

Like with policy gradient, you do a stochastic gradient ascent, but unlike policy gradient, you do not do so in θ, rather in μ space.

$$\nabla_\mu E_{\theta \sim P_\mu(\theta), \tau \sim \pi_\theta} \left[R(\tau) \right] = E_{\theta \sim P_\mu(\theta), \tau \sim \pi_\theta} \left[\nabla_\mu \log P_\mu(\theta) \cdot R(\tau) \right] \qquad (13\text{-}4)$$

The previous expression is similar to what you saw in policy gradients. However, there is a subtle difference. You are not doing a gradient step in θ. Hence, you do not worry about $\pi_\theta(a|s)$. You ignore most of the information about a trajectory, that is, states, actions, and rewards. We only worry about the policy parameter θ and the total trajectory reward $R(\tau)$. This in turn enables scalable distributed training similar to the A3C approach of running multiple workers as shown in Chapter 8.

Consider a concrete example. Suppose $\theta \sim P_\mu(\theta)$ is a Gaussian distribution with a mean of μ and a covariance matrix of $\sigma^2 \cdot I$. $\log P_\mu(\theta)$ given inside the "expectation" in Equation 13-4 can then be expressed as follows:

$$\log P_\mu(\theta) = -\frac{\|\theta - \mu\|^2}{2\sigma^2} + const$$

Taking the gradient of the previous expression with regard to μ, you have this:

$$\nabla_\mu \log P_\mu(\theta) = \frac{\theta - \mu}{\sigma^2}$$

Let's say you draw two parameter samples θ_1 and θ_2 and obtain two trajectories: τ_1 and τ_2.

$$E_{\theta \sim P_\mu(\theta), \tau \sim \pi_\theta} \left[\nabla_\mu \log P_\mu(\theta) R(\tau) \right] \approx \frac{1}{2} \left[R(\tau_1) \frac{\theta_1 - \mu}{\sigma^1} + R(\tau_2) \frac{\theta_2 - \mu}{\sigma^2} \right] \qquad (13\text{-}5)$$

This merely converts the expectation in Equation 13-4 to an estimate based on two samples. Can you interpret Equation 13-5? The analysis is similar to what you saw in Chapter 8. If a reward of trajectory is +ve, you adjust the mean μ to get closer to that θ. If the trajectory reward is -ve, you move μ away from that sampled θ. In other words, like with the policy gradient, you adjust μ to increase the probability of good trajectories and reduce the probability of bad trajectories. However, you do so by making adjustments directly to the distribution from which the parameters came and not by adjusting the θ on which the policy depends. The approach allows you to ignore the details of states and actions.

The referenced paper uses *antithetic sampling*. In other words, it samples a pair of policies with mirror noise $(\theta_+ = \mu + \sigma\epsilon, \theta_- = \mu - \sigma\epsilon)$ and then samples the two trajectories τ_+ and τ_- to evaluate Equation 13-5. Substituting these in Equation 13-5, the expression can be simplified as follows:

$$\nabla_\mu E\left[R(\tau)\right] \approx \frac{\epsilon}{2\sigma}\left[R(\tau_+) + R(\tau_-)\right]$$

The previous manipulation allows efficient parameter passing between workers and parameter servers. At the beginning, μ is known, and only ϵ needs to be communicated, which reduces the number of parameters that need to be passed back and forth. It brings significant scalability in making the approach parallel.

Figure 13-26 shows the parallelized evolution strategy pseudocode.

Parallelized evolution strategies

Input:
    Learning rate α, noise standard deviation σ, initial policy paramters θ_0

Initialize:
    n workers with known random seeds, and initial parameters θ_0

for $t = 0, 1, 2, \ldots$ **do**
    for each worker $i = 1, \ldots, n$ **do**
        Sample $\epsilon_i \sim N(0, I)$
        Compute returns $F_i = F(\theta_t + \sigma\epsilon_i)$
    end for
    Send all scalar returns F_i from each worker to ever other worker
    for each worker $i = 1, \ldots, n$ **do**
        Reconstruct all perturbations ϵ_j for $j = 1, \ldots, n$
        Set $\theta_{t+1} \leftarrow \theta_t + \alpha\frac{1}{n\sigma}\sum_{j=1}^{n} F_j\epsilon_j$
    end for
end for

Figure 13-26. *Parallelized evolution strategies algorithm 2*[38]

The authors of the paper reported the following:

- They found that the use of virtual batch normalization and other reparameterizations of the neural network policy greatly improved the reliability of the evolution strategies.

- They found the evolution strategies method to be highly parallelizable (as discussed earlier). In particular, using 1,440 workers, it was able to solve the MuJoCo 3D humanoid task in less than ten minutes.

---

[38] https://arxiv.org/pdf/1703.03864.pdf

- The data efficiency of evolution strategies was surprisingly good. One-hour ES results require about the same amount of computation as the published one-day results for asynchronous Advantage Actor-Critic (A3C). On MuJoCo tasks, they were able to match the learned policy performance of TRPO.

- ES exhibited better exploration behavior than policy gradient methods like TRPO. On the MuJoCo humanoid task, ES was able to learn a wide variety of gaits (such as walking sideways or backward). These unusual gaits were never observed with TRPO, which suggests a qualitatively different exploration behavior.

- They found the evolution strategies method to be robust.

If you are interested, you should read through the paper to get into the details and see how it compares with other methods.

Transfer Learning and Multitask Learning

In the previous chapters, you looked at using DQN and the policy gradient algorithms to train agents to play Atari games. If you check out the papers that showcase these experiments, you will notice that some Atari games are easier to train and some are harder. If you look at Atari Breakout versus Montezuma's Revenge, as shown in Figure 13-27, you will notice that it is easier to train on Breakout as compared to training on Montezuma. Why is this?

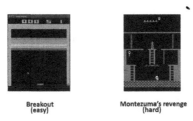

Breakout
(easy)   Montezuma's revenge
                       (hard)

Figure 13-27. *Easy and hard to learn Atari games*

Breakout has simple rules. However, Montezuma's Revenge has complex rules. They are not easy to learn. If you are playing it for the first time and have no prior knowledge of the exact rules, you know that the "keys" are something you usually pick to open new things and/or get big rewards. You know "ladders" can be used to climb up or down and that "skulls" are something to be avoided. In other words, human's past experience of having played other games or having read about treasure hunts or having watched movies gives them the context, or previous learning, to quickly perform new tasks that they might not have seen before.

Prior understanding of the problem structure can help you solve new complex tasks quickly. When an agent solves prior tasks, it acquires useful knowledge that can help the agent solve new tasks. But where is this knowledge stored? The following are some possible options:

- *Q-function*: They tell you what a good state and action are.

- *Policy*: They tell you which actions are useful and which are not.

- *Models*: They codify the learned knowledge about how the world operates, such as the laws of physics such as Newton's laws, friction, gravity, momentum, and so on.

- *Features/hidden states*: The hidden layers of neural networks abstract higher-level constructs and knowledge that can be generalized across different domains/tasks. You can see this in computer vision in supervised learning.

The ability to use the experience from one set of tasks to get faster and more effective performance on a new task is called *transfer learning*. Knowledge gained is transferred from past experiences to tackle new tasks. This is significantly used in supervised learning, especially in the domain of computer vision when popular convolutional network architectures like ResNet trained on an ImageNet dataset are used as pretrained networks to train on new vision tasks. You can also see these in Large Language Models where pre-trained models on languages can be further fine-tuned with a very small corpus of data on newer tasks. This section looks at how this technique can be applied to the field of reinforcement learning. Let's first define some terminology that is commonly found in transfer learning literature.

- *Task*: In RL, the MDP problem that you are trying to train the agent to solve.

- *Source domain*: The problem that the agent is trained on first.

- *Target domain*: The MDP that you hope to solve faster by leveraging the knowledge from the "source domain."

- *Shot*: The number of attempts in the target domain.

- *0-shot*: Running a policy trained on the source domain directly on the target domain.

- *1-shot*: Retraining the source domain's trained agent just once on the target domain.

- *Few-shot*: Retraining the source domain's trained agent a few times on the target domain.

Next let's look at how you can transfer the knowledge gained from the source domain to the target domain in the context of reinforcement learning. At this point, three broad classes of approaches have been tried.

- *Forward transfer*: Training on one task and transferring it to a new task.

- *Multitask transfer*: Training on many tasks and transferring it to a new task.

- *Transfer models and value functions*.

This section briefly talks about each of these approaches. *Forward transfer* is one of the most common ways to transfer knowledge in supervised learning, especially in computer vision. One model, like the popular architecture ResNet, is trained to classify images on the ImageNet dataset. This is called *pretraining*. The trained model is then changed by replacing the last layer or last few layers. The network is retrained on a new task called *fine-tuning*. However, forward transfer in reinforcement learning can face some issues of domain shift. In other words, the representations learned in the source domain may not work well in the target domain. In addition, there is a difference in the MDP. In other words, certain things are possible in the source domain that are not possible in the target domain. There are also fine-tuning issues. For example, the policies trained on the source domain could have sharp peaks in probability distribution, almost close to being deterministic. Such a peaked distribution could hamper exploration while fine-tuning on target domain.

The transfer learning in supervised learning seems to work well probably because the large set of varied images in the ImageNet dataset helps the network learn very good, generalized representation, which can then be fine-tuned to specific tasks. However, in reinforcement learning, the tasks are generally much less diverse, which makes it harder for the agent to learn the high-level generalizations. Further, there is the issue of policies being too deterministic, which hampers exploration for better convergence during fine-tuning. This issue can be handled by making the policies learn on a source domain with an objective that has an entropy regularizer term, something you saw in a few examples from previous chapters, such as Soft Actor-Critic (SAC) in Chapter 9. The entropy regularization ensures that the policies learned on the source system retain enough Stochasticity to allow for exploration during fine-tuning.

There is another approach to make learning on the source domain more generic. You can add some randomization to the source domain. Say you are training a robot to do some tasks; you could instantiate many versions of the source domain with a different mass of each arm for the robot, or the friction coefficients. This will induce the agent to learn the underlying physics instead of memorizing to do well in a specific configuration. In another real-world setup involving images, you can again borrow from the "image augmentation" practice from computer vision, wherein you augment the training images with some random rotation, scaling, and so on.

Next, consider *multitask transfer*, the second way to approach transfer learning. There are two key ideas to this: accelerate learning of all the tasks that are executed together and solve multiple tasks to provide better pretraining for the target domain.

One easy approach to train an agent on multiple tasks is to augment the state with a code/indicator signifying the specific task, for example, extending state S as $(S + Indicator)$. When an episode starts, the system chooses an MDP (task) at random and then chooses the initial state based on the initial state distribution. The training then runs like any other MDP. Figure 13-28 shows a schematic of this approach. This kind of approach can be tough at times. Imagine a policy getting better at solving a specific MDP; the optimization will start prioritizing the learning of that MDP at the cost of others.

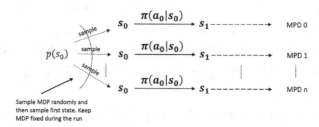

Sample MDP randomly and
then sample first state. Keep
MDP fixed during the run

Figure 13-28. *Solving multiple tasks together*

You could also train the agents to solve different tasks separately and then combine these learnings to solve a new task. You need to combine/distill the policies learned for different tasks to a single one in some way. There are various approaches for this, and I do not get into the details here.

Consider another variation of multitask transfer: *that of a single agent learning two different tasks in the same environment.* An example would be a robot learning to do laundry as well as wash utensils. In this approach, you augment the state to add the context of the task and train the agent. This is known as *contextual policies.* Using this approach, the state is represented as follows:

$$\tilde{s} = \begin{bmatrix} s \\ \omega \end{bmatrix}$$

You learn the policies as follows:

$$\pi_\theta \left(a|s,\omega \right)$$

Here, ω is the context, that is, the task.

Lastly, there is a third approach of transfer learning, that of *transferring models or value functions.* In this setting, you assume that the dynamics $p(s_{t+1}| s_t, a_t)$ are the same in both the source and target domains. However, the reward functions are different; for example, the autonomous car learns to drive to a small set of places (source domain). It then has to navigate to a new destination (target domain). You could transfer either of the three: the model, the value functions, or the policies. Models are a logical choice and simple to transfer, since the model $p(s_{t+1}| s_t, a_t)$ in principle is independent of the reward function. Transferring policies is usually done via contextual policies but otherwise not easy, as policies $\pi_\theta(a|s)$ contain the least information about the dynamics function. Value function transfer is also not so simple. Value functions couple dynamics, rewards, and policies.

You can do value transfer using successor features and representation, the details of which can be found in texts detailing transfer learning.

$$Q^\pi(s,a) = \underbrace{r(s,a)}_{reward} + \gamma E_{\underbrace{s' \sim p(s'|s,a)}_{dynamics}, \underbrace{a' \sim \pi(a|s)}_{policy}}[Q^\pi(s',a')]$$

Figure 13-29. *Transferring Q functions*

Meta-Learning

You have looked at various approaches to making agents learn from experiences. But is that the way humans learn? Current AI systems excel at mastering a single skill. Agents like AlphaGo have beaten the best human Go player. IBM Watson beat the best human Jeopardy player. But can AlphaGo play the card game of Bridge? Can the Jeopardy agent hold an intelligent chat? Can an expert helicopter controller for aerobatics be used to carry out a rescue mission? Comparatively, humans learn to act intelligently in many new situations, drawing on their past experience or current expertise.

If we want agents to be able to acquire many skills across different domains, we cannot afford to train agents for each specific task in a data-inefficient way—the way current RL agents are. To get true AI, we need agents to be able to learn new tasks quickly by leveraging their past experiences. This approach of *learning to learn* is called *meta-learning*. It is a key step to making RL agents human-like, always learning and improving based on experience and learning across varied tasks efficiently.

Meta-learning systems are trained on many different tasks (meta-training set) and then tested on new tasks. The terms *meta-learning* and *transfer-learning* can be confusing. These are evolving disciplines, and the terms are used inconsistently. However, one easy way to distinguish them is to think of meta-learning as learning to optimize hyperparameters of the model (e.g., number of nodes, architecture) and to think of transfer learning as fine-tuning an already tuned network.

During meta-learning, there are two optimizations at play—the learner, which learns (optimizes) the tasks, and the meta-learner, which trains (optimizes) the agents. There are three broad categories of meta-learning methods.

- *Recurrent models*: This approach trains a recurrent model like long short-term memory (LSTM) through episodes of tasks in the meta-training set.

- *Metric learning*: In this approach, the agent learns a new metric space in which learning is efficient.

- *Learning optimizers*: In this approach, there are two networks: the meta-learning and the learner. The meta-learner learns to update the learner network. The meta-learner can be thought of learning to optimize the hyperparameters of the model, and the learner network can be thought of as the regular network that is used to predict the action.

It is an interesting area of study with a lot of potential. Interested readers can refer to a great tutorial from the International Conference on Machine Learning (ICML-2019).[39]

Unsupervised Zero-Shot Reinforcement Learning

Following the idea of multitask learning, a good agent is one that can achieve many goals in a domain. The more chores they can do, the more useful household robots are; the more places they can go, the better self-driving cars are. Based on this idea and inspired by the recent success of unsupervised learning in language and vision—which has shown that a single model trained on data from the Internet can solve a variety of tasks without further training or fine-tuning—the authors of a recent paper titled "Unsupervised Zero-Shot Reinforcement Learning via Functional Reward Encodings"[40] explore a similar way to train a generalist agent from unlabeled offline data so that it can solve new tasks specified by users without training. This is called the *zero-shot reinforcement learning* (RL) problem. The hard challenge from this data is how to find, without labels, a task representation that is resilient to downstream objectives, avoiding the need for a human to provide well shaped reward functions before training.

The main contribution of the paper is to directly learn a latent representation that can capture any arbitrary reward functions based on their samples of state-reward pairs. The paper calls this concept *Functional Reward Encoding* (FRE). This is different from previous works in zero-shot RL or multi-task RL that use domain-specific task representations.

An FRE is trained using a prior distribution over reward functions. When the downstream tasks are unknown, a prior distribution is chosen that covers a wide range of possible objectives in a way that does not depend on a specific domain. In the paper's experiments, the authors demonstrate that a combination of random unsupervised reward functions, such as reaching goals and random MLP rewards, are a good option for the reward prior. They improve an FRE-conditioned policy for all rewards in this space. By doing this, they have already learned approximate solutions to many downstream tasks, and the zero-shot RL problem becomes simply finding the FRE encoding for the task, which the learned encoder does.

---

[39] https://sites.google.com/view/icml19metalearning
[40] https://arxiv.org/pdf/2402.17135.pdf

The framework in the paper presents a simple yet scalable method for training zero-shot RL agents in an unsupervised manner, as shown in Figure 13-30. The main idea is to (1) train an FRE network over random unsupervised reward functions, then (2) optimize a generalist FRE-conditioned policy toward maximizing said rewards, (3) after which, novel tasks can be solved by simply encoding samples of their reward functions, so that the FRE agent can immediately act without further training.

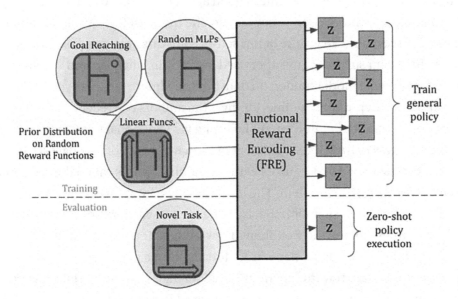

Figure 13-30. *FRE discovers latent representations over random unsupervised reward functions. At evaluation, user-given downstream objectives can be encoded into the latent space to enable zero-shot policy execution*
Source: Figure 1 of the paper[41]

Interested readers should refer to the paper. In this chapter, you have already seen a few examples of how advances from LLMs and diffusion models trained in an unsupervised way with generalist capabilities are being applied in RL. I am expecting a lot of these in the near future.

---

[41] https://arxiv.org/pdf/2402.17135.pdf

REINFORCE Learning from Human Feedback in LLMs

In Chapter 11 on PPL and RLHF, you saw the use of PPO-based RL fine-tuning on LLM. A 2024 paper titled "Back to Basics: Revisiting REINFORCE Style Optimization for Learning from Human Feedback in LLMs"[42] shows the use of REINFORCE-style optimization and demonstrates that it outperforms the PPO-based RLHF and RL-free Direct Preference Optimization (DPO) as well as Reward Ranked Fine Tuning (RAFT) approaches to AI alignment by finetuning LLMs .

The authors argue that the PPO approach has high computational cost and needs careful hyperparameter tuning. The authors claim that many of the reasons that motivated the development of PPO are less relevant in RLHF and propose a simpler method that saves computation and improves performance. They review the formulation of alignment from human preferences in RL. Following simplicity as a key principle, they show that many parts of PPO are not needed in an RLHF setting and that the basic REINFORCE-style optimization variants do better, suggesting that tailoring to LLMs alignment features allows you to use online RL optimization with low cost. This list summarizes the key findings of the paper:

1. **PPO is not the right tool for doing RL in RLHF**. The paper breaks apart PPO and shows that the most "basic" policy gradient algorithm, Vanilla Policy Gradient REINFORCE, is consistently outperforming PPO by 3.2 to 20.3 percent in terms of win rate.

2. **REINFORCE Leave-One-Out (RLOO) outperforms key baselines**. Built on top of REINFORCE, RLOO enables using multiple online samples, and the authors empirically show that it consistently outperforms baselines such as PPO, DPO, and RAFT. The paper shows that RLOO makes better use of online samples than RAFT, while presenting a higher robustness to noise and degree of KL penalty.

---

[42] https://arxiv.org/pdf/2402.14740v1.pdf

3. **Modeling partial completions is not necessary**. The paper effectively demonstrates that modeling partial sequences is an unnecessary undertaking for LLM preference training. Instead, modeling the full generations preserves performance while reducing complexity in the RL stage and significantly accelerating learning.

4. **RLOO is relatively robust to noise and KL penalty sensitivity**. The paper showcases RLOO robustness to noise and degree of KL penalty compared to RAFT.

How to Continue Studying

Deep reinforcement learning is seeing a lot of growth, and this is the most favored approach today to get agents to learn from experiences to behave intelligently. I hope that this book is just a start for you. I hope that you will continue to learn about this exciting field. You can use the following resources to continue learning.

Look for courses and online videos to extend your knowledge. MIT, Stanford, and UC Berkeley have many online courses that are the next logical way to dive deeper into this discipline. There are some YouTube videos from DeepMind and other experts in this field as well.

Make it a habit to visit the OpenAI and DeepMind websites regularly. They have a ton of material supplementing the research papers and are the basis of many of the algorithms you saw in this book.

Subscribe to Deep Learning, RL, and Generative AI-related newsletters and mailing lists. Use Notions (or something similar) to keep track of what you read.

Set up a Google alert for new papers in the field of reinforcement learning and try to follow along with the papers. Reading research papers is an art, and Professor Andrew Ng of Stanford has some useful tips on what it takes to master a subject.[43]

Finally, follow along with the algorithms covered in this book. Dive deep into the papers that form the basis of these algorithms. Try to reimplement the code yourself, either by following the code given in the notebooks accompanying this book or by looking at the implementations, especially in the Baselines3 and CleanRL libraries. Pick a domain and develop smaller weekend projects in application of RL.

---

[43] https://www.youtube.com/watch?v=733m6qBH-jI

At the end, each reader has a unique need and style. You should follow what already works for you and what you find interesting in these suggestions.

Summary

This chapter was a whirlwind tour of various emerging topics in and around reinforcement learning.

The first section presented some RL settings that are not the same as the ones used in the prior chapters, such as the MineRL, Donkey Car RL, FinRL, PySC2, and Godot RL agents. These settings pose various difficulties and possibilities for RL research and applications.

Next, you looked at additional methods in model-based RL in the context of deep learning, such as world models, imagination-augmented agents, model-based value expansion, and IRIS. These methods aim to take advantage of the sample efficiency of model-based methods and the scalability of model-free methods. Causal world models briefly mentioned a paper that establishes a formal link between causality and generalization in RL and suggests that causal reasoning is a crucial component for developing reliable and general AI.

This was followed by a quick introduction to imitation learning and inverse reinforcement learning, which are methods to learn from expert demonstrations or observations. The chapter discussed the differences between behavior cloning, direct policy learning, and inverse reinforcement learning, and have some examples of their applications.

The section on derivative-free methods showed some methods that optimize policies without taking derivatives, such as the cross-entropy method and the covariance matrix adaptation evolutionary strategy. These methods are based on evolutionary algorithms that select and improve the best policies from a set of candidates.

Next, the chapter covered the transfer learning and multitask learning methods that use the experience from one or more tasks to improve the learning of new tasks, such as forward transfer, multitask transfer, transfer models, and value functions. These methods aim to achieve faster and better learning by leveraging the similarities and generalizations across different tasks.

The section on meta-learning described the concept of meta-learning, which is learning to learn from past experiences. The chapter outlined three broad categories of meta-learning methods: recurrent models, metric learning, and learning optimizers. Meta-learning is a key step to making RL agents more human-like and adaptable to new situations.

The unsupervised zero-shot reinforcement learning section introduced a new approach to zero-shot reinforcement learning, which is the problem of learning a generalist agent from unlabeled offline data that can solve new tasks without further training. The approach is based on functional reward encodings, which are latent representations that capture any arbitrary reward functions from their samples.

The last section talked about a new approach to RLHF fine-tuning, that of replacing PPO with a simpler REINFORCE Leave-One-Out (RLOO) that's shown to be faster and performs better.

Lastly, the chapter covered some ways to continue exploring this field.

Index

A

Action, 6, 12, 15, 41, 44, 47, 48, 556
Action space, 50, 149, 311, 325
Action value functions, 66, 68, 69, 75, 178,
212, 283, 290
Activation function, 212, 228, 243
Actor-critic (AC) approach, 313, 369, 371,
415, 477, 553
A2C, 342, 344–346
pseudocode, 344
advantage, 341, 342
asynchronous advantage, 350, 351
implemeting A2C, 346, 348
Adaptability, 595
9.a-ddpg.ipynb file, 376–383, 385
Advantage actor-critic (A2C), 342–346
actor, 343
implementation, 346, 348, 349
MC approach, 343, 346
pseudocode, 344, 345
Advantage functions, 476
Agent, 6, 12
Agent Environment Cycle (AEC), 545, 546
Agent learning network, 233
Agent modeling, 539
AlphaGo, 451
branching factor, 451
general approaches, 451
MCTS, 453
neural network training, 452
policies, 452
RL network, 452

SL policy network, 453
standard search tree, 451
Antithetic sampling, 604
Arcade Learning Environment (ALE), 81,
251, 253
Aroon Indicator, 559
Artificial intelligence (AI)
definition, 2
Generative AI, 9, 10
Artificial neural networks, 212
Asynchronous advantage actor-critic
(A3C), 82, 350, 351, 572, 605
Asynchronous Reinforcement
Learning, 484
Asynchronous version, 99
Atari games, 253, 480
actions, 254
breakout, 255, 257
frameskipping, 254
observation spaces, 254
preprocessing, 257, 258
sticky actions, 254
Atari game simulator, 81
Auto-differential packages, 324
Auto-differentiation libraries, 315, 323
Autonomous car, 6, 49, 51, 609
Autonomous driving, 524
Autonomous vehicles (AVs), 14, 587
Autoregressive models, 10, 589
Average Directional
Index (ADX), 559
AWS Studio Lab, 38, 39

© Nimish Sanghi 2024
N. Sanghi, *Deep Reinforcement Learning with Python*, https://doi.org/10.1007/979-8-8688-0273-7

E

Printed in the United States
by Baker & Taylor Publisher Services